MICROSCALE and NANOSCALE HEAT TRANSFER

Analysis, Design, and Applications

MOURAD REBAY
SADIK KAKAÇ
RENATO M. COTTA

MICROSCALE and NANOSCALE HEAT TRANSFER

Analysis, Design, and Applications

CRC Press
Taylor & Francis Group
Boca Raton London New York

CRC Press is an imprint of the
Taylor & Francis Group, an **informa** business

CRC Press
Taylor & Francis Group
6000 Broken Sound Parkway NW, Suite 300
Boca Raton, FL 33487-2742

Printed on acid-free paper
Version Date: 20151103

International Standard Book Number-13: 978-1-4987-3630-5 (Hardback)

Visit the Taylor & Francis Web site at
http://www.taylorandfrancis.com

and the CRC Press Web site at
http://www.crcpress.com

Contents

Preface

The growing use of electronics, in both military and civilian applications, has led to the widespread recognition of the need for thermal packaging and management. The use of higher densities and frequencies in microelectronic circuits for computers is increasing day by day. Hence, the development of efficient cooling techniques for integrated circuits has become one of the important contemporary applications of microscale heat transfer, which has received significant attention for cooling of high power electronics and applications in biomechanical and aerospace industries. Microelectromechanical systems (MEMS) are the subject of increasing active research in a widening range of disciplines.

Designing effective cooling–heating devices is an important task in energy systems applications. There are various types of heat exchangers, including those used in chemical processing systems, nuclear reactor systems, and automotive engineering. By increasing the thermal conductivity of conventional heat transfer fluids by adding metallic or metal oxide in conventional fluids like water, oil, and ethylene glycol mixtures, heat transfer in cooling–heating is enhanced, resulting in smaller, energy-saving devices.

These topics and others are the main theme of this book. There are 18 chapters contributed by recognized individuals and active researchers in the field of micro- and nanoscale heat transfer and associated technologies. This book covers a wide range of analysis and applications of microscale and nanoscale convective heat transfer, which are organized into chapters as follows.

Chapter 1 deals with the solid–liquid interface phonon transfer at the molecular level and also studies the validity of the continuum hypothesis and Fourier law in nanochannels. The primary focus is on the conventional methods of using molecular dynamics (MD) for heat transport problems. For such cases, the effects of the applied thermostat and the intermolecular parameters employed at the interface are investigated. Next, the temperature dependence on the interface resistance for different wetting surfaces is characterized, and the temperature dependence on phonon transport and the near surface density structure is differentiated. Finally, using proper silicon–water interaction parameters to capture the correct wetting behavior and proper thermostat technique, the silicon–water interfacial thermal resistance (ITR) as a function of the surface temperature is determined with a proper thermostat technique. The reported resistance values show perfect agreement with the experimental measurements from the previous literature.

In Chapter 2, the molecular dynamics approach to calculate ITR is presented. The effect of bonding energy on liquid layering and ordering of the interfacial liquid on the effective thermal conductivity of the nanofluids, with some numerical exercises, is discussed. Simulations with a controlled increase in the hot wall temperature of the metal are presented to reveal that the interfacial water thermal resistance is not static but depends on the changes on the surface temperature. The presence of a thermal resistance at the interface between water and various metals is exhibited through a series of molecular dynamics simulations, including their basic software. The fundamentals of incorporation and derivation of multidimensional molecular dynamics modeling into simple thermal models are presented, which may prove useful in the practice of accurately modeling heat transfer in nanoscale devices with a range of operating conditions.

In Chapter 3, a state-of-the-art review is presented on convective heat transfer with nanofluids that have been investigated for their advantages and potentials in improving heat transfer rates since they can be used as working fluids in various thermal systems. This chapter provides details of experimental and theoretical investigations made on nanofluids. The experimental results present some measured viscosity and thermal conductivity values of nanofluids and important correlations obtained from experimental data. The theoretical part is presented under two main considerations: single-phase and two-phase modeling; the former is concerned with steady homogeneous properties of nanofluids and the latter distinguishes nanoparticle properties from base-fluid properties. Finally, applications using nanofluids are presented to show the feasibility of nanofluid developments with important correlations obtained from experimental data.

Chapter 4 reviews experimental results of commonly used nanofluids dispersed in both spherical and nonspherical nanoparticles, including Al_2O_3, CuO, Cu, Au, carbon nanotubes (CNTs), and graphite nanosheets. With the aid of a large amount of experimental data, an unbiased statistical analysis to trace the exact trends of thermal transport in nanofluids is developed. Effective medium theory (EMT) incorporated with the Kapitza resistance can be used to statistically explain the effect of volume fraction on the effective thermal conductivity of nanofluids containing Al_2O_3, CuO, CNTs, and graphite nanosheets. In contrast, abnormal enhancements, more than EMT predictions, in the nanofluids containing Cu nanoparticles are presented. To reveal the heat conduction mechanisms in nanofluids, systematic and benchmark studies are suggested and many aspects are to be taken into consideration, such as the morphology of nanoparticles, insulation of the metallic sensor and contact leads, stability of nanofluids, and the distribution of nanoparticles. Furthermore, the thermal transport at the nanoparticle–fluid interface is especially emphasized for understanding the effective thermal conductivity of nanofluids.

In Chapter 5, experimental and theoretical knowledge regarding nanofluid thermal conductivity models and viscosity is summarized. The effects of nanofluid parameters such as particle type, particle size, particle volume fraction, and temperature on thermal conductivity are discussed, providing insight into the mechanisms of enhancement and reasons for discrepancy among the reported values. Mechanisms of thermal conductivity enhancement, including Brownian motion, clustering, liquid layering, and ballistic transport are proposed, and the current understanding regarding the significance of these mechanisms, which are important in the enhancement of heat transfer with nanofluids, is explained.

Chapter 6 deals with the stability of nanofluids. The long-term stability of nanofluids is critical for sustainable heat transfer augmentation and enlarged application areas. In this study, a state-of-the-art review on the stability of nanofluids with an emphasis on thermal and hydrodynamic performance is presented. Stability assessment techniques and stability indicators are discussed in detail. Some future prospects are given as guidelines for contemporary researchers to take the lead in systematic research in this field.

Chapter 7 deals with the numerical approaches for computational fluid dynamics (CFD) simulation of nanofluids, which are discussed in detail and a review of the recent related studies in the literature presented. Single-phase and two-phase modeling characteristics and abilities are explained by theoretical and experimental studies. Two-phase flow models that consider fluids and nanoparticles as two different phases are defined in several forms for different flow regimes and mathematical approaches such as an Euler–Euler approximation (i.e., volume of fluids, mixture, and Eulerian models) or discrete phase model as an Euler–Lagrangian approximation. The main differences of two-phase modeling approaches for nanofluids by using CFD are presented and discussed in relation to recent numerical studies.

The aim of Chapter 8 is to present a review of some experimental results in the field of heat pipes and two-phase flows in thermosyphons. A new trend in the successful application of heat pipes and thermosyphons is related to nanotechnologies. Nanofluids and the invention of nanocoatings and nanocomposites have opened a new niche in heat pipe and thermosyphon design and use. Nanofluids are considered as the vacant working liquids for the transparent flat mini–heat exchangers heated by laser beams or solar radiation. The combination of nanofluids and small channels constitutes an innovative method for providing effectiveness, compactness, and low thermal resistance. Thin porous nanocoating in coaxial mini-channels or mini-grooves of heat pipe evaporators plays a role of additional centers for stable vapor generation, which do not require superheating of the surface to enhance heat transfer. Porous nanocoating (thickness 25–100 μm) on the generation of engineered nanostructured heat pipe heat exchangers represents a major breakthrough in material science and energy-saving technologies, which are discussed in this chapter.

In Chapter 9, nanofluids are treated as a porous media. Since nanofluids consist of a base fluid, a very small amount of nanoparticles with dimensions from 1 to 100 nm and high thermal conductivities, they can be considered as a large amount of base fluid flowing in a high-porosity media. This chapter presents a mathematical model that has been developed for the steady flow of the base fluid in the porous medium with Al_2O_3 nanoparticles. The flow is considered to be under a fully developed laminar flow condition through a rectangular pipe as in electronic circuit cooling applications. The governing equations written in terms of the primitive variables are solved through an in-house program using the finite volume method and the SIMPLE algorithm. The effects of the simulated porosity, permeability, and thermal conductivity models in the porous media are studied. A comparison of the simulated results with the experimental ones showed that the developed mathematical model demonstrates an excellent potential in nanofluid simulation in porous media.

Chapter 10 covers microscale convective heat transfer with gaseous flow in ducts. The fundamentals and current literature on microscale single-phase forced convection heat transfer phenomenon, focusing on gaseous slip flow and with particular attention to the literature published in the past five years, are also presented. Pertinent nondimensional parameters, such as Reynolds, Prandtl, Nusselt, Peclet, Knudsen, and Brinkman numbers, are defined and discussed with respect to their influence on flow and heat transfer characteristics. Physical phenomena, such as laminar-to-turbulent flow transition, rarefaction, viscous heat dissipation, axial conduction, variation of fluid thermophysical properties, flow geometry, and surface roughness, are also discussed. Analytical, numerical, and experimental research data are included and correlations recently introduced in literature are presented with respect to their limitations. Major conclusions in relation to the physical effects prevailing in convective heat transfer in microducts are given.

The aim of Chapter 11 is to present a numerical analysis of laminar flow of a Newtonian fluid in a parallel-plate microchannel subjected to a uniform heat flux or a uniform temperature on the external surface of the plates. The finite volume method is used for solving conservation equations in the problem of simultaneous developing of heat and fluid flows. Emphasis is placed on the Navier–Stokes numerical resolution method and energy equations, including combined microscale effects for liquid and gaseous flows. An in-house FORTRAN program is developed for integrating the effects of heat conduction in the wall, viscous dissipation, and rarefaction in the case of gaseous flow. Slip velocity and temperature jump boundary conditions are implemented at the fluid–solid interface. Separated and combined microscale effects are discussed. Some typical results of temperature profiles, Poiseuille number and Nusselt number evolutions, are given for water and air flows.

In Chapter 12, both experimental and numerical results of flow and heat transfer in microchannels and micronozzle mainly based on the studies of the authors' group are presented. After a short introduction, the experimental and numerical results of gas flow and deionized water flow and heat transfer in microchannels are provided. Then, numerical studies of several special cases, including liquid flow in wavy microchannels, gaseous flow in microscale porous media, gas flow and heat transfer in micronozzles, simultaneously developing flow and heat transfer in microtubes with thick walls, and electroosmotic flow in microchannels, are discussed. Finally, some important applications are presented, including a numerical design of a mini-microchannel heat sink with variable height that can provide a highly uniform temperature field (temperature difference less than 2°C) for heat fluxes as high as 2×10^6 W/m^2.

In Chapter 13, the generalized integral transform technique is introduced and employed in combination with a single domain reformulation strategy to provide hybrid numerical–analytical solutions to conjugated heat transfer in microfluidic systems composed of multiple fluid streams and walls, such as in direct contact micro heat exchangers and stratified flow microreactors with immiscible fluids. The governing energy equations for both the walls and fluid streams, within the thermally developing region and accounting for axial heat diffusion, are then combined into a single domain formulation, which is written in terms of space variable coefficients, representing the transitions of the thermophysical properties and velocity profiles. After illustrating the convergence characteristics of the proposed eigenfunctions, a couple of physical situations are explored more closely. Infrared thermography with microscopic lenses can be a powerful tool in nonintrusive thermal character[illegible] microsystems, especially in combination with an accurate direct problem solution appro[illegible] robust inverse problem analysis methodology, allowing for parametric and functional identifications in different classes of problems that are discussed in Chapter 14.

Chapter 14 illustrates the combination of integral transforms for direct problem solution and Bayesian inference for inverse problem handling, as applied to the thermal characterization of two distinct microsystems, namely, a microchannel with slip flow conditions and a micromodel of a porous medium.

Chapter 15 covers the application of the lattice Boltzmann method (LBM) for thermal microflows. In microchannel heat transfer, fluid flow at a certain range of Knudsen number can be strongly influenced by noncontinuum effects depending on flow conditions and the fluid phase (gas or liquid). This, in turn, poses limitations to conventional CFD, most notably when it comes to boundary conditions at the fluid–solid interface. In liquid flows, for example, there is a drastic increase in the surface area-to-volume ratio, and the microchannel walls can influence the fluid dynamics as a result of electrokinetic effects. For gas flows at a certain range of Knudsen number values ($0.01 < K_n < 0.1$), the gas enters a slip-flow regime where noncontinuum effects manifest in the form of a velocity slip and a temperature jump at the solid surface. In such cases, the Navier–Stokes equations remain applicable in the bulk flow but special boundary conditions are required at the solid surface to account for the noncontinuum effects. Then the problem can be treated using the LBM. This chapter presents an application of this method to thermal microflows. LBM, which operates at the mesoscale, finds its roots in lattice gas automata (LGA) and has been extensively used to solve the Navier–Stokes equations for microfluidic applications. Focus is given on the double distribution thermal LB model and various computational aspects that arise from implementing it. Furthermore, various LB boundary condition treatments that attempt to capture noncontinuum effects in the slip-flow regime are discussed.

In Chapter 16, pore-level heat and gas flows at the microscale are investigated in order to provide volume-averaged parameters for VAM transport equations in slip-flow regime. The velocity and temperature fields in various representative volumes of two-dimensional periodic and isotropic porous structures consisting of square rods at various conditions and porosity levels are determined. At the same time, the studied geometrical dimensions are gradually decreased in order to characterize size-dependent rarefaction effects for each case. Corresponding permeabilities, interfacial heat transfer coefficients, and effective thermal conductivity coefficients are calculated. The existing correlations for the prediction of volume-average transport properties for various forms of porous media at the microscale in slip-flow regime are revisited.

Chapter 17 covers a numerical method for the resolution of the problem of subcooled convective boiling flows in microchannel heat sinks. The modeling of a two-phase flow was achieved using the volume of fluid method. The geometric reconstruction scheme, which is based on the piecewise linear interface calculation method, is applied to reconstruct the liquid–vapor interface. A control volume method is used for the resolution of governing equations in conjugated heat transfer problems. An application of the numerical method for the cooling of an absorber of radiation in synchrotron light source (SOLEIL) is presented. The microchannel heat sink is integrated into the light absorber. Flow patterns during bubble generation and heat transfer coefficient are discussed.

Finally, Chapter 18 covers the two-phase boiling flow and condensation heat transfer in mini-/microchannels. The main physical properties, dimensionless numbers used in the evaluation of the prominent driving forces are defined. In particular, criteria defining transition between conventional and mini-/microchannels are given. In these two types of channels, flow regimes are described while noting their differences. Specific attention is paid to surface properties where wet ability plays an important role. The dimensionless numbers associated with flow boiling are presented.

The main flow boiling heat transfer results and heat transfer mechanisms are discussed. Current theoretical models are presented. Semiempirical correlations are presented to predict heat transfer coefficients. An analysis of pressure drop is presented by considering core friction and acceleration components. In condensation inside tubes, flow regimes are identified for different conditions and are shown to be different for those observed in conventional channels. Several transition criteria are defined for various fluids, operating conditions, and channel sizes. In particular, the flow pattern evolution from mini- to microchannels is described.

Acknowledgments

We express our sincere appreciation to all the authors for their valuable contributions, time, and efforts in preparing their chapters on time, which facilitated the early production of this book. Special thanks go to our colleagues at the University of Reims, at the TOBB University of Economics and Technology, and at Federal University of Rio de Janeiro for their encouragement and assistance during the preparation of this book. We have made every effort to minimize typographical errors during the editing of the chapters. We welcome feedback from the readers to improve subsequent editions of this important book. Thanks are also due to Arlene Kopeloff, editorial assistant, and Laurie Oknowsky, project coordinator. We are thankful to Jonathan W. Plant, executive editor, for initiating this project and for his close collaboration during the preparation of this book, and to other individuals at CRC Press who contributed their talents and energy to this book.

We are forever grateful to our spouses, Amel, Filiz, and Carolina, for their continuous support.

Mourad Rebay
Sadık Kakaç
Renato M. Cotta

Contributors

Murat Barisik
Department of Mechanical Engineering
Izmir Institute of Technology
Izmir, Turkey

Yildiz Bayazitoglu
Department of Mechanical Engineering
and
Department of Materials Science and Nanoengineering
Rice University
Houston, Texas

Ali Beskok
Department of Mechanical Engineering
Southern Methodist University
Dallas, Texas

André Bontemps
Paris Interdisciplinary Energy Research Institute
Denis Diderot University–Paris 7
Paris, France

Nadia Caney
Laboratoire des Écoulements Géophysiques et Industriels
University of Grenoble Alpes
Grenoble, France

Renato M. Cotta
Department of Mechanical Engineering
and
Department of Nanoengineering
Federal University of Rio de Janeiro
Rio de Janeiro, Brazil

Eskişehir B. Elçioğlu
Department of Mechanical Engineering
Middle East Technical University
Ankara, Turkey

and

Sivrihisar Vocational School
Eskisehir Osmangazi University
Eskişehir, Turkey

Nadim El Wakil
GRESPI Laboratory
University Institute of Technologie (IUT Reims-Chalons-Charleville)
University of Reims Champagne-Ardenne
Reims, France

Anis Hamza
GRESPI Laboratory
University of Reims Champagne-Ardenne
Reims, France

and

Synchrotron SOLEIL Facility
Gif sur Yvette, France

Ya-Ling He
Key Laboratory of Thermo-Fluid Science and Engineering
School of Energy and Power Engineering
Xi'an Jiaotong University
Xi'an, Shaanxi, People's Republic of China

Khaled M. Issa
Department of Mechanical and Manufacturing Engineering
University of Calgary
Calgary, Alberta, Canada

Yassine Kabar
Department of Mechanical Engineering
Jijel University
Jijel, Algeria

Sadık Kakaç
Department of Mechanical Engineering
TOBB University of Economics and Technology
Ankara, Turkey

Diego C. Knupp
Department of Mechanical Engineering and Energy
State University of Rio de Janeiro
Rio de Janeiro, Brazil

Hasna Louahlia-Gualous
LUSAC Laboratory
Department of Energy
University of Caen Basse Normandie
Saint Lô, France

Ghazali Mebarki
LESEI Laboratory
Department of Mechanical Engineering
University of Batna
Batna, Algeria

Moghtada Mobedi
Department of Mechanical Engineering
Izmir Institute of Technology
Izmir, Turkey

A.A. Mohamad
Department of Mechanical and Manufacturing Engineering
University of Calgary
Calgary, Alberta, Canada

A. Nakayama
Department of Mechanical Engineering
Shizuoka University
Johoku, Japan

Carolina P. Naveira-Cotta
Laboratory of Nano and Microfluidics and Microsystems—LabMEMS
Department of Mechanical Engineering
and
Department of Nanoengineering
Federal University of Rio de Janeiro
Rio de Janeiro, Brazil

Tuba Okutucu-Özyurt
Department of Mechanical Engineering
Middle East Technical University
Ankara, Turkey

S. Özerinç
Department of Mechanical Engineering
Middle East Technical University
Ankara, Turkey

Anchasa Pramuanjaroenkij
Faculty of Science and Engineering
Kasetsart University
Bangkok, Thailand

Mourad Rebay
GRESPI Laboratory
University Institute of Technologie (IUT Reims-Chalons-Charleville)
University of Reims Champagne-Ardenne
Reims, France

Ismail Ozan Sert
Department of Mechanical Engineering
TOBB University of Economics and Technology
Ankara, Turkey

Nilay Sezer-Uzol
Department of Mechanical Engineering
TOBB University of Economics and Technology
Ankara, Turkey

E. Şimşek
Department of Mechanical Engineering
Middle East Technical University
Ankara, Turkey

Wen-Quan Tao
Key Laboratory of Thermo-Fluid Science and Engineering
School of Energy and Power Engineering
Xi'an Jiaotong University
Xi'an, Shaanxi, People's Republic of China

Amarin Tongkratoke
Faculty of Science and Engineering
Kasetsart University
Sakon Nakhon, Thailand

Leonard L. Vasiliev
A.V. Luikov Heat and Mass Transfer Institute
National Academy of Sciences of Belarus
Minsk, Belarus

Leonid L. Vasiliev, Jr.
A.V. Luikov Heat and Mass Transfer Institute
National Academy of Sciences of Belarus
Minsk, Belarus

Jerry Vera
Department of Mechanical Engineering
Rice University
Houston, Texas

J.L. Wang
Jiangsu Key Laboratory for Design and Manufacture of Micro/Nano Biomedical Instruments
School of Mechanical Engineering
Southeast University
Nanjing, People's Republic of China

Almila Güvenç Yazicioğlu
Department of Mechanical Engineering
Middle East Technical University
Ankara, Turkey

X. Zhang
Key Laboratory for Thermal Science and Power Engineering
Department of Engineering Mechanics
Tsinghua University
Beijing, People's Republic of China

J.L.Z. Zotin
Federal Center of Technological Education Celso Suckow da Fonseca (CEFET/RJ)
Department of Mechanical Engineering
Itaguaí, Rio de Janeiro, Brazil

temperatures using a thermostat. However, imposing a constant temperature region in the solid domain results in an artificial thermal resistance previously overlooked in the literature. For such cases, we investigate the effects of applied thermostat [4]. Second, we question the intermolecular parameters employed at the interface. Previous works employed interaction parameters based on various mixing rules and did not validate the resulting surface wetting behavior. Hence, the previously reported interface resistance values are vague. By conducting a detailed water nano-droplet study, we relate the molecular interaction parameters to the macroscopic wetting behavior [5]. Next, we characterize the temperature dependence of the interface resistance for different wetting surfaces. By doing this, the temperature dependence of phonon transport and the near-surface density structure are distinguished [6]. Finally, using the proper silicon–water interaction parameters that capture the correct wetting behavior, we determine the silicon/water ITRs as a function of the surface temperature. Reported resistance values show perfect agreement with the experimental measurements in the literature.

Objectives of this chapter are

1. Demonstrations of length scales and conditions for onset of continuum hypothesis in nanoscale-confined liquids
2. Characterization of the artificial thermal resistance induced by the applied thermostat at the liquid/solid interface [4]
3. Revealing the relation between the surface wetting and interface phonon transport [5]
4. Exhibiting temperature dependence of ITR for different wetting surfaces [6]

1.2 ONSET OF CONTINUUM BEHAVIOR

Heat transfer is a physical process of energy flow developed as a result of a temperature difference. From the continuum view point, Fourier's law states that the heat flux is proportional to the developed temperature gradient while this proportionality is constant and independent of system size. Such consideration is questionable for nanoscales. Specifically in a fluid domain, ordering of liquid molecules formed due to surface forces creates complications in the definition of a local temperature. As a start, we explored the validity ranges of the continuum hypothesis for nano-confined fluids by performing MD simulations of argon liquid inside nanoscale graphite and silver channels of different heights.

We studied the validity of Fourier's law for one-dimensional liquid heat conduction inside 3.57, 7.6, and 14.9 nm height channels. Different temperatures are imposed on opposite channel surfaces to create a thermal gradient in the system. Fluid domain is divided into small bins where the local temperatures are measured and averaged through the simulation time. Resulted liquid number density profiles showed density layering near surfaces while the oscillation of density converges to the desired value in the center region of the channels, except the 3.57 nm case. Hence, fluid in a 3.57 nm confinement with such ordering is expected to show a non-continuum behavior as a result of the size effects.

In statistical mechanics sense, local temperature can be defined only if the local thermal equilibrium is established inside the local region of interest. First, we tested the local thermal equilibrium in our MD system by calculating the higher moments of velocity. At equilibrium, the velocity distribution at the point of measurement should have normal (Gaussian) distribution with zero skewness and kurtosis. We calculated higher moments of velocities in each local bin of the simulated domain. The results showed normal distribution of the thermal velocity, which validates that the local thermal equilibrium is maintained in our system. Time-averaged temperature profile exhibits a linear variation, typical of continuum predictions. Even for the smallest case of 3.57 nm height channel, temperature distribution agrees with the Fourier's law's expectations regardless of the observed density oscillations.

Next, we calculated the thermal conductivity of liquid argon. Generally, two methods can be applied for thermal conductivity calculations in MD simulations. One approach is the Green–Kubo method that is based on the fluctuation–dissipation theorem, and applicable for the equilibrium MD technique. The Green–Kubo method is derived from the linear response theory to extract thermal conductivity from the energy current correlation function. The second approach is the utilization of Fourier's law, which is based on a linear constitutive law, and may not be applicable in nano-confined geometries. First, we calculated the thermal conductivity of liquid argon using the Green–Kubo method. When there are no electronic contributions and net flow in the system, thermal motion becomes the sole contributor in the calculation of thermal conductivity. Our code successfully calculated the thermal conductivity of liquid argon using the Green–Kubo method, which matched with published values in the literature.

The thermal conductivity of liquid argon was also obtained using the Fourier's law with heat flux and a time-averaged temperature gradient predicted from MD. The thermal conductivity calculated from the Fourier's law matches well with the previously published thermal conductivity of liquid argon [7–13]. Independent of the developed heat flux and size of the computational domain, MD simulations successfully reproduce the thermal conductivity of liquid argon using the Fourier's law. Comparison of thermal conductivity obtained from the Green–Kubo method with that extracted from the Fourier's law of heat conduction enables the assessment of reliability of Fourier's law in nano-confined dimensions. Hence, in this study, we presented the emergence of continuum behavior and the validity of Fourier's law for liquid argon in channels with dimensions as small as 10 molecular diameters (≈3.5 nm).

1.3 BOUNDARY TREATMENT EFFECTS ON INTERFACE THERMAL RESISTANCE

Through the MD literature, the most widely used approach to create a heat flux in the domain is to employ thermostats to regulate the temperature in hot and cold regions, while utilizing a thermostat free zone in the middle. The thermostat applied regions often utilize the velocity rescaling, Nose–Hoover or Langevin thermostats [14]. Each of these algorithms constrains the temperature based on the kinetic energy of atoms. Many MD-based heat conduction studies for homogenous solid domains report emergence of an *artificial thermal resistance* (i.e., temperature jump) at the interface of the thermostat applied and thermostat free regions [15–24]. Phonon transport in solids arises from long-term correlated motion between neighboring atoms. Thermostat applies a random force to the atoms to keep them at the prescribed temperature. This de-correlates the motion between atoms at the interface of the thermostat applied and thermostat free regions, and leads to lower thermal conductivity with an *artificial interface resistance*. Although this phenomenon has been widely observed in solid systems, its significance at the liquid/solid interface is largely unknown.

In this study, we investigate heat conduction through the liquid argon–solid silver interface using non-equilibrium MD simulations by employing three different thermal boundary treatments. Hence, we investigate the effect of this *artificial thermal resistance* and distinguish it from the actual Kapitza resistance at the liquid/solid interface. In order to achieve these goals, we present heat conduction results in the solid–liquid–solid domains obtained by applying Langevin thermostat to all wall layers (Case 1); only to a small number of wall layers that are sufficiently away from the liquid/solid interface (Case 2); and by injecting and removing energy from the wall layers (Case 3). All results are organized to characterize the artificial thermal resistance, Kapitza resistance, and length using simple heat conduction analysis. Results of systematic studies obtained by varying the wall temperatures are also presented.

1.3.1 Three-Dimensional MD Simulation of Argon/Silver System

We studied liquid argon confined between two parallel solid silver walls by using LAMMPS (Large-scale Atomic/Molecular Massively Parallel Simulator), a MD code from Sandia National Laboratories [25]. Figure 1.1 illustrates a snapshot of the simulation domain. Total dimensions in longitudinal (x), vertical (y), and lateral (z) directions are 16.4, 3.7, and 3.7 nm, respectively. The domain was constructed using 1830 argon and 7776 silver molecules. The first silver domain lies between 0 and 48 Å, and the second silver domain is from 118 to 166 Å, while the liquid

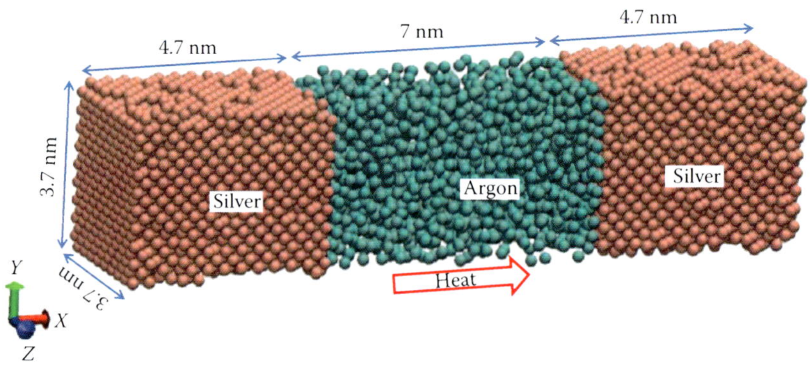

FIGURE 1.1 Snapshot of the MD simulation domain with domain size.

argon is confined to 70 Å spacing between the two walls. The corresponding number density of argon is $\rho = 0.8$ ($\#\sigma^3/nm^3$). The walls on each side of the domain contain 24 silver layers that are 0.2043 nm apart from each other. Each silver layer contains 162 molecules resulting in number density of 5.7 ($\#\sigma^3/nm^3$). Periodic boundary conditions are applied in the vertical (y) and lateral (z) directions. Silver walls are modeled as face-centered cubic (FCC) structures with (0, 0, 1) crystal plane facing the fluid. Atoms in the outmost layer of both silver walls are fixed to their original locations to maintain a fixed volume system, while the remaining atoms throughout the domain are free to move.

Mass for an argon molecule is $m = 6.63 \times 10^{-26}$ kg, its molecular diameter is $\sigma = 0.3405$ nm and the depth of the potential well for argon is $\varepsilon = 119.8 \times k_b$. Lennard-Jones (L-J) 6–12 potential was utilized to model the van der Waal's interactions between argon–argon and argon–silver molecules with a cutoff distance of 1 nm. The truncated (6–12) L-J potential is given as

$$\Phi_{truncated}(r_{ij}) = 4\varepsilon\left(\left(\left(\frac{\sigma}{r_{ij}}\right)^{12} - \left(\frac{\sigma}{r_{ij}}\right)^{6}\right) - \left(\left(\frac{\sigma}{r_c}\right)^{12} - \left(\frac{\sigma}{r_c}\right)^{6}\right)\right), \tag{1.1}$$

where

- r_{ij} is the intermolecular distance
- ε is the depth of the potential well
- σ is the molecular diameter
- r_c is the cutoff radius [14]

Lorentz–Berthelot mixing rule is employed to calculate the L-J parameters for argon–silver interactions [26]. Silver–silver interactions are calculated using the *embedded atom method* (EAM), which is a many-body potential for slightly more computational cost than the pair potential, and works especially well for FCC structures [27,28].

Simulations are started from the Maxwell–Boltzmann velocity distribution for all molecules at 115 K. Initial particle distribution is evolved 10^6 time steps (4 ns) to reach an isothermal steady state using 4 fs time steps. During this process, the Nose–Hoover thermostat is applied to the molecules to maintain the system's temperature at 115 K. Afterward, temperatures of the silver walls are modified to serve as heat baths to induce heat flux through the liquid/solid interfaces. At the same time, NVE ensemble (constant number of molecules, constant energy, and constant volume) is applied to liquid argon domain. Simulations are performed for an additional 10^6 time steps (4 ns) to ensure that the system attains equilibrium in presence of the heat flux. After which, time averaging of desired properties is performed for 2×10^6 additional time steps (8 ns). Longer time averaging has also been performed to confirm the convergence of density and temperature profiles to steady state.

Two different heat bath techniques are employed to investigate their effects. First, a constant surface temperature is maintained using Langevin thermostat. Langevin damping factor is taken to be 0.1 ps, which is prescribed by the phonon frequency of solid silver ($f_D = 0.223$ ps) at Debye temperature of $T_D = 215$ K [29,30]. Different wall temperatures are assigned to create heat flux through the liquid/solid interfaces. Langevin technique is employed as a global thermostat on the specified number of solid layers for each case. Second, a constant heat flux is obtained by injecting and removing energy through the heat baths to support a temperature gradient across the system. The simulated cases are summarized in Table 1.1. The computational domain is divided into 200 slab bins with the size of 0.083 nm. In order to interpret the temperature profiles obtained using fine bins, we also obtained temperature distribution using 20 slab bins. Temperature profiles from the fine and coarse bin cases were consistent and showed that local thermal equilibrium is reached in the simulations using fine bins. Coarse bin results are omitted for brevity of this work. Achievement of local thermal equilibrium in such nanoscale-confined domains was addressed earlier.

Irving–Kirkwood (I–K) expression is utilized to compute the heat flux vector for an N particle system using unity differential operator approximation as follows [31,32]:

$$J_k = \frac{1}{\text{Vol}}\left\langle \sum_{i}^{N} V_k^i\left(E^i + \Phi^i\right) + \sum_{i,j}^{N}\left(r_k^j + r_k^i\right)W^{i,j} \right\rangle, \tag{1.2}$$

$$E^i = \frac{1}{2}m^i\left(\left(V_x^i\right)^2 + \left(V_y^i\right)^2 + \left(V_z^i\right)^2\right), \tag{1.3}$$

$$W^{i,j} = \frac{1}{2}\left(V_x^i f_x^{i,j} + V_y^i f_y^{i,j} + V_z^i f_z^{i,j}\right), \tag{1.4}$$

where

the first term on the right-hand side of Equation 1.2 is the kinetic and potential energies carried by particle i

the second term is the energy transfer to particle i by force interactions with the surrounding particles

In the first term, V_x^i is the peculiar velocity component of particle i in k-direction, while k is the axes of the Cartesian coordinate system; E^i is the kinetic and Φ^i is the potential energy of particle i calculated using Equations 1.1 and 1.3, respectively. In the second component of Equation 1.2, $(r_k^j - r_k^i)$ is the kth component of the relative distance vector between particles i and j. The $W^{i,j}$ term is given in Equation 1.4, where $f_l^{i,j}$ is the intermolecular force exerted on particle i by particle j in the Cartesian coordinate direction l.

1.3.2 Constant Surface Temperature

Case 1 utilizes constant temperature surfaces, where all 23 silver layers of the left and right surfaces are maintained at 140 and 90 K, respectively, using a Langevin thermostat. The number density profile shown in Figure 1.2 illustrates the system consisting of three distinct domains. argon density exhibits strong near-wall layering due to the wall force field effects and local argon–argon interactions [33]. The density layering extends approximately 2 nm from each surface, after which the liquid argon density reaches nearly a constant value of 0.8 ($\#\sigma^3/\text{nm}^3$). Slight density variation in

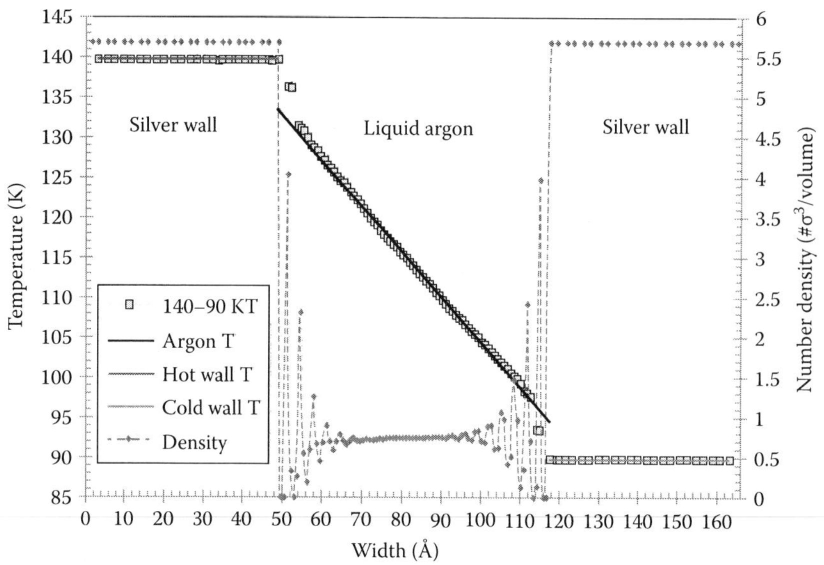

FIGURE 1.2 Temperature (left axis) and density (right axis) profiles of liquid argon confined between 24-layer silver walls. Langevin thermostat is applied onto 23 layers of each silver wall separately to maintain them at 140 and 90 K (Case 1).

this bulk liquid region is due to the temperature gradient, where the argon density increases with decreased temperature. Local argon density near the high temperature (left) wall is less than that near the low temperature (right) wall. Argon density fluctuates severely near the surface, resulting in nearly zero molecules within certain bins. Depletion of argon molecules near the surfaces has profound effects on the definition of local liquid temperature. Very few liquid molecules are present at these density minima, where local temperature cannot be defined accurately. In order to address this problem, we omitted presentation of temperature in these "empty" bins as can be observed from the near-wall temperature profiles.

Figure 1.2 also shows the temperature profiles obtained using 200 slab bins. While the silver surfaces are kept at constant temperatures, liquid argon temperature varies linearly in bulk of the nano-channel. Temperature jumps at the liquid/solid interfaces are apparent in the results. These temperature jumps can be associated with an *interfacial thermal resistance* (R_K), known as the Kapitza resistance [34], for given a heat flux $\vec{j}$ as

$$\Delta T = -R_K \vec{j} \cdot \vec{n}, \tag{1.5}$$

where $\vec{n}$ is the outward unit normal from the wall. Alternatively, it is possible to define a thermal resistance (Kapitza) length (L_K), by extrapolating the temperature profile from liquid into the solid, where the wall temperature is reached. The Kapitza length can be predicted using

$$\Delta T = L_K \left.\frac{\partial T}{\partial n}\right|_{liquid}, \tag{1.6}$$

where $\partial T/\partial n$ is the thermal gradient on the liquid side,

$$\Delta T = T_{liquid} - T_{wall}$$

In order to characterize the temperature distribution in the bulk of liquid argon, we performed linear least square fits to the MD data and measured sum of the squares of the residual, which is a measurement of how well the least square fit represents the data. The values are all below 1%, indicating that the temperature profile is linear in the bulk of the channel. Slight curvature in the temperature profile can be attributed to the temperature dependence of argon's thermal conductivity. Surprisingly, linear temperature profiles are observed up to the second argon-density peak, approximately 0.5 nm from the silver wall. Hence, strong density fluctuations do not significantly affect the temperature profile beyond 0.3 nm from each wall. In addition, the fine bin temperature distribution in liquid argon near the silver surface, especially within the first density peak near the walls, shows steep temperature variations compared to the bulk region. This behavior can be attributed to strong interactions of the silver–argon molecules.

Next, we calculate the Kapitza length, L_K, by using Equation 1.6. First, we obtained the local temperature gradient from the MD results and matched this with the slope of the linear curve fits to the temperature profiles in the bulk of the channel. Then, we extrapolated the liquid temperature onto the solid surface, which was assumed to be the center of the last silver layer facing the liquid, and calculated the local temperature jump ($\Delta T = T_{liquid} - T_{wall}$). As suggested by Equation 1.6, this temperature jump and the linear slope of argon temperature profile yield the value of L_K. In the current case, when thermostat is applied to all 23 silver layers, L_K of argon/silver interface is 11.1 and 9.4 Å at the hot (140 K) and cold (90 K) surfaces, respectively. These results clearly indicate the temperature dependence of L_K. In the mean time, we also calculated the heat flux, and utilized the temperature gradient obtained from MD result to calculate the thermal conductivity using the Fourier's law. Thermal conductivity for liquid argon at average temperature of 115 K is 103.9 mW/mK, which matches the published bulk values well [9,35].

In order to investigate the thermostat effects, we created Case 2 by applying the Langevin thermostat to the outer most second, third, and fourth layers of the left and right silver slabs (note that the outer first layer of each wall is fixed to define a constant simulation volume). As a result, the remaining 20 silver layers are left to interact freely to develop their own temperature profile based on phonon scattering of silver surfaces. By considering that the wall layers are 0.2043 nm apart from each other, 20 silver layers separate the thermostated molecules from the liquid/solid interface by approximately 15 atomic spacing. Such large distances minimize the thermostat effects at the liquid/solid interface. Figure 1.3 shows the temperature profile for Case 2. Sudden temperature drops at the interface of the thermostat applied and thermostat-free regions are observed inside the silver domains. This is due to the dynamic rescaling of thermostat, which results in phonon mismatch at this interface. Any thermostat altering particle motions corrupts this autocorrelation by reducing the thermal conductivity at the interface of the thermostat applied region. This implies that the Langevin thermostat leads to reduced conductivity between the thermostat applied and thermostat-free silver layers and creates an *artificial thermal resistance*. Similar behavior has also been consistently observed for Berendsen and Nose–Hoover thermostat techniques with different amounts of temperature jumps.

Figure 1.3b shows zoomed view of the temperature distribution in silver. After the temperature jump due to the artificial thermal resistance, silver temperature is reduced linearly to 136.8 K at the argon/silver interface. This reduction in silver temperature is due to heat flux through the solid wall. Similar to the hot wall, the cold wall also experiences a temperature jump, flowed with linear increase in silver temperature to 92.7 K. Linear least square fits to silver and argon temperatures result in good fits with measured sum of the squares of the residual lower than 1%. We must note that MD simulations here consider only the intramolecular vibrational energy transfer, while thermal transport in metals is dominated by electron interactions. As a result, calculated thermal conductivity of silver is less than the reported bulk material values. Despite this limitation, silver blocks serve as heat baths to induce heat flux on the argon/silver interfaces. Because of the definitions of the Kapitza resistance and Kapitza length given in Equations 1.5 and 1.6, heat transfer in silver domains has negligible effects on our investigation. Similar to the earlier calculations, L_K is

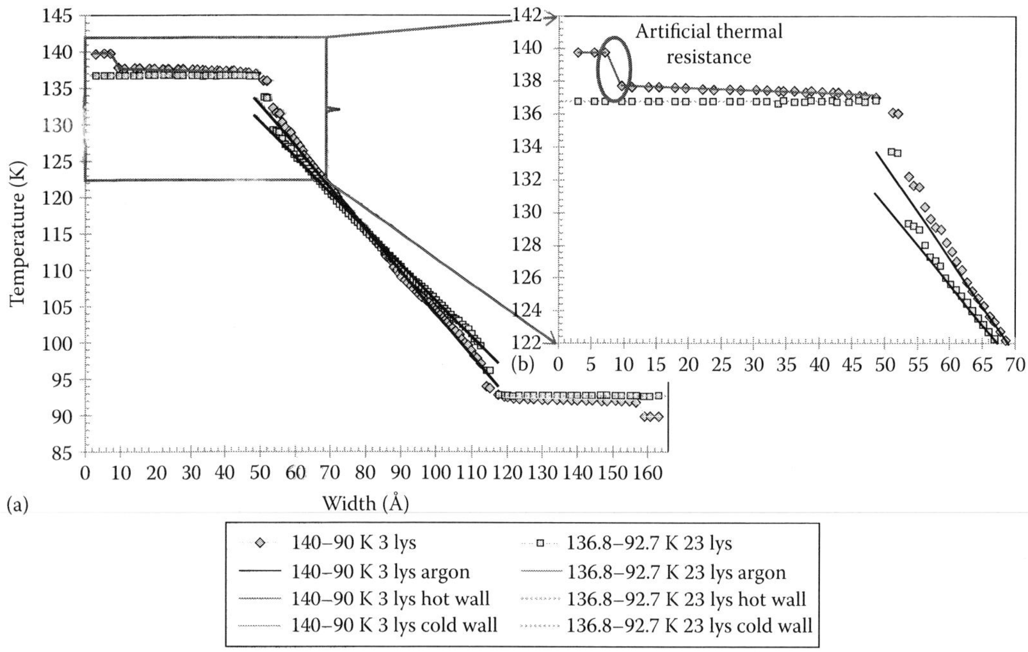

FIGURE 1.3 Temperature profile when Langevin thermostat is applied on three layers of silver walls to keep them at 140 and 90 K, respectively (Case 2). Result for Case 1 is also shown for comparison (a). Zoomed view of the hot surface shows presence of an artificial thermal resistance at the interface of the thermostat applied and thermostat-free regions of the solid domain (b).

found by using Equation 1.6. Extrapolation of argon temperature onto the center of the last silver layer facing liquid argon enabled us to calculate the local temperature jump ($\Delta T = T_{fluid} - T_{wall}$), thus the value for L_K. When thermostat is applied only to the outer three layers of silver walls, L_K at the argon/silver interface is found to be 5.8 and 3.1 Å on the hot (136.8 K) and cold (92.7 K) surfaces, respectively. Values of the Kapitza length are approximately half of the values calculated in Case 1. This proves the thermostat effect on the silver/argon interface. Interestingly, the calculated heat flux values for both cases are almost identical revealing that the total thermal resistance of the simulation system is approximately kept constant regardless of the location where the thermostat is applied. Slight reduction in the heat flux in Case 2 is due to heat conduction in silver slabs.

1.3.3 Constant Heat Flux

In an NVT ensemble (constant number of molecules, constant volume, and constant temperature), the main aim of the thermostat mechanism is keeping the temperature constant. Thermostat adding energy to the molecules at an incident when their instantaneous temperature is lower than the desired value, may remove energy from the same group of molecules if their temperature is higher at any following time step. This behavior corrupts the vibrational nature of the material and affects phonon transfer, resulting in a temperature jump between the thermostat applied and free regions. Instead of assigning a constant surface temperature using a thermostat, one can maintain heat flux in the system by simply injecting and removing energy from the left and right heat baths, respectively. Different than using a thermostat, non-translational kinetic energy is added to the molecules of hot reservoir (and removed from the molecules of the cold reservoir) at every time step continuously, regardless of their instantaneous temperatures. In such case, NVE ensemble is obtained where the *heat flow is constant* and the surface temperature fluctuates. This approach is identical to the solution of heat conduction problem by assigning heat flux (Neumann boundary condition) on both boundaries, which makes the continuum solution of heat conduction equation ill-posed, due to the undetermined integration constant that fixes the temperature value. MD simulation of such a case has similar difficulties. Since the energy is added and removed at a constant rate, the surface temperature is continuously adjusted, and hence, actual temperature of the surface cannot be assigned a priori. For such cases, the initial temperature of the system plays a key role to obtain the desired temperature. For example, in order to create a case with 140–90 K surface temperatures, simulation should initially be brought to thermal equilibrium at 115 K, which is the average temperature. Then, energy is added and removed from the left and right walls, respectively, while the heat flux value is adjusted to obtain the desired temperature difference of $\Delta T \approx 50$ K between the two reservoirs. The temperature profile in the system, obtained by adding and removing energy from outermost three layers of the left and right silver slabs (Case 3), is shown in Figure 1.4. Unlike the cases that utilized thermostat, adding energy does not create unphysical temperature jumps inside the solid domains. Similar to the previous cases, we calculated the L_K values at the liquid/solid interface, which are 5.9 and 3.0 Å on the 139.8 and 90.9 K surfaces, respectively. Consistent with our earlier observations, the Kapitza lengths obtained by adding heat are similar with Case 2, which utilized thermostat on outermost three layers of the walls. Since adding and removing energy from the solid domains eliminate the artificial thermal resistance observed in thermostat applied systems, this method seems to be the most proper way to investigate the ITR using non-equilibrium MD simulations.

Results clearly show that the Kapitza resistance and Kapitza length should be predicted either by using a large solid domain by applying thermostat to several wall layers away from the liquid/solid interface, or by adding and extracting energy from the surfaces. The latter approach can utilize smaller number of wall layers and minimizes the computational burden of simulating large number of wall molecules. It is also shown that the artificial thermal resistance varies between

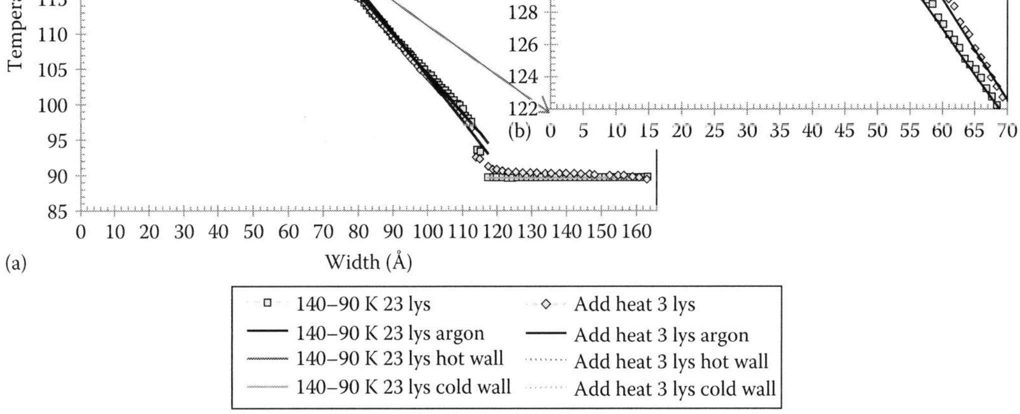

FIGURE 1.4 Temperature profile when energy is added and removed from three outermost layers of silver walls (Case 3). Result of Case 1 is plotted for comparison (a). Zoomed view of the hot surface shows elimination of the artificial thermal resistance in Case 3 (b).

3.8×10^{-9} and 4.2×10^{-9} km^2/W using a Langevin thermostat in 140–90 K range. The Kapitza resistance at liquid argon silver interface varies from 5.4×10^{-9} to 2.2×10^{-9} in the same temperature range.

1.4 TEMPERATURE DEPENDENCE OF THERMAL RESISTANCE AT THE WATER/SILICON INTERFACE

Silicon and silicon-based materials as semiconductors have been widely used in integrated circuits. Parallel to the technological developments, the transistor density on a single circuit has been increasing exponentially each year according to the prediction of Moore's law [36]. The main driving mechanism, enabling current developments, is the micro- and nano-fabrication technologies providing higher packing densities. It was also Richard Feynman's vision 50 years ago, when he was pointing out the necessity to understand the unique physics governing the matter in small scale [37]. Since then, micro- and nanoscale physics of silicon-based materials attracted greatest attention of scientists and engineers.

MD has been utilized to investigate phonon transport through the interfaces [38–55]. There are quite a number of studies revealing the dependence of ITR to various molecular parameters; but the effect of surface temperature on ITR is still a question. For a solid/solid interface, it was validated that the temperature dependence of ITR is similar to the theoretical predictions where increase of temperature enhances the phonon density of states and decreases the ITR [53]. However, in the case of a solid/liquid interface, liquid molecular structure formed near the surface plays a dominant role on the interface phonon transport. Opposite to the theoretical expectations, our earlier studies showed increase of ITR with increased temperature [4–8,40]. Such behavior develops as a result of *the temperature dependence of liquid density layering.* For the studied monatomic liquid argon in contact with silver [4,7] and solid argon surfaces [8,40], liquid density was found to be higher and better packed near the cold surface; as a result, the ITR was lower than the hot side. Hence, correct characterization of temperature dependence of ITR needs a detailed investigation of near-surface liquid structure whose dependence on temperature shows variation for surfaces having different wetting behavior.

A liquid near a solid surface exhibits layering on the molecular scale due to the ordering of the liquid molecules under the influence of surface forces [33]. This density layering is a specific property of a liquid/solid couple and develops mainly as a function of the surface wettability driven by *the intermolecular interactions.* However, the parameters determining the interactions of *the specific liquid/solid couples* are still under investigation. Frequently, the interaction parameters of non-identical molecules are calculated from parameters of the pair of identical molecules by using the Lorentz–Berthelot mixing rule. However, the interaction parameters between identical molecules are optimized for a bulk material system and may need to be re-parameterized for the interaction of non-identical pairs instead of using the simple Lorentz–Berthelot mixing rule. Therefore, the intermolecular interaction parameters should be properly specified in order to capture the correct behavior of ITR for a given solid/liquid couple.

Many authors indicate their concerns about the Lorentz–Berthelot mixing rule and proposed ways to calculate the interaction parameters of nonidentical molecules accurately [56,57]. One methodology is the measurement of water contact angle to define the interactions of water with carbon [58] and silica [59] surfaces for rigid solid surfaces.

The angle at which a liquid/vapor interface meets the solid surface is known as the contact angle. Even though the hydrophobic or hydrophilic nature of a surface should be defined by also considering other additional surface properties [60,61], contact angle can be utilized as a measure of the wetting phenomena, qualitatively. When a drop of liquid is on a solid surface, it may remain as a spherical drop or spread to cover the solid surface. Specifically, wetting occurs when the contact angle is less than 90°, while the liquid does not wet the surface for contact angle values greater than 90°. The wetting characteristic of silicon has been of great interest for a vast number of applications.

From the earliest experimental studies, it was reported that the measured contact angle value is strongly dependent on the surface conditions of silicon. A thin layer (less than 1 nm) of silicon dioxide forms naturally on the surface when silicon is exposed to the ambient air. Hermanson et al. measured contact angles between 35° and 96° by varying the surface treatment that removed the native oxide layers on the surface at different amounts [62]. Silicon and thermally grown silicon dioxide are studied by Martinez by removing the native oxide formed on silicon with etching [63]. Water contact angle on silicon and silicon dioxide are reported to be 89° and 48°, respectively. Similarly, Arkles reported the contact angle between water and etched silicon between 86° and 88° [64]. In a recent study, Cho et al. measured the water contact angle as 70.1° on bare silicon surface [65]. Our personal communications with the authors point out the difficulties of obtaining and maintaining a silicon surface without any chemical termination. The amount and time scale of oxide layer growth was mentioned in [66], where freshly etched silicon exposed to air at room temperature is rapidly covered by a thin oxide layer of thickness 0.5 nm in a minute, followed with a logarithmic growth rate. Hence, measured contact angle values in the literature differ with the thickness of the SiO_2 layer and experimental conditions. Bare silicon surface without any chemical termination is hydrophobic, while the oxidized layer adds hydrophilicity, based on the oxide layer's thickness. In this study, we use a contact angle value of 88° to compare with the MD calculated water contact angles on pure silicon surface.

The *objective* of this study is to predict the ITR between silicon and water for the case where the interaction parameters are properly tuned to capture the hydrophobic behavior of bare silicon surface. Starting with the classical mixing rule estimations, we varied the interaction parameters between silicon and water in order to match the microscopic contact angle calculated by MD with the experimentally measured macroscopic contact angle of 88°. In the meantime, an enhanced understanding of the near-wall density structure influences on ITR and its dependence on the surface temperature will be provided.

1.4.1 Three-Dimensional MD Simulation of Water/Silicon System

Liquid water is confined between two parallel solid silicon walls as illustrated in Figure 1.5. Total dimensions in longitudinal (x), vertical (y), and lateral (z) directions are 13.4, 2.7, and 2.7 nm, respectively. The domain was constructed using 2500 silicon and 5088 water molecules while the corresponding density of water is $\rho = 1.006$ g/cm^3. Walls on each side of the domain contain 25 silicon layers, which are 0.136 nm apart from each other, and the spacing between two silicon walls is 6.4 nm in x-direction. Periodic boundary conditions are applied in the vertical (y) and lateral (z) directions. Silicon surface is modeled as diamond cubic structure with (0, 0, 1) crystal plane facing the fluid. Atoms in the outmost layer of both silicon walls are fixed to their original locations to maintain a fixed volume system, while the remaining atoms throughout the domain are free to move.

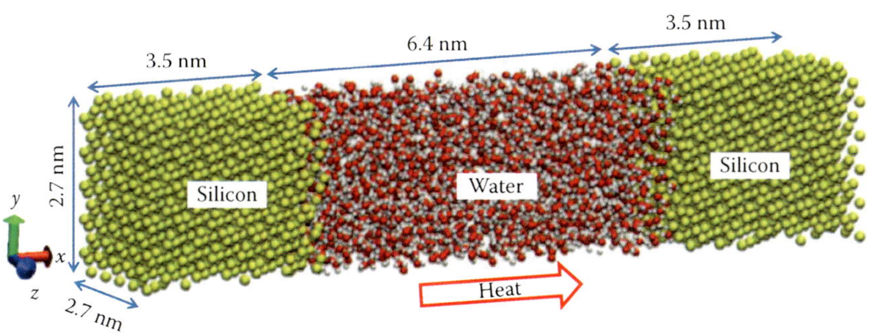

FIGURE 1.5 Snapshot of the simulated water–silicon system.

Many different models for liquid water have been developed by fitting some physical properties of water; as a result, different models show agreement with the different properties of water. A model requiring more fitting parameters gives better results, but becomes computationally inefficient due to its complexity. On the other hand, many studies showed that the thermodynamic properties of water models were most sensitive to the van der Waal's repulsive, short range Coulomb, and the polarization components of the potential [67]. Thus, the model explaining the van der Waal's and Coulombic interactions with a proper polarization of water structure provide a fair understanding of complex water behavior.

Water models can be classified by the number of points used to define the model, and whether the structure is treated as rigid or flexible. In our study, for the purpose of simplicity and computational efficiency, we chose SPC/E (Extended Simple Point Charge) model, which can be described as effective rigid pair potentials composed of Lennard-Jones and Coulombic terms [68]. This water model has three interaction sites, corresponding to the three atoms of the water molecules. Each atom gets assigned a point charge to model the long-range Coulombic interactions, and the oxygen atom also exhibits L-J potential to model van der Waal's forces. The truncated (6–12) L-J potential given in Equation 1.1 is utilized.

SPC/E is a tetrahedral model with an OH bond length of 0.1 nm and H–O–H angle of 109.47°. SHAKE algorithm [69] is used to constrain the bond lengths and angle, thus making the model rigid. A short-range L-J term with a cutoff distance of 1 nm acts on the oxygen, while a charge $q(0.4238e)$ is on each hydrogen, compensated by a charge $-2q(-0.8476e)$ on the oxygen. The long-range Coulombic part has been treated with a particle–particle particle-mesh (PPPM) solver, which can handle long-range interactions for periodic systems with slab geometry more efficiently than the Ewald sum algorithm [70]. PPPM invokes a particle–particle particle-mesh solver, which maps atom charge to a 3D mesh. It uses 3D fast Fourier transforms to solve Poisson's equation on the mesh, and then interpolates electric fields on the mesh points back to the atoms. Si–Si interactions were calculated by Stillinger–Weber potential, which considers two-body interactions with an additional many body dependence [71]. SPC/E water molecules interact with silicon surface only by van der Waal's interactions of silicon and oxygen using a cutoff distance of 1 nm. The molecular interaction parameters are given in Table 1.1. For the silicon–oxygen interactions, parameters can be calculated by the Lorentz–Berthelot (L–B) mixing rule given as,

$$\sigma_{Si-O} = \frac{\sigma_{Si-Si} + \sigma_{O-O}}{2}, \quad \varepsilon_{Si-O} = \sqrt{\varepsilon_{Si-Si} \times \varepsilon_{O-O}} \tag{1.7}$$

Using the corresponding parameters given in Table 1.1, the L–B mixing rule predicts the interaction parameters $\sigma^*_{Si-O} = 2.6305$ Å and $\varepsilon^*_{Si-O} = 0.12088$ eV.

Newton's equations of motion were integrated using the Verlet algorithm with a time step of 0.001 ps. We use LAMMPS [25] as our MD solver. Simulations were started from the Maxwell–Boltzmann velocity distribution for all molecules at 323 K. NVT ensemble was applied with Nose–Hover thermostat keeping the system at 323 K. Initial particle distribution was evolved 2×10^6 time steps (2 ns) to reach an isothermal steady state. Afterward, the Nose–Hover thermostat is applied only on the outmost

TABLE 1.1
Molecular Interaction Parameters Utilized

Molecule Pair	σ (Å)	ε (eV)	q (e)
O–O	3.166	0.006739	−0.8476
H–H	0	0	+0.4238
Si–Si	2.095	2.168201	0

second, third, and fourth layers of the both silicon slabs to induce heat flux through the liquid/solid interfaces by maintaining different wall temperatures. At the same time, NVE ensemble is applied to the remaining silicon and water molecules. Simulations are performed for an additional 4×10^6 time steps (4 ns) to ensure that the system attains equilibrium in presence of the heat flux. After which, time averaging of desired properties are performed through 24×10^6 additional time steps (24 ns).

Classically, energy fluctuation in an NVT system of a limited number of molecules is nonzero. In such case, increase of the averaging time decreases the fluctuations to a minimum value. This minimum value is a function of multiple molecular parameters, but mostly depends on the time scales of the molecular mechanisms that are modeled. We performed 20 ns averaging which is considered excessively long for a classical MD system. However, the phonon transport in our system occurs at different frequencies. Depicting time-averaged behavior of low frequency phonons requires excessively long simulation times that are computationally unfeasible. Further complexity develops due to the phonon scattering at the interface, which makes theoretical estimation of the time scale for the complete phonon spectra impossible. Overall, the fluctuations are mainly dependent on ε_{Si-O} and the specified temperature difference, where high ε_{Si-O} values and large temperature differences in the system result in lower fluctuations. Although the highest fluctuations in the temperature and density profiles are less than 2% and 1%, respectively, these variations result in higher fluctuations (≈9%) in the ITR calculations. Due to these fluctuations, we performed statistical analysis of our results and report the ITR behavior with corresponding standard deviations (see Section 1.3 for details).

The computational domain is divided into 100 slab bins with the size of 0.1343 nm for temperature and 200 slab bins with the size of 0.0672 nm for density profiles. Smaller bin size, 800 slab bins with the size of 0.0168 nm, is also employed in order to resolve the fine details of the near-wall water density distributions. Irving–Kirkwood (I–K) expression given in Equations 1.4 through 1.6 is utilized to compute the heat flux vector for an N particle system.

1.4.2 Influence of the Near-Wall Water Density Structure on ITR

In an earlier study [3], we investigated the effect of interaction parameters, specifically interaction strength, to the wetting angle of water on silicon surface. We decreased the interaction strength value starting from the classical mixing rules estimation, and observed transition from hydrophilic to hydrophobic behavior in the studied range. Decrease in the interfacial energy leads to lower wetting, and hence the water density distribution near the surface changes drastically. In this study, we further investigate the effect of interaction parameter on the ITR. For such purpose, we used the ε_{Si-O} values of 0.4, 0.2, 0.125, 0.1, and $0.05 \times \varepsilon^*_{Si-O}$, where the ε^*_{Si-O} is the value calculated by the L–B rule ($\varepsilon^*_{Si-O} = 0.12088$ eV). Our main purpose here is not only to understand the influence of the resulting near-surface fluid structure on the ITR, but also to reveal the importance of the interaction parameter values, since some of the interaction strength values used in this work are similar to the values utilized in previous investigations of ITR using silicon-SPC/E water model [49–51].

Temperature distributions obtained in a silicon–water–silicon system subjected to 353 and 293 K on its ends are shown in Figure 1.6. The temperature profiles vary with the change of interaction strength value. For silicon domain, we observed a temperature drop at the interface of the thermostat applied and thermostat free regions. This artificial ITR created by the thermostat is due to the dynamic rescaling of thermostat, which results in phonon mismatch at this interface. This artificial effect of the thermostat was investigated earlier [2]. Nose–Hoover thermostat creates temperature gradient inside the thermostat applied region different than the temperature gradient of the thermostat-free domain depending on the temperature and heat flux of the system, which was previously investigated in detail [2]. The remaining 21 silicon layers interacting freely develop their own temperature profile based on the phonon characteristic of silicon. Silicon temperature varies linearly sufficiently away from the interface. We observed nonlinear behavior near the interface extending ≈0.54 nm into the silicon

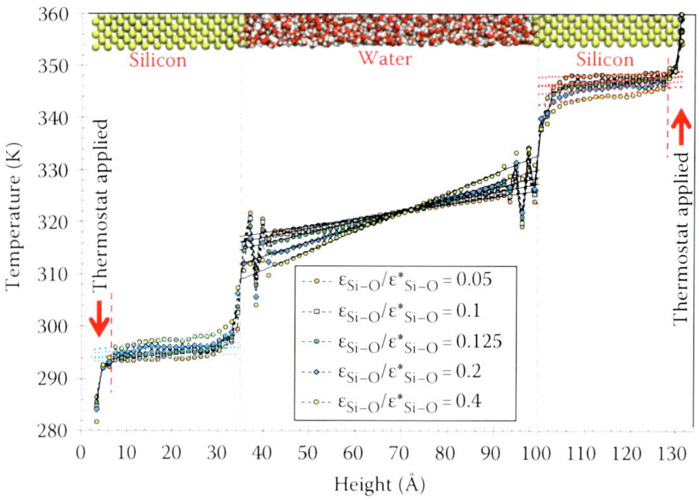

FIGURE 1.6 Temperature profiles of the silicon–water–silicon system at different ε_{Si-O} values while hot and cold reservoirs are kept at 353 and 293 K, respectively.

slab for the highest heat flux case. Such behavior is a strong evidence of phonon scattering mechanism from the interface, which needs further attention and will be investigated in detail in a separate study.

There are sudden temperature jumps at the liquid/solid interfaces. These temperature jumps are due to the mismatch in phonon spectrum of the water–silicon pair and results in the ITR. One can define a thermal resistance length (L_K), known as the Kapitza length, by extrapolating the temperature profile from liquid into the solid, where the wall temperature is reached. The Kapitza length can be predicted using Equation 1.6 where $\partial T/\partial n$ is the thermal gradient on the liquid side and $\Delta T = T_{liquid} - T_{wall}$. It is crucial to realize that Equation 1.6 requires onset of continuum behavior where a local temperature profile can be defined. In order to characterize the temperature distribution in the bulk of liquid water, we performed linear least square fits to the MD data, and measured sum of the squares of the residual, which is a measurement of how well the least square fit represents the data. The values are all below 1%, indicating that the temperature profile is linear in the bulk of the channel. Slight curvature in the temperature profile can be attributed to the temperature dependence of water's thermal conductivity. Water temperature varies linearly except the near-wall region where temperature fluctuates as a result of density layering formed due to the surface forces. Such behavior depends on the chosen bin size. While we presented fine distribution with 0.1343 nm bin size (100 bins), a slight increase in the bin size to 0.2686 nm (50 bins) resulted in linear behavior up to very near-surface region (≈0.15 nm away from surface). We discussed in an earlier study that in order to recover continuum behavior in such small environments, employed bin size should be equal to or bigger than one molecular diameter.

Next, we calculate the Kapitza length, L_K, by using Equation 1.6. First we obtained the local temperature gradient from the MD results, and matched this with the slope of the linear curve fits to the temperature profiles in the bulk of the channel. Then, we extrapolated the liquid temperature onto the solid surface, which was assumed to be the center of the last silicon layer facing the liquid; and calculated the local temperature jump ($\Delta T = T_{liquid} - T_{wall}$). As suggested by Equation 1.6, this temperature jump and the linear slope of water temperature profile yield the value of L_K. In the mean time, we also calculated the heat flux using Equation 1.3, and utilized the temperature gradient obtained from the MD result to calculate the thermal conductivity of confined water using the

TABLE 1.2
Kapitza Lengths at Hot (L_{K-H}) and Cold (L_{K-C}) Surfaces, Heat Flux, Temperature Gradients, and Thermal Conductivity of Liquid Water for Five Different Interaction Strength Values

Case	ε_{Si-O} (eV)	$\varepsilon_{Si-O}/\varepsilon^*_{Si-O}$	Contact Angle (°) [3]	q_x (W/m^2)	dT/dx (K/Å)	k_{water} (W/km)	T_H (K)	L_{K-H} (nm)	T_C (K)	L_{K-C} (nm)
1	0.048352	0.4	46.3	7.712	0.369	0.690	344.4	2.977	297.3	2.687
2	0.024176	0.2	57.8	5.691	0.281	0.67	345.8	5.263	296.1	5.119
3	0.01511	0.125	108.9	4.286	0.215	0.659	347	8.186	295.1	8.878
4	0.012088	0.1	129.7	3.588	0.173	0.684	347.1	10.597	295.1	11.57
5	0.006044	0.05	169.3	2.883	0.137	0.694	348.3	14.785	294.1	16.261

Notes: Hot and cold reservoirs are kept at 353 and 293 K, respectively. Contact angles for the corresponding interaction strengths are adopted from a water nano-droplet study.

Fourier's law. Calculated thermal conductivity for liquid water at average temperature of 323 K is ≈0.68 W/km, which matches the published bulk values well. The calculated Kapitza lengths at hot (L_{K-H}) and cold (L_{K-C}) surfaces, heat flux, temperature gradients, and thermal conductivity of liquid water are given in Table 1.2 for five different interaction strength values.

In the studied system, phonon transport that propagates inside the silicon domain at different wavelengths creates thermal fluctuations. As a result L_K cannot converge to a precise value. In order to have a fair judgment on L_K behavior, we performed a statistical investigation of L_K variation. Using a total simulation time of 24 ns, we obtained time averages at every 2, 4, 8, and 24 ns, resulting in 12, 6, 3, and 1 data sets, respectively. Using these total 22 differently averaged temperature profiles, 22 different L_K values are calculated. Later, the standard deviation of L_K and surface temperatures are calculated while the reported L_K value is the mean of the 22 L_K values. Standard deviations of the L_K values are given in Figure 1.7 as error bars. Overall, fluctuation of L_{K-H} is higher than L_{K-C}. Also standard deviation increases with decreased ε_{Si-O} while the highest value of SD is around 7%.

The increase of interaction strength creates better coupling between the water and silicon molecules, and hence, decreases the ITR and L_K values. Variation of L_K values for different $\varepsilon_{Si-O}/\varepsilon^*_{Si-O}$ at hot and cold surfaces are given in Figure 1.7a. Surface temperatures are also indicated on the plot. Since the temperature difference through the system is specified by the thermostat and fixed in the studied system, decrease of ITR results in increase of developed heat flux value. Regardless of this, resulted hot surface temperatures and cold surface temperatures for different ε_{Si-O} cases are very close to each other ($T_H \approx 347$ K and $T_C \approx 295$ K). Measured L_K values at cold and hot surface are found to be different. Since the L_K values of different ε_{Si-O} cases are different, we decided to normalize the L_{K-H} with L_{K-C} of the corresponding ε_{Si-O} case in order to understand the difference of L_{K-H} and L_{K-C} for each case (Figure 1.7b). While the L_K measured on the hot surface is lower than the cold surface values for $\varepsilon_{Si-O} < 0.2 \times \varepsilon^*_{Si-O}$, we observed higher L_{K-H} than L_{K-C} for larger ε_{Si-O} cases. The interesting behavior of L_{K-H} and L_{K-C} for different ε_{Si-O} cases clearly indicates that L_K at a liquid/solid interface is not dominated by the surface temperature.

Next, we investigate the water and silicon density profiles in Figure 1.8, which illustrates the system consisting of three distinct domains. A constant silicon density of 2.35 g/cm^3 is observed. Water density is constant and equal to 1.006 g/cm^3 in the bulk of the channel. Near-surface water density exhibits well-known density layering due to the surface forces and local water–water interactions [3].

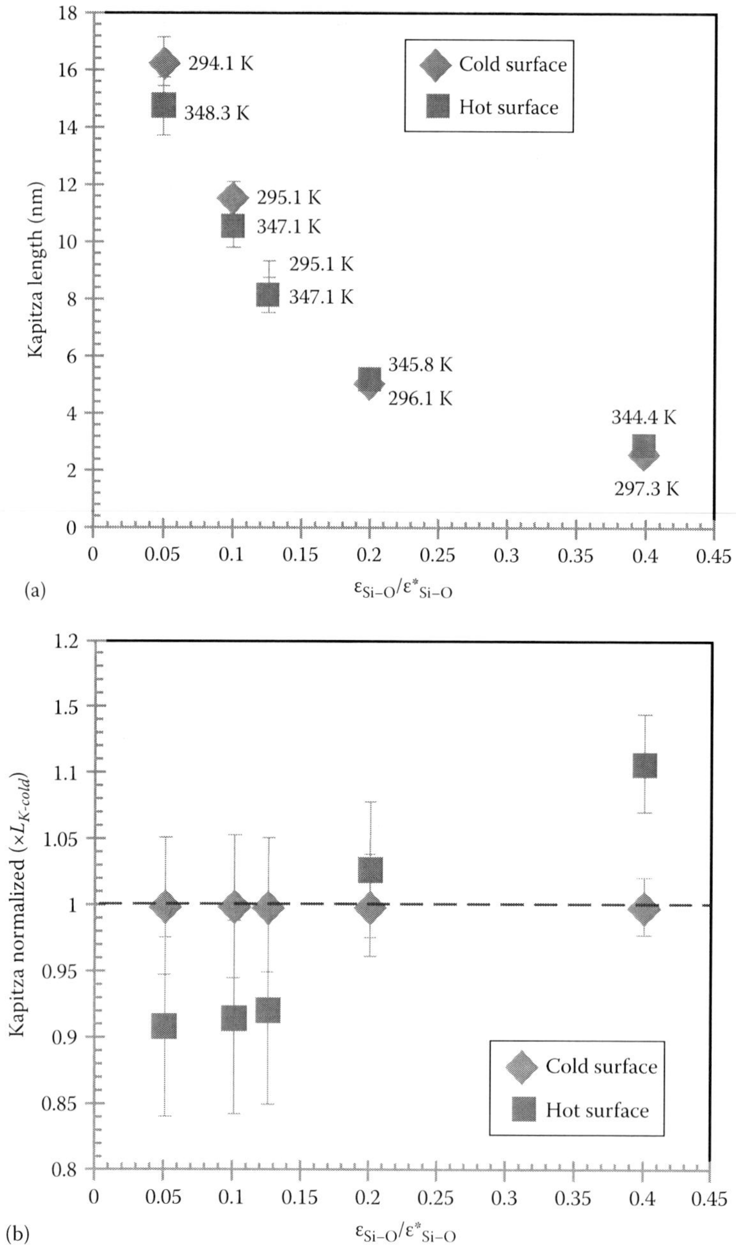

FIGURE 1.7 (a) L_K values for different $\varepsilon_{Si-O}/\varepsilon^*_{Si-O}$ ($\varepsilon^*_{Si-O} = 0.12088$ eV) cases obtained at hot and cold surfaces. Surface temperatures are also given on the plot. (b) Variation of the hot wall L_K value normalized with the corresponding cold surface L_{K-C} value for different $\varepsilon_{Si-O}/\varepsilon^*_{Si-O}$ cases.

Increase in the interaction strength develops stronger water layering for $\varepsilon_{Si-O} \geq 0.2 \times \varepsilon^*_{Si-O}$, while the increase in the near-wall water density results in slight decrease of density in the bulk. By simple examination of Figure 1.8, one can conclude that the water density peaks near the hot surface are higher than the cold side for $\varepsilon_{Si-O} \geq 0.2 \times \varepsilon^*_{Si-O}$. However, a more detailed investigation should be conducted to understand the behavior of water density near the surface and the effects of near-wall density distribution on the ITR.

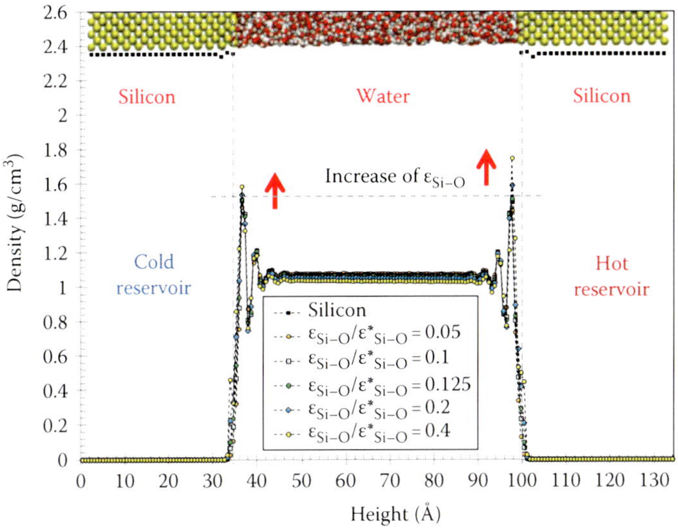

FIGURE 1.8 Density profiles of the silicon–water–silicon system at different ε_{Si-O} values while hot and cold reservoirs are kept at 353 and 293 K, respectively.

Water density variations near the hot and cold reservoirs are shown in Figure 1.9 using finer bin size of 0.0168 nm for various ε_{Si-O} values. Similar to our earlier study, water penetration into the gap between the first and second silicon layers is observed, and the water density in this region increases with increased ε_{Si-O}. There also forms an additional density layer between the first density peak of water and silicon surface while these two density behaviors become profound for $\varepsilon_{Si-O} \geq 0.2 \times \varepsilon^*_{Si-O}$. These two new density layers are indicated with arrows in Figure 1.9c. Simply, decrease of ITR with an increased ε_{Si-O} is the result of the stronger coupling between water and silicon molecules due to closer interactions. Comparisons of hot and cold side water densities of each case are also shown in Figure 1.9a through d. Water distributions near the hot and cold surfaces are identical for low ε_{Si-O} values ($\varepsilon_{Si-O} < 0.2 \times \varepsilon^*_{Si-O}$). Starting with the $0.2 \times \varepsilon^*_{Si-O}$ case, the additional density layer near the surface is more prominent at the cold side, which creates a better phonon coupling resulting in the L_{K-C} value being lower than L_{K-H}. While the densities for the $\varepsilon_{Si-O} < 0.2 \times \varepsilon^*_{Si-O}$ cases show no dependency on the surface temperature, density becomes more packed with decreased surface temperature for $\varepsilon_{Si-O} \geq 0.2 \times \varepsilon^*_{Si-O}$, which decreases the ITR. Thus, in Figure 1.7b, we observed not only the variation of ITR with temperature but also the temperature dependency of the water density, which affects the ITR value.

1.4.3 Temperature Dependence of ITR

Until now, we clarified that ITR strongly depends on the near-surface water density structure, which is a function of ε_{Si-O} value. Therefore, we need to identify the proper parameters for water–silicon interactions. For this purpose, we conducted a nano water droplet study where we characterized the interaction parameters by matching the microscopic wetting behavior of water on silicon with the macroscopic experimental measurements [3]. Figure 1.10a through e shows the snapshots of the nanoscale droplets formed on bare (1, 0, 0) silicon surface and the microscopic contact angle values for the 0.4, 0.2, 0.125, 0.1, and $0.05 \times \varepsilon^*_{Si-O}$ values. Increase of interaction strength decreases the contact angle so that the wettability changes from low to high in the studied range. By considering the line tension effects creating deviations between the nanoscale contact angle and the macro-scale value, an appropriate estimate for the interaction strength value

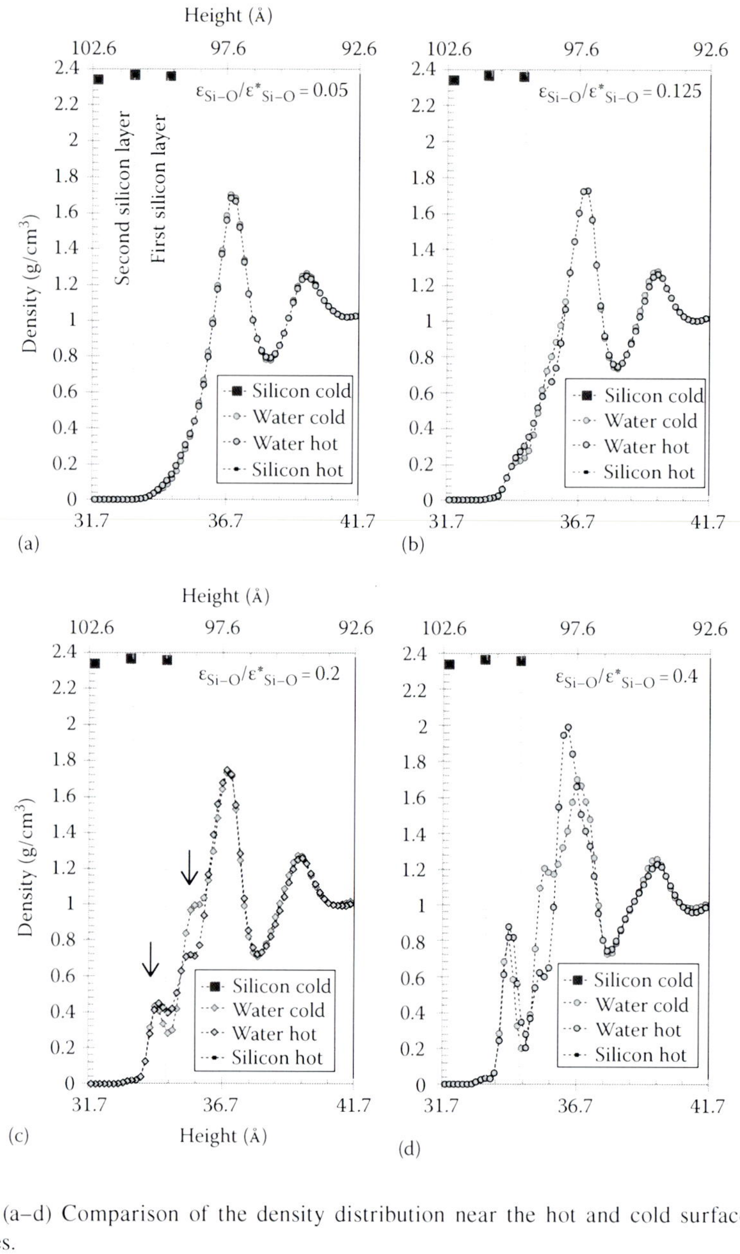

FIGURE 1.9 (a–d) Comparison of the density distribution near the hot and cold surfaces for various $\varepsilon_{Si-O}/\varepsilon^*_{Si-O}$ cases.

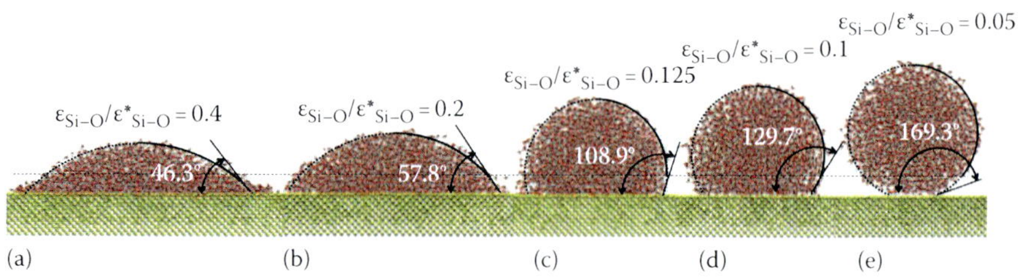

FIGURE 1.10 (a–e) Snapshot of the nanoscale water droplets formed on bare (1, 0, 0) silicon surface and microscopic contact angle values at the corresponding interaction strength values.

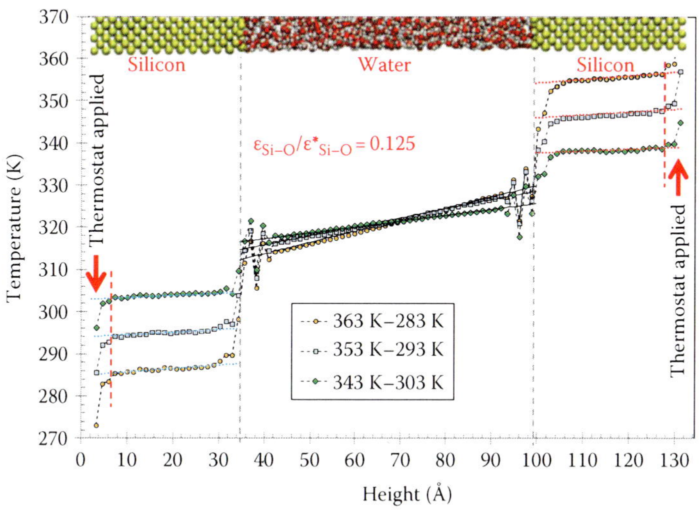

FIGURE 1.11 Temperature profiles of the $0.125 \times \varepsilon^*_{Si-O}$ case with various hot and cold reservoir temperatures.

was found as $\varepsilon_{Si-O} = 0.125 \times \varepsilon^*_{Si-O}$, which recovers the experimental *macroscopic* contact angle of water on pure silicon surface as 88° [3]. Details of this result and the methodology for calculation of the line tension effects to relate the wetting angles of nanoscale and macroscopic droplets can be found in [3].

Using the proper interaction strength value of $\varepsilon_{Si-O} = 0.125 \times \varepsilon^*_{Si-O}$, we investigated the temperature dependency of L_K at the water–silicon interface. For this purpose, we assigned different temperature values on the hot reservoir as 363, 353, and 343 K while the cold reservoir was kept at 283, 293, and 303 K, respectively. Temperature profile for each case is shown in Figure 1.11. As the temperature difference assigned through the heat baths decreases, resulting heat flux values are decreased. The calculated L_{K-H}, L_{K-C}, heat flux, temperature gradients, and thermal conductivity of liquid water are tabulated in Table 1.3.

Before focusing on the L_K variation at different surface temperatures, we should study the density structures near surface at each different temperature case. Density profiles of silicon–water–silicon system for different temperature differences are given in Figure 1.12. While Figure 1.12a shows slightly higher density peaks on the cold side for coarse bin (0.0672 nm) results, real behavior can only be characterized with a finer bin size (0.0168 nm) as shown in Figure 1.12b. Comparison of water densities near different temperature surfaces through Figure 1.12b validates that the near-wall water structure shows no dependency on the surface temperature, similar to our earlier results for low wetting surfaces.

TABLE 1.3
Kapitza Lengths at Hot (L_{K-H}) and Cold (L_{K-C}) Surfaces, Heat Flux, Temperature Gradients, and Thermal Conductivity of Liquid Water of $0.125 \times \varepsilon^*_{Si-O}$ Case

Case	T_H and T_C (K)	$\varepsilon_{Si-O}/\varepsilon^*_{Si-O}$	q_x (W/m^2)	dT/dx (K/Å)	k_{water} (W/km)	T_H (K)	L_{K-H} (nm)	T_C (K)	L_{K-C} (nm)
1	363–283	0.125	5.411	0.271	0.661	355.5	8.213	286.5	8.97
2	353–293	0.125	4.286	0.215	0.659	347	8.186	295.1	8.878
3	343–303	0.125	2.843	0.14	0.672	338.4	8.321	303.8	8.621

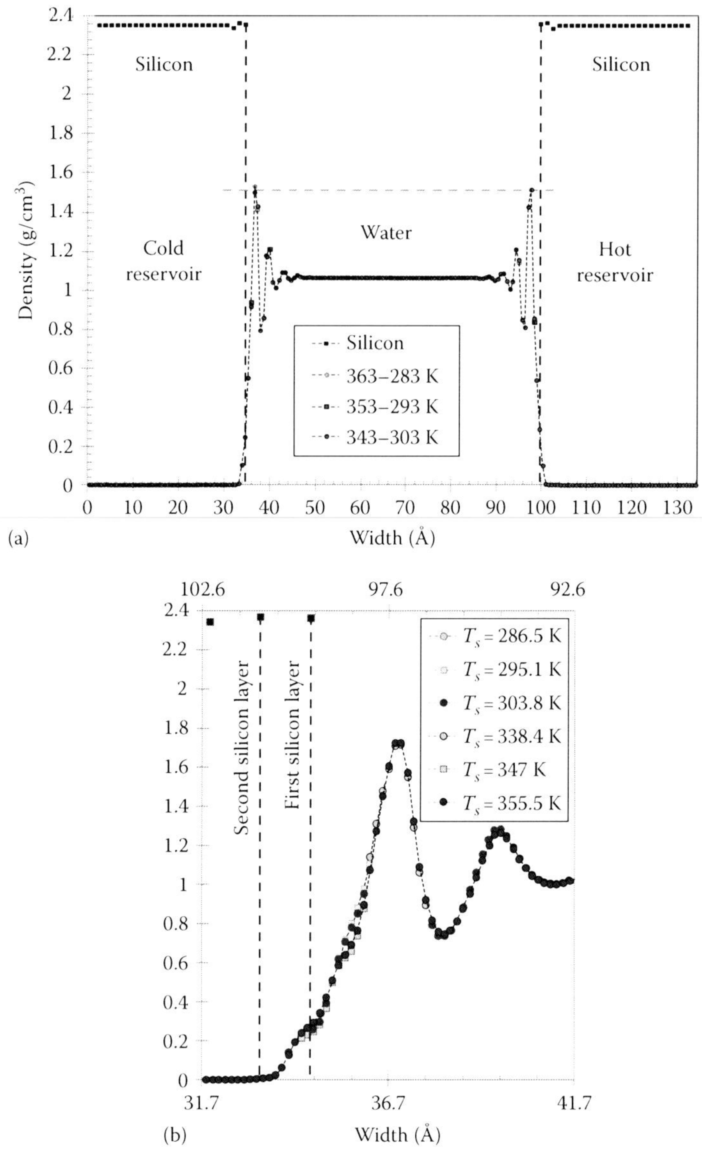

FIGURE 1.12 (a) Density profiles for different temperature cases. (b) Comparison of water distributions near the surfaces at different temperatures.

Since the near-surface water density is independent of the wall temperature, we can now properly characterize the temperature dependency of L_K. Figure 1.13 gives the L_K values at different temperature surfaces with the corresponding standard deviations, where lower temperature difference cases exhibit higher fluctuations. Similar to the theoretical predictions, ITR decreases with increased temperature since the phonon density of state is higher at high temperatures, which creates better coupling. Calculated L_K values (≈9 nm) are consistent with the experimental measurements, where 6–12 nm L_K values are found for hydrophilic and hydrophobic surfaces [72].

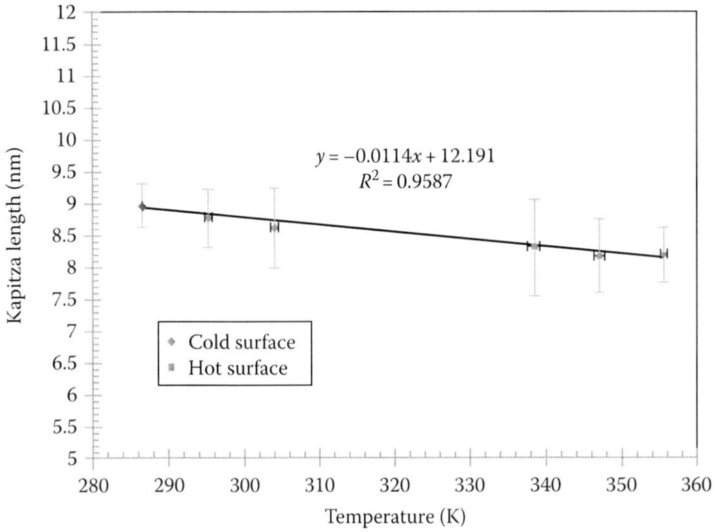

FIGURE 1.13 Variation of L_K with the change in the surface temperature for $0.125 \times \varepsilon^*_{Si-O}$ case. Standard deviations of L_K are given as error bars.

1.5 SUMMARY

In this chapter, we investigated molecular level interface phonon transfer between solids and liquids. The continuum hypothesis and Fourier's law in nano-channels were studied while they found valid in channels with dimensions as small as 10 molecular diameters (≈3.5 nm). Afterward, we focused on the MD techniques for heat transport problems. We started with questioning the current thermostat use for creating a heat flux with the non-equilibrium dynamics. Using three distinct boundary treatment methods, we investigated heat conduction in liquid argon-filled silver nano-channels with the objective of characterization of the Kapitza resistance. The first and most common approach utilizes constant temperature surfaces imposed via a thermostat (Case 1). However, the Kapitza resistance calculated with this method results in artificially large values, compared with Case 2, which thermostats the outermost three wall layers. In fact, Case 2 exhibits artificial resistance at the edge of the thermostat applied region as an unphysical temperature jump in the homogenous silver domain. Detailed thermal resistance analysis of both cases clearly show that the isothermal wall approach suffers from the artificial thermal resistance on the liquid/solid interface, where MD-predicted resistance value is a superposition of the physical Kapitza resistance and the artificial resistance due to the thermostat effects. The alternative approach of adding and extracting energy at three outermost layers of silver walls (Case 3) eliminates the artificial thermal resistance, and results in Kapitza resistance values consistent with Case 2. However, this approach has its own limitations, such as the inability to assign wall temperature, which is similar to the undetermined temperature based on integration of heat conduction equation using Neumann boundary conditions. Unlike the continuum-based approach, MD converges to a temperature distribution determined by the initial temperature of the system. Results clearly show that the Kapitza resistance and Kapitza length should be predicted either by using a large solid domain by applying thermostat to several wall layers away from the liquid/solid interface, or by adding and extracting energy from the surfaces.

Next, we studied temperature dependence of the water/silicon ITR with proper emphasize on the surface wetting properties and liquid layering near the surface. Results show that the ITR between a liquid and solid is dominated by the liquid structure formed on the surface under high wetting conditions. In such case, change of temperature not only alters the phonon transport, but also creates different coupling between liquid and solid molecules due to the variation of liquid ordering near the surface. As a result, interface resistance increases with increased temperature for high-wetting surfaces. The opposite behavior is true for low-wetting surfaces, where the density structure near the surface is independent of the surface temperature. As a result, change of surface temperature only affects phonon transport. Similar to the theoretical expectations, ITR decreases with increased temperature of a surface with wettability. In the light of these results, the wetting behavior of a liquid/solid couple should be properly considered in order to characterize the behavior of ITR. For such cases, we also clarified that the classical mixing rules frequently employed in the literature are not suitable to capture the correct interface behavior. The interaction strength value predicted by the Lorentz–Berthelot mixing rule creates highly hydrophilic silicon surface opposite to the experimental measurements. Hence, by conducting a nanoscale water droplet study, we properly tuned the interaction parameters in order to recover the real hydrophobic nature of bare silicon surface. Using the proper parameters, we characterized the temperature dependence of Kapitza length at the water/silicon interface, where MD-predicted Kapitza length values ($\approx$9 nm) agree with the experimental results for slightly hydrophobic surfaces.

REFERENCES

1. G.E. Karniadakis, A. Beskok, N. Aluru, *Micro Flows and Nano Flows: Fundamentals and Simulation*, Springer-Verlag, New York, 2005.
2. D.G. Cahill, W.K. Ford, K.E. Goodson, G.D. Mahan, A. Majumdar, H.J. Maris, R. Merlin, S.R. Phillpot, Nanoscale thermal transport, *Journal of Applied Physics*, 93, 793 (2003).
3. S. Maruyama, Molecular dynamics methods in microscale heat transfer. In G.P. Celata (ed.), *Heat Transfer and Fluid Flow in Microchannels*, Begell House Inc., New York, 2002.
4. M. Barisik, A. Beskok, Boundary treatment effects on molecular dynamics simulations of interface thermal resistance, *Journal of Computational Physics*, 231, 7881–7892 (2012).
5. M. Barisik, A. Beskok, Wetting characterization of silicon (1,0,0) surface, *Molecular Simulation*, 39(9), 700–709 (2013).
6. M. Barisik, A. Beskok, Temperature dependence of thermal resistance at the water/silicon interface, *International Journal of Thermal Sciences*, 77, 47–54 (2013).
7. Z. Shi, M. Barisik, A. Beskok, Interface thermal resistance between liquid argon and various solids, *International Journal of Thermal Sciences*, 59, 29–37 (2012).
8. B.H. Kim, A. Beskok, T. Cagin, Molecular dynamics simulations of thermal resistance at the liquid-solid interface, *Journal of Chemical Physics*, 129, 174701 (2008).
9. A.J.H. McGaughey, M. Kaviany, Thermal conductivity decomposition and analysis using molecular dynamics simulations: Part I. Lennard-Jones argon, *International Journal of Heat and Mass Transfer*, 47, 1783–1798 (2004).
10. P. Borgelt, C. Hoheisel, G. Stell, Exact molecular-dynamics and kinetic-theory results for thermal transport-coefficients of the Lennard-Jones argon fluid in a wide region of states, *Physical Review A*, 42, 789–794 (1990).
11. H.M. Roder, C.A.N. Decastro, U.V. Mardolcar, The thermal-conductivity of liquid argon for temperatures between 110-K and 140-K with pressures to 70 MPA international, *Journal of Thermophysics*, 8, 521–540 (1987).
12. B.J. Bailey, K. Kellner, Thermal conductivity of liquid and gaseous argon, *Physica*, 39, 444 (1968).
13. H. Ziebland, J.T.A. Burton, The thermal conductivity of nitrogen and argon in the liquid and gaseous states, *British Journal of Applied Physics*, 9, 52–59 (1958).
14. M.P. Allen, D.J. Tildesley, *Computer Simulation of Liquids*, Clarendon, Oxford, U.K., 1987.
15. K. Jolley, S.P.A. Gill, Modeling transient heat conduction in solids at multiple length and time scales: A coupled non-equilibrium molecular dynamics/continuum approach, *Journal of Computational Physics*, 228, 7412–7425 (2009).
16. P.K. Schelling, S.R. Phillpot, P. Keblinski, Comparison of atomic-level simulation methods for computing thermal conductivity, *Physical Review B*, 65, 144306 (2002).

17. Z.X. Huang, Z.A. Tang, Evaluation of momentum conservation influence in non-equilibrium molecular dynamics methods to compute thermal conductivity, *Physica B*, 373, 291–296 (2006).
18. A. Maitia, G.D. Mahana, S.T. Pantelidesa, Dynamical simulations of nonequilibrium processes: Heat flow and the Kapitza resistance across grain boundaries, *Solid State Communications*, 102, 517–521 (1997).
19. Q. Tang, A molecular dynamics simulation: The effect of finite size on the thermal conductivity in a single crystal silicon, *Molecular Physics*, 102, 1959–1964 (2004).
20. P. Jund, R. Jullien, Molecular-dynamics calculation of the thermal conductivity of vitreous silica, *Physical Review B*, 59, 13707 (1999).
21. H. Zhong, J.R. Lukes, Interfacial thermal resistance between carbon nanotubes: Molecular dynamics simulations and analytical thermal modeling, *Physical Review B*, 74, 125403 (2006).
22. C. Schröder, V. Vikhrenko, D. Schwarzer, Molecular dynamics simulation of heat conduction through a molecular chain, *Journal of Physical Chemistry A*, 113, 14039–14051 (2009).
23. Y.W. Yang, X.J. Liu, J.P. Yang, Nonequilibrium molecular dynamics simulation for size effects on thermal conductivity of Si nanostructures, *Molecular Simulation*, 34, 51–55 (2008).
24. J. Chen, G. Zhang, B. Li, Molecular dynamics simulations of heat conduction in nanostructures: Effect of heat bath, *Journal of the Physical Society of Japan*, 79, 074604 (2010).
25. S.J. Plimpton, Fast parallel algorithms for short–range molecular dynamics, *Journal of Computational Physics*, 117, 1–19 (1995).
26. P.M. Agrawal, B.M. Rice, D.L. Thompson, Predicting trends in rate parameters for self-diffusion on FCC metal surfaces, *Surface Science*, 515, 21–35 (2002).
27. S.M. Foiles, M.I. Baskes, M.S. Daw, Embedded-atom-method functions for the fcc Metals Cu, Ag, Au, Ni, Pd, Pt, and their alloys, *Physical Review B*, 33, 7983–7991 (1986).
28. X.W. Zhou, R.A. Johnson, H.N.G. Wadley, Misfit-energy-increasing dislocations in vapor-deposited CoFe/NiFe multilayers, *Physical Review B*, 69, 144113 (2004).
29. S. Qu, V. Shastry, W.A. Curtin, R.E. Miller, A finite temperature dynamic coupled atomistic/discrete dislocation method, *Modelling and Simulation in Materials Science*, 13, 1101 (2005).
30. P.A. Tipler, R. Llewellyn, *Modern Physics*, W. H. Freeman, New York, 2007.
31. J.H. Irving, J.G. Kirkwood, The statistical mechanical theory of transport processes. IV. The equations of hydrodynamics, *The Journal of Chemical Physics* 18, 817–829 (1950).
32. B.D. Todd, P.J. Daivis, D.J. Evans, Heat flux vector in highly inhomogeneous nonequilibrium fluids, *Physical Review E*, 51, 4362–4368 (1995).
33. M. Barisik, A. Beskok, Equilibrium molecular dynamics studies on nanoscale confined fluids, *Microfludics Nanofludics*, 11(3), 269–282 (2011).
34. G.L. Pollack, Kapitza resistance, *Reviews of Modern Physics*, 41, 48–81 (1969).
35. H.M. Roder, C.A. Nieto de Castro, U.V. Mardolcar, The thermal-conductivity of liquid argon for temperatures between 110-K and 140-K with pressures to 70 MPa, *International Journal of Thermophysics*, 8, 521–540 (1987).
36. G.E. Moore, Cramming more components onto integrated circuits, *Proceedings of the IEEE*, 86, 82–85 (1998).
37. R.P. Feynman, There's plenty of room at the bottom, *Engineering Sciences*, 23, 22–36 (1960).
38. E.T. Swartz, R.O. Pohl, Thermal boundary resistance, *Reviews of Modern Physics*, 61, 605–668 (1989).
39. M. Hu, P. Keblinski, P.K. Schelling, Kapitza conductance of silicon–amorphous polyethylene interfaces by molecular dynamics simulations, *Physical Review B*, 79, 104305–104312 (2009).
40. B.H. Kim, A. Beskok, T. Cagin, Thermal interactions in nanoscale fluid flow: Molecular dynamics simulations with solid-liquid interfaces, *Microfluidics and Nanofluidics*, 5, 551–559 (2008).
41. D. Torii, T. Ohara, K. Ishida, Molecular-scale mechanism of thermal resistance at the solid-liquid interfaces: Influence of interaction parameters between solid and liquid molecules, *Journal of Heat Transfer*, 132(1), 012402 (2009).
42. L. Xue, P. Keblinski, S.R. Phillpot, S.U.-S. Choi, J.A. Eastman, Effect of liquid layering at the liquid–solid interface on thermal transport, *International Journal of Heat and Mass Transfer*, 47, 4277–4284 (2004).
43. L. Xue, P. Keblinski, S.R. Phillpot, S.U.-S. Choi, J.A. Eastman, Two regimes of thermal resistance at a liquid–solid interface, *Journal of Chemical Physics*, 118, 337–339 (2003).
44. K.M. Issa, A.A. Mohamad, Lowering liquid-solid interfacial thermal resistance with nanopatterned surfaces, *Physical Review E*, 85, 031602 (2012).

45. P.A.E. Schoen, B. Michel, A. Curioni, D. Poulikakos, Hydrogen-bond enhanced thermal energy transport at functionalized hydrophobic and hydrophilic silica–water interfaces, *Chemical Physics Letters*, 476, 271–2[illegible]
46. J.L. Barrat, [illegible]ini, Kapitza resistance at the liquid–solid interface, *Molecular Physics*, 10, 1605–1610 (200[illegible]).
47. J.H. Cha, S. Chiashi, J. Shiomi, S. Maruyama, Generalized model of thermal boundary conductance between SWNT and surrounding supercritical Lennard-Jones fluid—Derivation from molecular dynamics simulations, *International Journal of Heat and Mass Transfer*, 55, 2008–2013 (2012).
48. S. Murad, I.K. Puri, Thermal transport across nanoscale solid-fluid interfaces, *Applied Physics Letters*, 92, 133105 (2008).
49. S. Murad, I.K. Puri, Molecular simulation of thermal transport across hydrophilic interfaces, *Chemical Physics Letters*, 467, 110–113 (2008).
50. S. Murad, I.K. Puri, Thermal transport through a fluid–solid interface, *Chemical Physics Letters*, 476, 267–270 (2009).
51. R.J. Stevens, L.V. Zhigilei, P.M. Norris, Effects of temperature and disorder on thermal boundary conductance at solid–solid interfaces: Nonequilibrium molecular dynamics simulations, *International Journal of Heat and Mass Transfer*, 50, 3977–3989 (2007).
52. J.C. Duda, C.J. Kimmer, W.A. Soffa, X.W. Zhou, R.E. Jones, Influence of crystallographic orientation and anisotropy on Kapitza conductance via classical molecular dynamics simulations, *Journal of Applied Physics*, 112, 09351 (2012).
53. T. Luo, J.R. Lloyd, Enhancement of thermal energy transport across graphene/graphite and polymer interfaces: A molecular dynamics study, *Advanced Functional Materials*, 22, 2495–2502 (2012).
54. J.C. Duda, T.S. English, E.S. Piekos, W.A. Soffa, L.V. Zhigilei, P.E. Hopkins, Implications of cross-species interactions on the temperature dependence of Kapitza conductance, *Physical Review B*, 84, 193301 (2011).
55. H.D. Young, *University Physics*, 7th edn., Addison Wesley, Boston, MA, 1992, Table 15-5.
56. A.K. Al-Matar, D.A. Rockstraw, A generating equation for mixing rules and two new mixing rules for interatomic potential energy parameters, *Journal of Computational Chemistry*, 15, 660–668 (2004).
57. M. Fyta, R.R. Netz, Ionic force field optimization based on single-ion and ion-pair solvation properties: Going beyond standard mixing rules, *Journal of Chemical Physics*, 136, 124103 (2012).
58. T. Werder, J.H. Walther, R.L. Jaffe, T. Halicioglu, P. Koumoutsakos, On the water-carbon interaction for use in molecular dynamics simulations of graphite and carbon nanotubes, *Journal of Physical Chemistry B*, 107, 1345–1352 (2003).
59. E.R. Cruz-Chu, A. Aksimentiev, K. Schulten, Water-silica force field for simulating nanodevices, *Journal of Physical Chemistry B*, 110, 21497–21508 (2006).
60. L. Gao, T.J. McCarthy, Teflon is hydrophilic: Comments on definitions of hydrophobic, shear versus tensile hydrophobicity, and wettability characterization, *Langmuir*, 24, 9183–9188 (2008).
61. T.A. Ho, D.V. Papavassiliou, L.L. Lee, A. Striolo, Liquid water can slip on a hydrophilic surface, *PNAS*, 108, 16170–16175 (2011).
62. K. Hermansson, U. Lindberg, B. Hok, G. Palmskog, Wetting properties of silicon surfaces, in *Proceedings of International Solid State Sensors and Actuators*, San Francisco, CA, 1991, pp. 193–196.
63. N. Martinez, Wettability of silicon, silicon dioxide, and organosilicate glass, Master of Science thesis, University of North Texas, Denton, TX, 2009.
64. B. Arkles, Hydrophobicity, hydrophilicity and silanes, *Paint & Coatings Industry Magazine*, 22, 114–135 (October 2006).
65. K.B. Seok, S. Sangwoo, S.S. Jae, K.K. Min, C.H. Hee, Micro-nano hybrid structures with manipulated wettability using a two-step silicon etching on a large area, *Nanoscale Research Letters*, 6, 333–343 (2011).
66. I.J. Haller, Covalently attached organic monolayers on semiconductor surfaces, *Journal of American Chemical Society*, 100, 8050–8055 (1978).
67. T.D. Iordanov, G.K. Schenter, B.C. Garrett, Sensitivity analysis of thermodynamic properties of liquid water: A general approach to improve empirical potentials, *Journal of Physical Chemistry A*, 110, 762–771 (2006).

68. H.J.C. Berendsen, J.R. Grigera, T.P.J. Straatsma, The missing term in effective pair potentials, *Journal of Physical Chemistry*, 91, 6269–6271 (1987).
69. M.P. Allen, D.J. Tildesley, *Computer Simulation of Liquids*, Oxford University Press, New York, 1989.
70. S. Miyamoto, P.A. Kollman, Settle: An analytical version of the shake and rattle algorithm for rigid water models, *Journal of Computational Chemistry*, 13, 952–962 (1992).
71. S.J. Plimpton, R. Pollock, M. Stevens, Particle-mesh ewald and rRESPA for parallel molecular dynamics simulations, in *Proceedings of the Eighth SIAM Conference on Parallel Processing for Scientific Computing*, Philadelphia, PA, 1997, pp. 8–21.
72. Z. Ge, D.G. Cahill, P.V. Braun, Thermal conductance of hydrophilic and hydrophobic interfaces, *Physical Review Letters*, 96, 186101 (2006).

2.1 INTRODUCTION

Devices such as nanoshells, nanorods, nanofins, and metallic nanofluids depend on the ability of a metallic nanomaterial to transmit kinetic energy in and out of the material surrounding them. With every particle, interface, and pair of interacting atoms, there are inefficiencies in the transfer of momentum and energy that manifest macroscopically as thermal resistance [1–14]. This may be due to a variety of factors. Such factors include (1) the atomic attraction between the two materials, (2) the lattice structure of the metal, (3) the curvature at the interface, (4) the frequency of oscillation for each component, (5) the temperature of the metallic surface, (6) the heat flux at the surface, and finally (7) the direction that the heat flux occurs. A high interfacial thermal resistance (ITR) could mean the difference between a surface that transmits heat well, and stays at a low steady temperature, and one that conducts poorly, rises in temperature and melts. It also impacts the amount of heating that takes place in the liquid medium. A low ITR results in good delivery of heat, allowing a nanoparticle to successfully warm up a medium such as water around it.

The ITR that takes place between the metallic surface and water is analyzed by using molecular dynamics (MD) simulations [15]. The results are compiled and used to create a series of empirical correlations that can be used as a predictive tool for estimating the interfacial resistances for metal/water interfaces within a range of surface temperatures and heat fluxes. In addition, it is shown how these correlations can be used in a range of practical applications to help thermal engineers design a nanoscale system that delivers heat from metal to water as desired.

2.2 BACKGROUND

MD simulations take the summation of the microscopic details of a system (the hundreds to thousands of particle forces, velocities, and trajectories in a system at the scale of atoms) and process this information to directly calculate macroscopic properties of experimental interest such as temperature, pressure, and thermal conductivity. The first MD simulations performed to gain insight of an atomic system took place in 1936, not on a computer, but physically in the form of gelatin balls and metal ball bearings representing molecules [16]. Despite their simplicity, these models produced a "surprisingly good" three-dimensional picture of the structure of a liquid, justifying the use of modeling. However, the effects of gravity limited their application, driving the need to produce computer-based mathematical models that accurately reproduced the physics without this limitation.

MD is a deterministic method that involves tracking the atomic trajectories of each atom by numerically solving Newton's equation of motion starting from specific initial conditions and boundary conditions and calculating forces between every pair of atoms using potentials. In general, the force corresponding to a potential $u(r)$ between two interacting particles i and j is

$$\mathbf{f}_{ij} = -\nabla u(r_{ij}) \tag{2.1}$$

Forces between particles are pair-additive, meaning the equations of motion follow from Newton's second law and the summation,

$$m\ddot{\mathbf{r}}_i = \mathbf{f}_i = \sum_{\substack{j=1 \\ (j\neq i)}}^{N_m} \mathbf{f}_{ij} \tag{2.2}$$

where

the sum is over all N_m atoms, excluding i

m is the atomic mass

MD employing the Monte Carlo method used highly idealized representations of molecules such as hard spheres and flat disks, along with Boltzmann probabilities of state, to determine the precise description of a particle system at thermodynamic equilibrium [17]. In 1957, Wood and Parker used Monte Carlo to derive thermodynamic data of liquid argon comparable to data from [18].

In 1957 and 1959, Alder and Wainwright took a different approach to molecular modeling and directly applied Newtonian equations to describe the trajectories of molecular motion for a system of hard spheres [1]. The particles moved at constant velocity between perfectly elastic collisions without the need to make any approximations of scattering interactions. The only limit to trajectory accuracy was the numerical truncation limits of the machine and the algorithms used to integrate their forces as the particles moved between steps.

Years later, in 1964, Rahman successfully applied the Lennard-Jones potential (often called L-J potential) to a set of Newtonian particles in motion [19]. The Lennard-Jones potential approximates higher-order van der Waal's electron-cloud dipole formation through a weak attractive potential at long distances and overlapping electron-cloud repulsion through a strong repulsive potential at short distances. Extensive studies using the Lennard-Jones model are followed [20–25].

2.2.1 Models of Molecular Water

The first models of molecular water followed soon thereafter by Barker and Watts using Monte Carlo [26] and then by Rahman and Stillinger using MD [27]. Models of flexible hydrocarbons followed soon by Ryckaert and Bellemans [28], with models of large proteins followed by McCammon et al. [29]. In 1981, Berendsen et al. developed a simple three-point charge model of water using a Lennard-Jones 6–12 potential on the oxygen positions and electrostatic point charge potentials between the oxygen and hydrogen atoms [30]. The Lennard-Jones potential parameters were adjusted to fit experimental data for the potential energy of water at 25°C [30]. They found that the simple point charge (SPC) model successfully reproduced the radial distribution function, the diffusion constant, and the hydrogen-bonding energy.

In 1984, Toukan and Rahman modified the rigid SPC model to allow for flexible molecules. In addition to the intermolecular potentials of the SPC model, they introduced a harmonic potential in quadratic form for the intramolecular bonds. This water model was able to reproduce the vibrational absorption spectra of water vapor with good agreement. The flexibility of the water molecule also provides a direct inclusion of polarizability, which affects the many-body potential, the structure of water clusters, and calculation of bulk thermodynamic properties [31]. Subsequent water models have been created since then, such as the TIP3P and TIP4P models (transferable intermolecular potential, 3-(or 4)-points) created by Jorgensen et al. in the 1980s [32], the SPC/E (extended simple point charge) created by Berendsen et al. in 1987 [33], and TIP4P-F (flexible TIP4P) created by Lawrence and Skinner in 2003 [34]. Each of these models was derived from the basic SPC model. They are all similar in nature, but the Lennard-Jones and point charge terms differ. The TIP4P models also have an extra point to keep track of that mark a charge point location slightly offset from the oxygen atom center.

2.2.2 Calculation of Thermal Transport Properties

The water models mentioned have been very successful in reproducing the bulk thermal conductivity of liquid water using MD. Generally, the thermal conductivity of homogenous material can be calculated by different techniques:

1. Equilibrium MD with the Green–Kubo formula
2. Ensemble averaging of NEMD
3. NEMD using the Irving–Kirkwood formula
4. NEMD using noncanonical linear response theory
5. NEMD heat transfer with direct temperature differences

The fourth option is based on Fourier's law, provided the system is large enough to have a temperature gradient. This option mostly mirrors the heat flux in macroscopic systems, and thus so, requires a boundary region acting as a heat source, a boundary region acting as a heat sink, and a set of atoms or molecules in-between whose temperatures average in time to generate a steady-state profile.

In some versions of the fourth method, the temperature is directly assigned to each of the boundary regions via a velocity-scaling thermostat. The velocity of the molecules on each side are scaled up or down in order to maintain a certain prescribed velocity, and the kinetic energy added and subtracted to the system on each side is tabulated to calculate an unknown heat flux [2,35]. The problem with this method is that the heat flux is a quantity with large fluctuations, so its average converges slowly and large temperature gradients are needed to induce a heat flow distinguishable from the noise.

A third method, created by Müller-Plathe in 1996, uses velocity switching between particles in the source and sink regions of a single-component system. In this method, the velocity of the fastest particle in the heat sink adopts the velocity of the slowest particle in the heat source, and vice versa [36]. Doing so, kinetic energy and momentum is exchanged between the two sides while each quantity remains conserved. While the heat flux and resulting temperature gradients are initially unknown, the temperature is averaged over time—and the heat flux is well-defined and exactly tabulated over time—leading to a rapid convergence of thermal conductivity. In 2000, this model was extended to include mixtures and molecular fluids such as water [37].

A second method bypasses this and imposes a prescribed heat flux by directly adding and subtracting a specific amount of kinetic energy from the heat source and sink regions, creating unknown temperatures at the boundaries and an unknown temperature slope. The drawback with this method is that it is sometimes difficult to know beforehand the magnitude of the heat flux in the system that is required to impose a reasonable temperature gradient.

2.3 INTERFACIAL THERMAL RESISTANCE EVALUATIONS

The measurement of thermal transport using MD extends beyond that of continuous media. In 1998, Maruyama and Kimura studied the thermal resistance over a solid–liquid interface using Lennard-Jones particles of liquid argon between platinum crystal walls [23]. They independently controlled the temperatures at the ends of the walls, allowing them to accurately calculate the energy flux. After periods of 1000–4000 ps, they considered the system at steady state, and they measured the temperatures of the walls and argon atoms, observing a distinctive jump at the solid–liquid interface. The temperature jump is regarded to be the direct result of the thermal resistance over the interface. The interaction potential parameters between the solid and liquid molecules were varied, covering a wide range of wettability. They found that the thermal resistance was lowest for strongly hydrophilic surface interactions (small contact angles) and sharply increased with the decrease of wettability (larger contact angles). They also found that the ITR was equivalent to 5–20 nm of thickness of additional liquid, meaning the amount of thermal resistance cannot be neglected in systems of similar size.

Although Maruyama et al. studied this phenomena using MD in the late 1990s, this distinctive solid/liquid temperature gap was first formally acknowledged and studied as far back as 1936 and 1941 by Pyotr Kapitza, who reported his measurements of the temperature jump at the interface between liquid helium and a solid wall [38].

During the following decades, the study of the ITR, also known as the Kapitza resistance, mainly focused on the field of cryogenics because of its importance in attaining ultra-low temperatures. However, with the decreased length scale and increased surface-to-volume ratio of nanotechnologies, the Kapitza resistance is important today even at room temperature.

2.3.1 Continuum-Based Models of Interfacial Thermal Resistance

The study of the Kapitza resistance left a legacy of continuum-based acoustic models, which attempted to explain the temperature gap in terms of phonon reflection and scattering at the interface. The first of these was the acoustic mismatch model (AMM), developed by Khalatnikov in 1952 [39], and the second one is in 1989, Swartz and Pohl attempted to correct this deviation and developed the diffusive mismatch model (DMM) to supplement the AMM [40].

The AMM defines the heat transmission probability at solid–liquid interfaces, modeling phonons as plane waves traveling through mediums with distinct acoustic impedances based on the phonon velocity and density of each medium. The AMM assumes both sides of an interface to be isotropic and the transmission probability to be independent of temperature, surface wettability, and intermolecular interaction. The gross heat current density from one medium to another medium is a function of the number of phonons with a given frequency, incident angle, phonon energy, and transmission probability, summed over all the frequencies and incident angles.

When compared to laboratory measurement, the AMM has been found to grossly overestimate ITR: temperature gaps were found to be between 10 and 100 times smaller than those predicted by the AMM. This indicated that phonons are able to cross the interface more easily.

DMM assumes that the scattering of phonon waves at the interface changes the energy transmission coefficient. This model assumes that all phonons incident upon the interface are elastically scattered in a random direction. The probability of scattering into one side of the interface or the other is proportional to the densities of states for phonons in the different materials. Temperature gaps in experiments, however, were found to be much larger than those values predicted with the DMM, meaning this model underestimates ITR.

Given the two extremes mentioned, the two models practically serve as upper and lower bounds for actual thermal resistance. Neither of these models fully explain the complexity of thermal resistance at all temperature ranges since predictions of thermal resistance on the interface are based on the faulty continuum assumption.

2.3.2 Use of Molecular Dynamics to Calculate Interfacial Thermal Resistance

As shown in the works of Maruyama et al., MD methods have been tapped recently to study the Kapitza resistance between two dissimilar media with success. Matsumoto et al. studied the thermal resistance between two species of Lennard-Jones solid crystals [9]. Using a method similar to Maruyama, 1D steady heat conduction was implemented using a pair of temperature-controlled heat baths. Matsumoto investigated four types of interfaces: (1) crystals with different masses, (2) crystals with different interaction parameters, (3) crystals with various surface interaction parameters, and (4) crystals with different size parameters. They found that each of the cases, except for the fourth case, exhibited temperature profiles with a distinct discontinuity at the interface. They also found that the reduction of energy flux due to the temperature jump was larger than that predicted with the AMM model. Xue et al. performed an analysis similar to Maruyama and performed MD simulations calculating the thermal resistance between solid and liquid Lennard-Jones particles of varying bonding energies [14].

2.3.3 Bonding Energy Effect on Layering and Interfacial Thermal Resistance

In 1990, Raghavan et al. performed MD simulations of water at the platinum interface to study its structure and dynamics, noting adsorption energy and molecular layering as a function of distance from the surface [41]. They noted that water too exhibited molecular layering near the surface, and that the first layer of water molecules arranged themselves into an ice-like hexagonal lattice. The second and third layers exhibited a similar structure due to hydrogen bonding influenced by the template effect of the first layer. Beyond the third layer, a transition to bulk behavior was noticed. They noted that at 300 K, the motions of the water molecules in the first layer are mostly oscillations about equilibrium positions, and each successive layer up to layer four showed the same behavior with progressively more movement. With these tests, they concluded that the structure of water is very sensitive to the nature of the bonding surface and its geometry. This study was a follow-up to a previous study they conducted in 1989 where they observed water on platinum at two distinct temperatures, T = 275 K and T = 360 K [6]. In that study, they observed greater ordering in three distinct layers at the colder temperature, and only two distinct layers at the same positions at the higher temperature. In the higher temperature simulation, the water molecules hop from one adsorption site to a neighboring vacant site, while the molecules are well localized on the adsorption sites at T = 275 K. Such results from both studies show a clear dependence of water structure on metal geometry and temperature.

In 1996, Ignaczak and Gomes used density functional theory to calculate the interaction energies of water molecules near Cu, Ag, and Au surfaces, finding copper to be the most hydrophilic metal with an adhesion energy of −31.8 kJ/mol, followed by the less attractive gold (−29.7 kJ/mol) and silver (−26.6 kJ/mol) [42]. As Keblinski found using Lennard-Jones liquids, they found that bonding energy plays a direct role in wetting behavior and ITR. Later Ge et al. [7] demonstrated this link experimentally by using time-domain thermo reflectance to measure the transport of vibrational energy across planar interfaces of water and a variety of chemically functionalized solids [30]. They tested four different surfaces: Al coated with octadecyltrichlorosilane to make it hydrophobic, Al coated with 2-[methoxy(polyethyleneoxy)-propyl]-trichlorosilane, to make it hydrophilic, Au coated with 1-octadecanethiol to make it hydrophobic, and Au coated with 11-mercapto-1-undecanol to make it hydrophilic. They found that the thermal conductance for the interfaces to be 60 and 50 MW/m^2-K for the hydrophobic Al and Au surfaces, respectively. The hydrophilic surfaces resulted in a significant increase in thermal conductance (decrease in thermal resistance) at 180 and 100 MW/m^2-K for the Al and Au surfaces, respectively, consistent with the trend observed in Lennard-Jones simulations. Shenogina et al. [43] repeated this same type of study using MD simulations of water on self-assembled monolayers (SAMs) alone that exhibited a broad range of surface chemistries from hydrophobic to hydrophilic. They, too, showed that thermal transport properties correlated directly with adhesion energy and with macroscopic wetting characterized by the thermal contact angle. Moreover, they demonstrated that with this knowledge, thermal transport measurements may be used alternatively as probes for interfacial bonding and adhesion.

2.3.4 Thermal Effects on Interfacial Thermal Resistance

Despite the insights that liquid wetting changes with bonding energy, that liquid layering increases at lower temperatures, and that wetting directly correlates with ITR, the question still remained: How does ITR change with temperature? Using solid and liquid Lennard-Jones particles, Merabia et al. [9,10] modeled the solid to liquid heat transfer from a heat-generating spherical nanoparticle using several bonding energies and five different dimensionless temperatures [10]. At low temperatures, they showed that the temperature profile around the spherical nanoparticle exhibited a clear $1/r$ drop off, as predicted from a macroscopic theory. As the temperature of the nanoparticle increased, deviations from this behavior were observed. Particularly, increasing the temperature of the nanoparticle resulted in larger temperature gaps at the interface. In addition, they found a steep temperature profile close to the nanoparticle surface, indicative of a decrease in the local thermal conductivity. This decreased conductivity acted like a "thermal shield" against conducting heat to the surrounding media. They found that, effectively, a particle could have a very high temperature while the liquid at a particle diameter away from the surface could have close to ambient temperature. They attribute this change in local thermal conductivity to decreased layering and decreased local density with increased wall temperature. They also examined the heat transfer from a planar geometry and discovered a relation between heat flux and wall temperature: they found that heat flux increases linearly with wall temperature until a kink temperature, T_k, beyond the boiling point, after which heat flux rapidly decreases with wall temperature. For the spherical nanoparticle, they noticed a much higher heat flux was possible than for the planar case, showing a linear increase in heat flux with wall temperature up until a different critical temperature, after which the increase in heat flux plateaus. Their model showed that this maximum, or "critical heat flux" in the spherical case is a factor of four times higher than the planar case.

In 2009, the same group followed up with MD simulations of more realistic MD cases rather than Lennard-Jones equivalents, modeling gold nanoparticles in octane and water [11]. For all heating cases, the resulting radial temperature field of the octane around the nanoparticle exhibited a $1/r$ drop off, as predicted by a macroscopic theory. A temperature gap was observed at the gold/octane interface, and this gap increased with increasing heat flux, indicating increasing ITR. At low heating powers, they observed an interfacial conductance of 100 MW/m^2-K (resistance of 10×10^{-9} m^2-K/W). At $P = 700$ nW, this conductance decreased to 50 MW/m^2-K (increased resistance of 20×10^{-9} m^2-K/W). They attribute this decreased conductance with a decrease in liquid density adjacent to the solid surface. For the gold in water case, they tested a similar-sized gold nanoparticle in 10,000 SPC/E water molecules. Using heating loads between 350 and 700 nW, they found radial temperature profiles much flatter than the octane case. They attributed this difference to water having five times the thermal conductivity as octane. On the other hand, they noted that the interfacial temperature jumps are smaller for the water/gold case. Over the range of heating power investigated, they measured interfacial conductances of 150–170 MW/m^2-K. This translates to resistance values between 6.66×10^{-9} and 5.88×10^{-9} m^2-K/W. This decreased interfacial resistance for water compared to octane is consistent with the trend noted in previous studies that interfacial conductance increases with the work of adhesion, which is higher for the gold–water case than for the gold–octane case. In both cases, these studies conclude that there exhibits an increased ITR with increased heating power.

2.3.5 Wall Temperature from Effects of Liquid Layering

Barisik and Beskok applied a similar rigor to models of planar surfaces of water on silicon [3,4]. After applying the temperatures and allowing the system to come to equilibrium, they measured the temperature gaps at the water/silicon interface located at the inner walls. They found that an increase in interaction energy is accompanied by a decrease in ITR, similar to the results of Keblinski et al. However, they also found that the two main interfacial hot and cold wall temperatures remained relatively constant in all ε_{Si-O} cases, indicating that the ITR is not dominated by the wall temperature [3]. For their next study [4], they altered the ε_{Si-O} interaction strength until the contact angle of a droplet of MD water on silicon equaled 88° to match experiment. At this interaction strength, considered low wetting, they observed that for each wall temperature, the water layer density remained nearly constant. With an unchanging water density structure for all cases, they compared the resultant ITR. They found that the thermal resistance actually decreased with higher temperatures. That is, with low-wetting interfaces the water structure is independent of wall temperature, and in such cases, the change in thermal contact becomes purely dependent on the phonon densities of state between the two interacting materials. They concluded that higher wall temperatures result in a higher density of state, which creates better coupling.

2.4 MOLECULAR MODELING FOR WATER–METAL INTERFACIAL THERMAL RESISTANCE

2.4.1 H_2O–H_2O Interactions: SPC-Flexible Water Model

Water molecules presented in Figures 2.1 and 2.2 can be modeled as flexible bodies using the simple point charge-flexible (SPC-flexible) model of water [31]. The SPC-flexible model was developed by Toukan and Rahman as a modified version of the SPC model originally developed by Berendsen et al. [30]. As originally devised, the SPC model assumes a rigid molecular structure

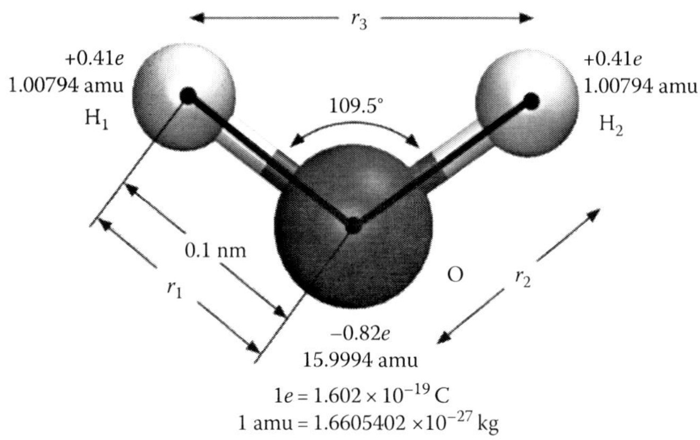

FIGURE 2.1 Water-molecule dimensions for SPC-flexible model.

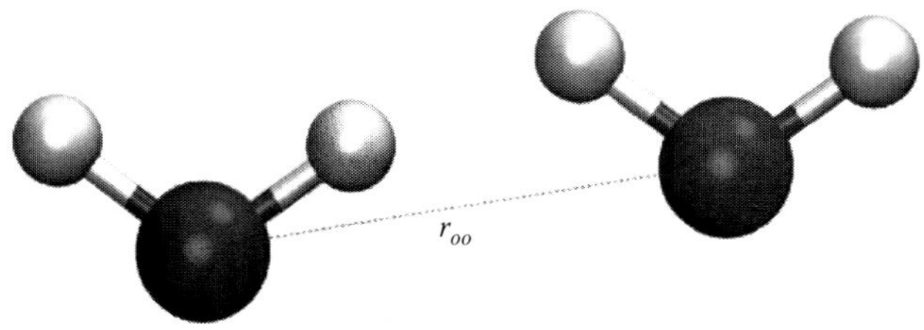

FIGURE 2.2 Oxygen–oxygen interaction pair in SPC-flexible water.

for the water molecule. The SPC-flexible adopts the same form for intermolecular interactions but allows for flexible molecules. The intermolecular interactions consist of three point charges. Lennard-Jones potentials interact between oxygen atoms of different molecules and an electrostatic coulomb potential interact between the molecules arising from the charges on the oxygens and hydrogens of each molecule. The three point charges are located at the oxygen and hydrogen sites. The water molecule is arranged with an H–O–H angle equal to the tetrahedral angle of 109.5° with O–H bond lengths of 0.1 nm.

2.4.1.1 Lennard-Jones Interactions

The oxygen atoms (O–O) of each molecule within the spherical cutoff distance of r_{cut} interact using the Lennard-Jones potential.

$$V_{oo} = \left(-\left(\frac{A}{r_{ij}}\right)^6 + \left(\frac{B}{r_{ij}}\right)^{12}\right) - \left(-\left(\frac{A}{r_{cut}}\right)^6 + \left(\frac{B}{r_{cut}}\right)^{12}\right) \tag{2.3}$$

where

$A = 0.37122$ nm-(kJ/mol)$^{1/6}$
$B = 0.3428$ nm-(kJ/mol)$^{1/12}$
r_{ij} is the distance between the oxygen atoms of molecules i and j

The $(1/r)^{12}$ terms represent a highly repulsive core resulting from Pauli repulsion at short ranges due to overlapping electron orbitals. The $(1/r)^6$ terms represent a long-range attraction that results from van der Waal's interaction (instantaneous dipoles) due to electron-cloud correlations as two atoms

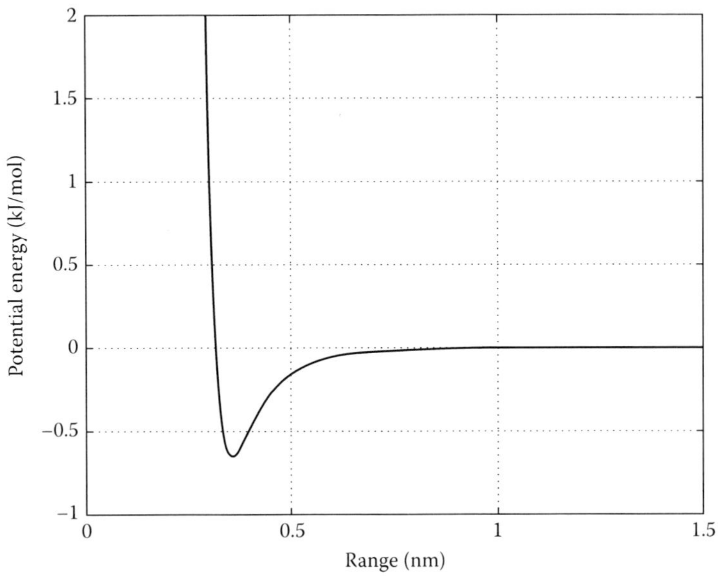

FIGURE 2.3 Oxygen–oxygen Lennard-Jones potential as a function of interatomic distance.

approach one another. The right-hand side of Equation 2.3 is a truncation term for a smooth cutoff at r_{cut} = 0.90 nm. This method slightly shifts the potential at all distances so that the potential can equal to zero at r_{cut} in order to minimize sudden shifts in energy between atoms entering and exiting the cutoff zone. The O–O Lennard-Jones interaction is shown in Figure 2.3.

2.4.1.2 Coulomb Interactions

The charges for the coulomb interaction sites are +0.41e at the hydrogen and −0.82e at the oxygen. The charge points of each molecule interact with those of the others to form nine interactions per molecular pair. The coulomb potential is of the form

$$V_{Coulomb} = \sum_{i}^{O,H_1,H_2} \sum_{j}^{O,H_1,H_2} \left(\frac{q_i q_j}{4\pi\varepsilon_0 r_{ij}} - \frac{q_i q_j}{4\pi\varepsilon_0 r_{cut}} \cdot \left(\frac{2(erf - 1)}{2erf + 1} \right) \right) \tag{2.4}$$

where long-range forces are handled using the error function and a smooth truncation at r_{cut} = 0.90 nm. The coulomb potential for each atomic pair as a function of distance between atoms is shown in Figure 2.4. Atom pairs are presented in Figure 2.5.

2.4.1.3 External Electric Field

In some special cases, the charge points can experience a force imposed by an external electric field from a source such as a laser, microwave, or surface plasmon near the surface of a plasmonic nanoparticle. Such a force can be modeled using the charge points. The force imposed on each atom is expressed by

$$F_{H_1,H_2,O} = q_{H_1,H_2,O} E_{external} \tag{2.5}$$

where

$q_{H_1,H_2,O}$ are the charges of the hydrogen and oxygen atoms
$E_{external}$ is the instantaneous magnitude of the electric field vector

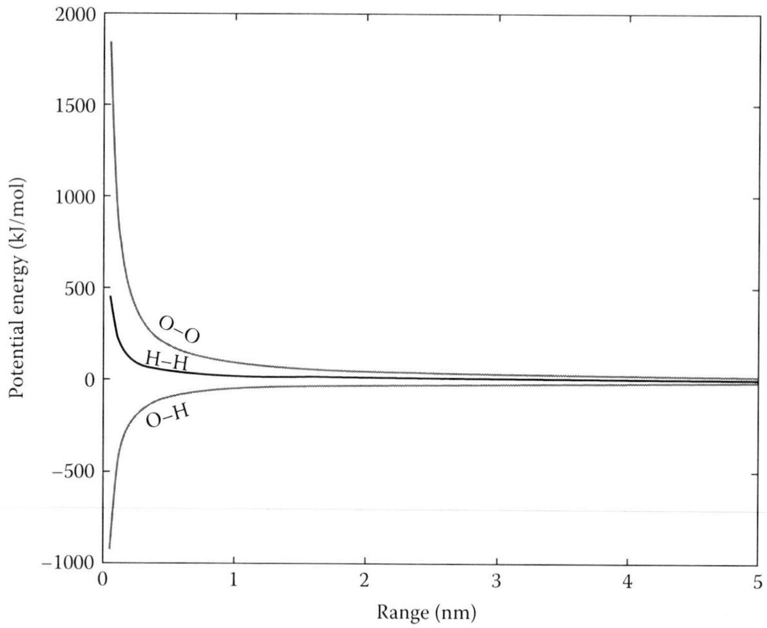

FIGURE 2.4 Oxygen–oxygen, oxygen–hydrogen, and hydrogen–hydrogen coulomb potentials between interacting water molecules.

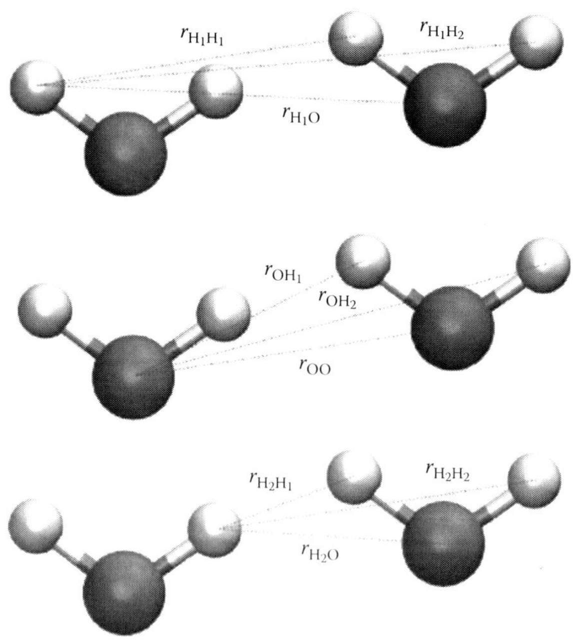

FIGURE 2.5 Atom pairs with coulomb interactions in SPC-flexible water.

2.3.1.4 Intramolecular Interactions

Internal bond flexibility can be handled using a quadratic formula with a combination of harmonic and Morse potentials:

$$V_{mol} = D_{\mathrm{OH}}\left\{1-\exp\left(\rho\left(\Delta r_1\right)\right)\right\}^2 + D_{\mathrm{OH}}\left\{1-\exp\left(\rho\left(\Delta r_2\right)\right)\right\}^2 + \frac{1}{2}b(\Delta r_3)^2 + c\left(\Delta r_1 + \Delta r_2\right)\Delta r_3 + d\left(\Delta r_1 + \Delta r_2\right) \quad (2.6)$$

where

Δr_1 and Δr_2 are stretch lengths in the O–H bond

Δr_3 is the stretch in the H–H separation vector (Figures 2.6 and 2.7)

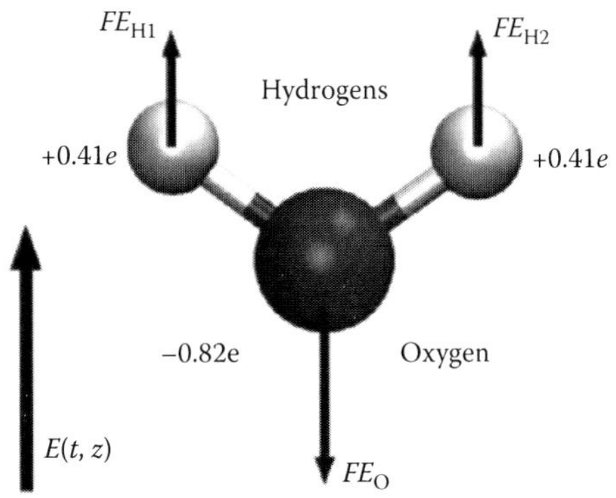

FIGURE 2.6 External electric field forces on water molecule atoms.

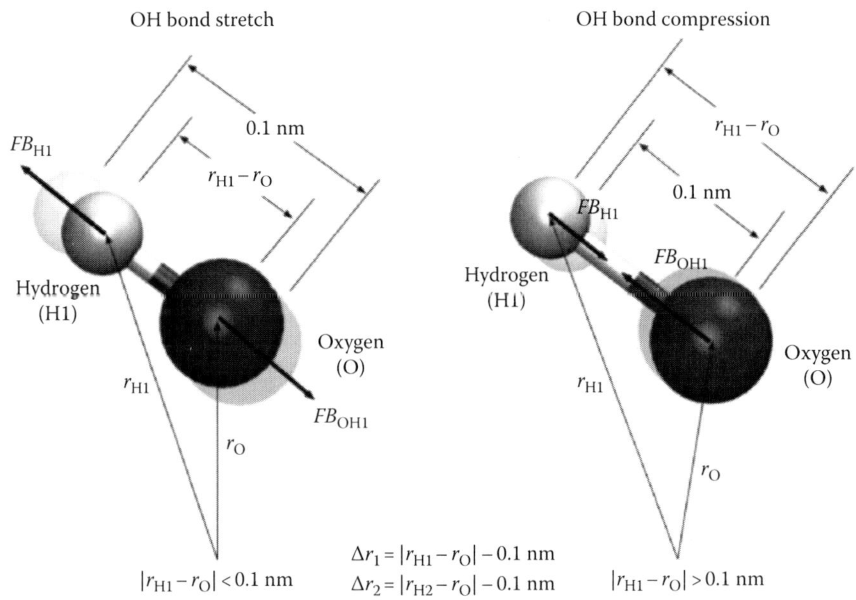

FIGURE 2.7 OH bond-stretching and compression due to application of Morse potential function resulting from changes in Δr_1 and Δr_2.

TABLE 2.1
Harmonic and Morse Constants

B	137485.5	(kJ/mol)/nm
C	−88465.26	(kJ/mol)/nm
D	46731.82	(kJ/mol)/nm
D_{OH}	426.37	kJ/mol
ρ	25.66	nm^{-1}

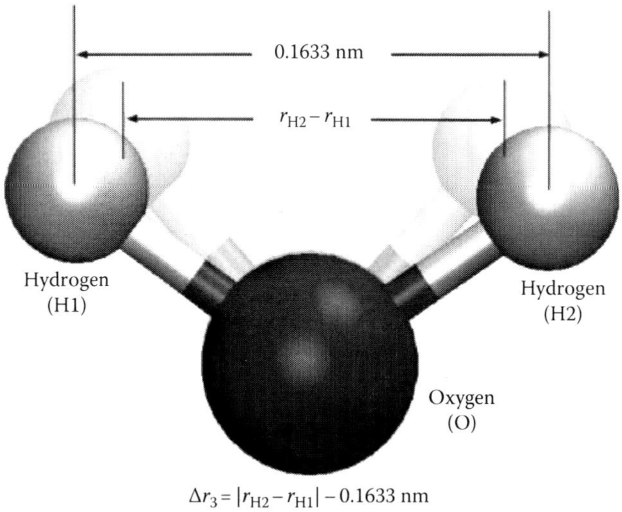

FIGURE 2.8 Angle-bending is captured by measuring changes in the Δr_3 term.

Values for the parameters in Equation 2.6 are listed in Table 2.1. The constant D_{OH} is taken as the experimentally determined dissociation energy of the O–H bond.

In contrast to simulations that use rigid molecules, the modeling of flexible hydrogen motion comes at a high computational cost, as the high frequency of vibrations necessitates the use of ultra-small time steps on the order of 5×10^{-4} ps in order to minimize computational error during the numerical integration of hydrogen trajectories (Figure 2.8).

2.4.2 Metallic Models

Various models to model the interatomic interactions within metals can be found in the literature. Among these are the Sutton-Chen potential [44], the embedded-atom model (EAM), and the simplest of these, which is the metallic Heinz Lennard-Jones potential [45].

2.4.2.1 Heinz Lennard-Jones Potential

In 2008, new Lennard-Jones parameters were developed by Heinz et al. that produced simulations of face-centered cubic metals that rival the accuracy of tight-binding and EAMs at a reported "million times lower" computational cost [45]. This new potentials could successfully "reproduce densities, surface tensions, interface properties with water and bioorganic molecules, as well as mechanical properties in quantitative (<0.1%) to good qualitative (25%) agreement with experiments under ambient conditions." The parameters use the density and surface tension as suitable reference points to experiment. Compared to parameterizations that rely on density and vaporization energy, these models reportedly increase the reliability of previous Lennard-Jones parameters by an order of magnitude. Additionally, these Lennard-Jones parameters were calibrated to reproduce the densities, surface tensions, mechanical properties, and interface properties with water.

The model employs the 12–6 functional form of the Lennard-Jones potential, where the parameter ε explicitly governs the strength of interactions and σ defines the distance between particles at which the interparticle potential is zero.

$$V_{ij} = 4\varepsilon\left[\left(\frac{\sigma}{r_{ij}}\right)^{12} - \left(\frac{\sigma}{r_{ij}}\right)^{6}\right] - 4\varepsilon\left[\left(\frac{\sigma}{r_{cut}}\right)^{12} - \left(\frac{\sigma}{r_{cut}}\right)^{6}\right] \tag{2.7}$$

where

r_{ij} is the distance between atom pairs i and j

r_{cut} is a truncation cutoff at 2.5σ

Figure 2.9 shows the geometrical relationship of an FCC (face-centered cubic) metal with lattice constant a and Lennard-Jones parameter σ_{ij}.

The Lennard-Jones parameters for some metals are listed in Table 2.2. A comparison between the Lennard-Jones potentials versus distance for some metals is graphed in Figure 2.10.

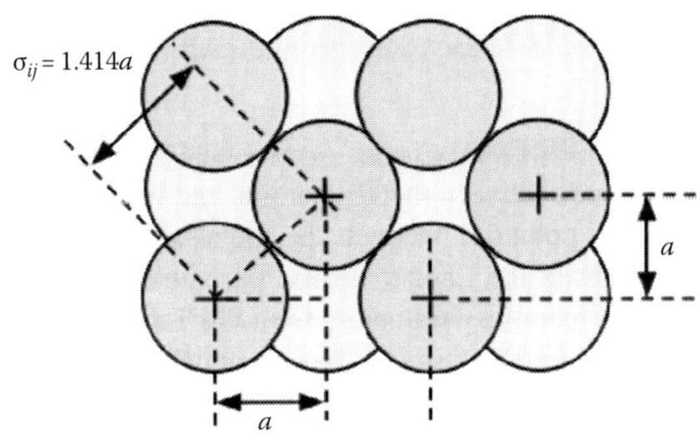

FIGURE 2.9 A metal with a lattice constant of a and a Lennard-Jones parameter of σ_{ij}.

TABLE 2.2
Lennard-Jones Parameters for FCC Metals

Metal	σ (nm)	ε (kJ/mol)
Au	0.2951	22.1334
Ag	0.2955	19.079
Cu	0.2616	19.7485
Al	0.2925	16.8197
Ni	0.2552	23.6396
Pb	0.3565	12.2591
Pd	0.2819	25.7316
Pt	0.2845	32.6352

2.5.4 Temperature Thermostat

When starting a simulation, it is convenient to place the atoms and molecules in a simple lattice arrangement to use for their initial positions. For liquid water, such a perfectly spaced arrangement is almost never seen in nature, even for water in its solid form, ice, which possesses a unique, hexagonally spaced formation. Systems in nature are typically found in their least energy configuration, that is, a configuration where the potential energy is at a minimum. Liquid water artificially placed in a cubic lattice arrangement is easy to write into a code that writes the initial starting positions of each molecule, however it will possess an amount of potential energy that may be greater or less than "normal" water found in nature. If the water molecules are placed in a cubic lattice closely together, the potential energy in the system will be larger than normal water, and the system, with greater potential energy, will move the molecules into an arrangement with less potential energy, converting the excess energy into kinetic energy in the process of creating motion. This placement of the water into an artificial lattice is like setting multiple springs in the system that can readily snap the system into rapid motion. The conversion of this excess potential energy into kinetic energy results in a rapid temperature rise in the system within fractions of a picosecond. For this reason, a temperature "thermostat" is needed to hold back this rise in temperature while the positions of the atoms and molecules evolve toward a conservative state of lowest energy. This process is called energy minimization. After energy minimization has finished, excess potential energy in the system will be lost, and a system starting at one temperature will possess a nearly constant amount of kinetic and potential energy, with values of each slightly oscillating around a constant value. Additionally, in accordance with the energy equipartition principal, the bond-stretching, angle-bending, rotational, and translational components of temperature will all be equal. Measuring these components of temperature for equality, thus, serves as a good check for minimization. After minimization, the system will retain a nearly constant average temperature (conservation of energy) as the simulation progresses over time, and this state allows for the accurate gathering of data to calculate material properties. The modifier "nearly" is used since slight numerical errors in the calculation of positions and velocities of several atoms over several time steps will continuously propagate over time, leading to an unnatural energy increase in the system often referred to as energy drift. However, this slight systematic drift can be ignored if data are gathered during a period short enough before the energy gains become too high, or if the drift is negated by the appropriate application of a universal thermostat that removes the artificial addition of kinetic energy.

Temperature thermostat in a monatomic system: The necessity to keep the average temperature in a system constant at a specified temperature is handled here using a scaling multiplier scheme in what is known as a velocity-scaling thermostat. In a velocity-scaling thermostat, energy is artificially added or taken away from the system so that the system's temperature matches a desired value. This scheme is simple to calculate and implement in any MD code and is rapidly effective. Given two temperatures, the temperature of the system at any instantaneous time, T_i, and the desired temperature of the system, T, a scaling term λ is calculated using the equation

$$\lambda = \sqrt{\frac{T}{T_i}} \tag{2.17}$$

For a simple monoatomic system such as argon, after the scaling constant is calculated, the velocity of every atom in a system is multiplied by λ:

$$v_{i,new} = \lambda v_{i,old} \tag{2.18}$$

where i spans from 1 to N, where N is the total number of atoms in the system. The total kinetic energy and the average temperature of the system are then reevaluated to reveal a new temperature that will match the temperature T set in Equation 2.17. It should be stressed that the temperature reflects the *average* temperature of the system, meaning that if the systems were divided into smaller regions, and their regional energy and temperatures were evaluated individually after their constituents were multiplied by the same scaling term, each regional temperature may not necessarily match; some would be larger, and some would be smaller. Furthermore, if every atom were to be evaluated individually, the use of an identical scaling term across the set would result in a set of individual temperatures whose individual values do not match the temperature T used by Equation 2.17, but whose average as a set does. The velocity-scaling thermostat may be used at every time step, which will result in a constant system temperature over time, or every few time steps, which will result in slight fluctuations in the temperature, resulting in a system that more closely matches that of a stochastic ensemble.

Temperature thermostat in molecular system (water): Since temperature in water is evaluated using the center of mass translational velocity, velocity scaling is handled a little differently. The basic algorithm proceeds as follows:

1. For every molecule, calculate the center of mass translational velocity.
2. Evaluate instantaneous temperature of the system using center of mass velocities, and use Equation 2.17 to find the scaling constant.
3. For every molecule, subtract the center of mass velocity from the velocities of each atom, H_1, H_2, and O.
4. For each molecule, scale the center of mass velocity by multiplying it by the scaling factor found in step 2.
5. For every molecule, add this newly scaled center of mass velocity to the individual velocities of atoms H_1, H_2, and O.

By utilizing the temperature thermostat, a system-wide "overheating" of molecules due to artificial lattice placement during initialization is mitigated. A thermostat can be applied while the system's potential energy is monitored, and as time progresses, the system will evolve toward a state of lower potential energy until the drop in potential energy flattens and the energy is minimized. At this point, if the system is used for the determination of physical properties, the thermostat should be turned off, and the properties should only be evaluated after the thermostat routine is inactive. In some cases, such as in the case of a heat source, a heat sink, or an intentionally hot or cold surface whose temperature must be specified, the thermostat may be used during the duration of a production run. However, thermal properties can only be taken using a subset of molecules or atoms whose velocities are not directly modified using a thermostat scheme.

2.5.5 Density

Density is a thermodynamic property of the substance that is equal to its mass per unit volume. The density of a molecular system is calculated simply by first counting the number of atoms in a specified region, then multiplying by the molecular or atomic mass, and then dividing by the region's total volume. For water, this is expressed as

$$\rho_{H_2O} = \frac{N_{mol} m_{H_2O}}{V_{tot}} \tag{2.19}$$

The density of a substance results from a balance between the kinetic energy giving translational motion to molecules moving them apart and the intermolecular potential that attracts them together. Thus, a difference in temperature—and kinetic energy—results in a change to this balance and thus a change to the average density. An increase in density results in molecules and atoms that are closer together, and based on the Lennard-Jones potential in Equation 2.3 and coulomb potential in Equation 2.4, smaller interatomic distances result in a higher potential energy. Likewise, a decrease in density results in atoms further apart, larger interatomic distances, and a lower potential energy.

2.5.6 Thermal Conductivity

The thermal conductivity of the system is directly measured using two methods. The first is called the Müller-Plathe method, which involves a method of swapping velocities between atoms in a heat source and heat sink, then measuring the difference in kinetic energy between them. The second involves the use of two imposed temperature thermostats between a heat source and a heat sink and tallying the amount of energy required to keep each side at their respective temperatures. Each method results in an imposed heat flux to the system, which results in a thermal gradient across a given length.

2.6 EVALUATION OF PROPERTIES ACROSS A LINEAR SYSTEM

While evaluating the bulk temperature or density in a volume that encompasses the entire simulation space may be useful in determining average information about a bulk material, the subject of nanoscale heat transfer may require the evaluation of properties *across* a system. In addition, a detailed evaluation of properties may be required near or at a nanoparticle surface/fluid interface. In order to accommodate this need, the simulation space may be divided up into a series of subvolumes labeled as "bins" where the properties of density, temperature, and pressure may be individually evaluated. The data from each bin may then be plotted against a position axis to describe how these properties change across the system. The equations for bin temperature and bin density can be modified as follows:

$$T_{bin} = \frac{2E_{bin}C_J}{3N_{bin}} \tag{2.20}$$

$$\rho_{bin} = \frac{N_{bin}m_{H_2O}}{V_{bin}} \tag{2.21}$$

Each bin occupies a slice of equal volume, and cross sections of each bin are uniform (Figure 2.14). The slices of the density bin are very fine—0.01 nm—while the slices for the temperature bin are more course at 0.1 nm. Since these bins are very small, the temperatures and densities tend to fluctuate very rapidly with every time step as molecules enter and leave the local system. Because of this, the values in each bin are averaged and recorded every 1000 time steps.

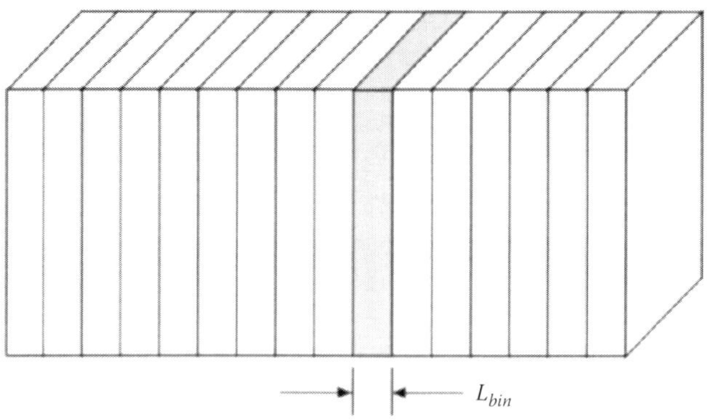

FIGURE 2.14 Volumetric bin slices of width L_{bin} along length of simulation space.

2.6.1 Determination of Thermal Conductivity

The thermal conductivity of a substance is a measurement of its ability to transmit heat from one location to another in the absence of convection or radiation. Conduction may come acoustically in the form of traveling waves of lattice vibrations (phonons) from one atom to another, or electrically where free valance electrons serve as carriers of thermal energy. In MD simulations, only thermal energy in the form of phonons is modeled.

2.6.1.1 Müller-Plathe Method

Müller-Plathe suggested a NEMD method for the calculation of thermal conductivity that imposed an exactly known heat flux on the system and then determined the resulting steady-state temperature gradient in the direction parallel to the heat flux. The thermal conductivity is given by the expression:

$$\lambda = \frac{\langle J_z(t) \rangle}{2tA\langle dT / dz \rangle} \tag{2.22}$$

where

$\langle J_z(t) \rangle$ is the time-averaged heat flux imposed during the simulation in the z direction
$\langle dT/dz \rangle$ is the resultant time-averaged temperature gradient in this direction
A is the surface area perpendicular to the z-axis
t is the total simulation time

The heat flux is created by exchanging the velocities of "cold" and "hot" slabs located at different ends of the simulation box. The cold slab donates its hottest particles (particles with the highest temperature, i.e., highest kinetic energy) to the hot slab, and the hot slab donates its coldest particles (particles with the lowest temperature and kinetic energy) to the cold slab. This exchange results in a transfer of kinetic energy from one side to the other, while keeping the total amount of energy and momentum conserved. This transfer also results in a heating of the warm slab and a cooling of the cold slab, while yielding a steady-state temperature gradient in the z direction due to thermal conduction across the medium. Using the series of temperatures pertaining to each of the bins between the hot and cold slabs, this temperature gradient can be directly measured.

The imposed heat flux is given by measuring the exchange in kinetic energy made by swapping the velocities pertaining to the hottest and coldest particles. In a water system, this heat exchange is given by the relationship

$$\langle J_z(t) \rangle = \sum_{transfers} \frac{m_{H_2O}}{2}\left(v_c^2 - v_h^2\right) \tag{2.23}$$

where

v_c and v_h are the center of mass velocities of the hottest and coldest molecules
m_{H_2O} is the water molecular mass

The total swapped energy is added with each transfer over a given time t. This heat exchange is applied to Equation 2.19 to find the thermal conductivity across the medium.

It should be noted that the "hottest" and "coldest" water molecules are found using the kinetic energies derived from the center of mass velocities. Once these molecules are identified, the atomic velocities from the individual atoms within these two molecules are swapped with one another. That is, the velocity for the H_1 atom in the hottest molecule in the heat sink is swapped with the velocity for the H_1 atom in the coldest molecule in the heat source. This same process applies for atoms H_2 and O.

It should also be noted that at any particular swap, the hottest and coldest molecules may differ, so the location of energy being added or removed from the sink and source will be scattered over time.

A swap may be programmed to occur at every time step, or it may be set to occur periodically every N time steps. The pertinent parameters are the total amount of energy exchanged and the total duration of time in which this exchange occurs for proper use in Equation 2.22.

The Müller-Plathe method is very effective at imposing a heat flux in the system that generates a sizeable thermal gradient. However, due to the random nature of finding the hottest and coldest particles to exchange, the exact rate of heat transfer is not explicitly set. Nor are the final temperatures of the heat source or heat sink. For a system starting at 300 K, the final source and sink temperatures may be 450 and 150 K, respectively. These temperatures may be too extreme for certain applications.

2.6.1.2 Imposed Thermostat Source and Sink

For finer control over these temperatures, a method using imposed thermostats may be employed. In this method, a thermostat is used to maintain a temperature for the heat source and heat sink, respectively. Acting as a thermostat, energy can be added or subtracted on both ends through use of the velocity-scaling multiplier. However, in this use of the thermostat, energy can only be added to the heat source, and energy can only be subtracted from the heat sink. For the heat source, this is implemented by only applying the thermostat when the average temperature of the source drops below a desired temperature T_{source} through contact with cooler atoms in the system. When this occurs, the following process takes place: First the sum of the squares of the velocities of all the atoms in the source is recorded. Next, the velocities are subsequently scaled using a scaling factor as previously shown, so the temperature of the source matches T_{source}.

After this scaling occurs, a new "sum of the squares" of the velocities in the source is recorded. These two amounts are then subtracted from one another and used in the following to calculate the kinetic energy added to the source as a result of the scaling:

$$\Delta E_{hot} = \sum_{transfers} \frac{m}{2}\left(\sum_{N_{source}} v_{source,new}^2 - \sum_{N_{source}} v_{source,old}^2\right) \tag{2.24}$$

where m is the atomic mass or molecular mass of the species in the heat source.

Similarly, kinetic energy is subtracted from the heat sink when the average temperature of the heat sink rises over a desired sink temperature T_{sink}, and scaling is applied so that the sink temperature matches T_{sink}. The total energy subtracted as a result of scaling is the following:

$$\Delta E_{cold} = \sum_{transfers} \frac{m}{2}\left(\sum_{N_{sink}} v_{sink,new}^2 - \sum_{N_{sink}} v_{sink,old}^2\right) \tag{2.25}$$

From the total heat added and total heat subtracted, the average energy flux is computed:

$$\Delta E_{ave} = \frac{\Delta E_{hot} + \Delta E_{cold}}{2} \tag{2.26}$$

From this value, the effective thermal conductivity is calculated using

$$k = \frac{\Delta E_{ave}}{A\tau(dT/dz)} \tag{2.27}$$

where

τ is the duration of time in which the source and sink thermostats are active

A is the cross-sectional area in the channel between the source and the sink

dT/dz is the linearized temperature gradient of the liquid region between the source and sink

After a steady-state temperature is reached, Equations 2.24 and 2.25 can also be used to directly calculate heat flux across the system:

$$q_z = \frac{\Delta E_{hot} + \Delta E_{cold}}{2\tau A} \tag{2.28}$$

Using q_z, the ITR, R_K, across an interface can also be computed:

$$R_K = \frac{\Delta T}{q_z} \tag{2.29}$$

where ΔT is the measured temperature gap at a liquid/solid interface.

2.7 SAMPLE MD SIMULATION FOR GOLD–WATER–GOLD AND SILVER–WATER–SILVER SYSTEMS

The simulations are performed for metal–water–metal systems for different solid metallic walls enclosing water molecules [15]. One snapshot of the MD simulation domain with gold walls is given in Figure 2.15. The walls are separated to hold 1088 water molecules at a density of 996 kg/m^3. The atoms in the outermost three layers serve as the heat source and the heat sink.

2.7.1 Temperature Distributions

Figures 2.16 and 2.17 show the temperature distributions for gold and silver nanochannels using two distinct hot wall temperatures: 350 and 625 K for Au, and 375 and 650 K for Ag. These distributions and a few more presented in [15] show the temperature profiles starting from the heat source on first three molecular layers on the left, across the hot side metal slab, through the hot-wall/water interface, across the water channel (approximately between x = 2–7 nm), across the cold-wall/water interface, through the cold-side metal slab, and into the heat sink in the last three molecular layers on the right. These figures are labeled using the temperature set by the thermostat at the source.

Although both metals are subject to similar boundary temperatures, they exhibit different temperature distribution characteristics as the wall temperatures increase. The slope of the temperature distribution inside the Au wall shows a slight increase between the 350 and 625 K cases. In stark contrast, the temperature distribution inside the Ag wall exhibits a comparatively steep slope in the 650 K case. The water temperature adjacent to the Ag wall is fairly close to that of the hot wall, that

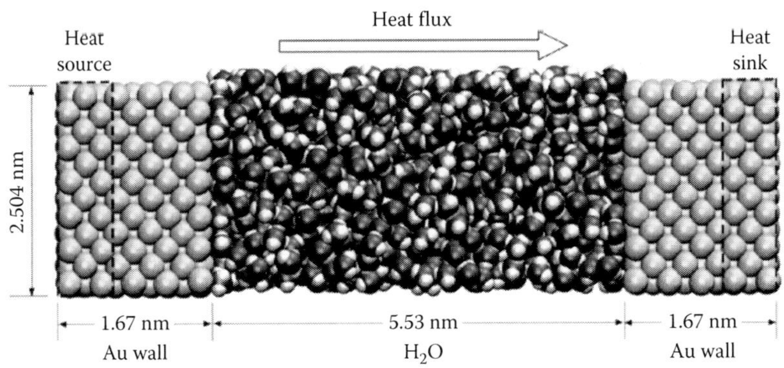

FIGURE 2.15 Snapshot of the MD simulation of water with gold walls.

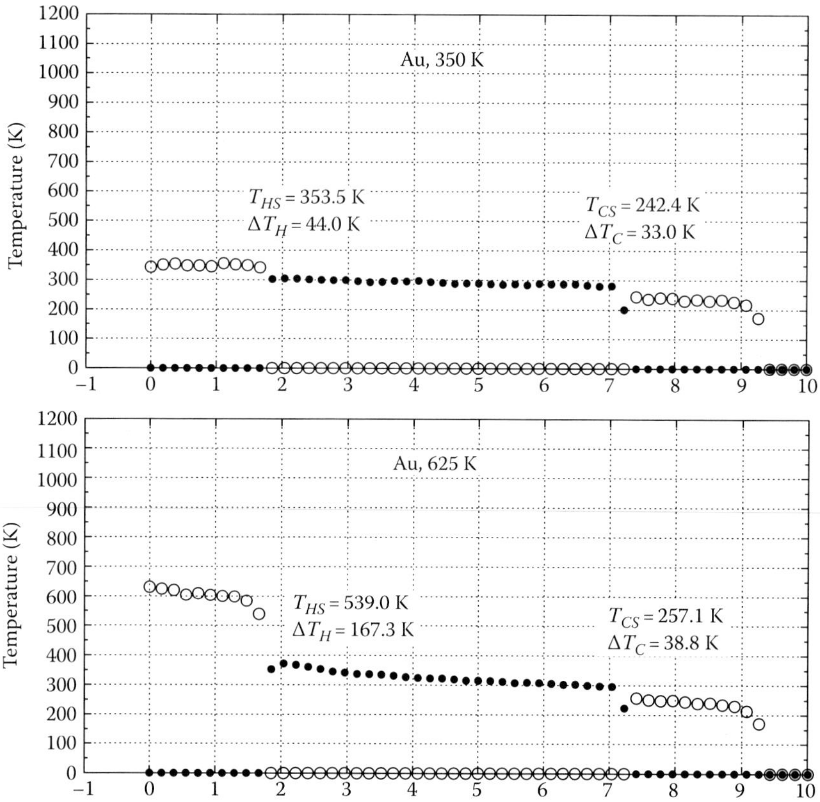

FIGURE 2.16 Au temperature profiles for the gold–water–gold system with source temperature of 350 and 625 K. (Adapted from Vera, J. and Bayazitoglu, Y., *Int. J. Heat Mass Transf.*, 86, 433, 2015.)

is, the temperature gap between the two materials is small, with a maximum temperature gap of 45 K. This small gap may be attributed to good vibrational coupling between the atoms in the Ag wall at the interface and the water molecules at the interface leading to a small ITR. On the contrary, the temperature gap between the water and Au wall is much higher, showing that Au exhibits a higher ITR. In the 625 K case, the gap is as high as 167.3 K.

Both Au and Ag cases show a very small temperature gap on the cold wall interfaces located on the right side of the channel. Between the 350 and 625 K cases, the cold Au interface only increases from 33.0 to 33.8 K. The cold Ag interfaces show an even smaller gap: 8.6 K for the 375 K case and 9.4 K for the 650 K case, results also indicative of a small ITR for silver and a larger ITR for gold.

Equations 2.24 and 2.25 may be used to calculate the amount of energy required to maintain the hot and cold temperature boundary regions to their respective temperatures. This energy is used to calculate the overall heat flux moving through the system. When used in conjunction with the measured temperature gaps noted at the interfaces, the exact boundary resistance values can be calculated using Equation 2.29.

2.7.2 Thermal Resistances

A two-dimensional least squares reduction analysis for determining a thermal resistance R_i at the interface was performed using the combined hot and cold wall data for several metals in contact with water using MD data given in [15] as follows:

$$R_i\left(T_w, q\right) = C_1 T_w + C_2 q + C_3 \tag{2.30}$$

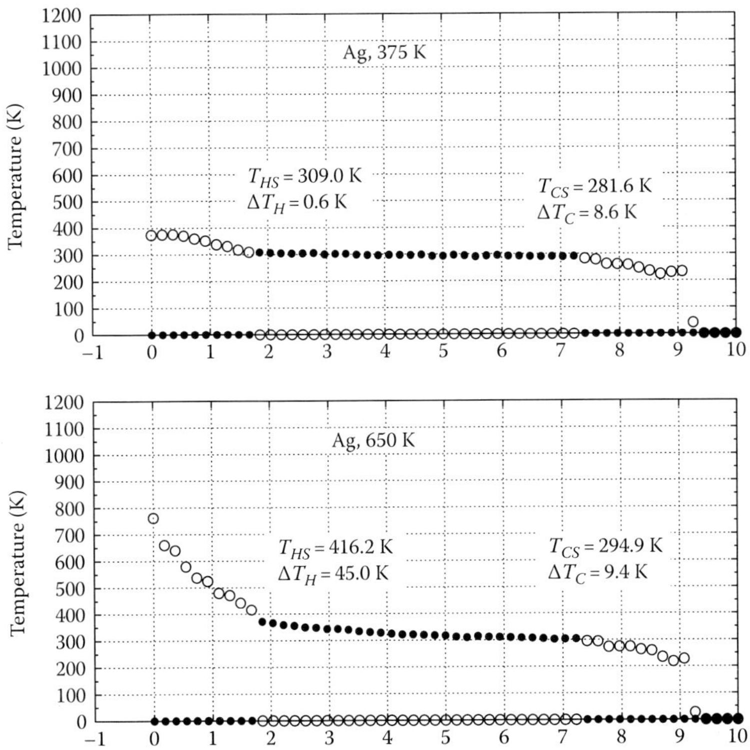

FIGURE 2.17 Ag temperature profiles for the silver–water–silver system with source temperature of 375 and 650 K. (Adapted from Vera, J. and Bayazitoglu, Y., *Int. J. Heat Mass Transf.*, 86, 433, 2015.)

In Equation 2.30 the constants for gold are

$$\begin{aligned} C_1 &= 0.044828843451793 \\ C_2 &= -0.950662757946684 \\ C_3 &= 2.206475560205140 \end{aligned} \tag{2.31}$$

And the constants for silver are

$$\begin{aligned} C_1 &= 0.033153897793913 \\ C_2 &= -1.332758403835610 \\ C_3 &= 0.700141267414531 \end{aligned} \tag{2.32}$$

Here

T_w is the temperature of a wall at the interface

q is the heat flux through the system

The interface thermal resistance, R_i from Equation 2.30, with coefficients given in Equations 2.31 and 2.32 for different metal water interfaces can simply be arranged to solve wall temperature or wall heat flux at the interface. For example, if the planer surface as illustrated by Figure 2.18 is considered, one could solve wall temperature, T_w for given T_L and q, or to solve wall heat flux, q for given T_w and T_L. A similar thermal resistance concept can be developed for one-dimensional cylindrical and spherical geometries, and will not be repeated here [15]. Use of these equations gives us

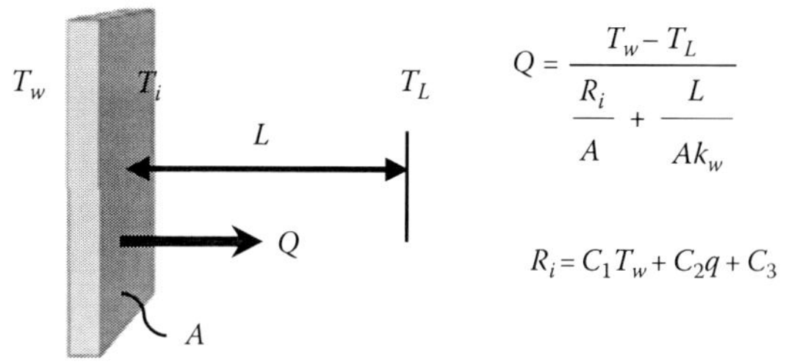

FIGURE 2.18 One-dimensional heat transfer equations for a planar heat conduction with interfacial thermal resistance.

some insight on the importance of including R_i in nanoscale heat transfer calculations. Considering the planer surface (Figure 2.18), we have

1. Planar surface with T_L and q known. Solving for T_w.

$$T_w = \frac{q(C_2 q + C_3 + (L/k_w)) + T_L}{(1 - C_1 q)} \tag{2.33}$$

2. Planar surface with T_w and T_L known. Solving for q. Must be solved iteratively.

$$q = \frac{T_w - T_L}{C_1 T_w + C_2 q + C_3 + (L/k_w)} \tag{2.34}$$

The incorporation of multidimensional MD-derived data into simple thermal models may prove useful in the practice of accurately modeling heat transfer in nanoscale devices with a range of operating conditions.

REFERENCES

1. B.J. Alder and T.E. Wainwright, Phase transition for hard sphere system, *Journal of Chemical Physics*, 27, 1208–1209 (1957).
2. M.P. Allen and D.J. Tildesley, *Computer Simulation of Liquids*, Clarendon Press, Oxford, UK (1987).
3. M. Barisik, Boundary treatment effects on molecular dynamics simulations on interface thermal resistance, *Journal of Computational Physics*, 231, 7881–7892 (2012).
4. M. Barisik and A. Beskok, Temperature dependence of thermal resistance at the water/silicon interface, *International Journal of Thermal Science*, 77, 47–54 (2014).
5. D.G. Cahill, W.K. Ford, K.E. Goodson, G.D. Mahan, A. Majumdar, H.J. Maris, R. Merlin, and S.R. Phillpot, Nanoscale thermal transport, *Journal of Applied Physics*, 93(2), 793–818 (2003).
6. K. Foster, K. Raghavan, and M. Berkowitz, A molecular dynamics study of the effect of temperature on the structure and dynamics of water between Pt walls, *Journal of Chemical Physics Letters*, 162(1 and 2), 32–38 (1989).
7. Z. Ge, D. Cahill, and P. Braun, Thermal conductance of hydrophilic and hydrophobic interfaces, *Physical Review Letters*, 96, 186101 (2006).
8. M. Matsumoto, H. Wakabayashi, and T. Makino, Thermal resistance of crystal interface: Molecular dynamics simulation, *Transactions of the JSME, Series B*, 68, 671, 1919–1925 (2002).
9. S. Merabia and P. Keblinski, Critical heat flux around strongly heated nanoparticles, *Physical Review E*, 79, 021404 (2009).
10. S. Merabia, S. Shenogin, L. Joly, P. Keblinski, and J.L. Barrat, Heat transfer from nanoparticles: A corresponding state analysis, *Proceedings of National Academy of Sciences of the United States of America*, 106(36), 15113–15118 (2009).
11. M. Shibahara and K. Takeuchi, A molecular dynamics study on the effects of nanostructural clearances on the thermal resistance at a liquid water-solid interface, *Third International Symposium on Atomic Technology IOP Publishing, Journal of Physics: Conference Series*, 191, 012008 (2009).

12. K. Toprak and Y. Bayazitoglu, Interfacial thermal resistance of Cu-SWCNT nanowire in water, *International Journal of Heat and Mass Transfer*, 79, 584–588 (2014).
13. K. Toprak and Y. Bayazitoglu, Numerical modeling of a CNT-Cu coaxial nanowire in a vacuum to determine the thermal conductivity, *International Journal of Heat and Mass Transfer*, 61, 172–175 (2013).
14. L. Xue, P. Keblinski, S.R. Phillpot, S. Choi, and J.A. Eastman, Two regimes of thermal resistance at a liquid-solid interface, *Journal of Chemical Physics*, 118(1), 337–339 (2003).
15. J. Vera and Y. Bayazitoglu, Temperature and heat flux dependence of thermal resistance of water/metal nanoparticle interfaces at sub-boiling temperatures, *International Journal of Heat and Mass Transfer*, 86, 433–442 (2015).
16. W.E. Morrell and J.H. Hildebrand, The distribution of molecules in a model liquid, *Journal of Chemical Physics*, 76, 650–660 (1936).
17. N. Metropolis, A.W. Rosenbluth, M.N. Rosnebluth, A.H. Teller, and E. Teller, Equation of state calculations by fast computing machines, *Journal of Chemical Physics*, 21, 1087–1092 (1953).
18. W.W. Wood and F.R. Parker, Monte Carlo equation of state of molecules interacting with the Lennard-Jones potential: A supercritical isotherm at about twice the critical temperature, *Journal of Chemical Physics*, 27, 720–733 (1957).
19. A. Rahman, Correlations in the motion of atoms in liquid argon, *Physical Review*, 136A, 405–411 (1964).
20. L. Verlet, Computer experiments on classical fluids. I. Thermodynamical properties of Lennard-Jones molecules, *Physical Review*, 159, 98–103 (1967).
21. L. Verlet, Computer experiments on classical fluids. II. Thermodynamical properties of Lennard-Jones molecules, *Physical Review*, 165, 201–214 (1968).
22. J.J. Nicholas, K.E. Gubbins, W.B. Streett, and D.J. Tildesley, Equation of state for the Lennard-Jones fluid, *Molecular Physics*, 37, 1429–1454 (1979).
23. S. Maruyama and T.A. Kimura, Study on thermal resistance over a solid-liquid interface by the molecular dynamics method, *Thermal Science and Engineering*, 7, 1, 63–68 (1999).
24. A. Bhansali, Y. Bayazitoglu, and S. Maruyama, Molecular dynamics simulation of an evaporating sodium droplet, *International Journal of Thermal Science* (*Reveue Generale de Thermique)*, 38, 66–74 (1999).
25. Y. Bayazioglu, S. Maruyama, P. Hos, Phase change studies with molecular dynamics: A computer simulation, inter. *Journal of Heat and Technology (Calorie e Tecnologia)*, 18, 1–2000 3–16 (2000).
26. J.A. Barker and R.O. Watts, Structure of water: A Monte Carlo calculation, *Chemical Physics Letters*, 3, 144–145 (1969).
27. A. Rahman and S.H. Stillinger, Molecular dynamics study of liquid water, *Journal of Chemical Physics*, 71, 129–132 (1971).
28. J.P. Ryckaert and A. Bellemans, Molecular dynamics of liquid n-butane near its boiling point, *Chemical Physics Letters*, 30, 123–125 (1975).
29. J.A. McCammon, B.R. Gelin, and M. Karplus, Dynamics of folded proteins, *Nature*, 267, 585–590 (1977).
30. H.J.C. Berendsen, J.P.M. Postma, W.F. van Genseteren, and J. Hermans, Interaction models for water in relation to protein hydration, *Jerusalem Symposium on Quantitative Chemistry and Biochemistry*, 14, 331–342 (1981).
31. K. Toukan and A. Rahman, Molecular-dynamics study of atomic motions in water, *Physical Review B*, 31(5), 2643–2648 (1985).
32. W. Jorgensen, J. Chandrasekhar, J. Madura, R. Impey, and M. Klein, Comparison of simple potential functions for simulating liquid water, *Journal of Chemical Physics*, 79, 926 (1983).
33. H.J.C. Berendsen, J.R. Grigera, and T.P. Straatsma, The missing term in effective pair potentials, *Journal of Physical Chemistry*, 91, 6269–6271 (1987).
34. C.P. Lawrence and J.L. Skinner, Flexible TIP4P model for molecular dyanmics simulation of liquid water, *Chemical Physics Letters*, 372, 842–847 (2003).
35. D.C. Rapaport, *The Art of Molecular Dynamics Simulation*, Cambridge University Press, Cambridge, U.K. (2004).
36. F. Müller-Plathe, A simple nonequilibrium molecular dynamics method for calculating the thermal conductivity, *Journal of Chemical Physics*, 106, 14 (1997).
37. D. Bedrov and G. Smith, Thermal conductivity of molecular fluids from molecular dynamics simulations: Applications of a new imposed-flux method, *Journal of Chemical Physics*, 113, 80 (2000).
38. P.L. Kapitza, Zhurnal Eksperimentalnoi i Teoreticheskoi Fiziki, *Journal of Physics* (*USSR*), 4, 181 (1941).

39. B. Kim, Thermal resistance at a liquid-solid interface dependent on the ratio of thermal oscillation frequencies, *Chemical Physics Letters*, 554, 77–81 (2012).
40. E.T. Schwartz and R.O. Pohl, Thermal boundary resistance, *Reviews of Modern Physics*, 61, 605–689 (1989).
41. K. Raghavan, K. Foster, and M. Berkowitz, Structure and dynamics of water at the Pt (111) interface: Molecular dynamics study, *Journal of Chemical Physics*, 94(3), 2110–2117 (1991).
42. G. Ignaczak, Simulations of liquid water in contact with a Cu(100) surface, *Journal of Elec. Chemistry*, 420, 209–218 (1997).
43. N. Shenogina, R. Godawat, P. Keblinski, and S. Garde, How wetting and adhesion affect thermal conductance of a range of hydrophobic to hydrophilic aqueous interfaces, *Physics Review Letters*, 102, 156101 (2009).
44. A.K. Shalek, J.T. Robinson, E.S. Karp, J.S. Lee, D.-R. Ahn, M.-H. Yoon, A. Sutton et al., Vertical silicon nanowires as a universal platform for delivering biomolecules into living cells, *Proceedings of National Academy of Sciences of the United States of America*, 107(5), 1870–5 (2010).
45. H. Heinz, R. Vaia, B. Farmer, and R. Naik, Accurate simulation of surfaces and interfaces of face-centered cubic metals using 12-6 and 9-6 Lennard-Jones Potentials, *Journal of Physical Chemistry C*, 112, 17281–17290 (2008).

3 Convective Heat Transfer Enhancement with Nanofluids

A State-of-the-Art Review

Sadık Kakaç and Anchasa Pramuanjaroenkij

CONTENTS

3.1 INTRODUCTION

Nanofluids have been utilized to enhance heat transfer rates by researchers and industrials as working fluids in various thermal systems. Base fluids, having low thermal conductivities, are mixed with nanoparticles—1–100 nm diameter particles, which have high thermal conductivities—to form the nanofluids for different applications. Metallic and nonmetallic materials have been used to produce the nanoparticles such as Al_2O_3, Ag, CuO, Cu, SiO_2, and TiO_2. These particles can enhance the thermal conductivities of their mixtures even at low concentrations [1,2]. Nanofluid selection is based on nanofluid properties, sizes, shapes, and volume fractions of the nanoparticles, as well as thermophysical properties of nanofluids, which play important roles on the heat transfer performance [3]. In nanofluid researches, nanofluids were investigated numerically and experimentally. The early studies on nanofluids were mainly focused on the measurements of thermal conductivity. Later, more experimental and numerical studies regarding the convective heat transfer of nanofluids had been developed continuously. Experimental researches have shown that the thermal conductivity enhancements obtained with nanofluids exceed the enhancements obtained by using conventional fluids [1,4].

Nanofluid preparation is a very important task with the use of nanoparticles for improving the thermal conductivity of base fluids [5]. Two methods were used for producing the nanofluids, (1) single-step method (2) two-step method. The single-step method is a process combining the preparation of nanoparticles with the synthesis of nanofluids, for which the nanoparticles are directly prepared by physical vapor deposition (PVD) technique or liquid chemical method. In this method the processes of drying, storage, transportation, and dispersion of nanoparticles are avoided, so the agglomeration of nanoparticles is minimized and

for 1–9.4 vol.%. CuO/water nanofluid

$$\mu_{nf} = (2.1275 - 0.0215T + 0.0002T^2)\mu_f \tag{3.10}$$

for 47 nm Al_2O_3/water nanofluid

$$\mu_{nf} = (0.904e^{0.148\phi})\mu_f \tag{3.11}$$

for 0–6 vol.%. Al_2O_3/water nanofluid [36,37]

$$\mu_{nf} = (1 + 7.3\phi + 123\phi^2)\mu_f \tag{3.12}$$

$$\mu_{nf} = (e^{4.91\phi/0.2092-\phi})\mu_f \tag{3.13}$$

and, for 40 nm MgO/water nanofluid [38]

$$\mu_{nf} = (1 + 11.61\phi + 109\phi^2)\mu_f \tag{3.14}$$

3.2.2 Numerical Investigation of Nanofluid Viscosity

Most of the viscosity studies obtained viscosity enhancement with the dispersion of nanoparticles in the base fluid and with the increase of volume fraction, no exact theoretical mechanism or common empirical correlations with effect of volume fraction, size, and temperature were available for nanofluid viscosity estimations [16]. The rheological behavior of nanofluids was categorized into four groups [15,35] as (1) dilute nanofluids (volume concentration less than 0.1 vol.%) whose viscosity fits the Einstein equation and there is no visible shear-thinning behavior; (2) semi-dilute nanofluids (with 0.1–5 vol.%) with aggregation of nanoparticles, whose viscosity fits the modified K–D equation and there is no obvious shear-thinning behavior; (3) semi-concentrated nanofluids (with 5–10 vol.%) with aggregation of nanoparticles whose viscosity fits the modified K–D equation and there is noticeable shear-thinning behavior; and (4) concentrated nanofluids (volume concentration more than 10 vol.%) with interpenetration of aggregation and this is out of the normal concentration range of nanofluids [18,28].

3.3 NANOFLUID THERMAL CONDUCTIVITY

Nanofluid thermal conductivities were investigated theoretically and experimentally to find the conductivity enhancements, which can also effect the convective enhancements. Many theoretical models were compared with experimental data available in their literatures [1,10,15,39–41].

The nanofluid thermal conductivity can be measured using techniques as summarized by Paul et al. [41] in three main measurement techniques: transient techniques (transient hot-wire, thermal constant analyzer, temperature oscillation, and 3ω technique), steady state technique (steady state parallel plate and cylindrical cell), and thermal comparator. The transient hot-wire technique was carried out dominantly. The most attractive advantages of this method were its capacity for experimentally eliminating the error due to natural convection, its fast results, and its simple conceptual design compared to other techniques [41]. Some experimental results obtained from these measuring techniques are concluded in Table 3.2 [10,15].

TABLE 3.2
The Experimental Thermal Conductivity of Nanofluids and Base Fluids

Base Fluids	Nanoparticles	Volume Fractions (%)	Enhancement (%)	Researchers
Water (pure, 0.613 W/m.K [1])	Al_2O_3	–0.30	1.44	Lee et al. [17]
	Al_2O_3	0.33–3.00	9.7	Chandrasekar et al. [18]
	TiO_2	0.20–2.00	7	Duangthongsuk and Wongwises [22]
	Fe	0.00–4.00	14.9/25	Li et al. [23]
	Ag	0.30/0.06/0.09	30	Godson et al. [24]
	CuO	1.00	5	Hwang et al. [44]
	SiO_2	1.00–4.00	23	Jahanshahi et al. [45]
Ethylene Glycol (EG) (pure, 0.253 W/m.K [1])	AIN	1.00–10.00	38.71	Yu et al. [27]
	Cu	0.30/0.50	46	Yu et al. [46]
	Al_2Cu	2.50	120	Chopkar et al. [47]
	CuO	1.00	9	Hwang et al. [44]
	Ag	1000–10000 ppm	18	Sharma et al. [48]
	Diamond	0.10–1.00	17.23	Yu et al. [30]

Sources: Based on Kakaç, S. and Pramuanjaroenkij, A., *Int. J. Heat Mass Transf.*, 52, 3187, 2009; Chandrasekar, M. et al., *Renew. Sustain. Energ. Rev.*, 16, 3917, 2012; Ghadimi, A. et al., *Int. J. Heat Mass Transf.*, 54, 4051, 2011.

The earlier empirical correlations of nanofluid thermal conductivity, Maxwell model, Hamilton and Crosser model, Brownian motion model, clustering model, liquid layering model, Yamada and Ota model for carbon nanotube were concluded in previous review works [1,10,15,39]. Maxwell offered the model that expressed an effective thermal conductivity equation for nanofluids containing spherical nanoparticles [39,42]:

$$k_{nf} = \frac{k_p + 2k_f + 2(k_p - k_f)\phi}{k_p + 2k_f - (k_p - k_f)\phi} k_f \tag{3.15}$$

Note that the effect of the size and shape of the particles and particle interactions were not included in Maxwell's equation.

Hamilton and Crosser model [43] represented the model for non-spherical particles with the shape factor (n) as experimental parameters for different types of nanoparticles [39,43]:

$$k_{nf} = \frac{k_p + (n-1)k_f - (n-1)\phi(k_f - k_p)}{k_p + (n-1)k_f + \phi(k_f - k_p)} k_f \tag{3.16}$$

where n is defined as

$$n = \frac{3}{\psi} \tag{3.17}$$

where ψ is the sphericity. Sphericity is the ratio of the surface area of a sphere with the same volume as the particle to the surface area of the particle, so $n = 3$ for a sphere [3].

Brownian motion model focused on Brownian motion which was the random motion of particles suspended in nanofluids. Brownian dynamics simulation was used to determine the effective thermal conductivity of nanofluids as [39,49]

$$k_{nf} = \phi k_p + (1-\phi)k_f \tag{3.18}$$

Note that k_p includes the effect of the Brownian motion of the nanoparticles on the thermal conductivity.

Koo and Kleinstreuer model [50] proposed the combination model of two thermal conductivity parts as

$$k_{nf} = k_{static} + k_{Brownian} \tag{3.19}$$

where

k_{static} represents the thermal conductivity enhancement due to the higher thermal conductivity of the nanoparticles such as the Maxwell model,

$k_{Brownian}$ represents the effect of Brownian motion, which the effect of nanoparticles moving around the nanofluid was taken into account as

$$k_{Brownian} = 5\times 10^4 \beta\phi\rho_f c_{p,f}\sqrt{\frac{\kappa_B T}{\rho_p d_p}}f \tag{3.20}$$

where

T represents the temperature in K

$c_{p,f}$ represents specific heat capacity of base fluid

β represents a constant obtained from their experiment

Both f and β for CuO nanofluids can be determined by utilizing available experimental data [51]:

$$f = (-134.63 + 1722.3\phi) + (0.4705 - 6.04\phi)T \tag{3.21}$$

For other nanofluids, f was taken as 1 [52]. Modified Koo and Kleinstreuer model [50] was represented as

$$k_{nf} = \frac{\left[k_p + (n-1)k_f - (n-1)\phi(k_f - k_p)\right]}{\left[k_p + (n-1)k_f + \phi(k_f - k_p)\right]}k_f + 5\times 10^4 \beta\phi\rho_f c_{p,f}\sqrt{\frac{\kappa_B T}{\rho_p d_p}}f \tag{3.22}$$

Since nanofluid clustering resulted in the fast heat transportation along relatively large distances, heat conduction occurred faster by solid particles; the thermal conductivity ratio model considering the clustering was [53–57]

$$\frac{k_{nf}}{k_f} = \frac{(k_{cl} + 2k_f) + 2\phi_{cl}(k_{cl} - k_f)}{(k_{cl} + 2k_f) - \phi_{cl}(k_{cl} - k_f)} \tag{3.23}$$

where

k_{cl} represents the thermal conductivity of the clusters

ϕ_{cl} represents the particle volume fraction of the clusters

Cylinder shaped nanoparticles were more effective in thermal conductivity enhancement when compared to spherical particles [39] but the excessive nanofluid clustering caused their sedimentation [10,15,40,41,55].

Xuan model [58] considered both Brownian motion and clustering of nanoparticles and was expressed

$$\frac{k_{nf}}{k_f} = \frac{k_p + 2k_f - 2\phi(k_f - k_p)}{k_p + 2k_f + \phi(k_f - k_p)} + \frac{\rho_p \phi c_{p,p}}{2k_f}\sqrt{\frac{\kappa_B T}{3\pi r_{cl}\mu_f}} \tag{3.24}$$

where, r_{cl} represents the apparent radius of the nanoparticle clusters obtained by experiments. The second term on the right-hand side of Equation 3.24 is the effect of the random motion of the nanoparticles taken into account.

Since liquid molecules formed as layered structures surrounding solid surfaces could enhance the effective thermal conductivity, the modified Maxwell model that considered the liquid layer was [59,60]

$$k_{nf} = \frac{k_{pe} + 2k_f + 2(k_{pe} - k_f)(1+\beta)^3\phi}{k_{pe} + 2k_f - (k_{pe} - k_f)(1+\beta)^3\phi} k_f \tag{3.25}$$

where k_{pe} is the thermal conductivity of the equivalent nanoparticle

$$k_{pe} = \frac{[2(1-\gamma) + (1+\beta)^3(1+2\gamma)]\gamma}{-(1-\gamma) + (1+\beta)^3(1+2\gamma)} k_p \tag{3.26}$$

and

$$\gamma = \frac{k_l}{k_p} \tag{3.27}$$

where k_l is the thermal conductivity of the nanolayer. β is defined as

$$\beta = \frac{t}{r_p} \tag{3.28}$$

where

t is nanolayer thickness

r_p the nanoparticle radius

Equation 3.26 is also known as Yu and Choi Model [59,60]. Xue and Xu model [61] considered Bruggeman's effective media theory [62], was expressed as

$$\left(1 - \frac{\phi}{\alpha}\right)\frac{k_{nf} - k_f}{2k_{nf} + k_f} + \frac{\phi}{\alpha}\frac{(k_{nf} - k_l)(2k_l + k_p) - \alpha(k_p - k_l)(2k_l + k_{nf})}{(2k_{nf} + k_l)(2k_l + k_p) + 2\alpha(k_p - k_l)(k_l - k_{nf})} = 0 \tag{3.29}$$

where subscript l refers to nanolayer. α is defined as

$$\alpha = \left(\frac{r_p}{r_p + t}\right)^3 \tag{3.30}$$

where t is the thickness of the nanolayer.

Yamada and Ota model [63] represented the normalized effective thermal conductivity for carbon nanotube (CNT) nanofluid as

$$k_{nf} = \frac{\left[k_p/k_f + n - n\phi\left(1 - k_p/k_f\right)\right]}{\left[k_p/k_f + n + \phi\left(1 - k_p/k_f\right)\right]} k_f \tag{3.31}$$

where n is the empirical shape factor defined as

$$n = 2\phi^{0.2}\left(L_p/d_p\right) \tag{3.32}$$

For carbon nanotube, L_p and d_p are length and diameter of the cylindrical particles.

Chon model [64] introduced a temperature-dependent empirical correlation to determine the thermal conductivity as

$$k_d = C(\rho c_p)_{nf} u_x \phi d_p r_0 \tag{3.33}$$

$$\frac{k_{nf}}{k_f} = 1 + 64.7\phi^{0.7460}\left(\frac{d_f}{d_p}\right)^{0.3690}\left(\frac{k_p}{k_f}\right)^{0.7476} Pr^{0.9955} Re^{1.2321} \tag{3.34}$$

where

$$Pr = \frac{\mu_f}{\rho_f \alpha_f} \tag{3.35}$$

$$Re = \frac{\rho_f V_{Br} d_p}{\mu_f} \tag{3.36}$$

$$V_{Br} = \frac{\kappa_B T}{3\pi\mu_f d_p \lambda_f} \tag{3.37}$$

They noted that V_{Br} is Brownian velocity of nanoparticles.

Corcione model [65] presented an empirical thermal conductivity correlation using the different data available in their literatures for several types of nanoparticles (10–150 nm) between 20°C and 50°C temperature range as [66]

$$\frac{k_{nf}}{k_f} = 1 + 4.4 Re^{0.4} Pr^{0.66}\left(\frac{T}{T_{fr}}\right)^{10}\left(\frac{k_p}{k_f}\right)^{0.03} \phi^{0.66} \tag{3.38}$$

where T_{fr} represents the freezing point of base fluid.

Elias et al. [13] found that by increasing the temperature, thermal conductivity were observed to be intensified; therefore, the nanofluid could show better performances at higher temperatures. They claimed that no experimental data of thermal conductivity of radiator coolant based nanofluids; therefore, they performed their experimental work to measure the values of the thermal conductivity of Al_2O_3–radiator coolant nanofluid at 30°C and compared their results with different empirical equations under different volume concentrations ranging from 0 to 1 vol.%; their results indicated higher thermal conductivities than the models developed by Maxwell [42], 1.92% higher, and Hamilton–Crosser [43], 2.2% higher, for all the volume concentrations.

From Hachey et al. work [67], the transient heating (Phase I) allowed a novel concurrent analysis of each circulated nanofluid sample and, in turn, an explicit correlation between viscosity and thermal conductivity. This feature also produced some unexpected results, such as most samples exhibiting thermal conductivity in an inverse proportion to concentration and the formation of parabolic regressions. Phase II proved successful in demonstrating the degradation of the samples by their exposure to heat, as their properties showed gradual alteration in isothermal heating. This alteration, hysteresis, was further observed in Phase III where it was also found to be highly specific to the composition, with some discernable patterns being observed in same base fluid species. The degradation of the nanoparticle suspensions was later confirmed with an additional dynamic light scattering (DLS) analysis of the spent experimental fluid samples. The DLS analysis of the spent samples indicated a shift of the particle-size distribution toward higher particle diameters, further evidencing that heating effects are a catalyst to nanofluid agglomeration. The particle agglomeration is thus believed to have been the cause of the hysteresis phenomenon observed for both the thermal conductivity and dynamic viscosity, as well as the nanofluid instability under heating effects.

Saleh et al. [68] investigated titanium dioxide nanofluids as working fluids in heat pipe and found that the relative thermal conductivity of distilled water was significantly increased with the addition of titanium dioxide nanoparticles and with temperature. The enhancement exhibited a nonlinear relationship with respect to the volume fraction nanoparticles and temperature.

Manikandan et al. [69] studied influence of nanoparticle concentrations and temperatures (10°C–50°C) on thermal conductivity and revealed the linear increase of thermal conductivity with nanoparticle concentration. They found 46.2% higher thermal conductivity enhancement at lower temperatures (at 10°C with 2 vol.% nanofluid) and concluded the role of ordered liquid layers toward thermal conductivity enhancement in well-dispersed nanosand–propylene glycol dispersion.

Singh et al. [70] presented that the thermal conductivity of the Cu/TH59 and Cu/TH66 nanofluids increases with the increasing copper loading. The enhancements were above the predictions of the Effective Medium Theory (EMT) [42]. Their conclusion was drawn that the Cu/TH66 nanofluids appeared to be more stable than the Cu/TH59 nanofluids, and stability improved as the particle size decreases, a figure of merit criterion was used to assess the heat transfer effectiveness of the suspensions as compared to the base fluid. Halelfadl et al. [71] studied the influence of particles concentrations and temperatures on the thermal conductivity of the CNTs/water nanofluids.

Esfe et al. [38] experimentally measured the thermal conductivity and found that the Hamilton–Crosser [43] and Yu and Choi [60] models underestimated the thermal conductivity ratio of MgO/water nanofluid due to the fact that both models do not take into account the effects of important factors such as the particle size, temperature, and interfacial layer on the thermal conductivity of nanofluids.

Ghanbarpour et al. [72] noticed that the thermal conductivity of Al_2O_3/water nanofluids increased nonlinearly with an increase in mass concentration while its increment was almost linear with temperature at a certain concentration. The thermal conductivity enhancement was in the range of 1.1%–87%, for mass concentration and temperature in the range of 3%–50% and 293–313 K. Based on limited available experimental results, they proposed a new and simple correlation based on nonlinear fitting of data as

$$\frac{k_{nf}}{k_f} = 1 + 3.5\phi + 2.5\phi^2 \tag{3.39}$$

The deviations of their experimental results from the proposed nonlinear pattern were less than 5%, which showed the capability of the proposed empirical correlation to predict the thermal conductivity of Al_2O_3/water nanofluid under the condition of their study.

Ghozatloo et al. [73] studied effects on the thermal conductivity of different surface modification methods; surfactant, acid, base, amide, sulfate on multiwalled CNTs. The basic treatment and SDS treated CNT/water nanofluids (SDS-KCNT/water) showed the highest conductivity of 0.765 W/m.K, which increases 24.9% of water as a base fluid conductor.

Lee et al. [74] introduced a new method to track the temporal changes of the particle volume fraction and size distribution, and the effects of the change of the nanofluid thermal conductivity. Their five different nanofluids were generated as Al_2O_3/DI water, SiO_2/ DI water, Ag/DI water, Al_2O_3/EG, and ZnO/EG. The influence of the volume concentration and the particle size variation on the thermal conductivity of the nanofluid was studied using both the transient hot-wire method and the DLS method.

Nine et al. [75] carried out a pre- and postexperimental analysis of copper–water and silver–water nanofluids in a single loop oscillating heat pipe (OHP) with 2.4 mm inner diameter. They found that color, deposition, dispersibility, propensity to be oxidized, disintegration, agglomeration, and thermal conductivity of metal nanofluids were strictly affected by heat transfer process and vice versa. They reported that metal nanofluids were found completely unstable in boiling heat transfer application. Nanofluid thermal properties were found different in heating up and cooling down modes and metal nanoparticles were more prone to be oxidized at even low temperature.

Karami et al. [76] experimentally observed significant enhancement of thermal conductivity: 32.2% by adding 150 ppm of f-CNTs to water as f-CNTs/water nanofluids as working fluid used in direct absorption solar collectors (DASCs). They saw that the thermal conductivity increased with increasing temperatures and f-CNT volume fractions, but the dependence of the thermal conductivity on temperature was much more significant.

Madhesh et al. [77] experimentally studied the heat transfer and rheological characteristics of Cu–TiO_2 hybrid nanofluids in a tube-in-tube counter flow heat exchanger. Their conclusions showed that the surface functionalized and highly crystalline nature of the hybrid nanofluids improved thermal conductivity and heat transfer potential of nanofluids.

Ahmed et al. [78] identified effects due to uncertainties of thermal conductivity and dynamic viscosity of nanofluid (Cu/water, Ag/water, CuO/water, and TiO_2/water) on boundary layer flow and heat transfer characteristics due to permeable stretching tube in the presence of heat source/sink with four models of thermal conductivity and dynamic viscosity depending on the shape of nanoparticles. Their results indicated that the skin friction coefficient decreased as the Reynolds number and the suction/injection parameter increased, while the local Nusselt number increased as the Reynolds number and the suction/injection parameter increased. These results were compared and found to be in excellent agreement with their literature.

Zerradi et al. [79] preformed a numerical study based on the Monte Carlo simulations combined with the new Nusselt number correlations to predict the thermal conductivity of several nanofluids: Al_2O_3/water, CuO/water, TiO_2/water, and CNT/water, in good agreement with their existing experimental data as

$$k_{static} = \frac{\dfrac{k_p}{k_f} + \psi + \psi\phi\left(1 - \dfrac{k_p}{k_f}\right)}{\dfrac{k_p}{k_f} + \psi + \phi\left(1 - \dfrac{k_p}{k_f}\right)} \quad \begin{cases} \psi = 2\phi^{0.2\frac{l_p}{d_p}} & \text{for cylindrical particles} \\ \psi = 2\phi^{0.2} & \text{for spherical particles} \end{cases} \tag{3.40}$$

$$k_{Brownian} = G\frac{k_f\varepsilon_f \sum_{i=1}^{N} \dfrac{A_i Pr^p}{d_i}\left(\alpha\phi + (\beta + \chi\phi)\left(\dfrac{1}{\nu}\sqrt{\dfrac{18k_bT}{\pi\rho d_i}}\right)^q + \delta\right)}{PrA_T} \tag{3.41}$$

where

G is empirical constant related to the thickness of the thermal boundary layer
α, β, and χ are thermophysic coefficients
ν is kinematic viscosity

They presented their results to show the thermal conductivity of Al_2O_3/water as function of temperature for three averaged nanoparticle diameters, 11, 47, and 150 nm, at 0.01 volume fractions and found a good agreement with the experimental values given by Chon et al. [64], for the case of nanoparticles with 47 and 150 nm as averaged diameter.

Lamas et al. [80] presented an exhaustive critical analysis, which was a statistical analysis of the different models to evaluate the predictive models currently available for thermal conductivity of CNT-based nanofluids. Their statistical study highlighted a lack of confidence on the models available since there was no convergence on the results because of poor quality data resulting in limited generic models. They concluded that more parametric experimental analysis might be given to specific control factors, not only particle geometry, volume fraction, temperature, and base fluid but also carbon nanotubes length and volume fraction.

Tso et al. [81] studied a semi-analytical model for calculating the enhanced thermal conductivity of nanofluids in spherical coordinates and focused on the effect of nanolayer thickness and the impact of the interfacial nanolayer on the effective thermal conductivity of nanofluids by using their model. They found that the effective thermal conductivities of nanofluids showed the same result as the Maxwell model when the nanolayer thickness equaled to zero and other results were in good agreement with the experimental results, and claimed that their prediction was much better than the Maxwell model and Bruggeman model.

3.4 NANOFLUID CONVECTIVE HEAT TRANSFER MODELING

Because of nanofluid nature as multicomponent fluids, it was also treated as either a two-phase homogeneous flow with no slip between nanoparticles and the fluid which was also in thermal equilibrium, or it was treated with a slip between the particles and the base fluid with thermal equilibrium. Most forced convection flows depended on both the Reynolds and Prandtl numbers, but for the case of nanofluids, additional parameters were included to take the thermal properties of all the constituents into account [39].

3.4.1 Empirical Correlation for Forced Convection with Nanofluids

Some experimental investigations in convection heat transfer of nanofluids, using water and propanol as base fluids, were collected in Table 3.3 [10,82].

TABLE 3.3
The Experimental Thermal Conductivity of Nanofluids: Water and Propanol Base Fluids

Base Fluids	Nanoparticles	Flow Regime	Experimental Condition	Researchers
Water	Al_2O_3	Laminar and developing flows	Constant wall temperature	Heris et al. [83]
	Al_2O_3	Laminar and developing flows	Constant heat flux	Anoop et al. [84]
	Al_2O_3	Turbulent	Constant wall temperature	Fotukian and Esfahany [85,86]
	Al_2O_3	Turbulent	Constant heat flux	Williams et al. [87] and Rea et al. [37]
Propanol	Al_2O_3	Laminar and Turbulent	Counter flow heat exchanger	Sommers and Yerkes [9]

Sources: Based on Chandrasekar, M. et al., *Renew. Sustain. Energ. Rev.*, 16, 3917, 2012; Godson, L. et al., *Renew. Sustain. Energ. Rev.*, 14, 629, 2010.

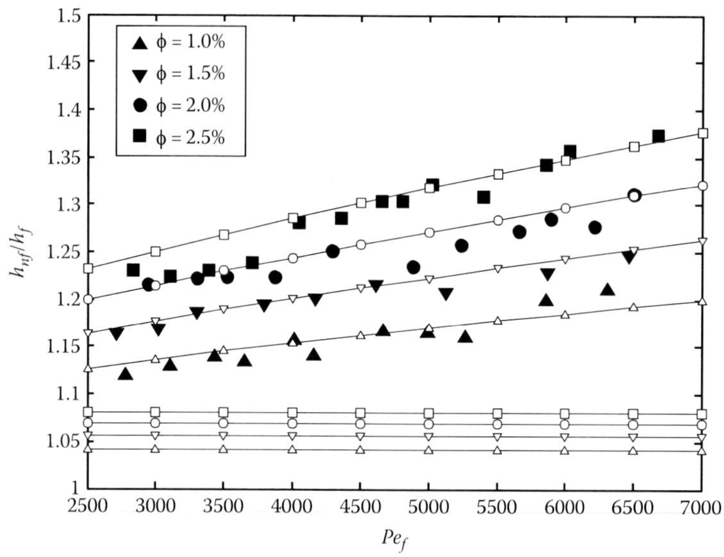

FIGURE 3.2 Comparison of the predictions of the correlations with experimental data as a function of Peclet number; the predictions of Li and Xuan correlation (solid lines), the predictions of Sieder-Tate correlation (dashed lines), and the experimental data of Heris et al. (markers). (Redrawn from Heris, S.Z. et al., *Int. J. Heat Fluid Flow*, 28(2), 203, 2007; Sieder, E.N. and Tate, G.E., *Ind. Eng. Chem.*, 28(12), 1429, 1936; Özerinç, S. et al., *Int. J. Therm. Sci.*, 62, 138, 2012; Li, Q. and Xuan, Y., *Sci. China E*, 45(4), 408, 2002.)

Sundar et al. [107] experimentally estimated the Nusselt number and friction factor as a function of particle concentration and Reynolds number for constant heat flux condition in forced convection apparatus with no phase change of the nanofluid flowing in a tube. The experiments were conducted for a Reynolds number range of 3,000–22,000 and for a particle concentration range from 0% to 0.6%. The results indicate that both Nusselt number and friction factor of the nanofluid increase with increasing particle volume concentration and Reynolds number. For 0.6% volume concentration, the enhancement of Nusselt number and friction factor is 39.18% and 19.12%, respectively, as compared to distilled water under the same flow conditions. It was verified that the classical Gnielinski and Notter–Rouse correlations underpredict the Nusselt number of the nanofluid; therefore, new generalized correlations are proposed for the estimation of the Nusselt number and friction factor based on the experimental data.

Halelfadl et al. [108] evaluated the convective heat transfer properties with constant wall temperature for four different multiwall CNTs based nanofluids at low concentration of 0.026 vol.% in a coaxial heat exchanger under laminar regime within the range of Reynolds numbers 500–2500. They claimed that the enhancement of the thermal conductivity and the average convective heat transfer of nanofluids increased with the aspect ratio of nanotubes and decreased when the thermal conductivity of the base fluid increases. This enhancement attained at least 10% in comparison to base fluid even with the low content of nanotubes used.

Saleh et al. [68] studied the transport properties of titanium dioxide nanofluids and their thermal performance as working fluids in a screen-mesh wick heat pipe investigated experimentally. The experimental results demonstrated that using nanofluid enhanced the thermal performance of screen-mesh wick heat pipes. Increasing the nanofluid volume fractions had a great effect in reducing the wall temperature of both types of heat pipes considered in their study. They also observed that an inclination angle 45% and 60% charge volume ratio of working fluid yielded the best thermal performance for heat pipes.

Halelfadl et al. [71] reported the efficiency of the nanofluids as coolants in both laminar and turbulent flow regimes as in energy systems and heat exchangers. They found that, in laminar regime with Reynold numbers ranging from 180 to 330, the nanofluids became efficient at 30°C and, by increasing the temperature to 40°C, the efficiency of nanofluids was increased to higher range of Reynold numbers. Their results in turbulent regime showed that the efficiency of the nanofluids was dependant to both the decrease in particle volume fractions and the increase in temperatures.

Esfe et al. [38] studied convective heat transfer characteristics and pressure drop of the (40 nm) MgO–water nanofluid flowing through a straight circular tube in fully developed turbulent flows. Their results showed that the maximum enhancement of heat transfer coefficient was about 35.93% for 1.0% volume fraction of nanofluid at $Re = 7331$. They compared their experimental results with existing correlations for Nusselt number of nanofluids and illustrated that experimental results were in good agreement especially with Gnielinski correlation at low concentration.

$$Nu = \frac{f/8\left(Re-1000\right)Pr}{1+12.7\sqrt{\frac{f}{8}}\left(Pr^{2/3}-1\right)}\left(\frac{Pr}{Pr_f}\right)^{0.11}\left[1+\left(\frac{d}{L}\right)^{2/3}\right] \tag{3.69}$$

They emphasized that the thermal performance factor for all the cases considered were greater than unity, which illustrated that the heat transfer enhancement was possible by utilizing nanofluid without huge penalty in pumping power. Their experimental friction factor was compared with the theoretical values obtained using Blasius relation [100] for distilled water as

$$f = 0.316RE^{-0.25} \tag{3.70}$$

Their friction factors for nanofluids according to Blasius prediction indicated that pumping power could be predicted with a well-known correlation when the solid volume fraction of nanofluid is less than 1%. They noted that the friction factor relation for the single-phase flow could be extended to MgO/water nanofluids. This trend is in general agreement with the results of Xuan and Li [90], Kayhani et al. [110], and Pak and Cho [11] under turbulent flow.

Ghanbarpour et al. [72] studied a theoretical analysis of thermal behavior of the nanofluid based on its experimental thermal properties in laminar and turbulent flow regimes. For thermally developing laminar flow, their results showed that the relative heat transfer coefficient increased almost 90% and 50% mass concentration at the same Reynolds number while it increased 50% and 20% at the same flow velocity and pumping power, respectively. For turbulent flow, the maximum increment achieved at the same Reynolds number while the relative heat transfer coefficient decreased at the same pumping power because of the higher increment in viscosity of the nanofluid, in comparison with thermal conductivity increment.

Karami et al. [76] investigated f-CNTs/water nanofluids as working fluid used in DASCs. The experimentally determined spectral transmission confirmed that f-CNT has a considerable role in raising the optical properties of the fluid, due to improvement of the light extinction level even at low volume fractions. The results show that the presence of 150 ppm f-CNT increases the extinction coefficient of pure water by about 4.1 cm^{-1}. With these promising properties, this nanofluid, f-carbon nanotube or f-CNT, is very interesting for increasing the overall efficiency of the DASCs.

Haghighi et al. [110] studied the convective heat transfer coefficients of water-based Al_2O_3, ZrO_2, and TiO_2 nanofluids at 9 wt.% and 20°C in a 1.5-m long, 3.7-mm diameter tube. They confirmed that classical correlations such as Shah for laminar flow and Gnielinski for turbulent flow could predict heat transfer coefficients within 15% accuracy when employing measured data and/or data from well accepted correlations for the thermophysical properties. They found that the average heat transfer enhancements at the same Reynolds number that were compared with base fluids was 23%, 8%, and 15% in laminar flow and 51%, 13%, and 41% in turbulent flow for Al_2O_3, TiO_2, and ZrO_2 nanofluids, respectively, while the results at the same pumping power showed average decrement of 63%, 17%, and 52% for Al_2O_3, TiO_2, and ZrO_2 nanofluids, respectively.

Hajipour and Dehkordi [111] investigated the Al_2O_3–water nanofluid mixed convective heat transfer inside a vertical channel partly filled with porous metal foam experimentally and numerically. They were interested in nanofluid flow in the porous metal foams as in porous matrix heat exchangers, solar collectors with porous absorber and chemical processing equipment because the nanofluid could improve the heat-transfer rate. They discovered that the nanofluids were stable and no sedimentation occurring in the porous foam and increasing the nanoparticle concentration did not show significant effect on the pressure drop in their work.

The nanofluid at 0.3% volume concentration produced 20% maximum nanofluid enhancement and 5% increase in pressure drop compared to distilled water.

Naik et al. [112] focused on the estimation of heat transfer and friction factor CuO/water nanofluid flowing in a plain tube with twisted tape and wire coil inserts. At 0.3% nanofluid and Reynolds number of 20,000, they met a maximum of 17.62% heat transfer enhancement, a maximum of 31.88% enhancement with inserting the twisted tape TT-2 and a maximum of 44.45% enhancement with inserting the wire coil WC-2. We noted that inserted tubes occupied higher friction factors in the order of 1.17 times and 1.19 times compared to water flowing in a tube at $Re = 20{,}000$, the thermal performance factor for nanofluids with wire coil inserts was more effective than twisted tape inserts under same particle loading and flow rate and the enhancement in friction factor was negligible in this study.

Abbassi et al. [113] observed the effect of TiO_2/water nanofluid in a vertical annulus with non-uniform heat flux and expressed that, for low volume fraction (less than 1%), there were no major changes in wall temperature, maximum wall temperature, heat transfer coefficient, and Nusselt number and, for higher volume fractions, wall temperature decreased and heat transfer and Nusselt number increased caused by the nanofluid dispersion.

Madhesh et al. [77] experimentally discovered that the convective heat transfer coefficient of nanofluids increased with Cu–TiO_2 hybrid nanofluid concentration and the Reynolds number, for the volume concentration up to 1.0%, the convective heat transfer coefficient, Nusselt number, and overall heat transfer coefficient were enhanced by 52%, 49%, and 68%, respectively. The new Nusselt number correlation as a function of the Reynolds number, Prandtl number, and volume concentration of the hybrid nanofluid was proposed as

$$Nu = 0.012 Re\, Pr^{0.333} \phi^{0.032} \tag{3.71}$$

which was valid under the transient flow regime with $4000 < Re < 8000$ and volume concentration ranging from 0.1% to 2.0%. They also proved the new correlation provided its results within the 17% average deviation respecting to the experimental data.

Esfe et al. [114] experimentally evaluated the effects of adding low concentrations of a new class of nanotubes; COOH-functionalized double-walled carbon nanotubes (DWCNTs), under turbulent flow and the nanofluid concentrations from 0.01% to 0.4%. Their results indicated that even using a very low quantity of 0.4 vol.% COOH-functionalized DWCNTs led to remarkable enhancement in the heat transfer coefficient (on average 25%), and the average Nusselt number increased on average by 18% but the pressure drop increased up to 20%; even using 0.01% of nanotubes was not advantageous due to the unfavorable effects of raising in pressure drop.

Esmaeilzadeh et al. [115] experimentally studied convective heat transfer performance and pressure drop of γ-Al_2O_3/water nanofluid in laminar flow through a circular tube fitted with twisted tape inserts with different thicknesses under constant heat flux and found that twisted tape inserts and the twisted tape thickness enhanced the average convective heat transfer coefficient. Ultimately, the convective heat transfer enhancement outweighed the effect of friction factor increase, leading enhanced thermal performance. They proposed two new empirical correlations for Al_2O_3/water nanofluid Nusselt number and thermal performance with volume fraction of 0.5% and 1%, with 15-nm diameter nanoparticles, through circular tube fitted with twisted tape inserts with different thicknesses (0.5, 1, and 2 mm), at constant twist ratio of 3.21, under laminar flow for $150 < Re < 1600$, from the experimental data as

$$Nu = 4.36 + a\left(x_*\right)^{-b} e^{\left(c \cdot x_*\right)}\left(0.001+\phi^{d}\right);\ \left(0.001+\left(\frac{t}{d}\right)^{f}\right) x_* > 0.00468 \quad (3.72)$$

where $a = 46.39$, $b = 1.142$, $c = 12.37$, $d = 1.48$, and $f = 0.1796$, which provided a maximum deviation of 15% from the experimental data and the thermal performance factor, η, given based on nanoparticle concentration, the dimensionless parameter (t/D) and Re, for $150 < Re < 1600$, as

$$\eta = aRe^{b}\left(1+\left(\frac{t}{D}\right)^{c}\right)\left(1+\phi^{d}\right) \quad (3.73)$$

where $a = 0.155$, $b = 0.3017$, $c = 0.2214$ and $d = 0.710$, which provided a maximum deviation of 15% from the experimental data.

Ghobadi and Muzychka [116] carried out an experimental study on the heat transfer inside mini scale coiled tubing for constant wall temperature condition with water and two low viscosity silicone oils and three categorized tube regions (1) low curvature region $De < 10$, where the heat transfer behavior resembled the one in straight tubes; (2) high curvature region $De > 50$; and (3) transitional region $10 < De < 50$. They found that an empirical method could predict the heat transfer in high curvature region. They proposed an asymptotic model using the experimental results for blending the rest two regions as

$$Nu = \left[3.66^{4} + \left(0.91375\sqrt{De}Pr^{-0.1}\right)^{4}\right]^{1/4} \quad (3.74)$$

The model predicted the Nusselt number in the Prandtl number range of $5 < Pr < 15$, and Dean number range of $40 < De < 700$, and the experimental results fell within the 10% range of the proposed model prediction. They mentioned that the Nusselt number in fully developed region was not affected by the length of coiled tubing.

Ho et al. [117] examined the cooling performance of a minichannel heat sink using the water based with Al_2O_3 nanoparticles and/or microencapsulated phase change material (MEPCM) particles and focused on the correlations of the averaged Nusselt number, the wall temperature control effectiveness and the averaged thermal resistance reduction effectiveness, respectively, as

$$Nu = 1.868\left(\frac{Pe_f}{L/D_h}\right)^{0.37}\left(1+1.72\omega_p-7.48\omega_p^2\right) \quad (3.75)$$

$$\varepsilon_{\tau_w} = 0.103\left(\frac{Pe_f}{L/D_h}\right)^{0.177}\left(1+5.782\omega_p-18.68\omega_p^2\right) \quad (3.76)$$

$$\varepsilon_{\tau_w} = 0.151\left(\frac{Pe_f}{L/D_h}\right)^{0.113}\left(1+5.774\omega_p-21.353\omega_p^2\right) \quad (3.77)$$

They presented that the heat transfer effectiveness of incorporating the Al_2O_3/water nanofluid and/or MEPCM suspension in the minichannel heat sink depended on the flow rate, the particle fraction dispersed in water, and the latent/sensible heat ratio of the MEPCM suspension, as well as, better effectiveness for the Al_2O_3/water nanofluid over the hybrid water-based suspensions of Al_2O_3 nanoparticles and MEPCM particles at higher Reynolds number.

Hussein et al. [118] measured the forced convection heat transfer enhancement by TiO_2 and SiO_2 suspended in water inside the flat copper tubes of an automotive cooling system and found the significant heat transfer enhancement, which associated with the concentration of the nanoparticles, 11% and 22.5% maximum Nusselt number enhancements for TiO_2 and SiO_2 nanofluids, respectively.

Utomo et al. [119] investigated the heat transfer coefficients of identical alumina, titania, and CNT nanofluids independently at two universities, University of Birmingham (UB), UK, and the Royal Institute of Technology (KTH), Sweden. They discovered that their investigated nanofluids behaved like homogenous mixtures and the experimental Nusselt numbers of all tested nanofluids in laminar tube flow followed classical correlations developed for single-phase fluids within ±10%, which was a typical uncertainty range for heat transfer correlations. They referred to the industrial point of view and commended that the heat transfer coefficients of nanofluids should be compared to that of the base fluid at the same pumping power since pumping power is related to the cost of transferring the heat. Unfortunately, their results showed that the viscosity ratios of nanofluids to base fluids were higher than the thermal conductivity ratios as in their literatures [11,37,120–122].

Gurav et al. [123] produced colloidal copper nanoparticles in a batch process from copper sulfate using as corbic acid as reducing agent and PVP as a particle stabilizer. After that, they found that the Nusselt number and heat transfer coefficient increased with an increase in the Reynolds number in the laminar flow regime, this relationship attributed to a significant reduction in particle agglomeration and intensification of mixing due to the dispersion of the nanoparticles.

Sarafraz and Hormozi [124] performed the experimental investigation on subcooled flow boiling to find the heat transfer coefficient of CuO/water and reported that, in two distinguishable heat transfer regions (forced convective and nucleate boiling), the heat transfer coefficient values in the nucleate boiling zone were much higher than those reported in forced convection zone because of two phase heat transfer occurred while single-phase heat transfer mechanism was the dominant phenomenon in forced convective region. Their experimental results also showed that increasing the heat flux and mass flux led to the significant increase of the flow-boiling heat transfer coefficient and the flow-boiling phenomena around the vertical cylinder led to the nanoparticle deposit on the heating section. The fouling rate was significantly increased which also occurred with increasing the wall temperature of cylinder and applied heat flux.

Sun and Yang [125] focused on flow boiling heat transfer characteristics of nano-refrigerants: Cu-R141b, Al-R141b, Al_2O_3-R141b, and CuO-R141b in a horizontal tube and found that adding nanoparticles to the refrigerant enhanced the heat transfer and the heat transfer coefficient increased with increased nanoparticle mass fraction. However, with increased mass velocities, the heat transfer coefficient of nano-refrigerant increased but the nanoparticle impact factor decreased causing the decrease in enhancement in heat transfer performance. Finally, they claimed that Cu nanoparticles had the best effect in enhancing the heat transfer performance of the nano-refrigerants because of Cu high thermal conductivity.

Waghole et al. [126] conducted an experimental study to investigate heat transfer enhancement of Ag/water nanofluid in an absorber by means of twisted tape inserts (inserted in absorber/receiver of parabolic trough collector). They concluded that, for absorber with tape inserts, the Nussult number varied from 1.25 to 2.10 times and the friction factor varied from 1.0 to 1.750 times in comparison with those of plain absorber for the comparable Reynolds number and the enhancement factor varied from 135% to 205% at constant pumping power. They endorsed that there was no significant increase in pressure drop or friction factor for silver nanofluid in comparison to water at the same twist ratio. They introduced the generalized correlation functions of Nussult number and friction factor for turbulent flow of Ag/water nanofluid in plain absorber and with twisted tape inserts as, for water with twisted tape inserts

$$Nu = 1.2546 Re^{0.2843} Pr^{0.4} \left(1 + H/D\right)^{0.004} \tag{3.78}$$

$$f = 0.0841 Re^{-0.2729} \left(1 + H/D\right)^{-0.1701} \tag{3.79}$$

and, for silver nanofluid with twisted tape inserts

$$Nu = 1.3167 Re^{0.2843} Pr^{0.4} \left(1 + H/D\right)^{0.004} \left(1 + \phi\right)^{-0.008} \tag{3.80}$$

$$f = 0.0841 Re^{-0.2729} \left(1 + H/D\right)^{-0.1701} \left(1 + \phi\right)^{-0.5} \tag{3.81}$$

which were validated for the range $500 \leq Re \leq 6000$, $0 \leq \phi \leq 0.1\%$, and $0.577 \leq H/D \leq 1.732$ where $\phi = 0$ for water and H/D = 0 for plain absorber.

Karami and Rahimi [127] worked on an experimental system, which was a hybrid of microchannels cooling system, photovoltaic (PV)-thermal module, and applying Boehmite/water nanofluid with concentrations of 0.01, 0.1, and 0.3 wt.% as the working fluid. They wanted to enhance heat transfer rate for the PV cell to achieve higher electrical efficiency and found that the flow regime in the microchannel is laminar. They noted that the presence of nanoparticles has a remarkable effect on decreasing the average temperature of the PV cell compared with that of pure water. The PV efficiency was increased by decreasing the operating temperature of PV surface, and the highest electrical efficiency (27%) was achieved by employing 0.01 wt.% Boehmite nanofluid.

Ho et al. [128] examined the forced convection heat transfer effectiveness with using water-based suspensions of alumina nanoparticles and MEPCM particles (PCM suspension) to replace the pure water as the working fluids in a minichannel heat sink (Mini-CHS). They summarized that the heat dissipation effectiveness of the nanofluid and PCM suspension depended significantly on their flow rates through the heat sink and the friction factor was significantly increased for a Mini-CHS with larger mass fractions of both particles. They also investigated the forced convective cooling efficacy of the hybrid Al_2O_3/water nanofluid and PCM suspension in a Mini-CHS and found that the efficacy depended on the flow rate, particle fractions, and latent/sensible heat ratio, additionally, Al_2O_3/water nanofluid effectiveness was better than the hybrid suspensions at higher Reynolds number.

Sohel et al. [129] investigated the heat transfer performances of a rectangular minichannel heat sink using Al_2O_3/water nanofluid at different volume fractions and Reynolds number; a great performance and a significant improvement in the heat transfer coefficient (18%) of using Al_2O_3/water nanofluid as a coolant was obtained when they were compared to the pure distilled water, as well as, the 15.72% lower convective thermal resistance was found when it was compared to the distilled water.

Vermahmoudi et al. [130] applied Fe_2O_3/water nanofluid in a compact air-cooled heat exchanger to measure the overall heat transfer coefficient using the logarithmic mean temperature difference (LMTD) technique under laminar flow regime. Their measured overall heat transfer coefficient of the nanofluid demonstrated a very good agreement with the correlation of Xuan et al. [58] with average error of 3%. They conducted that the overall heat transfer coefficient and the heat transfer rate of nanofluid were improved with the enhancement in the nanofluid and air flow Reynolds numbers; with 0.65 vol.% nanoparticle concentration, the overall heat transfer coefficient was enhanced up to 13% and the heat transfer rate was enhanced up to 11.5% comparing with the distilled water.

Reddy and Rao [131] focused their present study on the investigation of heat transfer and friction factor of ethylene glycol/water based TiO_2 nanofluid flowing in a circular tube with and without helical coil inserts. The experimental investigations concluded that the use of helical coil inserts was advantageous to enhance heat transfer on one hand and on the other hand it introduces undesirable pressure drop in the circuit. Heat transfer enhancements in the presence of helical coil inserts with *P/d* ratio of 2.5 were from 7.85% to 16.11% at Reynolds number of 4,000 and the enhancement was increased to 17.71% when Reynolds number increased to 15,000. While, without the inserts, 0.02% TiO_2 nanofluid occupied heat transfer enhancements from 7.85% to 10.73% compared to base fluid when the Reynolds number varied from 4,000 to 15,000. They introduced their Nusselt number and friction factor correlations from their investigation as

$$Nu = 0.007523 Re^{0.8} Pr^{0.5} \left(1+\phi\right)^{7.6} \left(1+\frac{P}{d}\right)^{0.037} \tag{3.82}$$

$$f = 0.3250 Re^{-0.2377} \left(1+\phi\right)^{2.723} \left(1+\frac{P}{d}\right)^{0.041} \tag{3.83}$$

Khairul et al. [132] focused on the benefits of using CuO/water nanofluids in a corrugated plate heat exchanger and revealed that CuO/water nanofluids at 0.5%, 1.0%, and 1.5 vol.% increased the heat transfer coefficient about 17.70%, 21.80%, and 24.7%, respectively, as compared to water. Besides, higher particle volume fraction and volume flow rate of nanofluids higher than the friction factor, their analytical results revealed that CuO/water nanofluids reduced the energy destruction by 24%, 16.25%, and 8% for above concentrations, respectively, compared to water. At the end, they concluded that the average 34%, 22%, and 12% enhanced energetic heat transfer effectiveness was found above concentrations, respectively, compared to water.

Buschmann and Franzke [133] aimed to make nanofluids applicable for thermosyphons. They employed water-based titanium dioxide and gold nanofluids with different concentrations as working fluids in a vertical thermosyphon. They found that a maximal reduction of the thermal resistance of about 24% was achieved and an optimum condition was reached at concentrations between 0.2 and 0.3 vol.%, whereas at higher concentrations the thermal resistance remains either unchanged or increases again.

Heris et al. [134] studied the effect of electric field on thermal performance of two-phase closed thermosyphon as external modifications by adding two different water-based nanofluids; Al_2O_3 and CuO nanoparticles, into the pure water using ultrasonic vibration system without adding any dispersant or stabilizer to prevent any possible change of chemical properties of the nanofluid. Their results showed that the thermal performance and the Nusselt number ratio of thermosyphon gradually increased by increase in electric field intensity and nanofluids concentration. Al_2O_3/water nanofluid improved thermal performance of thermosyphon more than CuO/water nanofluid.

3.4.2 Theoretical Analysis of Convective Heat Transfer of Nanofluids

Since there are phenominals dealing with interactions between the fluid and the solid particles as well as among the particles themselves such as Brownian motion, researchers have investigated these factors in the enhancement of heat transfer. The governing equations under the specified boundary conditions can be solved. If the micro convective and micro diffusion of the suspended particles (hydrodynamic dispersion) are neglected, the general conservation equations can be written as [8,39] follows.

Conservation of mass

$$div\left(\rho\bar{v}\right) = 0 \tag{3.84}$$

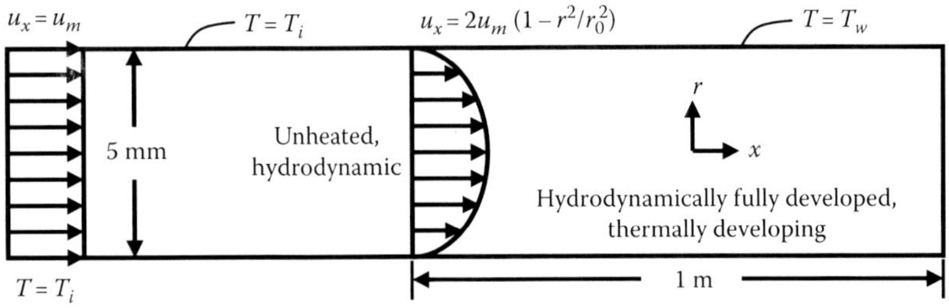

FIGURE 3.3 Schematic of hydrodynamically fully developed flow. (Redrawn from Özerinç, S., Heat transfer enhancement with nanofluids, Master thesis, Middle East Technical University, Ankara, Turkey, 2010.)

Conservation of momentum

$$div(\rho\bar{v}\ \bar{v}) = -grad\,P + \mu\nabla^2\bar{v} \tag{3.85}$$

Conservation of energy

$$div(\rho\bar{v}\ c_p T) = div(k \cdot grad\,T) \tag{3.86}$$

The simple domain as the axisymmetric flow was shown in Figure 3.3 and the energy equation for nanofluids can be considered as [52]

$$(\rho c)_{nf}\left(\frac{\partial T}{\partial t} + u\frac{\partial T}{\partial x}\right) = \frac{1}{r}\frac{\partial}{\partial r}\left(k_{eff}\,r\frac{\partial T}{\partial r}\right) + \frac{\partial}{\partial x}\left(k_{eff}\frac{\partial T}{\partial x}\right) + \mu_{nf}\left(\frac{\partial u}{\partial r}\right)^2 \tag{3.87}$$

When the isotropical flow is assumed, D represents the effective apparent thermal diffusivity, for laminar flow, as [1]

$$\bar{D} = \alpha^*_{eff} = \alpha_{eff} + \frac{D}{(\rho c_p)_{eff}} \tag{3.88}$$

and the apparent thermal conductivity becomes [1]

$$\bar{k} = k_{eff} + D \tag{3.89}$$

For fully developed laminar flow in a tube, the velocity distribution is given as [39]

$$u = 2u_m\left(1 - \frac{r^2}{r_0^2}\right) \tag{3.90}$$

The energy equation can be solved under constant wall temperature and constant wall heat flux boundary condition as [52]

$$\frac{\partial T}{\partial t} + u\frac{\partial T}{\partial x} = \frac{1}{r}\frac{\partial}{\partial r}\left(\bar{D}r\frac{\partial T}{\partial r}\right) + \frac{\partial}{\partial x}\left(\bar{D}\frac{\partial T}{\partial x}\right) + \left(\frac{\mu}{\rho c}\right)_{nf}\left(\frac{\partial u}{\partial r}\right)^2 \tag{3.91}$$

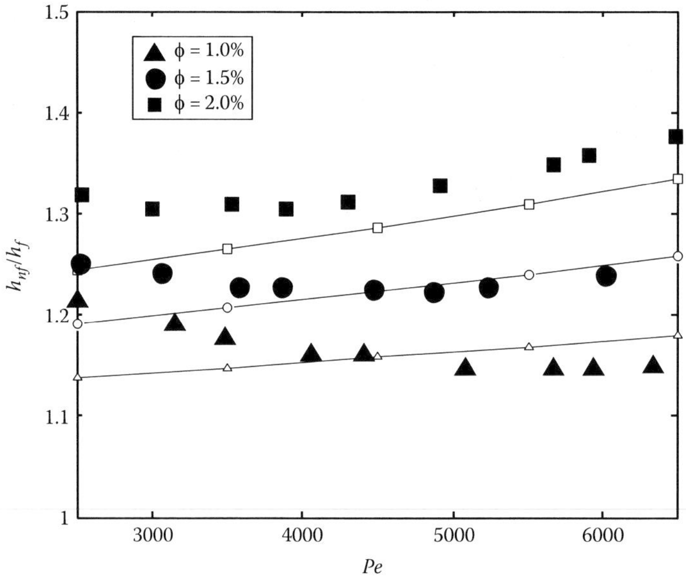

FIGURE 3.5 Comparison of the numerical results with numerical data of Heris et al. [146]. (Redrawn from Özerinç, S. et al., *Int. J. Therm. Sci.*, 62, 138, 2012.)

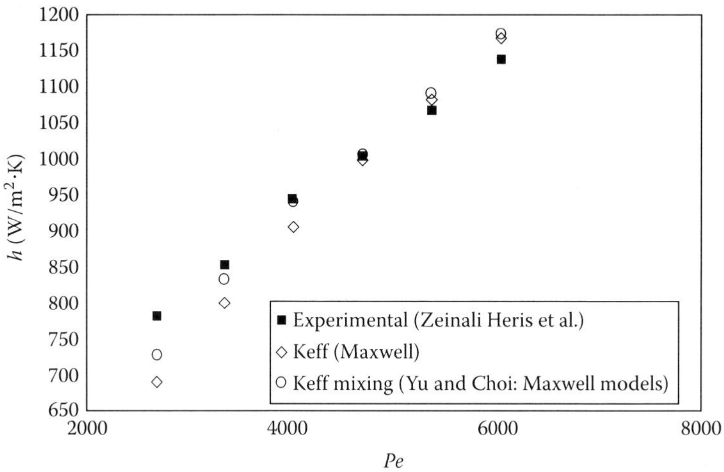

FIGURE 3.6 Simulation results obtained from single and mixing thermal conductivity models and experimental data. (Redrawn from Tongkratoke, A. et al., *Comput. Therm. Sci.*, 6(1), 1, 2014.)

Hussein et al. [142] were interested in the friction factor and forced convection heat transfer enhancement of SiO_2/water nanofluid. They performed a simulation and found that the maximum friction factor increased to 22% and the highest Nusselt number enhanced up to 40% for the 2.5% volume concentration SiO_2 nanofluid. Their simulation results showed that the friction factor and Nusselt number of the nanofluids were highly depended on the volume concentration and Reynolds number. They mentioned that there was a significant increasing in Nusselt number in the experimental study and agreeing with the numerical study. They claimed that the friction factor decreased with increasing of volume flow rate and the inlet temperature and the SiO_2 nanofluid had high potential for hydrodynamic flow and heat transfer enhancement and were highly appropriate to industrial and practical applications.

Hussein et al. [118] brought their input and output parameters from their experimental work to develop statistical models of cooling system components using multiple linear regression methods and factorial methodology (FM) or the analysis of variance (ANOVA) for Nusselt number for pure water, TiO_2/water, and SiO_2/water, respectively, as

$$Nu = 1.69414 + 0.32529Q + 0.0058446T_{\infty} \tag{3.141}$$

$$\begin{aligned} Nu = 6.38776 + 0.06018Q - 0.0610291T \\ - 285.668\phi + 0.0176852(Q \cdot T) + 51.2556(Q \cdot \phi) \end{aligned} \tag{3.142}$$

$$Nu = -3.0125 - 1.31736Q + 0.083777T - 1071.89\phi \tag{3.143}$$

Malvandi and Ganji [143] theoretically investigated force convective heat transfer of alumina/water nanofluid inside a cooled parallel-plate channel in the creeping flow regime and the presence of heat generation. They developed a modified two-component four-equation nonhomogeneous equilibrium model with adiabatic bottom wall and no-slip condition of the fluid–solid interface by using Fortran. Their results indicated that nanoparticles move from the adiabatic wall (nanoparticles depletion) toward the cold wall (nanoparticles accumulation) and construct a nonuniform nanoparticle distribution.

Nield and Kuznetsov [144] conducted an analytical study of fully developed laminar forced convection in a parallel-plate channel occupied by a nanofluid or by a porous medium saturated by a nanofluid, which was subjected to uniform-flux boundary conditions, constant heat flux boundaries, and constant temperature boundaries. Their model incorporated the effects of Brownian motion and thermophoresis and found that the combined effect of these two agencies had the effect of reducing the Nusselt number.

Selimefendigil and Öztop [145] introduced their numerical analysis of the heat transfer enhancement and fluid flow characteristics of a rotating cylinder under the influence of magnetic dipole in the backward facing step geometry using finite element method provided by COMSOL, a commercial software. They observed that the length and size of the recirculation zones were controlled with magnetic dipole strength ($0 \leq \gamma \leq 16$) and cylinder rotation angles ($-75 \leq \Omega \leq 75$), as well as, local Nusselt number increased and number of peaks in the presence of the magnetic field decreased as the Reynolds number ($10 \leq Re \leq 200$) increased.

Abdolbaqi et al. [146] observed the nanofluids (CuO/water, TiO_2/water, and Al_2O_3/water) heat transfer enhancement under turbulent flow through a straight square channel with constant heat flux condition at the upper and lower walls by using the finite volume method in the CFD analysis by FLUENT software with boundary conditions; under constant heat flux of 5000 W/m^2, Reynolds numbers from 104 to 106 and a constant volume concentration from 1% to 4%. Their results showed that the heat transfer rates and wall shear stress were increased by increasing the nanofluid volume concentration, the CuO nanofluid significantly enhanced heat transfer, and their results showed good agreement with their literature by 10% deviation.

Sokhansefat et al. [147] numerically studied three-dimensional fully developed turbulent mixed convection heat transfer of Al_2O_3/synthetic oil nanofluid in a trough collector tube and absorber tube with a nonuniform heat flux obtained from the Monte Carlo ray tracing technique. Their governing equations were solved by using the FLUENT software, and their results showed that the convection heat transfer coefficient had a direct dependency on the volumetric concentration of the nanofluid and the heat transfer enhancement due to the nanofluids reduced as the absorber operational temperature was increased. Then their results were compared with their existing experimental literatures.

Vanaki et al. [148] numerically investigated the effects of the SiO_2/EG nanofluids on the thermal and flow fields through transversely wavy wall channels with different phase shifts between the upper and lower wavy walls within the turbulent range of $6,000 \leq Re \leq 18,000$ and a uniform wall temperature of 400 K. In this work, the CFD software, FLUENT 6.3, was used to solve the governing equations by using finite volume method. Their simulation results indicated that the wavy channel performance was greatly influenced by changing the phase shift and the wavy amplitude; the highest performance evaluation criterion was obtained for the phase shift of 30° with wavy amplitude of 0.5 at $Re = 6000$ and the nanofluid with platelets nanoparticle shape gave the highest heat transfer enhancement compared with other tested nanofluids.

Selimefendigil and Öztop [149] performed a numerical study of pulsating rectangular jet with nanofluids to investigate the effects of pulsating frequency, Reynolds number, nanoparticle volume fraction on the fluid flow, and heat transfer characteristics by using FLUENT finite volume based code. From their results in the pulsating flow case, the combined effect of pulsation and inclusion of nanoparticles was not favorable for the augmentation of the stagnation point Nusselt number, at Re equals to 200 and 400 and volume fractions of 1% and 3%, when compared to the steady case.

Kherbeet et al. [150] simulated the laminar mixed convective flow of SiO_2/EG nanofluid over a 3-D horizontal microscale backward-facing step (MBFS) to explore the effect of step height on the flow and heat transfer characteristics by using a finite volume method. Their results showed that the Nusselt number and skin friction coefficient increased, while the Reynolds number and pressure drop decreased, with the increase of the step.

Malvandi and Ganji [151] theoretically studied mixed convective heat transfer of alumina/water nanofluid inside a vertical microchannel by using a modified Buongiorno's model to investigate the effect of the nanoparticles migration. Their results indicated that nanoparticles moved from the heated walls causing nanoparticles depletion toward the core region of the channel causing nanoparticles accumulation and constructed a nonuniform nanoparticles distribution.

Xu and Pop [152,153] presented their works on mixed convection flow of a nanofluid over a stretching surface with uniform free stream [152] and in a horizontal channel [153]. Their nanofluid contained both nanoparticles and gyrotactic microorganisms. The recursively analytical solutions with high precision in both works were obtained by the improved homotopy analysis method.

Ting et al. [154] performed field-synergy analysis on the viscous dissipative alumina/water nanofluid flow in circular microchannel heat sinks to scrutinize the synergetic relation between the flow and temperature fields for both heating and cooling processes. For the heating process, under the effect of viscous dissipation, they found that the degree of synergy between the velocity and temperature fields of nanofluid flow deteriorated, which led to a dwindled heat transfer performance of the nanofluid because of the presence of two different heat sources, the synergy angle and field synergy number for viscous dissipative flow. For nanofluid cooling process, they found that the viscous dissipation augmented the field synergy of the flow.

Azimi and Kalbasi [155] simulated the forced convective heat transfer of nanofluid in the developing laminar flow (entrance region) in a circular tube and found that the dynamic part of nanofluid thermal conductivity due to the Brownian motion had a minor effect on the heat transfer coefficients, however, static part of thermal conductivity including nanolayer around nanoparticle played an important role in heat transfer.

Elias et al. [156] aimed to study the effect of different particle shapes (cylindrical, bricks, blades, and platelets) on the overall heat transfer coefficient, heat transfer rate, and entropy generation of shell and tube heat exchanger with different baffle angles and segmental baffle. They found that cylindrical shape nanoparticles showed best performance in respect to overall heat transfer coefficient and heat transfer rate.

Minea [157] numerically analyzed the heat transfer performance for laminar fully developed forced convection of Al_2O_3/water in horizontal tubes with two zones: one adiabatic and one with uniform wall heat flux. In this work, the effect of the uncertainty in the values of the physical properties of γAl_2O_3/water nanofluid on their thermohydraulic performance for laminar fully developed forced convection in modeling nanofluids properties was noticed.

Aly [158] numerically investigated the heat transfer and pressure drop characteristics of Al_2O_3/water nanofluid flowing inside coiled tube-in-tube heat exchangers by using the three dimensional realizable k–ε turbulent model with enhanced wall treatment from a computational fluid dynamics. In this work, the friction factor increased with the increase in curvature ratio and pressure drop penalty was negligible with increasing the nanoparticles volume concentration increasing the coil diameter and nanoparticles volume concentration. This work showed that conventional correlations for predicting average heat transfer and friction factor in turbulent flow regime such as Gnielinski correlation and Mishra and Gupta correlation, respectively, for helical tubes were also valid for the tested nanofluids, which suggested that nanofluids behaved like a homogeneous fluid.

Numerical studies of nanofluids with magnetic field influence increased gradually. Goharkhah and Ashjaee [159] simulated forced convective heat transfer of Fe_3O_4/water nanofluid or ferrofluid in an alternating nonuniform magnetic field produced by eight line source dipoles imposed on several parts of the channel. They compared the results with zero magnetic field case and found that the heat transfer enhancement increased with the Reynolds number and reached a maximum of 13.9% at Re equals to 2000 and frequency equals to 20 Hz, nevertheless, the heat transfer enhancement due to the magnetic field was accompanied by a pressure drop penalty, a maximum pressure drop increase of 6% at Re equals to 2000 and frequency equals to 5 Hz. They concluded that the pressure drop increase was not as significant as the heat transfer enhancement.

Servati et al. [160] utilized the Lattice Boltzmann method to investigate the effects of uniform vertical magnetic field on the flow pattern and Al_2O_3/water nanofluid heat transfer in a channel that is partially filled with porous medium and constant temperatures on top and bottom walls. They found that the increase in the magnitude of magnetic field or Hartmann number led to the slow growth in the average Nusselt number with little drops of the outlet average temperature and velocity.

Sert et al. [161] numerically studied forced convection heat transfer characteristics of nanofluid incompressible transient laminar flow in a circular duct under step change in wall temperature and wall heat flux by assuming Al_2O_3/water nanofluid as a homogenous single-phase fluid. They obtained temperature distributions across the tube for a step jump of wall temperature and wall heat flux for various times during the transient calculations at a given location for a constant value of Peclet number and a particle diameter. They found that the heat transfer coefficient values for a step change wall heat flux condition was higher than a step change wall temperature condition; however, the enhancement of heat transfer coefficient was almost same for both conditions. They concluded that the variations resulted from the step change in the wall temperature, and heat flux were still significant only in short time interval. The results are given under transient and steady-state conditions.

Said et al. [162] theoretically analyzed entropy generation, heat transfer enhancement capabilities, and pressure drop for a flat-plate solar collector operated with single-walled carbon nanotubes (SWCNTs) based nanofluids as an absorbing medium using measured data; nanofluid specific heat (Cp) obtained from a PerkinElmer DSC 4000 and the nanofluid density obtained from a density meter, then the second law based energy analysis was carried out to evaluate the efficiency of the flat plate collector. They observed that the SWCNTs nanofluid reduced the entropy generation by 4.34% and enhanced the heat transfer coefficient by 15.33% theoretically compared to water as an absorbing fluid, contrarily, nanofluid pumping power penalty of solar collector was found to be 1.20% higher than that of the water.

Nasrin and Alim [163] focused on a numerical study to investigate the forced convection through a flat plate solar collector using the alumina/water nanofluid as the working fluid inside the riser pipe of the solar collector. Their governing differential equations with boundary conditions were solved by finite element method using Galerkin's weighted residual scheme. They proposed semi-empirical relation, which was established from this correlation with experimental data found in the literature for Re from 200 to 1100 as

$$Nu = 2.2275 + \left(0.8849 + 0.2233\,Pr\right)\left(Re\right)^{0.2120} \tag{3.144}$$

and confirmed that there was a good agreement between the experimental result and that obtained by the semi-empirical relation.

Lelea and Laza [164] presented their numerical simulations for nanofluid heat transfer and fluid flow in micro-heat sink with straight microtubes and multiple tangential inlet jets, constant heat sink of 50 W/cm^2 on the bottom surface and a fixed pumping power by using a finite volume method from the Ansys-Fluent commercial solver.

Li et al. [165] numerically studied the efficiency of compact heat exchanger and microchannel heat sinks and flow and heat transfer characteristics of Al_2O_3/water nanofluids in the microchannel with dimple and protrusion surface, which was claimed to be the first time. From their result, flow separation and reattachment in the dimple did not contain two symmetrical focuses or single focus any more when inlet velocity and volume fraction were larger, and the locations of separation and reattachment showed the trend of drawing close, so the size of separation bubble decreased.

Parvin et al. [166] numerically investigated the heat transfer performance and entropy generation of forced convection through a DASC with Cu/water nanofluid by using penalty finite element method with Galerkins weighted residual technique. Their results provided a reference for enhancing the forced convection heat transfer performance while simultaneously reducing the entropy generation.

Hatami and Ganji [167] investigated heat transfer of a fin-shaped microchannel heat sink (MCHS), which was cooled by Cu/water nanofluid and temperature distribution in solid section (fin) and nanofluid section, numerically by porous media approach and least square method. Their conclusions were as, by increasing the nanoparticles volume fraction, Brownian motion of the particles increased and the optimum design of MCHS which the minimum point of the friction factor was obtained when channel aspect ratio equaled to 2.45.

Tongkratoke et al. [135] compared results obtained from three turbulence models—Standard, RNG, and Realizable combined with the wall function equations—to distinguished differences among them. In their work, the experimental data of Lee and Mudawar [168] were used for comparison in the case of the single phase heat transfer in a channel. The comparison of the predicted heat transfer coefficient using the standard, RNG, and Realizable k–ε turbulence models combined with standard wall function was presented in Figure 3.7. It was seen that the Realizable k–ε models can predict the heat transfer coefficient more accurately than the standard and RNG k–ε turbulence models. Moreover, the computational time used by the RNG k–ε turbulence model was about 1–1.5

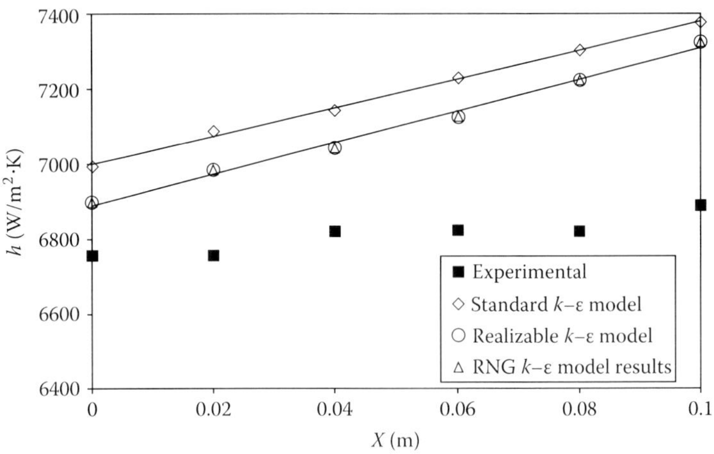

FIGURE 3.7 Comparison between the k–ε turbulence models and experimental data in prediction of the heat transfer coefficient at the nanofluid volume fraction at 0.01. (Redrawn from Tongkratoke, A. et al., Numerical study of turbulence nanofluid flow to distinguish models for in-house programming, in *Proceedings of the Third International Advances in Applies Physics and Material Science Congress*, Antalya, Turkey, 2013.)

times less than that of the Realizable k–ε turbulence model. They indicated average error percentages between the simulated and experimental results of Lee and Mudawar [168] as 5.57, 4.57, and 4.26, for standard, RNG k–ε, and realizable turbulence models the percentages less than 10% which can be acceptable. One may notice that the percentages also implied that the longer flow distance, more errors in the numerical results.

3.5 APPLICATION WITH NANOFLUIDS AS WORKING FLUIDS

There were numerous practical applications dealing with nanofluids. As we mentioned in the previous parts of our review work, many researchers have targeted their works to the practical applications. Wu and Zhao [169] reviewed the state-of-the-art nanofluid studies and critical heat flux (CHF) enhancement nanofluid heat transfer and CHF enhancement as gap to engineering application. They indicated that some nanofluid experimental data of thermal properties were neither sufficient nor reliable for engineering applications. No comprehensive theory explained the energy transfer processes in nanofluids.

Nanofluids appeared to be used in microchannel heat sink as in several literatures. Saidur et al. [170] presented advantages in using nanofluids with heat exchangers by referring conclusions from their literatures [171,172]. They proposed the enhanced heat transfer rate of flue gases [173]. They claimed that, in micro-scale equipment with increased thermal duties and using laminar flow, the use of a nanofluid instead of a conventional fluid seemed advantageous [121]. However, they mentioned that practical disadvantages brought into question the overall merit of using nanofluids in microchannel heat sinks [174]. The efficiency of heat exchanger with nanofluid utilization for waste heat recovery purpose was increased but the cogeneration efficiency of diesel electric generators was reduced [175]. Mohammed et al. review work [176] focused on the important aspects of nanofluids such as nanofluids types, properties, and heat transfer characteristics and limitations toward heat exchanger applications. Hussein et al. [177] expressed their review work in forced convection heat transfer in heat exchangers and other shapes of tubes, which was also carried on experimentally and numerically by other research groups.

Choi [178] introduced nanoparticles as an innovative utility, which were dispersed in conventional heat transfer fluids to produce a new class of high efficiency heat exchange media [170]. Nanofluids appeared promising for the use in electronics devices to cool electronics, transformers, and in chillers, in domestic refrigerator and thermal absorption systems, in machining and grinding processes [170]. Nagarani et al. [179] reviewed how extended surfaces with heat exchanger have been used in the field of heat transfer including an experimental study [180], which was conducted to determine the heat removal ability of multiwalled carbon nanotubes grown in a silicon minichannel with Al_2O_3/H_2O nanofluid as the cooling medium. Murshed et al. [181] performed investigations on boiling, droplet spreading, and convective heat transfer in order to exploit nanofluids as the next generation coolants. Saidur et al. [170] reported that nanofluids gave rise to an increase in the coefficient of performance (COP) by 5.15%, relative to a condition without nanofluids at the standard rating conditions, but the pressure drop penalty of the addition of nanofluids was almost negligible and, only adding 0.2 wt.% Al_2O_3 nanoparticles into water, the total freezing time can be saved by 20.5%. Saidur et al. [170] concluded that nanofluids in refrigerating systems were examined and shown nanofluid enhancements; using CNTs with R-113, using TiO_2 with HFC-134a, using Al_2O_3 with R-22, using TiO_2 with R-11, using CNTs with R-123, using CNTs with HFC-134a, using CuO with R-113 and using TiO_2 with HFC-134a and mineral oil. Bi et al. [182] claimed that the nanofluids could be used safely in the refrigerator. Choi et al. [183] used Al_2O_3 and AIN powder with transformer oil and found nanofluid heat transfer enhancement. There were works of nanofluids used in that thermal management in space applications, weapons, and ship hull heating [170].

Saidur et al. [170] reported that nanofluids were applied to cool engine as coolants; water and ethylene glycol are commonly used for vehicles to increase thermal signal variations in detecting knock occurrence in a gas SI engine, in diesel combustion by shorting ignition delay and increasing energy density, and high burn rates [170]. The increased workload of the pump, which was occupied by nanofluid in an engine cooling system, was acceptable according to the literatures [184]. Since hybrid vehicle industry has been growing, there was a project targeting to reduce the size and weight of this vehicle cooling system by using nanofluids as the working fluids, the nanofluids enabled to higher heat rejection in the vehicle [185].

3.6 NANOFLUID FUTURE APPROACHES

New nanoparticles can play an important role in improving nanofluid efficiency. Nanoparticle fabrications should be simple to reduce nanoparticle initial costs. The lower nanoparticle cost, the lower nanofluid investment. Reactions between nanoparticles and traditional base fluids must be evaluated for their potentials in retrofitting the nanoparticles into the base fluids or in replacing the base fluids with the nanofluids. To be able to utilize nanofluids commercially, some topics should be solved and suggested as soon as possible such as nanofluid stabilities and viscosities, which affect nanofluid pump powers. Nanofluid enhancements including the Nusselt number and thermophysical property relationships must be investigated and collected for industrial applications in turbulent and laminar regimes. Data mining may be occupied for a large amount of experimental data since some nanofluids were often investigated. More experimental works have to be carried under the same conditions as the commercial and industrial applications to discover the real affections in the applied conditions. When the experiment works can cause serious damages or side effects to the applications, the numerical and simulation works take their significant parts into the investigations because the numerical works consume less budgets and times with suitable reliances.

REFERENCES

1. S. Kakaç, A. Pramuanjaroenkij, Review of convective heat transfer enhancement with nanofluids, *International Journal of Heat and Mass Transfer*, 52, 3187–3196 (2009).
2. A. Tongkratoke, A. Pramuanjaroenkij, A. Chaengbamrung, S. Kakaç, Numerical study of nanofluid heat transfer enhancement with mixing thermal conductivity models, *Computational Thermal Sciences*, 6(1), 1–12 (2014).
3. S. Kakaç, A. Pramuanjaroenkij, Single-phase and two-phase treatments of convective heat transfer enhancement with nanofluids—A state-of-the-art review, *International Journal of Thermal Sciences*. THESCI-D-14-00978 (Submitted for Initial Review).
4. Y. Li, J. Zhou, S. Tung, E. Schneider, S. Xi, A review on development of nanofluid preparation and characterization, *Powder Technology*, 196, 89–101 (2009).
5. L.S. Sundar, M.K. Singh, Convective heat transfer and friction factor correlations of nanofluid in a tube and with inserts: A review, *Renewable and Sustainable Energy Reviews*, 20, 23–35 (2013).
6. W. Yu, D.M. France, J.L. Routbort, S.U.S. Choi, Review and comparison of nanofluid thermal conductivity and heat transfer enhancements, *Heat Transfer Engineering*, 29(5), 432–460 (2008).
7. M.M. Kostic, Critical issues in nanofluids research and application potentials, in: *Nanofluids: Research, Development Applications* (ed. Y. Zhang), Nova Science Publishers, Hauppauge, NY, 2013.
8. D. Wen, Y. Ding, Experimental investigation into convective heat transfer of nanofluids at the entrance region under laminar flow conditions, *International Journal of Heat and Mass Transfer*, 47, 5181–5188 (2004).
9. A.D. Sommers, K.L. Yerkes, Experimental investigation into the convective heat transfer and system-level effects of Al_2O_3-propanol nanofluids, *Journal of Nanoparticle Research*, 12, 1003–1014 (2010).
10. M. Chandrasekar, S. Suresh, T. Senthilkumar, Mechanisms proposed through experimental investigations on thermophysical properties and forced convective heat transfer characteristics of various nanofluids—A review, *Renewable and Sustainable Energy Reviews*, 16, 3917–3938 (2012).
11. B.C. Pak, Y.I. Cho, Hydrodynamic and heat transfer study of dispersed fluids with submicron metallic oxide particles, *Experimental Heat Transfer*, 11, 151–170 (1998).
12. Y. Xuan, W. Roetzel, Conceptions for heat transfer correlation of nanofluids, *International Journal of Heat and Mass Transfer*, 43, 3701–3707 (2000).

13. M.M. Elias, I.M. Mahbubul, R. Saidur, M.R. Sohel, I.M. Shahrul, S.S. Khaleduzzaman, S. Sadeghipour, Experimental investigation on the thermo-physical properties of Al_2O_3 nanoparticles suspended in car radiator coolant, *International Communications in Heat and Mass Transfer*, 54, 48–53 (2014).
14. O. Mahian, A. Kianifar, C. Kleinstreuer, M.A. Al-Nimr, I. Pop, A.Z. Sahin, S. Wongwises, A review of entropy generation in nanofluid flow, *International Journal of Heat and Mass Transfer*, 65, 514–532 (2013).
15. A. Ghadimi, R. Saidur, H.S.C. Metselaar, A review of nanofluid stability properties and characterization in stationary conditions, *International Journal of Heat and Mass Transfer*, 54, 4051–4068 (2011).
16. L.S. Sundar, K.V. Sharma, M.T. Naik, M.K. Singh, Empirical and theoretical correlations on viscosity of nanofluids: A review, *Renewable and Sustainable Energy Reviews*, 25, 670–686 (2013).
17. J.H. Lee, K.S. Hwang, S.P. Jang, B.H. Lee, J.H. Kim, S.U.S. Choi, Effective viscosities and thermal conductivities of aqueous nanofluids containing low volume concentrations of Al_2O_3 nanoparticles, *International Journal of Heat and Mass Transfer*, 51, 2651–2656 (2008).
18. M. Chandrasekar, S. Suresh, B.A.C. Bose, Experimental investigations and theoretical determination of thermal conductivity and viscosity of Al_2O_3/water nanofluid, *Experimental Thermal and Fluid Science*, 34, 210–216 (2010).
19. C.T. Nguyen, F. Desgranges, N. Galanis, G. Roy, T. Mare, S. Boucher, Viscosity data for Al_2O_3/water nanofluid-hysteresis: Is heat transfer enhancement using nanofluids reliable, *International Journal of Thermal Sciences*, 47, 103–111 (2008).
20. K.B. Anoop, S. Kabelac, T. Sundararajan, S.K. Das, Rheological and flow characteristics of nanofluids: Influence of electroviscous effects and particle agglomeration, *Journal of Applied Physics*, 106, 034909 (2009).
21. M.J. Pastoriza-Gallego, C. Casanova, R. Páramo, B. Barbés, J.L. Legido, M.M. Piñeiro, A study on stability and thermophysical properties (density and viscosity) of Al_2O_3 in water nanofluid, *Journal of Applied Physics*, 106, 064301 (2009).
22. W. Duangthongsuk, S. Wongwises, Measurement of temperature-dependent thermal conductivity and viscosity of TiO_2-water nanofluids, *Experimental Thermal and Fluid Science*, 33, 706–714 (2009).
23. Q. Li, Y. Xuan, J. Wang, Experimental investigations on transport properties of magnetic fluids, *Experimental Thermal and Fluid Science*, 30, 109–116 (2005).
24. L. Godson, B. Raja, L.D. Mohan, S. Wongwises, Experimental investigation on the thermal conductivity and viscosity of silver-deionized water nanofluid, *Experimental Heat Transfer*, 23, 317–332 (2010).
25. P.K. Namburu, D.P. Kulkarni, D. Misra, D.K. Das, Viscosity of copper oxide nanoparticles dispersed in ethylene glycol and water mixture, *Experimental Thermal and Fluid Science*, 32, 397–402 (2007).
26. E.V. Timofeeva, W. Yu, D.M. France, D. Singh, J.L. Routbort, Base fluid and temperature effects on the heat transfer characteristics of SiC in ethylene glycol/H_2O and H_2O nanofluids, *Journal of Applied Physics*, 109, 014914 (2011).
27. W. Yu, H. Xie, Y. Li, L. Chen, Experimental investigation on thermal conductivity and viscosity of aluminum nitride nanofluid, *Particuology*, 9, 187–191 (2001).
28. H. Chen, Y. Ding, A. Lapkin, Rheological behaviour of nanofluids containing tube/rod-like nanoparticles, *Powder Technology*, 194, 132–141 (2009).
29. W. Yu, H. Xie, L. Chen, Y. Li, Investigation of thermal conductivity and viscosity of ethylene glycol based ZnO nanofluid, *Thermochimica Acta*, 491, 92–96 (2009).
30. W. Yu, H. Xie, Y. Li, L. Chen, Q. Wang, Experimental investigation on the thermal transport properties of ethylene glycol based nanofluids containing low volume concentration diamond nanoparticles, *Colloids and Surfaces A: Physicochemical and Engineering Aspects*, 380, 1–5 (2001).
31. A. Einstein, A new determination of the molecular dimensions, *Annals of Physics*, 324(2), 289–306 (1906).
32. D.A. Drew, S.L. Passman, *Theory of Multicomponent Fluids*, Springer, Berlin, Germany, 1999.
33. H.C. Brinkman, The viscosity of concentrated suspensions and solutions, *Journal of Chemical Physics*, 20, 571–581 (1952).
34. G.K. Batchelor, The effect of Brownian motion on the bulk stress in a suspension of spherical particles, *Journal of Fluid Mechanics*, 83, 97–117 (1977).
35. C. Nguyen, F. Desgranges, G. Roy, N. Galanis, T. Maré, S. Boucher, H.A. Mintsa, Temperature and particle-size dependent viscosity data for water-based nanofluids—Hysteresis phenomenon, *International Journal of Heat and Fluid Flow*, 28(6), 1492–1506 (2007).
36. S.E.B. Maïga, C.T. Nguyen, N. Galanis, G. Roy, Heat transfer behaviours of nanofluids in a uniformly heated tube, *Superlattices and Microstructures*, 35, 543–557 (2004).

37. U. Rea, T. McKrell, L.W. Hu, J. Buongiorno, Laminar convective heat transfer and viscous pressure loss of alumina–water and zirconia–water nanofluids, *International Journal of Heat and Mass Transfer*, 52, 2042–2048 (2009).
38. M.H. Esfe, S. Saedodin, M. Mahmoodi, Experimental studies on the convective heat transfer performance and thermophysical properties of MgO–water nanofluid under turbulent flow, *Experimental Thermal and Fluid Science*, 52, 68–78 (2014).
39. S. Kakaç, Y. Yener, A. Pramuanjaroenkij, *Convective Heat Transfer*, CRC Press, Boca Raton, FL, 2013.
40. S. Özerinç, S. Kakaç, S., A.G. Yazıcıoğlu, Enhanced thermal conductivity of nanofluids: A state-of-the-art review, *Microfluidics and Nanofluidics*, 8(2), 145–170 (2010).
41. G. Paul, M. Chopkar, I. Manna, P.K. Das, Techniques for measuring the thermal conductivity of nanofluids: A review, *Renewable and Sustainable Energy Reviews*, 14, 1913–1924 (2010).
42. J.C. Maxwell, *Treatise on Electricity and Magnetism*, Oxford University Press, London, U.K., 1904.
43. R.L. Hamilton, O.K. Crosser, Thermal conductivity of heterogeneous two-component systems, *Industrial & Engineering Chemistry Fundamentals*, 1, 182–191 (1962).
44. Y. Hwang, H.S. Park, J.K. Lee, W.H. Jung, Thermal conductivity and lubrication characteristics of nanofluids, *Current Applied Physics*, 6(S1), 67–71 (2006).
45. M. Jahanshahi, S.F. Hosseinizadeh, M. Alipanah, A. Dehghani, G.R. Vakilinejad, Numerical simulation of free convection based on experimental measured conductivity in a square cavity using Water/SiO_2 nanofluid, *International Communications in Heat and Mass Transfer*, 37, 687–694 (2010).
46. W. Yu, H. Xie, L. Chen, Y. Li, Investigation on the thermal transport properties of ethylene glycol-based nanofluids containing copper nanoparticles, *Powder Technology*, 197, 218–221 (2010).
47. M. Chopkar, S. Kumar, D.R. Bhandari, P.K. Das, I. Manna, Development and characterization of Al_2Cu and Ag_2Al nanoparticle dispersed water and ethylene glycol based nanofluids, *Materials Science and Engineering B*, 139, 141–148 (2007).
48. P. Sharma, I.H. Baek, T. Cho, S. Park, K.B. Lee, Enhancement of thermal conductivity of ethylene glycol based silver nanofluids, *Powder Technology*, 208, 7–19 (2011).
49. P. Bhattacharya, S.K. Saha, A. Yadav, P.E. Phelan, R.S. Prasher, Brownian dynamics simulation to determine the effect thermal conductivity of nanofluids, *Journal of Applied Physics*, 95, 6492–6494 (2004).
50. J. Koo, C. Kleinstreuer, A new thermal conductivity model for nanofluids, *Journal of Nanoparticle Research*, 6(6), 577–588 (2004).
51. S.K. Das, N. Putra, P. Thiesen, W. Roetzel, Temperature dependence of thermal conductivity enhancement of nanofluids, *Journal of Heat Transfer: Transactions of the ASME*, 125, 567–574 (2003).
52. S. Özerinç, Heat transfer enhancement with nanofluids, Master thesis, Middle East Technical University, Ankara, Turkey, 2010.
53. C. Nan, R. Birringer, D.R. Clarke, H. Gleiter, Effective thermal conductivity of particulate composites with interfacial thermal resistance, *Journal of Applied Physics*, 81(10), 6692–6699 (1997).
54. B. Wang, L. Zhou, X. Peng, A fractal model for predicting the effective thermal conductivity of liquid with suspension of nanoparticles, *International Journal of Heat and Mass Transfer*, 46(14), 2665–2672 (2003).
55. R. Prasher, P.E. Phelan, P. Bhattacharya, Effect of aggregation kinetics on the thermal conductivity of nanoscale colloidal solutions (Nanofluid), *Nano Letters*, 6(7), 1529–1534 (2006).
56. Y. He, Y. Jin, H. Chen, Y. Ding, D. Cang, H. Lu, Heat transfer and flow behaviour of aqueous suspensions of TiO_2 nanoparticles (nanofluids) flowing upward through a vertical pipe, *International Journal of Heat and Mass Transfer*, 50(11–12), 2272–2281 (2007).
57. W. Evans, R. Prasher, J. Fish, P. Meakin, P. Phelan, P. Keblinski, Effect of aggregation and interfacial thermal resistance on thermal conductivity of nanocomposites and colloidal nanofluids, *International Journal of Heat and Mass Transfer*, 51(5–6), 1431–1438 (2008).
58. Y. Xuan, Q. Li, W. Hu, Aggregation structure and thermal conductivity of nanofluids, *AIChE Journal*, 49(4), 1038–1043 (2003).
59. C. Yu, A.G. Richter, A. Datta, M.K. Durbin, P. Dutta, Observation of molecular layering in thin liquid films using x-ray reflectivity, *Physical Review Letters*, 82(11), 2326–2329 (1999).
60. W. Yu, S.U.S. Choi, The role of interfacial layers in the enhanced thermal conductivity of nanofluids: A renovated Maxwell model, *Journal of Nanoparticle Research*, 5, 167–171 (2003).
61. Q. Xue, W. Xu, A model of thermal conductivity of nanofluids with interfacial shells, *Materials Chemistry and Physics*, 90(2–3), 298–301 (2005).

62. D.A.G. Bruggeman, The calculation of various physical constants of heterogeneous substances. I. The dielectric constants and conductivities of mixtures composed of isotropic substances, *Annals of Physics*, 416(7), 636–664 (1935).
63. E. Yamada, T. Ota, Effective thermal conductivity of dispersed materials, *Wärme- und Stoffübertragung*, 13, 27–37 (1980).
64. C.H. Chon, K.D. Kihm, S.P. Lee, S.U.S. Choi, Empirical correlation finding the role of temperature and particle size for nanofluid (Al_2O_3) thermal conductivity enhancement, *Applied Physics Letters*, 87(15), 153107 (2005).
65. M. Corcione, Empirical correlating equations for predicting the effective thermal conductivity and dynamic viscosity of nanofluids, *Energy Conversion and Management*, 52, 789–793 (2011).
66. I.O. Sert, N. Sezer-Uzol, S. Kakaç, Numerical analysis of nanofluids convective heat transfer with mixture model approaches, in: *Proceeding of the International Symposium on Convective Heat and Mass Transfer*, Kusadasi, Turkey, 2014.
67. M.A. Hachey, C.T. Nguyen, N. Galanis, C.V. Popa, Experimental investigation of Al_2O_3 nanofluids thermal properties and rheology—Effects of transient and steady-state heat exposure, *International Journal of Thermal Sciences*, 76, 155–167 (2014).
68. R. Saleh, N. Putra, R.E. Wibowo, W.N. Septiadi, S.P. Prakoso, Titanium dioxide nanofluids for heat transfer applications, *Experimental Thermal and Fluid Science*, 52, 19–29 (2014).
69. S. Manikandan, A. Shylaja, K.S. Rajan, Thermo-physical properties of engineered dispersions of nanosand in propylene glycol, *Colloids and Surfaces A: Physicochemical and Engineering Aspects*, 449, 8–18 (2014).
70. D. Singh, E.V. Timofeeva, M.R. Moravek, S. Cingarapu, W. Yu, T. Fischer, S. Mathur, Use of metallic nanoparticles to improve the thermophysical properties of organic heat transfer fluids used in concentrated solar power, *Solar Energy*, 105, 468–478 (2014).
71. S. Halelfadl, T. Maré, P. Estellé, Efficiency of carbon nanotubes water based nanofluids as coolants, *Experimental Thermal and Fluid Science*, 53, 104–110 (2014).
72. M. Ghanbarpour, E.B. Haghigi, R. Khodabandeh, Thermal properties and rheological behavior of water based Al_2O_3 nanofluid as a heat transfer fluid, *Experimental Thermal and Fluid Science*, 53, 227–235 (2014).
73. A. Ghozatloo, A.M. Rashidi, M. Shariaty-Niasar, Effects of surface modification on the dispersion and thermal conductivity of CNT/water nanofluids, *International Communications in Heat and Mass Transfer*, 54, 1–7 (2014).
74. J. Lee, K. Han, J. Koo, A novel method to evaluate dispersion stability of nanofluids, *International Journal of Heat and Mass Transfer*, 70, 421–429 (2014).
75. M.J. Nine, H. Chung, M.R. Tanshen, N.A.B.A. Osman, H. Jeong, Is metal nanofluid reliable as heat carrier?, *Journal of Hazardous Materials*, 273, 183–191 (2014).
76. M. Karami, M.A. Akhavan Bahabadi, S. Delfani, A. Ghozatloo, A new application of carbon nanotubes nanofluid as working fluid of low-temperature direct absorption solar collector, *Solar Energy Materials & Solar Cells*, 121, 114–118 (2014).
77. D. Madhesh, R. Parameshwaran, S. Kalaiselvam, Experimental investigation on convective heat transfer and rheological characteristics of Cu–TiO_2 hybrid nanofluids, *Experimental Thermal and Fluid Science*, 52, 104–115 (2014).
78. S.E. Ahmed, A.K. Hussein, H.A. Mohammed, S. Sivasankaran, Boundary layer flow and heat transfer due to permeable stretching tube in the presence of heat source/sink utilizing nanofluids, *Applied Mathematics and Computation*, 238, 149–162 (2014).
79. H. Zerradi, S. Ouaskit, A. Dezairi, H. Loulijat, S. Mizani, New Nusselt number correlations to predict the thermal conductivity of nanofluids, *Advanced Powder Technology*, 25(3), 1124–1131 (2014), http://dx.doi.org/10.1016/j.apt.2014.02.020.
80. B. Lamas, B. Abreu, A. Fonseca, N. Martins, M. Oliveira, Critical analysis of the thermal conductivity models for CNT based nanofluids, *International Journal of Thermal Sciences*, 78, 65–76 (2014).
81. C.Y. Tso, S.C. Fu, Christopher, Y.H. Chao, A semi-analytical model for the thermal conductivity of nanofluids and determination of the nanolayer thickness, *International Journal of Heat and Mass Transfer*, 70, 202–214 (2014).
82. L. Godson, B. Raja, D.M. Lal, S. Wongwises, Enhancement of heat transfer using nanofluids—An overview, *Renewable and Sustainable Energy Reviews*, 14, 629–641 (2010).
83. S.Z. Heris, M.N. Esfahany, S.G. Etemad, Experimental investigation of convective heat transfer of Al_2O_3/water nanofluid in circular tube, *International Journal of Heat and Fluid Flow*, 28(2), 203–210 (2007).

84. K.B. Anoop, T. Sundararajan, S.K. Das, Effect of particle size on the convective heat transfer in nanofluid in the developing region, *International Journal of Heat Mass Transfer*, 52, 2189–2195 (2009).
85. S.M. Fotukian, M.N. Esfahany, Experimental study of turbulent convective heat transfer and pressure drop of dilute CuO/water nanofluid inside a circular tube, *International Communications in Heat and Mass Transfer*, 37, 214–219 (2010).
86. S.M. Fotukian, M.N. Esfahany, Experimental investigation of turbulent convective heat transfer of dilute-Al_2O_3/water nanofluid inside a circular tube, *International Journal of Heat and Fluid Flow*, 31, 606–612 (2010).
87. W. Williams, J. Buongiorno, L.W. Hu, Experimental investigation of turbulent convective heat transfer and pressure loss of alumina/water and zirconia/water nanoparticle colloids (nanofluids) in horizontal tubes, *Journal of Heat Transfer*, 130, 042412 (2008).
88. L.G. Asirvatham, B. Raja, D.M. Lal, S. Wongwises, Convective heat transfer of nanofluids with correlations, *Particuology*, 9(6), 626–631 (2011).
89. Q. Li, Y. Xuan, Convective heat transfer and flow characteristics of Cu–water nanofluid, *Science in China. Series E. Technological Sciences*, 45(4), 408–416, (2002).
90. Y. Xuan, Q. Li, Investigation on convective heat transfer and flow features of nanofluids, *Journal of Heat Transfer*, 125, 151–155, (2003).
91. S. Suresh, K.P. Venkitaraj, P. Selvakumar, M. Chandrasekar, Effect of Al_2O_3-Cu/water hybrid nanofluid in heat transfer, *Experimental Thermal and Fluid Science*, 38, 54–60 (2012).
92. S. Suresh, M. Chandrasekar, S.C. Sekhar, Experimental studies on heat transfer and friction factor characteristics of CuO/water nanofluid under turbulent flow in a helically dimpled tube, *Experimental Thermal and Fluid Science*, 35(3), 542–549 (2011).
93. A.R. Sajadi, M. Kazemi, Investigation of turbulent convective heat transfer and pressure drop of TiO_2/water nanofluid in circular tube, *International Communications in Heat and Mass Transfer*, 38(10), 1474–1478 (2011).
94. L.S. Sundar, M.T. Naik, K.V. Sharma, M.K. Singh, T.C.S. Reddy, Experimental investigation of forced convection heat transfer and friction factor in a tube with Fe_3O_4 magnetic nanofluid, *Experimental Thermal and Fluid Science*, 37, 65–71 (2012).
95. R.S. Vajjha, D.K. Das, D.P. Kulkarni, Development of New Correlations for Convective Heat Transfer and Friction Factor in Turbulent Regime for Nanofluids, *International Journal of Heat and Mass Transfer*, 53, 4607–4618 (2010).
96. E.N. Sieder, G.E. Tate, Heat transfer and pressure drop of liquids in tubes, *Industrial & Engineering Chemistry*, 28(12), 1429–1435 (1936).
97. R.K. Shah, A.L. London, *Laminar Flow Forced Convection in Ducts*, Supplement 1 to Advances in Heat Transfer, Academic Press, New York, 1978.
98. O. Abouali, G. Ahmadi, Computer simulations of natural convection of single phase nanofluids in simple enclosures: A critical review, *Applied Thermal Engineering*, 36, 1–13 (2012).
99. A. Bejan, *Convection Heat Transfer*, 3rd edn., John Wiley & Sons, Inc., Hoboken, NJ, 2004, p. 270.
100. F.P. Incropera, D.P. Dewitt, T.L. Bergman, A.S. Lavine, *Fundamentals of Heat and Mass Transfer*, 6th edn., John Wiley & Sons, Inc., Hoboken, NJ, 2007, p. 589.
101. R.A. Kuyper, T.H. Van Dermeer, C.J. Hoogendorn, R.A.W.M. Henkes, Numerical study of laminar and turbulent natural convection in an inclined square cavity, *International Journal of Heat and Mass Transfer*, 36, 2899–2911 (1993).
102. T.H. Kuehn, R.J. Goldstein, Correlating equations for natural convection heat transfer between horizontal circular cylinders, *International Journal of Heat and Mass Transfer*, 19, 1127–1134 (1976).
103. M. Itoh, T. Fujita, N. Nishiwaki, M. Hirata, A new method of correlating heat transfer coefficients for natural convection in horizontal cylindrical annuli, *International Journal of Heat and Mass Transfer*, 13, 1364–1368 (1970).
104. R. Kumar, M.A. Kalam, Laminar thermal convection between vertical coaxial isothermal cylinders, *International Journal of Heat and Mass Transfer*, 34, 513–524 (1991).
105. T.Y. Chu, R.J. Goldstein, Turbulent convection in a horizontal layer of water, *J. Fluid Mech.*, 60, 141–159 (1973).
106. S. Özerinç, A.G. Yazıcıoğlu, S. Kakaç, Numerical analysis of laminar forced convection with temperature-dependent thermal conductivity of nanofluids and thermal dispersion, *International Journal of Thermal Sciences*, 62, 138–148 (2012).
107. L.S. Sundar, M.K. Singh, I. Bidkin, A.C.M. Sousa, Experimental investigations in heat transfer and friction factor of magnetic Ni nanofluid flowing in a tube, *International Journal of Heat and Mass Transfer*, 70, 224–234 (2014).

108. S. Halelfadl, P. Estellé, T. Maré, Heat transfer properties of aqueous carbon nanotubes nanofluids in coaxial heat exchanger under laminar regime, *Experimental Thermal and Fluid Science*, 55, 174–180 (2014).
109. M.H. Kayhani, H. Soltanzadeh, M.M. Heyhat, M. Nazari, F. Kowsary, Experimental study of convective heat transfer and pressure drop of TiO_2/water nanofluid, *International Communications in Heat and Mass Transfer*, 39, 456–462 (2012).
110. E.B. Haghighi, M. Saleemi, N. Nikkam, R. Khodabandeh, M.S. Toprak, M. Muhammed, B. Palm, Accurate basis of comparison for convective heat transfer in nanofluids, *International Communications in Heat and Mass Transfer*, 52, 1–7 (2014).
111. M. Hajipour, A.M. Dehkordi, Mixed-convection flow of Al_2O_3–H_2O nanofluid in a channel partially filled with porous metal foam: Experimental and numerical study, *Experimental Thermal and Fluid Science*, 53, 49–56 (2014).
112. M.T. Naik, Syed Sha Fahad, L. Syam Sundar, Manoj K. Singh, Comparative study on thermal performance of twisted tape and wire coil inserts in turbulent flow using CuO/water nanofluid, *Experimental Thermal and Fluid Science*, 57, 65–76 (2014).
113. Y. Abbassi, M. Talebi, A.S. Shirani, J. Khorsandi, Experimental investigation of TiO_2/Water nanofluid effects on heat transfer characteristics of a vertical annulus with non-uniform heat flux in non-radiation environment, *Annals of Nuclear Energy*, 69, 7–13 (2014).
114. M.H. Esfe, S. Saedodin, O. Mahian, S. Wongwises, Heat transfer characteristics and pressure drop of COOH-functionalized DWCNTs/water nanofluid in turbulent flow at low concentrations, *International Journal of Heat and Mass Transfer*, 73, 186–194 (2014).
115. E. Esmaeilzadeh, H. Almohammadi, A. Nokhosteen, A. Motezaker, A.N. Omrani, Study on heat transfer and friction factor characteristics of γ-Al_2O_3/water through circular tube with twisted tape inserts with different thicknesses, *International Journal of Thermal Sciences*, 82, 72–83 (2014).
116. M. Ghobadi, Y.S. Muzychka, Fully developed heat transfer in mini scale coiled tubing for constant wall temperature, *International Journal of Heat and Mass Transfer*, 72, 87–97 (2014).
117. C.J. Ho, W.C. Chen, W.M. Yan, Correlations of heat transfer effectiveness in a minichannel heat sink with water-based suspensions of Al_2O_3 nanoparticles and/or MEPCM particles, *International Journal of Heat and Mass Transfer*, 69, 293–299 (2014).
118. A.M. Hussein, R.A. Bakar, K. Kadirgama, K.V. Sharma, Heat transfer enhancement using nanofluids in an automotive cooling system, *International Communications in Heat and Mass Transfer*, 53, 195–202 (2014).
119. A.T. Utomo, E.B. Haghighi, A.I.T. Zavareh, M. Ghanbarpourgeravi, H. Poth, R. Khodabandeh, B. Palm, A.W. Pacek, The effect of nanoparticles on laminar heat transfer in a horizontal tube, *International Journal of Heat and Mass Transfer*, 69, 77–91 (2014).
120. H. Masuda, A. Ebata, K. Teramae, N. Hishinuma, Alteration of thermal conductivity and viscosity of liquid by dispersing ultra-fine particles (dispersion of c-Al_2O_3, SiO_2 and TiO_2 ultra-fine particles), *Netsu Bussei*, 7(4), 227–233 (1993).
121. M.N. Pantzali, A.A. Mouza, S.V. Paras, Investigating the efficacy of nanofluids as coolants in plate heat exchangers (PHE), *Chemical Engineering Science*, 64, 3290–3300 (2009).
122. E.V. Timofeeva, D.S. Smith, W. Yu, D.M. France, D. Singh, J.L. Routbort, Particle size and interfacial effects on thermo-physical and heat transfer characteristics of water-based a-SiC nanofluids, *Nanotechnology*, 21(21), 215703–215713 (2010).
123. P. Gurav, S.S. Naik, K. Ansari, S. Srinath, K.A. Kishore, Y.P. Setty, S. Sonawane, Stable colloidal copper nanoparticles for a nanofluid: Production and application, *Colloids and Surfaces A: Physicochemical and Engineering Aspects*, 441, 589–597 (2014).
124. M.M. Sarafraz, F. Hormozi, Scale formation and subcooled flow boiling heat transfer of CuO–water nanofluid inside the vertical annulus, *Experimental Thermal and Fluid Science*, 52, 205–214 (2014).
125. B. Sun, D. Yang, Flow boiling heat transfer characteristics of nano-refrigerants in a horizontal tube, *International Journal of Refrigeration*, 38, 206–214 (2014).
126. D.R. Wagholе, R.M. Warkhedkar, V.S. Kulkarni, R.K. Shrivastva, Experimental investigations on heat transfer and friction factor of Silver nanofluid in absorber/receiver of parabolic trough collector with twisted tape inserts, *Energy Procedia*, 45, 558–567 (2014).
127. N. Karami, M. Rahimi, Heat transfer enhancement in a hybrid microchannel-photovoltaic cell using Boehmite nanofluid, *International Communications in Heat and Mass Transfer*, 55, 45–52 (2014), http://dx.doi.org/10.1016/j.icheatmasstransfer.2014.04.009.
128. C.J. Ho, W.C. Chen, W.M. Yan, Experiment on thermal performance of water-based suspensions of Al_2O_3 nanoparticles and MEPCM particles in a minichannel heat sink, *International Journal of Heat and Mass Transfer*, 69, 276–284 (2014).

129. M.R. Sohel, S.S. Khaleduzzaman, R. Saidur, A. Hepbasli, M.F.M. Sabri, I.M. Mahbubul, An experimental investigation of heat transfer enhancement of a minichannel heat sink using Al_2O_3–H_2O nanofluid, *International Journal of Heat and Mass Transfer*, 74, 164–172 (2014).
130. Y. Vermahmoudi, S.M. Peyghambarzadeh, S.H. Hashemabadi, M. Naraki, Experimental investigation on heat transfer performance of Fe_2O_3/water nanofluid in an air-finned heat exchanger, *European Journal of Mechanics B/Fluids*, 44, 32–41 (2014).
131. M.C.S. Reddy, V.V. Rao, Experimental investigation of heat transfer coefficient and friction factor of ethylene glycol water based TiO_2 nanofluid in double pipe heat exchanger with and without helical coil inserts, *International Communications in Heat and Mass Transfer*, 50, 68–76 (2014).
132. M.A. Khairul, M.A. Alim, I.M. Mahbubul, R. Saidur, A. Hepbasli, A. Hossain, Heat transfer performance and energy analyses of a corrugated plate heat exchanger using metal oxide nanofluids, *International Communications in Heat and Mass Transfer*, 50, 8–14 (2014).
133. M.H. Buschmann, U. Franzke, Improvement of thermosyphon performance by employing nanofluid, *International Journal of Refrigeration*, 40, 416–428 (2014).
134. S.Z. Heris, F. Mohammadpur, A. Shakouri, Effect of electric field on thermal performance of thermosyphon heat pipes using nanofluids, *Materials Research Bulletin*, 53, 21–27 (2014).
135. A. Tongkratoke, A. Pramuanjaroenkij, A. Chaengbamrung, S. Kakaç, Numerical study of turbulence nanofluid flow to distinguish models for in-house programming, in: *Proceedings of the Third International Advances in Applies Physics and Material Science Congress*, Antalya, Turkey, 2013.
136. J. Boussinesq, Theory de L'eco Reynolds Stresses Ulment Tourbillant, Memoires Presentes Par Divers Savants Sciences Mathematique at Physiques, 46–50 (1877).
137. B.E. Launder, D.B. Spalding, Application of the energy-dissipation model of turbulence to the calculation of flow near a spinning disc, *Letters in Heat and Mass Transfer*, 1(2), 131–138 (1972).
138. D. Choudhury, Introduction to the renormalization group method and turbulence modeling, Fluent Inc. Technical Memorandum TM-107 (1993).
139. A. Behzadmehr, M. Saffar-Avval, N. Galanis, Prediction of turbulent forced convection of a nanofluid in a tube with uniform heat flux using a two phase approach, *International Journal of Heat and Fluid Flow*, 28, 211–219 (2007).
140. V. Bianco, F. Chiacchio, O. Manca, S. Nardini, Numerical investigation of nanofluids forced convection in circular tubes, *Applied Thermal Engineering*, 29(17–18), 3632–3642 (2009).
141. S.Z. Heris, M.N. Esfahany, G. Etemad, Numerical investigation of nanofluid laminar convection heat transfer through a circular tube, *Numerical Heat Transfer, Part A*, 52(11), 1043–1058 (2007).
142. A.M. Hussein, R.A. Bakar, K. Kadirgama, Study of forced convection nanofluid heat transfer in the automotive cooling system, *Case Studies in Thermal Engineering*, 2, 50–61 (2014).
143. A. Malvandi, D.D. Ganji, Effects of nanoparticle migration on force convection of alumina/water nanofluid in a cooled parallel-plate channel, *Advanced Powder Technology*, 25(4), 1369–1375 (2014), http://dx.doi.org/10.1016/j.apt.2014.03.017.
144. D.A. Nield, A.V. Kuznetsov, Forced convection in a parallel-plate channel occupied by a nanofluid or a porous medium saturated by a nanofluid, *International Journal of Heat and Mass Transfer*, 70, 430–433 (2014).
145. F. Selimefendigil, H.F. Öztop, Effect of a rotating cylinder in forced convection of ferrofluid over a backward facing step, *International Journal of Heat and Mass Transfer*, 71, 142–148 (2014).
146. M.K.H. Abdolbaqi, C.S.N. Azwadi, R. Mamat, Heat transfer augmentation in the straight channel by using nanofluids, *Case Studies in Thermal Engineering*, 3, 59–67 (2014), http://dx.doi.org/10.1016/j.csite.2014.04.001.
147. T. Sokhansefat, A.B. Kasaeian, F. Kowsary, Heat transfer enhancement in parabolic trough collector tube using Al_2O_3/synthetic oil nanofluid, *Renewable and Sustainable Energy Reviews*, 33, 636–644 (2014).
148. S.M. Vanaki, H.A. Mohammed, A. Abdollahi, M.A. Wahid, Effect of nanoparticle shapes on the heat transfer enhancement in a wavy channel with different phase shifts, *Journal of Molecular Liquids*, 196, 32–42 (2014).
149. F. Selimefendigil, H.F. Öztop, Pulsating nanofluids jet impingement cooling of a heated horizontal surface, *International Journal of Heat and Mass Transfer*, 69, 54–65 (2014).
150. A.S. Kherbeet, H.A. Mohammed, K.M. Munisamy, B.H. Salman, The effect of step height of microscale backward-facing step on mixed convection nanofluid flow and heat transfer characteristics, *International Journal of Heat and Mass Transfer*, 68, 554–566 (2014).
151. A. Malvandi, D.D. Ganji, Mixed convective heat transfer of water/alumina nanofluid inside a vertical microchannel, *Powder Technology*, 263, 37–44 (2014), doi:10.1016/j.powtec.2014.04.084.

152. H. Xu, I. Pop, Mixed convection flow of a nanofluid over a stretching surface with uniform free stream in the presence of both nanoparticles and gyrotactic microorganisms, *International Journal of Heat and Mass Transfer*, 75, 610–623 (2014).
153. H. Xu, I. Pop, Fully developed mixed convection flow in a horizontal channel filled by a nanofluid containing both nanoparticles and gyrotactic microorganisms, *European Journal of Mechanics B/Fluids*, 46, 37–45 (2014).
154. T.W. Ting, Y.M. Hung, N. Guo, Field-synergy analysis of viscous dissipative nanofluid flow in Microchannels, *International Journal of Heat and Mass Transfer*, 73, 483–491 (2014).
155. S.S. Azimi, M. Kalbasi, Numerical study of dynamic thermal conductivity of nanofluid in the forced convective heat transfer, *Applied Mathematical Modelling*, 38, 1373–1384 (2014).
156. M.M. Elias, I.M. Shahrul, I.M. Mahbubul, R. Saidur, N.A. Rahim, Effect of different nanoparticle shapes on shell and tube heat exchanger using different baffle angles and operated with nanofluid, *International Journal of Heat and Mass Transfer*, 70, 289–297 (2014).
157. A.A. Minea, Uncertainties in modeling thermal conductivity of laminar forced convection heat transfer with water alumina nanofluids, *International Journal of Heat and Mass Transfer*, 68, 78–84 (2014).
158. W.I.A. Aly, Numerical study on turbulent heat transfer and pressure drop of nanofluid in coiled tube-in-tube heat exchangers, *Energy Conversion and Management*, 79, 304–316 (2014).
159. M. Goharkhah, M. Ashjaee, Effect of an alternating non uniform magnetic field on ferrofluid flow and heat transfer in a channel, *Journal of Magnetism and Magnetic Materials*, 362, 80–89 (2014).
160. A.A. Servati V., K. Javaherdeh, H.R. Ashorynejad, Magnetic field effects on force convection flow of a nanofluid in a channel partially filled with porous media using Lattice Boltzmann Method, *Advanced Powder Technology*, 25, 666–675 (2014).
161. I.O. Sert, N.S. Uzol, S. Kakaç, Numerical analysis of transient laminar forced convection of nanofluids in circular ducts, *Heat and Mass Transfer: Wärme- und Stoffübertragung*, 49(10), 1405–1417 (2013), doi: 10.1007/s00231-013-1184-1.
162. Z. Said, R. Saidur, N.A. Rahim, M.A. Alim, Analyses of exergy efficiency and pumping power for a conventional flat plate solar collector using SWCNTs based nanofluid, *Energy and Buildings*, 78, 1–9 (2014).
163. R. Nasrin, M.A. Alim, Semi-empirical relation for forced convective analysis through a solar collector, *Solar Energy*, 105, 455–467 (2014).
164. D. Lelea, I. Laza, The water based Al_2O_3 nanofluid flow and heat transfer in tangential microtube heat sink with multiple inlets, *International Journal of Heat and Mass Transfer*, 69, 264–275 (2014).
165. P. Li, D. Zhang, Y. Xie, Heat transfer and flow analysis of Al_2O_3–water nanofluids in microchannel with dimple and protrusion, *International Journal of Heat and Mass Transfer*, 73, 456–467 (2014).
166. S. Parvin, R. Nasrin, M.A. Alim, Heat transfer and entropy generation through nanofluid filled direct absorption solar collector, *International Journal of Heat and Mass Transfer*, 71, 386–395 (2014).
167. M. Hatami, D.D. Ganji, Thermal and flow analysis of microchannel heat sink (MCHS) cooled by Cu–water nanofluid using porous media approach and least square method, *Energy Conversion and Management*, 78, 347–358 (2014).
168. J. Lee, I. Mudawar, Assessment of the effectiveness of nanofluids for single-phase and two-phase heat transfer in micro-channels, *International Journal of Heat and Mass Transfer*, 50, 452–463 (2006).
169. J.M. Wu, Jiyun Zhao, A review of nanofluid heat transfer and critical heat flux enhancement Research gap to engineering application, *Progress in Nuclear Energy*, 66, 13–24 (2013).
170. R. Saidur, K.Y. Leong, H.A. Mohammad, A review on applications and challenges of nanofluids, *Renewable and Sustainable Energy Reviews*, 15, 1646–1668 (2011).
171. Y. Yang, Carbon nanofluids for lubricant application, PhD thesis, University of Kentucky, Lexington, KY, 2006.
172. D. Elcock, Potential impacts of nanotechnology on energy transmission applications and needs, Environmental Science Division, Argonne National Laboratory, Lemont, IL, November 2007.
173. R. Saidur, J.U. Ahamed, H.H. Masjuki, Energy, exergy and economic analysis of industrial boilers, *Energy Policy*, 38(5), 2188–2197 (2010).
174. X.Q. Wang, A.S. Mujumdar, Review on nanofluids, Part II: Experiments and applications, *Brazilian Journal of Chemical Engineering*, 25(October–December (04)), 631–648 (2008).
175. D.P. Kulkarni, R.S. Vajjha, D.K. Das, D. Oliva, Application of aluminum oxide nanofluids in diesel electric generator as jacket water coolant, *Applied Thermal Engineering*, 28, 1774–1781 (2008).
176. H.A. Mohammed, G. Bhaskaran, N.H. Shuaib, R. Saidur, Heat transfer and fluid flow characteristics in microchannels heat exchanger using nanofluids: A review, *Renewable and Sustainable Energy Reviews*, 15, 1502–1512 (2011).

Despite numerous papers have been published on the thermal conduction of nanofluids since 1993 [14], both experimental findings and fundamental mechanisms are still lacking consistency. For instance, it is debated whether the thermal conductivity enhancement is anomalously beyond the prediction from the effective medium theory (EMT). Earlier experimental results showed that the thermal conductivity increases significantly compared to the base fluid. Masuda et al. [14] found that the thermal conductivity of water can be increased by 30% after dispersing 13 nm Al_2O_3 with volume fraction of 4.3% (4.3 vol.%). Eastman et al. [15] obtained similar results in Al_2O_3/water and CuO/water nanofluids; the enhancements of thermal conductivity were 30% and 58% (5 vol.%), respectively. Later, the thermal conductivity was reported to be increased by up to 40% for EG-based nanofluid containing only approximately 0.3 vol.% Cu nanoparticles [16]. Much larger improvement was observed in nanofluids containing multiwalled carbon nanotubes (MWCNTs); up to 160% enhancement in thermal conductivity was reported at only 1 vol.% [3]. On the other hand, some experiments found no significant enhancement in thermal conductivity compared with the prediction from the EMT theory. Zhang et al. [17] measured the thermal conductivity of different nanofluids, including Al/toluene, Al_2O_3/water, TiO_2/water, CuO/water, and carbon nanotube (CNT)/water; no anomalous enhancements were obtained, and the results can be well predicted by the Hamilton–Crosser model for spherical nanoparticles and by Yamada and Ota formula for CNTs. Putnam et al. [18] also observed no enhancement in the thermal conductivity for nanofluids containing C60–C70 fullerenes and Au nanoparticles; similar conclusions were drawn for Au/water [19], Bi_2Te_3/oil, and Bi_2Te_3/FC72 [20] nanofluids. Recently, a benchmark study on the thermal conductivity of nanofluids was carried out by over 30 organizations worldwide. The tested nanofluids were comprised of Al_2O_3 (nanorods)/water, Al_2O_3 (nanorods)/polyalphaolefins lubricant (PAO), Au/water, SiO_2/water, and $Mn_{1/2}$–$Zn_{1/2}$–Fe_2O_4/water. It was found that the effective thermal conductivities were in good agreement with the EMT theory combined with the effects of particle aspect ratio and thermal resistance at the particle/fluid interface [21]. The thermal conductivities of aqueous nanofluids containing surfactant-stabilized MWCNTs, and graphite nanosheets can be also well predicted by the EMT theory when the interfacial thermal resistance is taken into consideration [22,23]. Other than the inconsistency in the anomalous enhancement in thermal conductivity, the controversies in experimental data concerning the effects of nanoparticle size and temperature on the thermal conductivity of nanofluids are also hotly debated. In the case of particle size, some reports showed that thermal conductivity increases with decreasing particle size [24–30], while an opposite tendency was presented by others [22,31–35]. Temperature-dependent behaviors are more complicated. Thermal conductivity may increase linearly [28,29,36–39] or non-linearly [25,40–42] with increasing temperature, follow the same behavior of the base fluid [14,17,23,33,43–46], or even decrease as temperature increases [20,47].

In view of the dispersivity or even inconsistency in experimental results, many mechanisms are speculated, such as Brownian motion-induced microconvection [48–55], liquid layering [3,56–59], nanoparticle aggregation [44,60–65], radiation heat transfer [66,67], percolation theory for nanoparticles [68–71], and thermal wave theory [72]. Some models are proposed to explain the particular findings, but none of these mechanisms are able to predict the exact behavior of the thermal conductivity of nanofluids. The controversy regarding these mechanisms is far from over, and the experimental proof of these mechanisms is difficult, more systematic, and innovative experiments are demanded in the future study. In the case of Brownian motion, Wang et al. [73] claimed that the increase in thermal conductivity is small due to the small Peclet number, which was also confirmed by some works [60,74]. Later, based on the experimental findings that thermal conductivity increases abnormally as temperature increases and particle size decreases [24,36]. Jang and Choi [48] suggested a model considering the Brownian motion of nanoparticles, which was improved by other researchers [50–52], and was supported by some experiments [53–55]. Though the nanoparticle diffusion was visualized by Krishnamurthy et al. [54], the mass diffusion and thermal conductivity enhancement did not match with each other. Another ingenious experiment was carried out by Gao et al. [44], where Al_2O_3 nanoparticles were dispersed into hexadecane and hog fat, and the thermal conductivity in both liquid and solid state were measured. The anomalous increase in thermal conductivity was also observed in solid state when the Brownian motion was frozen out, indicating that Brownian motion is not the main cause of the thermal conductivity enhancement. The benchmark study [21] also ruled out the influence of Brownian motion on thermal conductivity.

Despite the large controversies as mentioned before, most reports addressed that the thermal resistance at particle/fluid interface plays an important role in heat transfer [75]. The interfacial thermal resistance (Kapitza resistance) R_{bd} relates the temperature drop ΔT at the interface to the heat flux across the interface q, which is defined by $R_{bd} = q/\Delta T$. Kapitza resistance depends on the dissimilarity of contact materials, and its value ranges from 10^{-10} to 10^{-7} m^2-K W^{-1} at room temperature [76]. R_{bd} at metal–metal interface is relatively small, of about 2.5×10^{-10} m^2-K W^{-1} [77] and 0.7×10^{-10} m^2-K W^{-1} [78] at Al/Cu and Pd/Ir interfaces respectively. A large value of about 10^{-7} m^2-K W^{-1} was obtained at the interface comprised of bismuth and hydrogen-terminated diamond [76]. Other data for solid–solid interfaces fall between these two limits. Compared with the solid–solid interface, the studies on the Kapitza resistance in nanofluids are scarcely reported [79–81]. The magnitude of R_{bd} ranges from 0.55×10^{-8} m^2-K W^{-1} (Al/water) to 2×10^{-7} m^2-K W^{-1} (AuPd/toluene). Surprisingly large R_{bd}, over 10^{-7}m^2-K W^{-1}, of CNT/liquid interface and fullerene/liquid interface was obtained, which was in good agreement with the molecular dynamic simulation [80]. For a typical Au/water nanofluids, the Kapitza length, defined by $a_k = R_{bd}\lambda_b$, is about 4 nm, which is comparable to the diameter of Au nanoparticle ($\lambda_b \sim 0.6$ W m^{-1} K^{-1} and $R_{bd} \sim 7 \times 10^{-8}$ m^2-K W^{-1}).

In this chapter, we summarize the previous experimental results of most widely used nanofluids containing both spherical and non-spherical nanoparticles, including Al_2O_3, CuO, Cu, Au, CNT, and graphite sheets, and analyze the experimental data by the EMT theory incorporated with the Kapitza resistance. The thermal conductivity of nanofluids depends on various parameters, such as the component thermal properties, temperature, volume fraction, particle size distribution, particle geometry, pH value, hydrophobicity, etc. Due to the great difference existing in samples and experiments, it is impossible to make quantitative predictions of all the thermal conductivities of nanofluids. To avoid this problem, with the aid of a large amount of experimental data, we employ an unbiased statistical analysis to reveal the concentration-dependent behavior of the thermal transport in nanofluids. The geometry parameters from statistical information of nanoparticles are used, and the Kapitza resistance is the only adjusting parameter to get the upper and lower bounds of thermal conductivity.

4.2 THEORETICAL MODELS

4.2.1 MAXWELL'S EQUATION

Based on the assumption that a nanofluid is a mixture consisting of a continuous base fluid with a discontinuous solid component, the properties of these heterogeneous mixtures are described by the EMT theory. Using a mean field approach, Maxwell and Garnett derived the expression for effective dielectric properties of a two-component mixture, then it has been extended to determine the thermal conductivity of a binary mixture in terms of its corresponding component properties and relative volume [82]. Neglecting the interactions among the spherical fillers, Maxwell [82] derived a formula to determine the effective thermal conductivity

$$\frac{\lambda_{eff}}{\lambda_b} = 1 + 3f\frac{\lambda_p - \lambda_b}{\lambda_b + 2\lambda_p - f(\lambda_p - \lambda_b)} \tag{4.1}$$

where

f is the volume fraction

λ_b and λ_p are the thermal conductivities of base fluid and particle, respectively

It is noted that the Maxwell–Garnett formula only considers the effect of volume fraction and can be solely applicable in the mixtures at low concentration. Based on this model, many studies have extended to consider the effects of particle distribution [83,84], particle geometry [85,86], and Kapitza resistance [87,88].

4.2.2 Effect of Kapitza Resistance

The Kapitza resistance is ignored in the conventional models for macroscopic objects, but may play an important role in the thermal transport at nanoscale [89]. As a comparison with nanofluids, the effective thermal conductivity of solid nanocomposites shows only modest improvement over matrix materials, addressing the importance of the Kapitza resistance [90,91]. For nanofluids dispersed with sphere nanoparticles, taking the Kapitza resistance into consideration, the effective thermal conductivity is given by [87,88]

$$\frac{\lambda_{eff}}{\lambda_b} = 1 + 3f\frac{\lambda_p(1-\alpha)-\lambda_b}{\lambda_p(1+2\alpha)+2\lambda_b - f\left[\lambda_p(1-\alpha)-\lambda_b\right]} \tag{4.2}$$

where

$\alpha = 2R_{bd}\lambda_b/d_p$

d_p is the average particle diameter

For the randomly oriented ellipsoidal inclusions with radii a_1, a_2, and a_3, Nan et al. [88] yield the following expression

$$\frac{\lambda_{eff}}{\lambda_b} = \frac{3+f\left[2\beta_{11}(1-L_{11})+\beta_{33}(1-L_{33})\right]}{3-f\left(2\beta_{11}L_{11}+\beta_{33}L_{33}\right)} \tag{4.3}$$

where

$$L_{11} = \begin{cases} \dfrac{p^2}{2(p^2-1)} - \dfrac{p}{2(p^2-1)^{3/2}}\cosh^{-1} p, & \text{for } p > 1 \\[2ex] \dfrac{p^2}{2(p^2-1)} + \dfrac{p}{2(p^2-1)^{3/2}}\cosh^{-1} p, & \text{for } p < 1 \end{cases} \tag{4.4}$$

$$L_{33} = 1 - 2L_{11} \tag{4.5}$$

$$\beta_{ii} = \frac{\lambda_{ii}^c - \lambda_b}{\lambda_b + L_{ii}\left(\lambda_{ii}^c - \lambda_b\right)} \tag{4.6}$$

$$\lambda_{ii}^c = \frac{\lambda_p}{1+\gamma L_{ii}\lambda_p/\lambda_b} \tag{4.7}$$

with subscript i (i = 1, 3) representing parameters along 1 and 3 axes, and

$$\gamma = \begin{cases} (2+1/p)R_{bd}\lambda_b/a_1, & \text{for } p>1 \\ (2+2p)R_{bd}\lambda_b/a_3, & \text{for } p<1 \end{cases} \tag{4.8}$$

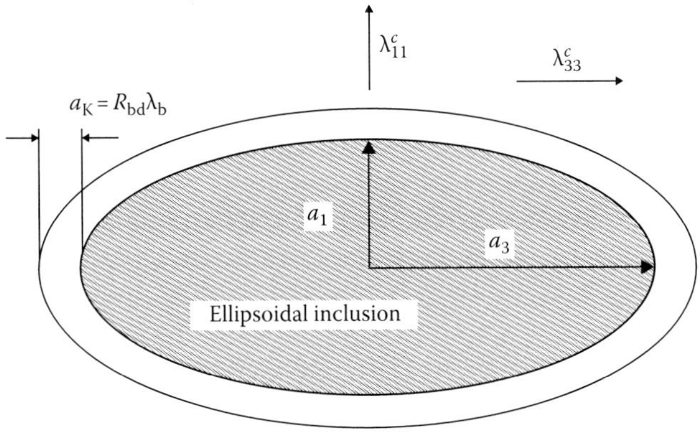

FIGURE 4.1 Schematic of an ellipsoidal inclusion coated with a thin interfacial thermal barrier layer.

Here, p is the aspect ratio of the ellipsoid, $p = a_3/a_1$, and $p > 1$ and $p < 1$ are for a prolate ($a_1 = a_2 < a_3$) and an oblate ($a_1 = a_2 > a_3$) ellipsoidal inclusion, respectively. λ_{11}^c and λ_{33}^c are the equivalent thermal conductivities along a_1 and a_3 axes of the ellipsoidal inclusion coated with a thin interfacial thermal barrier layer, seen in Figure 4.1. Obviously, for $R_{bd} = 0$ and $p = 1$, this equation reduces to the Maxwell–Garnett model.

For nanoparticles with large aspect ratio, such as CNTs, if $p > 1000$, L_{11} is reduced to 0.5 and L_{33} is zero, so the effective thermal conductivity of nanofluids is given by

$$\frac{\lambda_{eff}}{\lambda_b} = \frac{3 + f(\beta_{11} + \beta_{33})}{3 - \beta_{11} f} \tag{4.9}$$

with

$$\beta_{11} = 2\frac{\lambda_{11}^c - \lambda_b}{\lambda_{11}^c + \lambda_b}, \quad \beta_{33} = \frac{\lambda_{33}^c}{\lambda_b} - 1 \tag{4.10}$$

$$\frac{1}{\lambda_{11}^c} = \frac{1}{\lambda_p} + \frac{1}{d_p \lambda_b / 2a_K} \tag{4.11}$$

$$\frac{1}{\lambda_{33}^c} = \frac{1}{\lambda_p} + \frac{1}{L_p \lambda_b / 2a_K} \tag{4.12}$$

and a_K is the Kapitza length. For flat sheets, $p \ll 1$, $L_{11} \sim 0$ and $L_{33} \sim 1$, so the resulting effective thermal conductivity is expressed by

$$\frac{\lambda_{eff}}{\lambda_b} = \frac{3 + 2f\beta_{11}}{3 - \beta_{33} f} \tag{4.13}$$

with

$$\beta_{11} = \frac{\lambda_{11}^c}{\lambda_b} - 1, \quad \beta_{33} = 1 - \frac{\lambda_b}{\lambda_{33}^c} \tag{4.14}$$

$$\frac{1}{\lambda_{11}^{c}} = \frac{1}{\lambda_{\mathrm{p}}} + \frac{1}{W_{\mathrm{p}}\lambda_{\mathrm{b}} / 2a_{\mathrm{K}}} \tag{4.15}$$

$$\frac{1}{\lambda_{33}^{c}} = \frac{1}{\lambda_{\mathrm{p}}} + \frac{1}{t_{\mathrm{p}}\lambda_{\mathrm{b}} / 2a_{\mathrm{K}}} \tag{4.16}$$

where W_{p} and t_{p} are the width and thickness of flat sheets.

4.2.3 Effect of Aggregation

Previous models assume that nanofluids are well dispersed; however, the aggregation of nanoparticles takes place even at low concentration [92]. These aggregates can be considered as new particles, and the individual aggregates occupy more space than the corresponding nanoparticles that make up the aggregates; therefore, the aggregation is believed to enhance the thermal transport prominently [44,60,61,63]. It is worth noting that the thermal conductivity enhancement will decrease if the aggregates continue to agglomerate to settle down from the base fluid [93]. Prasher et al. [63] developed a three-level clustering model to evaluate the effect of the clustering process. In this model, some particles are assumed to form the backbone of the cluster, which is embedded within the whole aggregate with a radius of gyration, R_{g}, and the other particles form dead ends, seen in Figure 4.2. At the first level, the thermal conductivity of the aggregate with the presence of dead ends only is calculated by the Bruggeman model [85]

$$\left(1 - f_{\mathrm{nc}}\right)\frac{\lambda_{\mathrm{b}} - \lambda_{\mathrm{nc}}}{\lambda_{\mathrm{b}} + 2\lambda_{\mathrm{nc}}} + f_{\mathrm{nc}}\frac{\lambda_{\mathrm{b}} - \lambda_{\mathrm{nc}}}{\lambda_{\mathrm{p}} + 2\lambda_{\mathrm{nc}}} = 0 \tag{4.17}$$

Here

- λ_{nc} is the thermal conductivity of the aggregate sphere with the dead-end particles only
- f_{nc} is the volume fraction of dead-end particles in the aggregate sphere, which is calculated by $f_{\mathrm{nc}} = f_{\mathrm{int}} - f_{\mathrm{c}}$
- f_{int} is the fraction of the nanoparticles in the aggregate
- f_{c} is the fraction of the nanoparticles in backbone of the aggregate

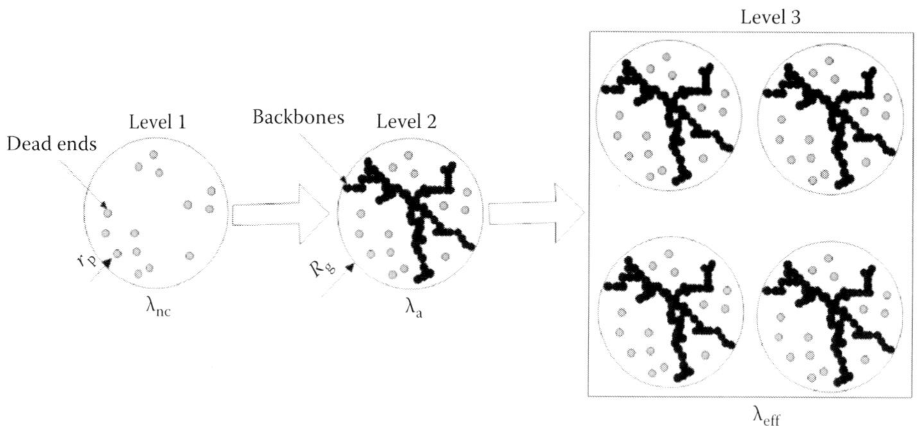

FIGURE 4.2 Schematic of the three-level clustering model. (From Prasher, R. et al., *Appl. Phys. Lett.*, 89, 143119, 2006.)

According to the fractal theory, $f_{int} = (R_g/r_p)^{d_f-3}$, $f_c = (R_g/r_p)^{d_l-3}$, d_f, and d_l are the fractal and chemical dimensions of the aggregate, respectively [61]. Bruggeman model [85] rather than MG model [82] is used since the former has no limitation in the concentration of inclusions. At the second level, the modified EMT [88] is used to calculate the thermal conductivity of the aggregate, λ_a, with both backbones and dead ends. In Equations 4.3 to 4.8, the aspect ratio is given by $p = R_g/r_p$, the thermal conductivity of base fluid, λ_b, and fraction of nanoparticle, f, are replaced by λ_{nc} (from Equation 4.17) and f_c, respectively. At the third level, combining the fraction and the thermal conductivity of the aggregates, f_a and λ_a, the effective thermal conductivity of nanofluid, λ_{eff}, which contains a large amount of aggregates, is calculated by using the MG model (Equation 4.1). Due to the number conservation of the particles, f_a is calculated by $f_a = f/f_{int}$.

4.2.4 Effect of Percolation

Unlike some dynamic models that are proposed to take the effect of random motion of nanoparticle, the empirical models based on the EMT theory are applicable in both solid nanocomposites and stationary nanofluids. The effective thermal conductivity increases rapidly with volume fraction above a certain concentration (threshold) in solid nanocomposites [94,95], similar phenomena are also observed in nanofluids containing CNTs and graphite (or graphene) nanosheets [69–71]. The nanoparticles in solid composites are likely to form highly conductive percolating network, resulting in large thermal conductivity enhancement [96], and the effective thermal conductivity in the vicinity of the percolation threshold can be described as

$$\frac{\lambda_{eff}}{\lambda_0} = \left(f - f_{cr}\right)^t \tag{4.18}$$

where

λ_0 is a pre-exponential factor that depends on the intrinsic thermal conductivity of nanoparticles and their contacts as well as the topology of the percolation cluster
f_{cr} is the critical fraction volume for percolation
t is the exponential index accounting for the percolating network characteristics

In those theoretical models mentioned earlier, Prasher's model [63] considers both thermal interface resistance and aggregate of nanoparticles; however, large uncertainties exist in determining the empirical parameters, including R_g, f_{int}, f_c, d_f, d_l, and f_a, which greatly limit its accuracy. For the dilute nanofluids, the aggregation can be assumed to be isolated from each other, the effect of percolation paths is not yet prominent. As a consequence, in this work, the upper ($R_{bd} = 0$) and lower bounds (R_{bd} is sufficiently large) of the effective thermal conductivity are calculated based on Equations 4.2 and 4.3 for nanofluids containing the spherical and the longitudinal nanoparticles, respectively.

4.3 EXPERIMENTAL TECHNIQUES

Different techniques have been used to measure the effective thermal conductivity of nanofluids, including the transient hot wire method [97], short hot wire method [98], 3ω method [99], steady-state method [37], temperature oscillation method [36], optical beam deflection technique [18], optical thermal lensing method [100], and others [101], and the majority of experiments are carried out by the homemade or commercialized apparatus (KD2) based on the transient hot wire method. In general, the thermal conductivity enhancement is reproducible among various experimental techniques, which was also verified by recent benchmark studies [21,29]. To ensure the measurement accuracy, some experimental details are addressed here:

1. By using the transient hot wire method, short hot wire method, and 3ω method, the metallic sensor as well as the contact leads should be well insulated to prevent the current

amazingly follow the EMT theory, when the Kapitza resistance is the only free parameter to fit. Ge et al. [108] found that the Kapitza resistance of the hydrophobic interface is a factor of 2–3 larger than that of hydrophilic interface, so the Kapitza resistance is believed to be tunable by changing the functionalization condition. The Kapitza resistances of 5×10^{-7} to 2×10^{-8} m^2-K W^{-1} are obtained, which is coarsely of the same order as the reported values [80,108].

Graphite nanosheet, consisting of a few layer of carbon in two-dimensional lattice, exhibits a number of intriguing electronic, optical, and thermal properties. As listed in Table 4A.6, the thickness of the graphite nanosheets (about few nanometers) is three orders smaller than its width (few micrometers); therefore, the effective thermal conductivity of nanofluids containing graphite nanosheets can be predicted from Equation 4.13. For a suspended single-layer nanosheet (graphene), the in-plane thermal conductivity ranges from 3000 to 5000 W m^{-1} K^{-1} [120]; as the number of layers increases, the thermal conductivity decreases dramatically, and saturates to a constant value beyond four layers, which is identical to that of buck graphite [111]. Here, the in-plane thermal conductivity of graphite nanosheets is taken to be 1000 W m^{-1} K^{-1}, with width 2 μm and thickness 1 nm; and the predictions are plotted in Figure 4.8. The calculation results agree reasonably well with the experimental data, and the Kapitza resistance ranges from 1×10^{-8} to 5×10^{-8} m^2-K W^{-1}. The cross-plane thermal conductivity in graphite nanosheets is about 6 W m^{-1} K^{-1} but has a negligible effect on the thermal conductivity of nanofluids due to the presence of the Kapitza resistance. Taking the Kapitza resistance to be 2×10^{-8} m^2-K W^{-1} and thermal conductivity to be 1000 W m^{-1} K^{-1}, the equivalent thermal conductivity along the cross-plane direction, λ_{33}^c, is only about 0.025 W m^{-1} K^{-1}, much smaller than the in-plane equivalent value, λ_{11}^c, which is about 47 W m^{-1} K^{-1}; therefore, the anisotropic behavior of thermal conductivity of graphite nanosheets is ignored in the earlier analysis.

The EMT theory can be used to statistically explain the effect of volume fraction on the thermal conductivity of nanofluids containing Al_2O_3, CuO, CNTs, and graphite nanosheets; however, it fails in explaining some other experimental findings. For instance, the thermal conductivity of

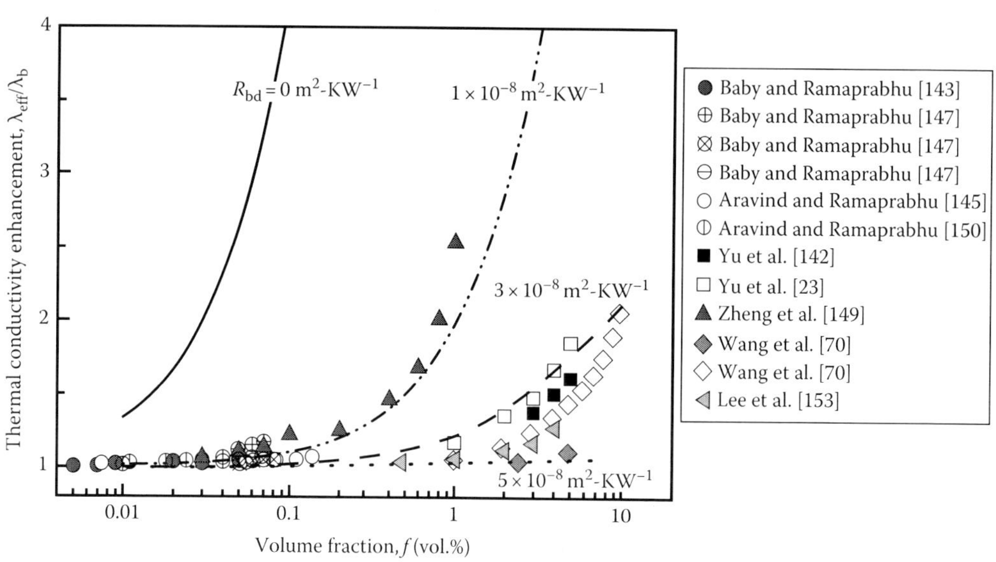

FIGURE 4.8 Comparison of experimental data on the effective thermal conductivity of graphite nanosheets/water nanofluids and predictions based on the EMT theory with different Kapitza resistances.

graphite nanosheets shows unexpectedly higher thermal conductivities than the EMT predictions especially at large volume fraction, as shown in Figure 4.8 indicating the formation of percolated structures [69–71]. Moderate enhancements in thermal conductivity beyond the EMT predictions were observed in nanofluids containing Al_2O_3 and CuO nanoparticles, which might be attributed to the nanoparticle aggregation [44,65]. In the case of nanofluids containing metallic nanoparticles, especially for Cu nanofluids, the EMT theory is probably not sufficient to describe the effective thermal conductivity, and its abnormal enhancement is still an open question. In contrast to the nonmetallic nanoparticles, the heat transport in metal is dominated by electrons, so the coupling between electrons within metallic nanoparticles and phonons within the matrix may play an important role. In addition, the mean free path of electrons is comparable or even larger than the particle size, which further complicates the modeling of the effective thermal conductivity. Consequently, to clarify these factors, with the aid of good characterization of nanoparticles and reliable experimental techniques, more systematic and benchmark studies of thermal property of nanofluids are demanded.

4.5 CONCLUSIONS

We perform a brief review of the experimental results of the effective thermal conductivity of the most commonly used nanofluids containing both spherical and non-spherical nanoparticles, including Al_2O_3, CuO, Cu, Au, CNTs, and graphite nanosheets. With the aid of a large amount of experimental data, we employ an unbiased statistical analysis to find the concentration-dependent behavior of thermal conductivity in nanofluids. Based on the statistical analysis, the effective thermal conductivity of nanofluids containing Al_2O_3, CuO, CNTs, and graphite nanosheets roughly agrees with the EMT predictions. Due to the nanoparticle aggregation, moderate increments of the thermal conductivity in nanofluids containing Al_2O_3 and CuO nanoparticles are observed. The nanoparticles with large aspect ratio are prone to form percolated structures to enhance the effective thermal conductivity especially at high volume fraction. In addition, the abnormal enhancement in the nanofluids containing Cu nanoparticles also indicates that the EMT theory is probably not sufficient to describe the heat conduction in some nanofluids.

The effective thermal conductivity of nanofluids depends not only on volume fraction, but also on some other parameters, such as temperature, particle size, particle geometry, pH value, hydrophobicity, and particle distribution. Consequently, by measuring the same nanofluids with different measurement techniques, the benchmark studies are demanded to reveal the exact heat conduction mechanisms in nanofluids. From the literature survey, a few suggestions are pointed out here. (1) The current leakage may increase the effective thermal conductivity; therefore, the metallic sensor as well as the contact leads should be well insulated. (2) The thermal conductivity is found to decrease with time after ultrasonication, so the stability of nanofluids must be ensured before the thermal conductivity measurement. (3) Due to the sedimentation in nanofluids, the nanoparticle may not be randomly distributed, so the reliable experimental techniques that test small-volume liquid may be preferred. (4) If the Kapitza resistances at nanoparticle/fluid and film/fluid are similar, the thermal transport at nanoparticle/fluid interface can be extensively studied by using the well-developed thermoreflectance techniques.

TABLE 4A.4
Summary of Experimental Data on Thermal Conductivity of Cu Nanofluids

	Base Fluid	*f* (vol.%)	d_p (nm)	λ_{eff}/λ_b
Xuan et al. [2]	Water	2.6–7.5	<100	1.22–1.76
Eastman et al. [15]	EG	0.10–0.28	<10	1.020–1.1417
Xuan et al. [4]	Water	0.3–2	<100	1.030–1.186
Assael et al. [125]	EG	0.09, 0.48	25	1.01, 1.03
Liu et al. [119]	Water	0.1	70–100	1.007
		0.1	100–200	1.009
		0.2	>250	1.005
Jana et al. [114]	Water	0.05–0.3	35–50	1.17–1.74
Garg et al. [126]	EG	0. 4–2	~200	1.02–1.126
Li et al. [116]	Water	0.002–0.09	~100	1.082–1.175
Sinha et al. [127]	EG	0.55, 1	30–45	1.34, 1.58
Patel et al. [29]	Water	0.1–3	80	1.035–1.210
	EG	0.1–3	80	1.045–1.260
Yu et al. [104]	EG	0.3, 0.5	5–10	1.11, 1.16
Saterlie et al. [118]	Water	0.55, 1	70	1.24, 1.55

TABLE 4A.5
Summary of Experimental Data on Thermal Conductivity of CNT nanofluids

	Base Fluid/Particle	*f* (vol.%)	d_p (nm), L_p(μm)	λ_{eff}/λ_b
Choi et al. [3]	PAO/MWCNT	0.05–1	25, 50	1.030–2.605
Xie et al. [128]	Water/MWCNT	0.4, 1	15, 30	1.040, 1.067
	EG/MWCNT	0.20–1		1.022–1.131
Wen et al. [62]	Water/MWCNT	0–0.84	20–60, >10	1.005
Assael et al. [102]	Water/MWCNT	0.6	100, 50	1.07–1.38
Assael et al. [129]	Water/DWCNT	0.75, 1	5, 20	1.020, 1.076
Liu et al. [130]	PAO/MWCNT	1, 2	20, —	1.085, 1.303
	EG/MWCNT	0.2–1.0		1.016–1.124
Ding et al. [5]	Water/MWCNT	0.05–0.5	20–30, >10	1.064–1.275
Hwang et al. [131]	Water/MWCNT	0.25–1	10–30, 10–50	1.013–1.070
	Oil/MWCNT	0.2–0.5		1.005–1.086
Jana et al. [114]	Water/MWCNT	0.3–0.8	10, 5–10	1.243–1.340
Shaikh et al. [132]	PAO/MWCNTs	0.1–1	—, —	1.336–2.160
Zhang et al. [17]	Water/MWCNT	0.1–0.9	150, 10	1.034–1.420
Glory et al. [133]	Water/MWCNT	0.006–1.73	40, 1.75	1.005–1.444
Jha et al. [134]	Water/MWCNT	0.007–0.384	20–30, —	1.033–1.149
	EG/MWCNT	0.007–0.384		1.032–1.068
Chen et al. [43]	Water/MWCNT	0.080–1.0	15, 30	1.004–1.118
	EG/MWCNT	0.2–1		1.037–1.175
Nanda et al. [135]	PAO/SWCNT	0.5–1.5	4–6, 0.5	1.043–1.210
	EG/SWCNT	0.27–1.09		1.090–1.344
Amrollahi et al. [136]	EG/SWCNT	0.5–2.5	—, —	1.150–1.220
Choi et al. [137]	Water/SWCNT	~0.57	1.33, —	1.11
Meibodi et al. [138]	Water/MWCNT	0.56, 1.11	~20, —	1.013, 1.103
Cherkasova et al. [22]	Water/MWCNT	0.01–0.1	30–50, 0.5–2	1.022–1.082
Walvekar et al. [139]	Water/MWCNT	0.005–0.056	20, 30	1.04–1.374
Wang et al. [140]	Water/MWCNT	0.3–1.26	20–30, 5–30	1.030–1.127
Halelfadl et al. [141]	Water/MWCNT	0.026	9.2, 1.5	1.113
		0.026	11.4, 1	1.096
	EG/MWCNT	0.026	9.2, 1.5	1.151

TABLE 4A.6
Summary of Experimental Data on Thermal Conductivity of Nanofluids Containing Graphite Nanosheets

	Base Fluid/Particle	f (vol.%)	W_p (μm), t_p (nm)	λ_{eff}/λ_b
Yu et al. [142]	EG/graphene oxide nanosheet	1–5	1–3, 0.8–1.2	1.105–1.610
Baby and Ramaprabhu [143]	Water/graphene	0.005–0.056	—, —	1.010–1.140
	EG/graphene	0.005–0.05		1.006–1.041
Yu et al. [144]	Water/graphene oxide nanosheet	1–5	0.5–3, 0.6–1.1	1.073–1.300
Yu et al. [23]	EG/graphene	1–5	0.2–3, 0.7–1.3	1.168–1.854
Aravind and Ramaprabhu [145]	EG/graphene	0.007–0.14	—, —	1.024–1.064
	Water/graphene	0.007–0.14		1.037–1.136
Baby and Sundara [146]	EG/graphene decorated with CuO	0.01–0.07	—, —	1.018–1.170
	Water/graphene decorated with CuO	0.005–0.05		1.090–1.280
Baby and Ramaprabhu [147]	EG/graphene	0.05–0.08	—, —	1.024–1.052
	Water/graphene	0.005–0.05		1.055–1.155
Baby and Ramaprabhu [42]	EG/graphene decorated with Ag	0.01–0.07	—, —	1.020–1.076
	Water/graphene decorated with Ag	0.005–0.05		1.07–1.28
Gupta et al. [148]	Water/graphene	0.05–0.2	0.005–1.5, —	1.018–1.100
Zheng et al. [149]	EG/graphite	0.03–1	—, —	1.064–2.534
Aravind and Ramaprabhu [150]	EG/graphene-MWCNT nanocomposite	0.011–0.04	1, —	1.032–1.068
	Water/graphene-MWCNT nanocomposite	0.011–0.04		1.011–1.092
Kole and Dey [151]	Water/graphene	0.041–0.395	—, —	1.055–1.150
Ghozatloo et al. [152]	Water/graphene	0.0038–0.0192	—, —	1.009–1.135
Wang et al. [70]	EG/graphite	2.41, 4.83	<0.03, —	1.041, 1.114
	EG/graphene oxide nanosheet	1–9.6	1–5, 0.8–1.2	1.055–2.063
Lee et al. [153]	EG/graphene	0.5–4	0.1, few layers	1.032–1.270
Mehrali et al. [154]	Water/graphene	0.011–0.045	2, 2	1.12–1.23
Ijam et al. [105]	Glycerol-water/graphene nanosheet	0.009–0.45	0.5–2, —	1–1.03

ACKNOWLEDGMENTS

This work was supported by the National Natural Science Foundation of China (Grant Nos. 51476033, 51327001, 51106029, 51356001), and the PhD Programs Foundation of Ministry of Education of China (Grant No. 20110092120006).

REFERENCES

1. S.U.S. Choi and J.A. Eastman. Enhancing thermal conductivity of fluids with nanoparticles. *ASME Tech Rep FED* 231, 99–103 (1995).
2. Y. Xuan and Q. Li. Heat transfer enhancement of nanofluids. *Int J Heat Fluid Flow* 21, 58–64 (2000).
3. S.U.S. Choi, Z.G. Zhang, W. Yu, F.E. Lockwood, and E.A. Grulke. Anomalous thermal conductivity enhancement in nanotube suspensions. *Appl Phys Lett* 79, 2252–2254 (2001).

4. Y. Xuan and Q. Li. Investigation on convective heat transfer and flow features of nanofluids. *ASME J Heat Transfer* 125, 151–155 (2003).
5. Y. Ding, H. Alias, D. Wen, and R.A. Williams. Heat transfer of aqueous suspensions of carbon nanotubes (CNT nanofluids). *Int J Heat Mass Transfer* 49, 240–250 (2006).
6. S.M. You, J.H. Kim, and K.H. Kim. Effect of nanoparticles on critical heat flux of water in pool boiling heat transfer. *Appl Phys Lett* 83, 3374–3376 (2003).
7. I.C. Bang and S. Heung Chang. Boiling heat transfer performance and phenomena of Al_2O_3–water nanofluids from a plain surface in a pool. *Int J Heat Mass Transfer* 48, 2407–2419 (2005).
8. P. Keblinski, J.A. Eastman, and D.G. Cahill. Nanofluids for thermal transport. *Mater Today* 8, 36–44 (2005).
9. X.-Q. Wang and A.S. Mujumdar. Heat transfer characteristics of nanofluids:, A review. *Int J Therm Sci* 46, 1–19 (2007).
10. W. Yu, D.M. France, J.L. Routbort, and S.U.S. Choi. Review and comparison of nanofluid thermal conductivity and heat transfer enhancements. *Heat Transfer Eng* 29, 432–460 (2008).
11. K. Khanafer and K. Vafai. A critical synthesis of thermophysical characteristics of nanofluids. *Int J Heat Mass Transfer* 54, 4410–4428 (2011).
12. K.V. Wong and O. De Leon. Applications of nanofluids, current and future. *Adv Mech Eng* 2010, 519659 (2010).
13. R. Saidur, K.Y. Leong, and H.A. Mohammad. A review on applications and challenges of nanofluids. *Renew Sustain Energ Rev* 15, 1646–1668 (2011).
14. H. Masuda, A. Ebata, K. Teramae, and N. Hishinuma. Alteration of thermal conductivity and viscosity of liquid by dispersing ultra-fine particles. *Jpn J Thermophys Prop* 4, 227–233 (1993).
15. J.A. Eastman, U.S. Choi, S. Li, L.J. Thompson, and S. Lee. Enhanced thermal conductivity through the development of nanofluids, *Proceedings of the Symposium on Nanophase and Nanocomposite Materials II*, Vol. 457, Materials Research Society, Boston, MA, 1997, pp. 3–11.
16. J.A. Eastman, S.U.S. Choi, S. Li, W. Yu, and L.J. Thompson. Anomalously increased effective thermal conductivities of ethylene glycol-based nanofluids containing copper nanoparticles. *Appl Phys Lett* 78, 718–720 (2001).
17. X. Zhang, H. Gu, and M. Fujii. Effective thermal conductivity and thermal diffusivity of nanofluids containing spherical and cylindrical nanoparticles. *Exp Therm Fluid Sci* 31, 593–599 (2007).
18. S.A. Putnam, D.G. Cahill, P.V. Braun, Z. Ge, and R.G. Shimmin. Thermal conductivity of nanoparticle suspensions. *J Appl Phys* 99, 084308 (2006).
19. D.C. Venerus, M.S. Kabadi, S. Lee, and V. Perez-luna. Study of thermal transport in nanoparticle suspensions using forced rayleigh scattering. *J Appl Phys* 100, 094310 (2006).
20. B. Yang and Z.H. Han. Temperature-dependent thermal conductivity of nanorod-based nanofluids. *Appl Phys Lett* 89, 083111 (2006).
21. J. Buongiorno, D.C. Venerus, N. Prabhat, T. McKrell, J. Townsend, R. Christianson, Y.V. Tolmachev et al. A benchmark study on the thermal conductivity of nanofluids. *J Appl Phys* 106, 094312 (2009).
22. A.S. Cherkasova and J.W. Shan. Particle aspect-ratio and agglomeration-state effects on the effective thermal conductivity of aqueous suspensions of multiwalled carbon nanotubes. *J Heat Transfer* 132, 082402–082402 (2010).
23. W. Yu, H. Xie, X. Wang, and X. Wang. Significant thermal conductivity enhancement for nanofluids containing graphene nanosheets. *Phys Lett A* 375, 1323–1328 (2011).
24. H.E. Patel, S.K. Das, T. Sundararajan, A. Sreekumaran Nair, B. George, and T. Pradeep. Thermal conductivities of naked and monolayer protected metal nanoparticle based nanofluids, manifestation of anomalous enhancement and chemical effects. *Appl Phys Lett* 83, 2931–2933 (2003).
25. C.H. Chon, K.D. Kihm, S.P. Lee, and S.U.S. Choi. Empirical correlation finding the role of temperature and particle size for nanofluid (Al_2O_3) thermal conductivity enhancement. *Appl Phys Lett* 87, 153107 (2005).
26. S.H. Kim, S.R. Choi, and D. Kim. Thermal conductivity of metal-oxide nanofluids, particle size dependence and effect of laser irradiation. *J Heat Transfer* 129, 298–307 (2006).
27. C.H. Li and G.P. Peterson. The effect of particle size on the effective thermal conductivity of Al_2O_3-water nanofluids. *J Appl Phys* 101, 044312 (2007).
28. H.A. Mintsa, G. Roy, C.T. Nguyen, and D. Doucet. New temperature dependent thermal conductivity data for water-based nanofluids. *Int J Therm Sci* 48, 363–371 (2009).
29. H. Patel, T. Sundararajan, and S. Das. An experimental investigation into the thermal conductivity enhancement in oxide and metallic nanofluids. *J Nanopart Res* 12, 1015–1031 (2010).

30. T.-P. Teng, Y.-H. Hung, T.-C. Teng, H.-E. Mo, and H.-G. Hsu. The effect of alumina/water nanofluid particle size on thermal conductivity. *Appl Therm Eng* 30, 2213–2218 (2010).
31. E.V. Timofeeva, A.N. Gavrilov, J.M. McCloskey, Y.V. Tolmachev, S. Sprunt, L.M. Lopatina, and J.V. Selinger. Thermal conductivity and particle agglomeration in alumina nanofluids, experiment and theory. *Phys Rev E* 76, 061203 (2007).
32. M. Beck, Y. Yuan, P. Warrier, and A. Teja. The effect of particle size on the thermal conductivity of alumina nanofluids. *J Nanopart Res* 11, 1129–1136 (2009).
33. M. Beck, Y. Yuan, P. Warrier, and A. Teja. The thermal conductivity of alumina nanofluids in water, ethylene glycol, and ethylene glycol + water mixtures. *J Nanopart Res* 12, 1469–1477 (2010).
34. E.V. Timofeeva, D.S. Smith, W. Yu, D.M. France, D. Singh, and J.L. Routbort. Particle size and interfacial effects on thermo-physical and heat transfer characteristics of water-based A-SiC nanofluids. *Nanotechnology* 21, 215703 (2010).
35. P. Warrier and A. Teja. Effect of particle size on the thermal conductivity of nanofluids containing metallic nanoparticles. *Nanoscale Res Lett* 6, 247 (2011).
36. S.K. Das, N. Putra, P. Thiesen, and W. Roetzel. Temperature dependence of thermal conductivity enhancement for nanofluids. *J Heat Transfer* 125, 567–574 (2003).
37. C.H. Li and G.P. Peterson. Experimental investigation of temperature and volume fraction variations on the effective thermal conductivity of nanoparticle suspensions (nanofluids). *J Appl Phys* 99, 084314 (2006).
38. H. Xie and L. Chen. Adjustable thermal conductivity in carbon nanotube nanofluids. *Phys Lett A* 373, 1861–1864 (2009).
39. M. Abareshi, E.K. Goharshadi, S. Mojtaba Zebarjad, H. Khandan Fadafan, and A. Youssefi. Fabrication, characterization and measurement of thermal conductivity of Fe_3O_4 nanofluids. *J Magn Magn Mater* 322, 3895–3901 (2010).
40. D. Wen and Y. Ding. Effective thermal conductivity of aqueous suspensions of carbon nanotubes (carbon nanotube nanofluids). *J Thermophys Heat Transfer* 18, 481–485 (2004).
41. P. Garg, J.L. Alvarado, C. Marsh, T.A. Carlson, D.A. Kessler, and K. Annamalai. An experimental study on the effect of ultrasonication on viscosity and heat transfer performance of multi-wall carbon nanotube-based aqueous nanofluids. *Int J Heat Mass Transfer* 52, 5090–5101 (2009).
42. T.T. Baby and S. Ramaprabhu. Synthesis and nanofluid application of silver nanoparticles decorated graphene. *J Mater Chem* 21, 9702–9709 (2011).
43. L. Chen, H. Xie, Y. Li, and W. Yu. Nanofluids containing carbon nanotubes treated by mechanochemical reaction. *Thermochim Acta* 477, 21–24 (2008).
44. J.W. Gao, R.T. Zheng, H. Ohtani, D.S. Zhu, and G. Chen. Experimental investigation of heat conduction mechanisms in nanofluids: Clue on clustering. *Nano Lett* 9, 4128–4132 (2009).
45. N. Shalkevich, W. Escher, T. Bürgi, B. Michel, L. Si-Ahmed, and D. Poulikakos. On the thermal conductivity of gold nanoparticle colloids. *Langmuir* 26, 663–670 (2010).
46. D. Singh, E. Timofeeva, W. Yu, J. Routbort, D. France, D. Smith, and J.M. Lopez-Cepero. An investigation of silicon carbide-water nanofluid for heat transfer applications. *J Appl Phys* 105, 064306 (2009).
47. W. Duangthongsuk and S. Wongwises. Measurement of temperature-dependent thermal conductivity and viscosity of TiO_2-water nanofluids. *Exp Therm Fluid Sci* 33, 706–714 (2009).
48. S.P. Jang and S.U.S. Choi. Role of Brownian motion in the enhanced thermal conductivity of nanofluids. *Appl Phys Lett* 84, 4316–4318 (2004).
49. D.H. Kumar, H.E. Patel, V.R.R. Kumar, T. Sundararajan, T. Pradeep, and S.K. Das. Model for heat conduction in nanofluids. *Phys Rev Lett* 93, 144301 (2004).
50. J. Koo and C. Kleinstreuer. A new thermal conductivity model for nanofluids. *J Nanopart Res* 6, 577–588 (2004).
51. R. Prasher, P. Bhattacharya, and P.E. Phelan. Thermal conductivity of nanoscale colloidal solutions (nanofluids). *Phys Rev Lett* 94, 025901 (2005).
52. B. Yang. Thermal conductivity equations based on Brownian motion in suspensions of nanoparticles (nanofluids). *J Heat Transfer* 130, 042408 (2008).
53. C.H. Chon and K.D. Kihm. Thermal conductivity enhancement of nanofluids by Brownian motion. *J Heat Transfer* 127, 810 (2005).
54. S. Krishnamurthy, P. Bhattacharya, and P.E. Phelan. Enhanced mass transport in nanofluids. *Nano Lett* 6, 419–423 (2006).
55. R. Azizian, E. Doroodchi, and B. Moghtaderi. Effect of nanoconvection caused by brownian motion on the enhancement of thermal conductivity in nanofluids. *Ind Eng Chem Res* 51, 1782–1789 (2011).

56. C.J. Yu, A.G. Richter, A. Datta, M.K. Durbin, and P. Dutta. Observation of molecular layering in thin liquid films using x-ray reflectivity. *Phys Rev Lett* 82, 2326–2329 (1999).
57. W. Yu and S.U.S. Choi. The role of interfacial layers in the enhanced thermal conductivity of nanofluids: A Renovated Maxwell model. *J Nanopart Res* 5, 167–171 (2003).
58. L. Xue, P. Keblinski, S.R. Phillpot, S.U.S. Choi, and J.A. Eastman. Effect of liquid layering at the liquid–solid interface on thermal transport. *Int J Heat Mass Transfer* 47, 4277–4284 (2004).
59. H. Xie, M. Fujii, and X. Zhang. Effect of interfacial nanolayer on the effective thermal conductivity of nanoparticle-fluid mixture. *Int J Heat Mass Transfer* 48, 2926–2932 (2005).
60. P. Keblinski, S.R. Phillpot, S.U.S. Choi, and J.A. Eastman. Mechanisms of heat flow in suspensions of nano-sized particles (nanofluids). *Int J Heat Mass Transfer* 45, 855–863 (2002).
61. B.-X. Wang, L.-P. Zhou, and X.-F. Peng. A fractal model for predicting the effective thermal conductivity of liquid with suspension of nanoparticles. *Int J Heat Mass Transfer* 46, 2665–2672 (2003).
62. S.M.S. Murshed, K.C. Leong, and C. Yang. Enhanced thermal conductivity of TiO_2—Water based nanofluids. *Int J Therm Sci* 44, 367–373 (2005).
63. R. Prasher, W. Evans, P. Meakin, J. Fish, P. Phelan, and P. Keblinski. Effect of aggregation on thermal conduction in colloidal nanofluids. *Appl Phys Lett* 89, 143119 (2006).
64. H.T. Zhu, C.Y. Zhang, Y.M. Tang, and J.X. Wang. Novel synthesis and thermal conductivity of Cuo nanofluid. *J Phys Chem C* 111, 1646–1650 (2007).
65. P.D. Shima, J. Philip, and B. Raj. Influence of aggregation on thermal conductivity in stable and unstable nanofluids. *Appl Phys Lett* 97, 153113 (2010).
66. G. Domingues, S. Volz, K. Joulain, and J.-J. Greffet. Heat transfer between two nanoparticles through near field interaction. *Phys Rev Lett* 94, 085901 (2005).
67. P. Ben-Abdallah. Heat transfer through near-field interactions in nanofluids. *Appl Phys Lett* 89, 113117 (2006).
68. J. Philip, P.D. Shima, and B. Raj. Evidence for enhanced thermal conduction through percolating structures in nanofluids. *Nanotechnology* 19, 305706 (2008).
69. R. Zheng, J. Gao, J. Wang, S.-P. Feng, H. Ohtani, J. Wang, and G. Chen. Thermal percolation in stable graphite suspensions. *Nano Lett* 12, 188–192 (2012).
70. B. Wang, J. Hao, and H. Li. Remarkable improvements in the stability and thermal conductivity of graphite/ethylene glycol nanofluids caused by a graphene oxide percolation structure. *Dalton Trans* 42, 5866–5873 (2013).
71. P. Dhar, S.S. Gupta, S. Chakraborty, A. Pattamatta, and S.K. Das. The role of percolation and sheet dynamics during heat conduction in poly-dispersed graphene nanofluids. *Appl Phys Lett* 102, 163114 (2013).
72. L. Wang and X. Wei. Nanofluids, synthesis, heat conduction, and extension. *J Heat Transfer* 131, 033102 (2009).
73. X. Wang and X. Xu. Thermal conductivity of nanoparticle—Fluid mixture. *J Thermophys Heat Transfer* 13, 474–480 (1999).
74. J. Eapen, R. Rusconi, R. Piazza, and S. Yip. The classical nature of thermal conduction in nanofluids. *J Heat Transfer* 132, 102402–102402 (2010).
75. D.G. Cahill, P.V. Braun, G. Chen, D.R. Clarke, S. Fan, K.E. Goodson, W.P. King et al. Nanoscale thermal transport. II. 2003–2012. *Appl Phys Rev* 1, 011305 (2014).
76. H.-K. Lyeo and D. Cahill. Thermal conductance of interfaces between highly dissimilar materials. *Phys Rev B* 73, 144301 (2006).
77. B. Gundrum, D. Cahill, and R. Averback. Thermal conductance of metal-metal interfaces. *Phys Rev B* 72, 245–426 (2005).
78. R.B. Wilson and D.G. Cahill. Experimental validation of the interfacial form of the Wiedemann-Franz law. *Phys Rev Lett* 108, 255901 (2012).
79. O.M. Wilson, X. Hu, D.G. Cahill, and P.V. Braun. Colloidal metal particles as probes of nanoscale thermal transport in fluids. *Phys Rev B* 66, 224301 (2002).
80. S.T. Huxtable, D.G. Cahill, S. Shenogin, L. Xue, R. Ozisik, P. Barone, M. Usrey et al. Interfacial heat flow in carbon nanotube suspensions. *Nat Mater* 2, 731–734 (2003).
81. Z. Ge, D.G. Cahill, and P.V. Braun. Aupd metal nanoparticles as probes of nanoscale thermal transport in aqueous solution. *J Phys Chemi B* 108, 18870–18875 (2004).
82. J.C. Maxwell. *A Treatise on Electricity and Magnetism*, Clarendon Press, Oxford, U.K., 1891.
83. L. Rayleigh. On the influence of obstacles arranged in rectangular order upon the properties of a medium. *Philos Mag* 34, 481–502 (1892).

84. D.J. Jeffrey. Conduction through a random suspension of spheres. *Proc R Soc Lond A Math Phys Sci* 335, 355–367 (1973).
85. D.A.G. Bruggeman. Berechnung Verschiedener Physikalischer Konstanten Von Heterogenen Substanzen. I. Dielektrizitätskonstanten Und Leitfähigkeiten Der Mischkörper Aus Isotropen Substanzen. *Ann Phys* 416, 665–679 (1935).
86. R.L. Hamilton and O.K. Crosser. Thermal conductivity of heterogeneous two-component systems. *Ind Eng Chem Fund* 1, 187–191 (1962).
87. D.P.H. Hasselman and L.F. Johnson. Effective thermal conductivity of composites with interfacial thermal barrier resistance. *J Compos Mater* 21, 508–515 (1987).
88. C.-W. Nan, R. Birringer, D.R. Clarke, and H. Gleiter. Effective thermal conductivity of particulate composites with interfacial thermal resistance. *J Appl Phys* 81, 6692–6699 (1997).
89. D.G. Cahill, W.K. Ford, K.E. Goodson, G.D. Mahan, A. Majumdar, H.J. Maris, R. Merlin, and S.R. Phillpot. Nanoscale thermal transport. *J Appl Phys* 93, 793–818 (2003).
90. S.A. Putnam, D.G. Cahill, B.J. Ash, and L.S. Schadler. High-precision thermal conductivity measurements as a probe of polymer/nanoparticle interfaces. *J Appl Phys* 94, 6785–6788 (2003).
91. M. Moniruzzaman and K.I. Winey. Polymer nanocomposites containing carbon nanotubes. *Macromolecules* 39, 5194–5205 (2006).
92. K. Kwak and C. Kim. Viscosity and thermal conductivity of copper oxide nanofluid dispersed in ethylene glycol. *Korea Aust Rheol J* 17, 35–40 (2005).
93. R. Prasher, P.E. Phelan, and P. Bhattacharya. Effect of aggregation kinetics on the thermal conductivity of nanoscale colloidal solutions (nanofluid). *Nano Lett* 6, 1529–1534 (2006).
94. R. Haggenmueller, C. Guthy, J.R. Lukes, J.E. Fischer, and K.I. Winey. Single wall carbon nanotube/polyethylene nanocomposites, thermal and electrical conductivity. *Macromolecules* 40, 2417–2421 (2007).
95. S. Kumar, M.A. Alam, and J.Y. Murthy. Effect of percolation on thermal transport in nanotube composites. *Appl Phys Lett* 90, 104105 (2007).
96. M. Foygel, R.D. Morris, D. Anez, S. French, and V.L. Sobolev. Theoretical and computational studies of carbon nanotube composites and suspensions, electrical and thermal conductivity. *Phys Rev B* 71, 104201 (2005).
97. Y. Nagasaka and A. Nagashima. Absolute measurement of the thermal conductivity of electrically conducting liquids by the transient hot-wire method. *J Phys E Sci Instrum* 14, 1435–1440 (1981).
98. X. Zhang and M. Fujii. Simultaneous measurements of the thermal conductivity and thermal diffusivity of molten salts with a transient short-hot-wire method. *Int J Thermophys* 21, 71–84 (2000).
99. D.-W. Oh, A. Jain, J.K. Eaton, K.E. Goodson, and J.S. Lee. Thermal conductivity measurement and sedimentation detection of aluminum oxide nanofluids by using the 3ω method. *Int J Heat Fluid Flow* 29, 1456–1461 (2008).
101. A. Kusiak, P. Ch, and J.L. Battaglia. Measuring the thermal conductivity of liquids using photo-thermal radiometry. *Meas Sci Technol* 21, 015403 (2010).
102. M.J. Assael, C. Chen, I. Metaxa, and W.A. Wakeham. Thermal conductivity of suspensions of carbon nanotubes in water. *Int J Thermophys* 25, 971–985 (2004).
103. N.R. Karthikeyan, J. Philip, and B. Raj. Effect of clustering on the thermal conductivity of nanofluids. *Mater Chem Phys* 109, 50–55 (2008).
104. W. Yu, H. Xie, L. Chen, and Y. Li. Investigation on the thermal transport properties of ethylene glycol-based nanofluids containing copper nanoparticles. *Powder Technol* 197, 218–221 (2010).
105. A. Ijam, A. Moradi Golsheikh, R. Saidur, and P. Ganesan. A glycerol–water-based nanofluid containing graphene oxide nanosheets. *J Mater Sci* 49, 5934–5944 (2014).
106. S.R. Choi, J. Kim, and D. Kim. 3ω Method to measure thermal properties of electrically conducting small-volume liquid. *Rev Sci Instrum* 78, 084902 (2007).
107. A. Plech, V. Kotaidis, S. Grésillon, C. Dahmen, and G. von Plessen. Laser-induced heating and melting of gold nanoparticles studied by time-resolved x-ray scattering. *Phys Rev B* 70, 195423 (2004).
108. Z. Ge, D.G. Cahill, and P.V. Braun. Thermal Conductance of Hydrophilic and Hydrophobic Interfaces. *Phys Rev Lett* 96, 186101 (2006).
109. D.R. Lide. *CRC Handbook of Chemistry and Physics*, CRC Press, Boca Raton, FL, 2004.
110. M. Fujii, X. Zhang, H. Xie, H. Ago, K. Takahashi, T. Ikuta, H. Abe, and T. Shimizu. Measuring the thermal conductivity of a single carbon nanotube. *Phys Rev Lett* 95, 065502 (2005).
111. S. Ghosh, W. Bao, D.L. Nika, S. Subrina, E.P. Pokatilov, C.N. Lau, and A.A. Balandin. Dimensional crossover of thermal transport in few-layer graphene. *Nat Mater* 9, 555–558 (2010).

112. C.A. Nieto de Castro, S.F.Y. Li, A. Nagashima, R.D. Trengove, and W.A. Wakeham. Standard reference data for the thermal conductivity of liquids. *J Phys Chem Ref Data* 15, 1073–1086 (1986).
113. H. Xie, J. Wang, T. Xi, Y. Liu, F. Ai, and Q. Wu. Thermal conductivity enhancement of suspensions containing nanosized alumina particles. *J Appl Phys* 91, 4568–4572 (2002).
114. S. Jana, A. Salehi-khojin, and W.-H. Zhong. Enhancement of fluid thermal conductivity by the addition of single and hybrid nano-additives. *Thermochim Acta* 462, 45–55 (2007).
115. B. Wang, X. Wang, W. Lou, and J. Hao. Gold-ionic liquid nanofluids with preferably tribological properties and thermal conductivity. *Nanoscale Res Lett* 6, 259 (2011).
116. X.F. Li, D.S. Zhu, X.J. Wang, N. Wang, J.W. Gao, and H. Li. Thermal conductivity enhancement dependent ph and chemical surfactant for Cu-H_2O nanofluids. *Thermochim Acta* 469, 98–103 (2008).
117. X. Wang, D. Zhu, and S. Yang. Investigation of pH and SDBS on enhancement of thermal conductivity in nanofluids. *Chem Phys Lett* 470, 107–111 (2009).
118. M. Saterlie, H. Sahin, B. Kavlicoglu, Y. Liu, and O. Graeve. Particle size effects in the thermal conductivity enhancement of copper-based nanofluids. *Nanoscale Res Lett* 6, 217 (2011).
119. M.-S. Liu, M.C.-C. Lin, C.Y. Tsai, and C.-C. Wang. Enhancement of thermal conductivity with Cu for nanofluids using chemical reduction method. *Int J Heat Mass Transfer* 49, 3028–3033 (2006).
120. A.A. Balandin, S. Ghosh, W. Bao, I. Calizo, D. Teweldebrhan, F. Miao, and C.N. Lau. Superior thermal conductivity of single-layer graphene. *Nano Lett* 8, 902–907 (2008).
121. T. Borca-Tasciuc, A.R. Kumar, and G. Chen. Data reduction in 3ω method for thin-film thermal conductivity determination. *Rev Sci Instrum* 72, 2139–2147 (2001).
122. B. Barbés, R. Páramo, E. Blanco, M. Pastoriza-Gallego, M. Piñeiro, J. Legido, and C. Casanova. Thermal conductivity and specific heat capacity measurements of Al_2O_3 nanofluids. *J Therm Anal Calorim* 111, 1615–1625 (2013).
123. J.R.V. Peñas, J.M.O.D. Zárate, and M. Khayet. Measurement of the thermal conductivity of nanofluids by the multicurrent hot-wire method. *J Appl Phys* 104, 044314 (2008).
124. M. Liu, M.C. Lin, and C. Wang. Enhancements of thermal conductivities with Cu, Cuo, and carbon nanotube nanofluids and application of MWNT/water nanofluid on a water chiller system. *Nanoscale Res Lett* 6, 297 (2011).
125. M.J. Assael, I.N. Metaxa, K. Kakosimos, and D. Constantinou. Thermal conductivity of nanofluids—Experimental and theoretical. *Int J Thermophys* 27, 999–1017 (2006).
126. J. Garg, B. Poudel, M. Chiesa, J.B. Gordon, J.J. Ma, J.B. Wang, Z.F. Ren et al. Enhanced thermal conductivity and viscosity of copper nanoparticles in ethylene glycol nanofluid. *J Appl Phys* 103, 074301 (2008).
127. K. Sinha, B. Kavlicoglu, Y. Liu, F. Gordaninejad, and O.A. Graeve. A comparative study of thermal behavior of iron and copper nanofluids. *J Appl Phys* 106, 064307 (2009).
128. H. Xie, H. Lee, W. Youn, and M. Choi. Nanofluids containing multiwalled carbon nanotubes and their enhanced thermal conductivities. *J Appl Phys* 94, 4967–4971 (2003).
129. M.J. Assael, I.N. Metaxa, J. Arvanitidis, D. Christofilos, and C. Lioutas. Thermal conductivity enhancement in aqueous suspensions of carbon multi-walled and double-walled nanotubes in the presence of two different dispersants. *Int J Thermophys* 26, 647–664 (2005).
130. M.-S. Liu, M.C.-C. Lin, I.T. Huang, and C.-C. Wang. Enhancement of thermal conductivity with carbon nanotube for nanofluids. *Int J Heat Mass Transfer* 32, 1202–1210 (2005).
131. Y. Hwang, H.S. Park, J.K. Lee, and W.H. Jung. Thermal conductivity and lubrication characteristics of nanofluids. *Curr Appl Phys* 6S1, e67–e71 (2006).
132. S. Shaikh, K. Lafdi, and R. Ponnappan. Thermal conductivity improvement in carbon nanoparticle doped pao oil: An experimental study. *J Appl Phys* 101, 064302 (2007).
133. J. Glory, M. Bonetti, M. Helezen, M.M.-L. Hermite, and C. Reynaud. Thermal and electrical conductivities of water-based nanofluids prepared with long multiwalled carbon nanotubes. *J Appl Phys* 103, 094309 (2008).
134. N. Jha and S. Ramaprabhu. Synthesis and thermal conductivity of copper nanoparticle decorated multiwalled carbon nanotubes based nanofluids. *J Phys Chem C* 112, 9315–9319 (2008).
135. J. Nanda, C. Maranville, S.C. Bollin, D. Sawall, H. Ohtani, J.T. Remillard, and J.M. Ginder. Thermal conductivity of single-wall carbon nanotube dispersions: Role of interfacial effects. *J Phys Chem C* 112, 654–658 (2008).
136. A. Amrollahi, A.A. Hamidi, and A.M. Rashidi. The effects of temperature, volume fraction and vibration time on the thermo-physical properties of a carbon nanotube suspension (carbon nanofluid). *Nanotechnology* 19, 315701 (2008).

137. T.Y. Choi, M.H. Maneshian, B. Kang, W.S. Chang, C.S. Han, and D. Poulikakos. Measurement of the thermal conductivity of a water-based single-wall carbon nanotube colloidal suspension with a modified 3-Ω method. *Nanotechnology* 20, 315706 (2009).
138. M.E. Meibodi, M. Vafaie-Sefti, A.M. Rashidi, A. Amrollahi, M. Tabasi, and H.S. Kalal. The role of different parameters on the stability and thermal conductivity of carbon nanotube/water nanofluids. *Int Commun Heat Mass Transfer* 37, 319–323 (2010).
139. R. Walvekar, I.A. Faris, M. Khalid, and K. Lumpur. Thermal conductivity of carbon nanotube nanofluid—Experimental and theoretical study. *Heat Transfer Asia Res* 41, 145–163 (2012).
140. J. Wang, J. Zhu, X. Zhang, and Y. Chen. Heat transfer and pressure drop of nanofluids containing carbon nanotubes in laminar flows. *Exp Therm Fluid Sci* 44, 716–721 (2013).
141. S. Halelfadl, P. Estellé, and T. Maré. Heat transfer properties of aqueous carbon nanotubes nanofluids in coaxial heat exchanger under laminar regime. *Exp Therm Fluid Sci* 55, 174–180 (2014).
142. W. Yu, H. Xie, and D. Bao. Enhanced thermal conductivities of nanofluids containing graphene oxide nanosheets. *Nanotechnology* 21, 055705 (2010).
143. T.T. Baby and S. Ramaprabhu. Investigation of thermal and electrical conductivity of graphene based nanofluids investigation of thermal and electrical conductivity of graphene based nanofluids. *J Appl Phys* 108, 124308 (2010).
144. W. Yu, H. Xie, and W. Chen. Experimental investigation on thermal conductivity of nanofluids containing graphene oxide nanosheets. *J Appl Phys* 107, 094317 (2010).
145. S.S.J. Aravind and S. Ramaprabhu. Surfactant free graphene nanosheets based nanofluids by in-situ reduction of alkaline graphite oxide suspensions. *J Appl Phys* 110, 124326 (2011).
146. T.T. Baby and R. Sundara. Synthesis and transport properties of metal oxide decorated graphene dispersed nanofluids. *J Phys Chem C* 115, 8527–8533 (2011).
147. T.T. Baby and S. Ramaprabhu. Enhanced convective heat transfer using graphene dispersed nanofluids. *Nanoscale Res Lett* 6, 289 (2011).
148. S.S. Gupta, V.M. Siva, S. Krishnan, T.S. Sreeprasad, and P.K. Singh. Thermal conductivity enhancement of nanofluids containing graphene nanosheets thermal conductivity enhancement of nanofluids containing graphene nanosheets. *J Appl Phys* 110, 084302 (2011).
149. R. Zheng, J. Gao, J. Wang, and G. Chen. Reversible temperature regulation of electrical and thermal conductivity using liquid–solid phase transitions. *Nat Commun* 2, 286–289 (2011).
150. S.S.J. Aravind and S. Ramaprabhu. Graphene–multiwalled carbon nanotube-based nanofluids for improved heat dissipation. *RSC Adv* 3, 4199–4206 (2013).
151. M. Kole and T.K. Dey. Based nanofluids investigation of thermal conductivity, viscosity, and electrical conductivity of graphene based nanofluids. *J Appl Phys* 113, 084307 (2013).
152. A. Ghozatloo, M. Shariaty-Niasar, and A.M. Rashidi. Preparation of nanofluids from functionalized graphene by new alkaline method and study on the thermal conductivity and stability. *Int Commun Heat Mass Transfer* 42, 89–94 (2013).
153. G.-J. Lee and C.K. Rhee. Enhanced thermal conductivity of nanofluids containing graphene nanoplatelets prepared by ultrasound irradiation. *J Mater Sci* 49, 1506–1511 (2014).
154. M. Mehrali, E. Sadeghinezhad, S.T. Latibari, S.N. Kazi, M. Mehrali, M. Nashrul, B. Mohd, H. Simon, and C. Metselaar. Investigation of thermal conductivity and rheological properties of nanofluids containing graphene nanoplatelets. *Nanoscale Res Lett* 9, 15 (2014).

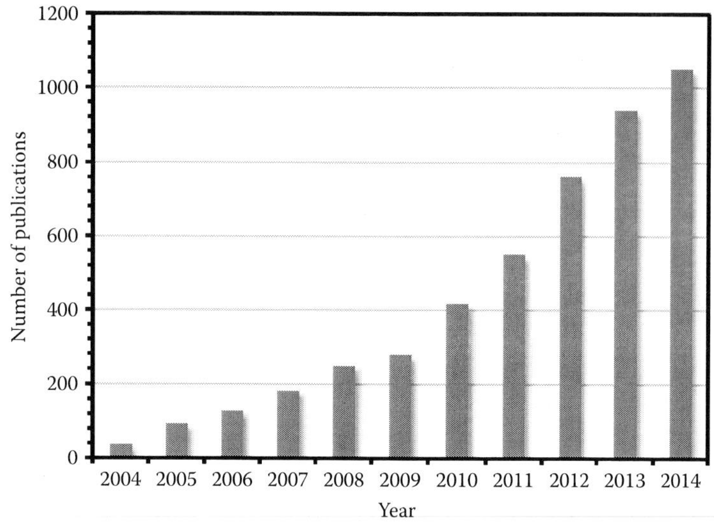

FIGURE 5.1 Number of publications related to nanofluid research per year. (Based on Taylor, R. et al., *J. Appl. Phys.*, 113(1), 011301, 2013.)

and governing mechanisms of thermal conductivity enhancement and viscosity increase in nanofluids. In order to give an idea about the rate of expansion of this field, the number of publications about nanofluids per year in the past decade is shown in Figure 5.1 [2].

Experimental studies regarding nanofluids have mostly utilized water, ethylene glycol, and engine oil as the base fluid, and metals and metal oxides as the nanoparticles with sizes ranging from ten nanometers to hundreds of nanometers. High aspect ratio particles such as carbon nanotubes and graphite flakes have also been investigated. There have been studies investigating both thermal conductivity and viscosity of nanofluids. Researchers reported different thermal conductivity enhancement values for similar nanofluids ranging from incremental enhancements that can be directly explained by the mean field theory to anomalous enhancements of more than 100% at particle volume fractions below 1%. Contradictory results have also been observed for the case of viscosity measurements, where different studies reported different dependencies of the viscosity on particle volume fraction, particle size, and temperature.

The experimental observation of the large thermal conductivity increase obtained using nanofluids led to many theoretical studies investigating the enhancement mechanisms of thermal transport. Proposed mechanisms include Brownian motion, clustering, liquid layering, ballistic transport, thermophoresis, and near-field radiation.

In this chapter, experimental and theoretical knowledge regarding nanofluid thermophysical properties is summarized. Effects of nanofluid parameters such as particle type, particle size, particle volume fraction, and temperature on the thermal conductivity and viscosity are discussed, providing insight to the mechanisms of thermal conductivity enhancement and viscosity increase as well as reasons of discrepancy among the reported values. Proposed mechanisms of thermal conductivity enhancement, including Brownian motion of nanoparticles, liquid layering around nanoparticles, and clustering of nanoparticles are described, and the current understanding regarding the significance of these mechanisms is explained.

5.1.1 Advantages of Nanofluids over Conventional Suspensions

The idea of adding high thermal conductivity particles into a liquid to improve its thermal conductivity is not new [3–5]. Until recently, such an approach has been limited to particles of sizes on the order of microns or millimeters. For such particle sizes, sedimentation, clogging, pressure drop, and tube wall erosion are severe; therefore, the application of these suspensions to commercial systems has not been feasible. With the recent advances in nanotechnology [6–8], mass production of nanoparticles has become possible, which enabled the possibility of utilizing these nanoparticles for improving heat transport in liquids. Due to the extremely small size of the nanoparticles, the associated liquid suspensions (nanofluids) do not have the problems of sedimentation, clogging, and tube wall erosion [9].

5.1.2 Nanofluid Preparation Techniques

Nanofluid preparation techniques can be divided into two main categories: single-step and two-step [10]. In the more commonly used two-step method, the nanoparticles are synthesized separately from the liquid matrix, and then they are added to the base liquid. Most nanofluids used in research activities are prepared by using the two-step approach. The two-step approach is compatible with mass production of nanofluids for future commercial applications.

In the two-step method, the nanoparticle synthesis is separate from the nanofluid preparation. Various techniques can be utilized for the nanoparticle synthesis such as mechanical grinding, inert-gas-condensation, chemical precipitation, chemical vapor deposition, spray pyrolysis, and thermal spraying [11]. Inert-gas-condensation [12] is one of the most popular techniques among these alternatives. The main disadvantage of the two-step techniques is the uncontrolled agglomeration of particles, which results in poor stability of the nanofluid [13].

Single-step methods are the methods where the nanoparticles are formed inside the liquid; therefore, the nanofluid is obtained in a single step. One of the most common single-step methods is the direct-evaporation technique [9,14–17]. Single-step methods usually result in better dispersion and stability characteristics when compared to the two-step method [11]. However, most of these single-step techniques are either not suitable for mass production, or are restricted to the synthesis of a specific type of nanofluid. Yu and Xie [10] and Haddad et al. [18] review the nanofluid preparation techniques in detail.

5.1.3 Stability and Dispersion Characteristics of Nanofluids

Stability of the nanoparticles inside the base liquid is critical for the practical application of nanofluids. Overall, nanofluids are superior in stability to suspensions of micron- or millimeter-sized particles in a base liquid. Researchers have reported the synthesis of nanofluids with excellent stability, with negligible settling after weeks of static storage [19,20]. However, especially in the nanofluids prepared by the two-step method, excessive agglomeration might take place resulting in large particles, which might eventually settle out of the nanofluid. This can be prevented by enhancing the dispersion characteristic of the nanofluid through adjusting the pH value [21] or using surfactants [22]. Existing agglomerates can be broken into smaller parts using ultrasonic vibration [20].

Hwang et al. [23] investigated the effect of different stirring techniques on the effectiveness of breaking the agglomerates into smaller particles. Carbon black/water and Ag/water nanofluids were prepared, and stirrer, ultrasonic vibrator, ultrasonic disrupter, and high pressure homogenizer were applied on the nanofluids. The smallest particle size was observed for the case of high pressure homogenizer. The average diameter of the carbon black particles for each stirring technique is summarized in Table 5.1. [23].

Surfactants are also very effective in improving the dispersion characteristics and, therefore, stability. Surfactants should be selected based on the nanoparticle type and the base liquid type. Some examples are oleylamine for carbon nanotube/decene [24], sodium dodecyl sulfate for CNT/water [25], salt and oleic acid for Cu/water [20], oleic acid and cetyltrimethylammonium-bromide for TiO_2/water [26] nanofluids. One disadvantage of the surfactants is that they might

TABLE 5.1
Dispersion Characteristics of Carbon Black/Water Nanofluid as a Result of the Application of Different Stirring Techniques

Stirring Technique	Average Particle Size (nm)
High pressure homogenizer	45
Ultrasonic disrupter	66
Ultrasonic bath	147
Stirrer	182
No physical treatment	585

Source: Hwang, Y. et al., *Powder Technol.*, 186(2), 145, 2008.

form a layer between the nanoparticle and the base liquid, which increases the thermal contact resistance between the particle and the base liquid.

The dispersion characteristics also affect the clustering characteristics of the nanoparticles, which in turn affects the thermal conductivity enhancement of the nanofluid (see Section 5.5.4). Nanofluids with finely dispersed nanoparticles provide better stability; on the other hand, suppressed clustering decreases the amount of thermal conductivity enhancement. Therefore, there is a compromise between the stability and high thermal conductivity. More detailed reviews of nanofluid preparation techniques and their stability characteristics are available in the literature [10,13,18,27].

5.2 DENSITY

Density of a nanofluid can be estimated by using the expression

$$\rho_{nf} = \phi\rho_p + (1-\phi)\rho_f, \tag{5.1}$$

where
- ρ is the density
- ϕ is the volume fraction of the nanoparticles
- Subscripts *nf* and *f* correspond to the nanofluid and to the base liquid, respectively

Pastoriza-Gallego et al. [28] indicate that the main source of error in nanofluid density calculations comes from the uncertainty in the density of the nanoparticles. Furthermore, the interaction between the nanoparticle and the liquid molecules might result in a deviation from the ideal colloid assumption [28], especially for very small nanoparticle sizes, that is, very large interface area. On the other hand, density measurements on Al_2O_3/water nanofluids [28] have shown that the error resulting from the aforementioned factors are below 0.008 g/cm^3. As discussed in the subsequent sections, there are much larger uncertainties regarding the thermal conductivity and viscosity of nanofluids in the literature. Therefore, Equation 5.1 is appropriate for all practical calculations regarding the thermofluidic applications of nanofluids.

5.3 SPECIFIC HEAT CAPACITY

Specific heat capacity of a nanofluid can be determined by the following expression [29,30]:

$$(\rho c_p)_{nf} = \phi(\rho c_p)_p + (1-\phi)(\rho c_p)_f, \tag{5.2}$$

where c_p is the specific heat capacity. Equation 5.1 should be substituted for the density of the nanofluid. Zhou and Ni [31] measured the specific heat capacity of Al_2O_3/water nanofluids and verified that Equation 5.2 is in good agreement with the experimental data. It should be noted that the volume fraction based averaging of specific heat does not predict the specific heat capacity of solid–liquid mixtures accurately.

5.4 THERMAL CONDUCTIVITY

Addition of nanoparticles into a base liquid dramatically increases the thermal conductivity of the liquid. Therefore, thermal conductivity is the key thermophysical property that makes nanofluids attractive for thermal applications.

There are many parameters that affect the thermal conductivity of nanofluids. These include but are not limited to particle material, base fluid, particle volume fraction, particle size, particle shape, and temperature. Furthermore, the parameters related to the preparation of nanofluids with well-dispersed nanoparticles discussed in Section 5.1.3 might affect the clustering behavior (see Section 5.5.4), which in turn affects the thermal conductivity of the nanofluid. In this section, some of the important nanofluid parameters and their effect on the thermal conductivity are discussed with an emphasis on their relationship to the clustering characteristics of the nanofluids.

5.4.1 Measurement Techniques

Transient hot-wire technique is the most commonly used approach for measuring the thermal conductivity of nanofluids [32–35]. For conducting fluids such as nanofluids prepared using metallic particles, the hot wire is required to be coated with a thin layer of electrically insulating layer of preferably high thermal conductivity [19]. In case of using an uninsulated wire, Zhang et al. [36] have observed an artificial increase of ~8% in the thermal conductivity of Au/toluene nanofluid. Das et al. [37] argued that the electrical field around the hot wire might increase the concentration of ions around the wire, creating a measurement artifact.

Some other techniques used for measuring the thermal conductivity of nanofluids are 3-ω method [38,39], steady-state coaxial cylinders method [40], steady-state parallel plate method [41], optical beam deflection technique [42], and temperature oscillation technique [37,43].

Schmidt et al. [44] measured the thermal conductivity of Al_2O_3/decane nanofluids and showed that the transient hot-wire technique and an optical technique [45] provide consistent results. Also, the temperature oscillation technique has been shown to be in agreement with the hot-wire technique [19,37]. Buongiorno et al. [46] performed a benchmark study by comparing transient hot-wire method, steady state methods, and optical methods. All the considered techniques provided consistent results, especially when normalized by the base liquid conductivity measurements.

5.4.2 Effect of Particle Material and Base Liquid

Table 5.2 lists the thermal conductivities of several materials that are commonly used in nanofluid preparation [13]. In almost all of the possible combinations of nanoparticles and base liquids, the nanoparticle thermal conductivity is significantly higher than the thermal conductivity of the base liquid. This implies that in the case of a temperature gradient, the temperature inside a nanoparticle will be practically constant, whereas the majority of the temperature variation will take place through the liquid medium [13,47]. Therefore, especially for nanofluids with well dispersed particles, the thermal conductivities of the base liquid and the nanoparticle are not expected to be the main determining factors for the thermal conductivity enhancement ratio. An increase or decrease in the particle's thermal conductivity is not going to significantly change the temperature distribution in the domain (unless the particle has comparable thermal conductivity to that of the base liquid).

TABLE 5.2
Room Temperature Thermal Conductivity of Some Liquids Used as Base Liquid and Some Solids Used as Nanoparticles for Nanofluid Preparation

Material	Thermal Conductivity (W/m·K)
Water	0.613
Ethylene glycol	0.253
Engine oil	0.145
Silver	429
Copper	401
Gold	318
Aluminum	237
Iron	80
Diamond	3300
Carbon nanotube	3500 [53]
Graphene	3000–5300 [54,55]
Silicon	148
Alumina	40

Source: Based on Eastman, J.A. et al., *Annu. Rev. Mater. Res.*, 34(1), 219, 2004.

On the other hand, nanoparticles are known to form clusters [17,48,49], and these clusters are very effective at further improving the thermal conductivity of the nanofluid [50,51]. In the case of clustering, high thermal conductivity paths are present, which are formed by chains of nanoparticles [52]. In this case, nanoparticles with very high thermal conductivity will be able to provide much more effective heat transfer along the chains, improving the overall heat transport performance of the nanofluid. There are also secondary effects due to the fact that the nanoparticle material and the base fluid determine the zeta potential and pH of the nanofluid, which affect the clustering dynamics [21], which in turn alters the thermal transport properties of the nanofluid.

5.4.3 Effect of Particle Volume Fraction

Volume fraction of the nanoparticles in a nanofluid is the primary parameter that determines the thermal conductivity enhancement. Most research findings show a linear relationship between the particle volume fraction and the thermal conductivity [17,19,37,56]. These findings imply that there is no interaction between individual nanoparticles (or between nanoparticle clusters) in the associated nanofluid samples. On the other hand, there are studies that show a decrease in the rate of thermal conductivity enhancement with particle volume fraction at higher particle volume fractions [34,49]. One possible explanation for the nonlinearity is the overlapping and interaction of the thermal fields around the individual nanoparticles. In addition, a change in the volume fraction might nonlinearly affect the clustering behavior. Since cluster size and morphology dominate the thermal conductivity of nanofluids (see Section 5.5.4), this might result in a nonlinear relationship between the particle volume fraction and the thermal conductivity. Especially the nanofluids that have large aspect ratio particles such as carbon nanotubes (CNT) show the nonlinear behavior more frequently, and at relatively smaller particle volume fractions [25,57,58]. This might be explained by the fact that the threshold particle volume fraction for forming interacting percolation networks of carbon nanotubes is much smaller than that required for forming similar networks by spherical nanoparticles [13,59].

5.4.4 Effect of Particle Size

Particle size is another important parameter that affects the thermal conductivity of a nanofluid. Because of the phonon scattering at the solid–liquid interface, thermal conductivity of solids decreases with decreasing particle size for sizes below 50 nm [35].

When experimental data of similar nanofluids with different particle sizes [9,60–65] are compared, it is seen that the thermal conductivity enhancement of nanofluids usually increase with decreasing particle size, indicating that there are other mechanisms that overcome the effect of size on solid thermal conductivity. Mechanisms of Brownian motion (see Section 5.5.2) and liquid layering around nanoparticles (see Section 5.5.3) predict increasing enhancement with decreasing particle size, which is in qualitative agreement with these experimental findings. However, as discussed in Section 5.5, these mechanisms are not very successful at explaining the high thermal conductivity enhancements obtained with nanofluids in a physically meaningful way, and therefore they cannot dominate the particle size effect directly. However, Brownian motion might interact with the clustering mechanism (see Section 5.5.6) to increase the thermal conductance of the cluster branches. This secondary effect might explain the particle dependence since Brownian motion becomes more effective with decreasing particle size. In addition, Feng et al. [66] observed that nanofluids with smaller nanoparticles have more pronounced clustering, which is in agreement with the aforementioned experiments regarding particle size dependence.

There are also experimental studies that showed the opposite particle size dependence, that is, decreasing thermal conductivity enhancement with decreasing particle size [35,67]. Assuming a nanofluid with well dispersed nanoparticles, such a trend can be explained by the decrease of thermal conductivity of nanoparticles with decreasing diameter [35]. A similar trend is expected when the interfacial thermal resistance between the nanoparticle and the liquid matrix is high and becomes the limiting factor in the thermal conductivity of a nanofluid. In such a case, decreasing particle size increases the total interfacial area for given particle volume fraction, decreasing the overall thermal transport efficiency of the nanofluid.

5.4.5 Effect of Particle Shape

Based on the effective medium theory [68,69] and Maxwell model [70], Hamilton and Crosser [5] proposed the following expression for calculating the thermal conductivity of suspensions of non-spherical particles in a liquid.

$$k_{nf} = \frac{k_p + (n-1)k_f - (n-1)\phi(k_f - k_p)}{k_p + (n-1)k_f + \phi(k_f - k_p)} k_f, \tag{5.3}$$

where

k is the thermal conductivity

n is the empirical shape factor defined as

$$n = \frac{3}{\psi}, \tag{5.4}$$

where ψ is the sphericity. Sphericity is the surface area of a sphere that has the same volume as the particle divided by the surface area of the particle.

Hamilton and Crosser model [5] predicts that high aspect ratio particles are thermally more efficient than spherical particles of the same volume fraction. For both rod-shaped and disc-shaped particles, the heat can be transported more efficiently over longer distances when compared to spherical particles; therefore Hamilton and Crosser model predicts a similar enhancement for both cases. However, for disc-shaped particles, the interfacial area increases more dramatically with increasing aspect ratio than that of rod-shaped particles. Interfacial thermal resistance is not taken into account in the Hamilton and Crosser model [5] since the model is originally developed for micron- and

increases in enhancement with increasing temperature is questionable, although some change in enhancement is possible due to the variations in the thermal conductivity of the liquid matrix and the nanoparticle with temperature. On the other hand, if clustering is taking place, then the severity of clustering as well as the general morphology of the clusters might change with temperature. Since the parameters associated with clustering affect the heat transport efficiency drastically, the temperature becomes a parameter that can significantly alter the thermal conductivity enhancement of nanofluids.

Another parameter that might induce temperature dependence is the interfacial thermal resistance between the nanoparticles and the liquid matrix. Currently no complete model that can predict the thermal behavior at a solid–liquid interface exists [74].

5.4.7 Conclusions

Classical models [5,70] for conventional suspensions of particles in liquids predict the thermal conductivity of the mixture as a function of the thermal conductivity of the constituents, the particle volume fraction, and the shape of the particles. For the case of nanofluids, many other parameters such as particle size, temperature, and pH value affect the thermal conductivity through the changes in the clustering behavior. Therefore, the controversies regarding the effects of these parameters can be resolved by focusing on the effects of these parameters on the clustering characteristics. Extensive reviews of the effects of different parameters on thermal conductivity are available in the literature [11,13,56,67,73,83–96].

5.5 MECHANISMS OF THERMAL CONDUCTIVITY ENHANCEMENT

High thermal conductivity of nanofluids triggered many studies that investigate the heat transport mechanisms in the nanofluids. In this section, a summary of the associated findings is presented.

5.5.1 Classical Models for Suspensions of Particles in a Liquid

There are many theoretical models for predicting the thermal conductivity of suspensions of solid particles in liquids. These models are developed for particle sizes on the order of microns or for larger particles; therefore, they do not take the effect of some nanoscale phenomenon such as Brownian motion, liquid layering, and Kapitza resistance into account.

One of the most useful expressions for the thermal conductivity of a suspension of solid particles in a liquid is developed by Maxwell [69] based on effective medium theory. The thermal conductivity is given as

$$k_{nf} = \frac{k_p + 2k_f + 2(k_p - k_f)\phi}{k_p + 2k_f - (k_p - k_f)\phi} k_f, \tag{5.5}$$

where

k is the thermal conductivity

ϕ is the particle volume fraction

Subscripts nf, p, and f correspond to the nanofluid, the particle, and the base liquid, respectively

Many experimental studies discussed before have illustrated that the Maxwell model [69] of thermal conductivity severely underestimates the thermal conductivity of nanofluids. As a consequence, there have been extensive efforts toward understanding the underlying mechanisms responsible for the high thermal conductivity enhancement obtained by the presence of nanoparticles in a liquid. The major mechanisms that have been proposed are Brownian motion of nanoparticles, liquid layering around nanoparticles, and clustering of nanoparticles. In this part of the chapter, these mechanisms are discussed in terms of relevant experimental, theoretical, and numerical research available in the literature.

5.5.2 Brownian Motion of Nanoparticles

Brownian motion is the random motion of particles suspended inside a liquid. By using the Einstein diffusion theory, the velocity of a particle can be written as [97]:

$$V_{Br} = \frac{k_b T}{3\pi\mu_f d_p l_f}, \tag{5.6}$$

where

k_b is the Boltzmann constant
T is temperature in K
μ_f is the base liquid viscosity
d_p is the particle diameter
l_f is the mean free path of base liquid molecules
Subscript f indicates the base liquid

Since particle diameter is on the order of nanometers in nanofluids, the Brownian motion is much more significant when compared to conventional suspensions of micron-sized particles. Therefore, researchers suggested that the Brownian motion might be partly responsible for the high thermal conductivity enhancement in nanofluids [98].

There are two possible contributions of the Brownian motion to thermal transport. First, the nanoparticle itself can enhance the heat transport through its own motion. Second, as the nanoparticle moves, it can drag fluid molecules around it which can create a micro-convection effect [98].

The first mechanism, that is the direct effect of the motion of the nanoparticles on thermal diffusion, can be analyzed by considering the relative significance of the diffusion of the liquid molecules and the diffusion of the nanoparticles [47]. Diffusion coefficient for a particle can be written as [47]

$$D = \frac{k_b T}{3\pi\mu_f d_p}. \tag{5.7}$$

One can compare the time required for the particles and the liquid molecules to move a distance of particle diameter, τ_p, and τ_f, respectively [47]:

$$\tau_p = \frac{d^2}{6D} = \frac{3\pi\mu_f d^3}{6k_b T}, \tag{5.8}$$

and

$$\tau_f = \frac{d^2}{6\alpha_f} = \frac{d^2\rho_f c_{p,f}}{6k_f}, \tag{5.9}$$

where

ρ is the density
α is the thermal diffusivity
k is the thermal conductivity

By considering water as the base liquid and the particle size as 10 nm, Keblinski et al. [47] have shown that the liquid diffusion is about 500 times more effective than the particle diffusion. Therefore, the direct contribution of the Brownian motion is a negligible effect when compared to the thermal conductivity of the base liquid [74]. The insignificance of the first mechanism of Brownian motion has also been illustrated by molecular dynamics (MD) simulations [47]. A thermal conductivity model has been developed based on the first mechanism of Brownian motion enhanced thermal transport by Kumar et al. [99]. Although this model predicts some experimental data well, an empirical constant of unphysical magnitude has been employed to obtain the presented results.

The fact that the aforementioned mechanism is not effective does not directly imply that the secondary mechanism of micro-convection is also negligible. Even though the particle itself is not efficient in transporting heat through its motion, it can drag many fluid molecules during its motion and create a mixing effect in the liquid. Jang and Choi [98,100] have developed a model based on this micro-convection concept. In this model, the heat conduction in the base liquid, heat conduction inside the nanoparticle, collisions between the nanoparticles, and the micro-convection effect are all taken into account. The model's prediction for the nanofluid thermal conductivity is [98]

$$k_{nf} = k_f(1-\phi) + k_p^*\phi + 3C_l\frac{d_f}{d_p}k_f Re_d^2 Pr_f\phi, \quad (5.10)$$

where

- C_l is an empirical constant
- d_f is the diameter of the fluid molecules
- Pr is the Prandtl number

k_p^* takes the effect of the Kapitza resistance into account empirically, that is, $k_p^* < k_p$. Reynolds number of the Brownian motion of the nanoparticle, Re_d, is defined as

$$Re_d = \frac{V_{Br}d_p}{\nu_f}, \quad (5.11)$$

where

- ν_f is the kinematic viscosity of the base fluid
- V_{Br} is the average velocity of the particles due to Brownian motion, which can be determined by using Equation 5.6

There are many other theoretical models based on the micro-convection effect in the literature [56,62,101–103]. More importantly, there have been many studies that investigated the significance of the contribution of the micro-convection effect. Sarkar and Selvam [104] performed MD simulations and claimed that the micro-convection is an important component of heat transfer enhancement obtained with nanofluids. Babaei et al. [105] has investigated the same problem through MD simulations and agreed that some components of the heat current autocorrelation function can actually be significant. However, they also proved that these significant terms cancel each other when treated properly and the overall effect of the micro-convection is negligible. Evans et al. [106] considered a limiting case in which all fluids molecules are affected by the micro-convection, which is an obvious overestimation of the real phenomenon. As a result of the analysis, they determined a thermal conductivity enhancement of 0.032% and concluded that the micro-convection effect is negligible.

The strength of the Brownian motion related effects (both the direct effect and the micro-convection) increase with increasing Brownian velocity. Then, based on Equation 5.6, one can predict that increasing temperature and decreasing particle size increase the Brownian motion effect on the thermal conductivity. As discussed in Sections 5.4.4 and 5.4.6, there are many experimental studies that are in qualitative agreement with this prediction. However, most of these studies do not provide insight to the effects of particle size and temperature on other enhancement mechanisms such as nanoparticle clustering and ballistic transport. On the other hand, there are a limited number of experimental studies, which provide direct insight to the possible contribution of Brownian motion. Eapen et al. [50] have analyzed the effect of the particle material density on the thermal conductivity of nanofluids by using silica particles and particles of a copolymer of tetrafluoroethylene and perfluoro-methylvinylether. The micro-convection argument implies a strong density dependence of the thermal conductivity enhancement (see Equation 5.6). Such an enhancement has not been observed in this work, which is considered as an indication of the insignificance of the Brownian motion.

A number of studies focused on obtaining nanofluids with well-dispersed particles. Since these investigations eliminated the possibility of thermal conductivity enhancement due to nanoparticle clustering, the main enhancement is expected to come from the Brownian motion. All of these studies [36,42,106–110] have indicated enhancement values that are in agreement with the Maxwell model [70], indicating that the micro-convection mechanism is not effective. Another parameter that is supposed to affect the enhancement values through Equation 5.6 is the viscosity of the base fluid. Phuoc et al. [111] investigated this possibility by systematically varying the viscosity of MWCNT/water nanofluid by adjusting the surfactant amount in the samples. They observed that the thermal conductivity enhancement is independent of the base fluid viscosity showing that Brownian motion is a negligible component.

A very effective way of eliminating Brownian motion is to perform experiments on frozen nanofluids [74,112]. This provides a controlled experiment where most of the other proposed mechanisms still continue to operate whereas Brownian motion is completely removed from the picture. The experiments on Al_2O_3/animal oil and Al_2O_3/hexadecane indicated a slight decrease and some increase in thermal conductivity upon freezing, respectively [74,112]. Researchers explained these trends through the percolation dynamics during freezing, showing that Brownian motion does not dominate the thermal conductivity enhancement.

5.5.3 Liquid Layering around Nanoparticles

Research has shown through x-ray reflectivity experiments [113] that liquid molecules form a layered structure around nanoparticles with an order level in between that of the solid and the liquid. Yu and Choi [114] argued that the thermal conductivity of this structured layer might have a value in between the thermal conductivity of the liquid and the solid. They developed a theoretical model based on the Maxwell model [70] to predict the additional enhancement that would be obtained through this mechanism. The nanoparticle and the structured layer were transformed into a larger particle with an effective thermal conductivity by using effective medium theory [115]. The expression for the nanofluid thermal conductivity is then as follows

$$k_{nf} = \frac{k_{pe} + 2k_f + 2(k_{pe} - k_f)(1+\beta)^3\phi}{k_{pe} + 2k_f - (k_{pe} - k_f)(1+\beta)^3\phi} k_f, \tag{5.12}$$

where k_{pe} is the thermal conductivity of the equivalent nanoparticle

$$k_{pe} = \frac{[2(1-\gamma) + (1+\beta)^3(1+2\gamma)]\gamma}{-(1-\gamma) + (1+\beta)^3(1+2\gamma)} k_p, \tag{5.13}$$

where

$$\gamma = \frac{k_l}{k_p}, \tag{5.14}$$

and k_l is the thermal conductivity of the nanolayer. β is defined as

$$\beta = \frac{t}{r_p}, \tag{5.15}$$

where
t is the nanolayer thickness
r_p is the nanoparticle radius

Yu and Choi [116] extended this approach to non-spherical particles by utilizing the Hamilton and Crosser [5] model. There have been many other studies that developed a theoretical model by taking the effect of the liquid layering into account [117–121]. In all these models, the key parameters are the thickness and the thermal conductivity of the layer. Eastman et al. [13] have shown that for a nanofluid with particle size of 10 nm, and assuming liquid layer thermal conductivity being equal to the particle thermal conductivity (an extreme scenario), the layer thickness required for 100% enhancement is about 2.5 nm. The experimental work [113] shows that the liquid layers around solid particles have a thickness of only a few atoms (~1 nm), whereas many theoretical modeling studies have utilized much thicker layers in order to be able to fit the experimental data. Since thermal conductivity increase due to the layering has not been quantified, these models also use mostly arbitrary values of layer thermal conductivity. Using molecular dynamics simulations, Xue et al. [122] have shown that the ordering around a nanoparticle does not significantly affect the thermal conductivity in the associated region indicating that the liquid layer's thermal conductivity is very close to the thermal conductivity of the base liquid, and that the layering cannot be responsible for the high thermal conductivity enhancement obtained with nanofluids. Wilson et al. [123] investigated the interfacial resistance for Au, Pt, and AuPd nanoparticles in water and in toluene and found no indication of a highly conductive layer surrounding the particles.

To summarize, although the concept of liquid layering around nanoparticles has been experimentally observed, the extent of this layering and the thermal conductivity of the ordered region do not seem to be able to explain the high thermal conductivity enhancements obtained in nanofluids.

5.5.4 Clustering of Nanoparticles

The most important mechanism proposed for explaining the thermal conductivity enhancement in nanofluids is clustering of nanoparticles. According to this mechanism, the nanoparticles in a liquid aggregate to form fractal-shaped clusters, which provide numerous paths of high conductivity over relatively long distances when compared to the particle size. This behavior is schematically illustrated in Figure 5.4 [51,124]. Along the high conductivity paths, the heat transport is very efficient, and these paths are separated by a relatively small number of liquid molecules within a cluster. Therefore, one can consider each cluster as a much larger particle having a thermal conductivity in between that of the nanoparticle and the base liquid.

As discussed by Wen et al. [86], this mechanism predicts a specific level of clustering for optimum thermal conductivity enhancement. Figure 5.5 qualitatively illustrates this behavior [13,86]. In the well-dispersed limit, that is, no clustering, each nanoparticle is surrounded by many liquid molecules; therefore, chains of particles that form the high thermal conductivity paths are not present. Effective medium theory [68–70] can be utilized to estimate the thermal conductivity enhancement for this configuration. When the particle concentration is small, and $k_p \gg k_f$, the effective medium theory predicts [51,69]

$$k_{nf,low} = k_f(1+3\phi). \tag{5.16}$$

It should be noted that this case corresponds to the prediction of the Maxwell model [70] for $k_p \gg k_f$ (see Equation 5.5).

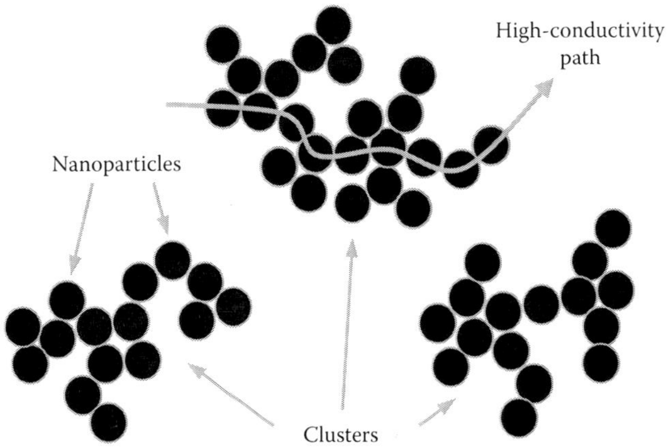

FIGURE 5.4 Illustration that shows the clustering of the nanoparticles, and a high thermal conductivity path that is formed as a result of the clustering. (Based on Eapen, J. et al., *J. Heat Transfer*, 132(10), 102402, 2010; Özerinç, S., *Heat Transfer Enhancement with Nanofluids*, Middle East Technical University, Ankara, Turkey, 2010.)

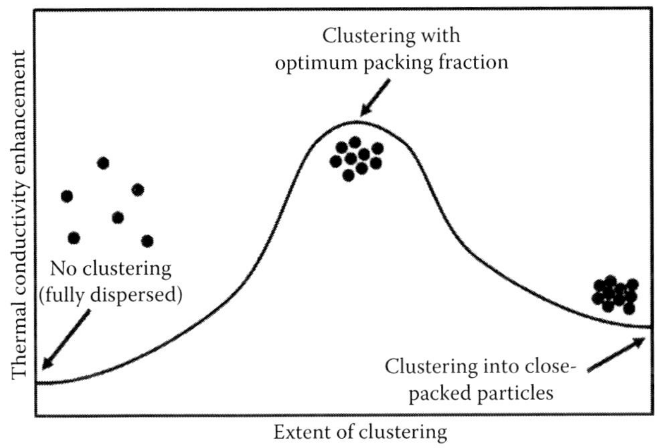

FIGURE 5.5 A qualitative plot showing the relationship between the extent of clustering and the thermal conductivity enhancement. (Adapted from Eastman, J.A. et al., *Annu. Rev. Mater. Res.*, 34(1), 219, 2004; Wen, D. et al., *Particuology*, 7(2), 141, 2009.)

At the other extreme of the clustering behavior, there is the fully aggregated case: The particles are close-packed, and they behave similar to a case of a smaller number of larger particles. This is somewhat similar to the well dispersed scenario; there are clusters (which can be considered as larger particles) and the overall solid volume fraction of the system has not changed [70]. However, it should be noted that even if the particles are close-packed, there are fluid molecules in between the particles, and the heat transport is very efficient within the cluster since the thermal paths over fluid molecules are very short. Keblinski et al. [69] note that for the case of randomly close-packed spherical particles, the *packing fraction* is about 60%, which is defined as the ratio of the volume of the solid particles to the volume formed by the boundaries of the cluster. Therefore, about 40% of the volume is occupied by the liquid molecules, which implies that there is more than 50% increase in the volume that has a very high *effective* thermal conductivity. Consequently, the enhancement for the full aggregation case is expected to be higher than that of the fully dispersed scenario.

In between these two extremes, the clustering is present with a smaller packing fraction, with more loosely aggregated nanoparticles. This further increases the *effective* particle volume fraction of the nanofluid (i.e., volume fraction of the clustered structures whose boundaries are defined by the outermost particles of the cluster), increasing the thermal conductivity beyond that of the full aggregation case. For the same volume fraction, now there are longer paths of high thermal conductivity which improves the thermal conductivity of the nanofluid beyond that is predicted by the Maxwell model [70].

A comparison between the predictions of the effective medium theory and the experimental data available in the literature provides further insight into the aforementioned levels of clustering [69]. Figure 5.6 illustrates this comparison in terms of the variation of the thermal conductivity enhancement with particle volume fraction for water-based nanofluids with Al_2O_3, SiC, and CuO nanoparticles [19,35,37,38,62,75,125–127]. The lower dashed line indicates the prediction of the effective medium theory for well dispersed particles (no clustering limit of Figure 5.5). The line in the middle indicates the estimation of the same theory for randomly close-packed particles (full aggregation limit of Figure 5.5). The upper dashed line corresponds to a more loosely packed case of 40% packing fraction. It is seen that there is good agreement between the predictions of the effective medium theory and the data for most of the experimental results [69].

Theoretically, there is more space for further enhancement without going beyond the predictions of the effective medium theory, which might explain some experimental data in Figure 5.6 with higher enhancement values. For given volume fraction and same packing fraction, a clustering with fractal-like branches can further improve the enhancement when compared to a loose aggregation case with spherical-shaped clusters [51]. In the limiting case of forming branches, the clusters can assume the shape of a percolation network [51]. Eapen et al. [51] modeled the thermal transport in this limiting case by considering the nanofluid as a composite of liquid regions inside a solid matrix, and calculated the associated thermal conductivity enhancement. The size of the liquid regions

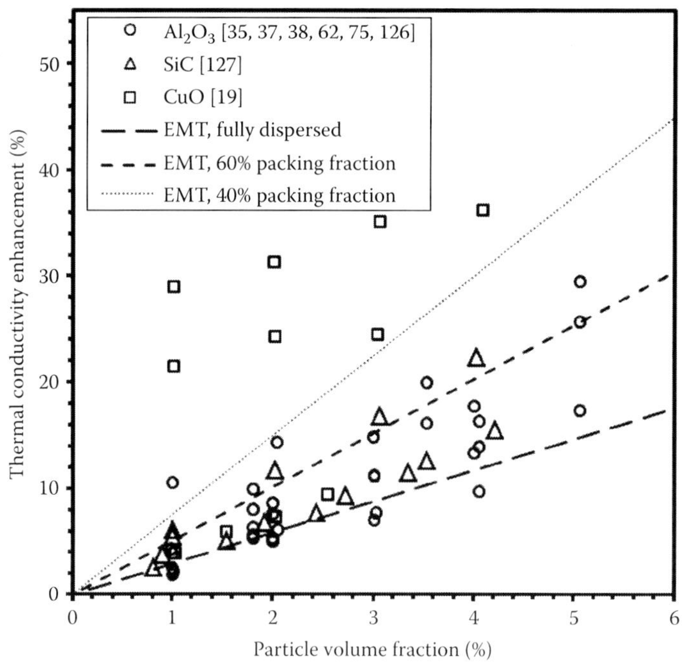

FIGURE 5.6 Comparison of the predictions of the effective medium theory (EMT) and some experimental data for water-based nanofluids from the literature. (Based on Özerinç, S. et al., *Microfluid. Nanofluid.*, 8(2), 145, 2010; Keblinski, P. et al., *Mater. Today*, 8(6), 36, 2005.)

corresponds to the size of the nanoparticle chains that form the percolation network. The associated expression for thermal conductivity enhancement is [51]

$$k_{nf} = k_f\left(1+\frac{2}{3}\frac{k_p}{k_f}\right). \tag{5.17}$$

Eapen et al. [51] compares the upper limit predicted by Equation 5.17 and the lower limit predicted by Equation 5.16 with the experimental data in the literature, and shows that almost all data lies in between the well-dispersed limit and the percolation limit. This finding indicates that the principles of the effective medium theory are sufficient to explain the "anomalous" thermal conductivity enhancements obtained with nanofluids.

One convincing aspect of the mechanism of clustering of nanoparticles is that the clustering of the nanoparticles has already been observed in many studies through scanning electron microscopy and transmission electron microscopy [17,48,49]. Furthermore, the proposed mechanism is fully capable of explaining the high thermal conductivity enhancement in the nanofluids by using the well-established effective medium theory [68–70] by simply considering the rearrangement of the nanoparticles in a heat transport-effective form. Therefore, the mechanism does not need any further assumptions and it does not propose questionable empirical parameters to match experimental data, which is the case for the other proposed mechanisms (see Sections 5.5.2 and 5.5.3).

There are many experimental studies that support the significance of the clustering mechanism. In specific circumstances, the clustering may actually result in a percolated network similar to the configuration modeled by Eapen et al. [51]. Zheng et al. [128] has shown that for graphite flake-based nanofluids, the volume fraction threshold for the formation of the percolation networks can be determined by the slope change in the thermal conductivity data. They reported a percolation threshold of 0.07 vol.% for the graphite/ethylene glycol nanofluid [129].

Further insight regarding the effect of clustering on thermal conductivity has been gained by altering the clustering behavior of the nanofluids. Lee et al. [130] studied the effect of particle surface charge on the thermal conductivity and they observed that the particle surface charge dramatically affects the thermal conductivity. Since the biggest effect of the surface charge is on the clustering dynamics of nanoparticles, it can be argued that the finding indicates the importance of the clustering on the thermal conductivity. Putnam et al. [42] considered well dispersed fullerenes in toluene and Au nanoparticles in ethanol and toluene, that is, no clustering. They did not observe any anomalous enhancement in the thermal conductivity and the results were in agreement with the predictions of the effective medium theory assuming well dispersed particles.

The particle shape might also affect the significance of the clustering mechanism. Keblinski et al. [69] have indicated that the spherical particles are the most effective in enhancing the thermal conductivity through clustering. Particles with high aspect ratios, such as carbon nanotubes provide very long conduction paths without clustering, and clustering of such particles might actually decrease the overall heat transport efficiency when compared to the well dispersed case [69].

Many experimental studies indicate that the thermal conductivity enhancement of nanofluids increase with increasing temperature and decreasing particle size (see Sections 5.4.4 and 5.4.6). It is known that clustering increases with increasing temperature and decreasing particle size [131], therefore, the clustering mechanism can also explain such dependencies of thermal conductivity [132].

5.5.5 Combined Effects of Ballistic Phonon Transport and Clustering

Although ballistic phonon transport alone cannot explain the high thermal conductivity enhancement obtained with nanofluids, it might improve the impact of the clustering mechanism on the thermal conductivity. The effectiveness of heat transport along the branches of the cluster might decrease when the nanoparticles are not in direct contact. On the other hand, the phonon mean free path in liquids is on the order of a few nanometers [13], and therefore, the heat might get ballistically transported from a particle to an adjacent particle efficiently over small distances [13]. Therefore, a cluster that has an average particle separation on the order of a few nanometers might thermally behave similarly to a cluster of particles with solid contact.

5.5.6 Combined Effects of Brownian Motion and Clustering

As discussed in Section 5.5.2, most of the experimental and computational studies indicate that Brownian motion and associated micro-convection effects are not sufficient to explain the thermal conductivity enhancement of the nanofluids. However, similar to the case of ballistic phonon transport, Brownian motion might contribute to heat transport through clustering, especially when the nanoparticles are separated from each other by a small distance [13]. Brownian motion helps the particles to have some relative motion with respect to each other with occasional solid contact, which would enhance the heat transport. This might partly explain the experimental trends of increasing thermal conductivity enhancement with increasing temperature and decreasing particle size.

5.6 VISCOSITY

The main advantage of the nanofluids is their high thermal conductivity, which provides an effective approach to improving the efficiency of thermal systems. Adding particles into a liquid also increases the effective viscosity [133], which would increase the pressure drop in systems based on forced convection heat transfer. Since high pressure drop increases the running cost of a system due to the increase in the pumping power, viscosity of nanofluids is a very critical parameter for determining the feasibility of nanofluids for heat transfer applications. Depending on the relative significance of the thermal conductivity enhancement and the viscosity increase, the application of a nanofluid might become disadvantageous.

5.6.1 Classical Models for Suspensions of Particles in a Liquid

Dynamic viscosity of dilute suspensions of spherical particles in a base liquid can be estimated by using the following expression provided by Einstein [134]:

$$\mu_{nf} = (1 + 2.5\phi)\mu_f, \tag{5.18}$$

where μ is the dynamic viscosity. This expression assumes rigid particles and no interaction between the particles [135]. Brinkman [136] proposed an alternative formula that converges to the case of a solid at 100% volume fraction:

$$\mu_{nf} = \frac{1}{(1-\phi)^{2.5}}\mu_f. \tag{5.19}$$

For higher particle volume fractions ($\phi > 1\%$), the hydrodynamic interactions between the particles affect the viscosity [135]. Batchelor [137] proposed a modified version of Equation 5.18, which takes these interactions into account:

$$\mu_{nf} = (1 + 2.5\phi + 6.2\phi^2)\mu_f. \tag{5.20}$$

As the particle volume fraction increases, the higher order interactions become more pronounced. Krieger and Dougherty [135,138] formulated the following semi-empirical relationship for these cases,

$$\frac{\eta_{nf}}{\eta_f} = \left(1 - \frac{\phi}{\phi_m}\right)^{-2.5\phi_m}, \tag{5.21}$$

where
η is the shear viscosity
ϕ_m is the maximum particle packing fraction [135]

5.6.2 Effects of Nanofluid Parameters

In almost all of the experimental studies [30,39,79,82,111,139–143], the nanofluid viscosity has been observed to be higher than that is predicted by the classical models. Figure 5.7 shows the viscosity ratio (μ_{nf}/μ_f) as a function of particle volume fraction [85] for water-based Al_2O_3, TiO_2, and SiC nanofluids [39,41,60,79,144,145]. It is seen that the measured viscosities are significantly higher than the prediction of the Einstein model [134] (Equation 5.18).

Particle volume fraction is the major parameter that affects the viscosity of a nanofluid. Regarding this dependence, there is agreement in the fact that viscosity increases monotonically with increasing particle volume fraction [146]. Some of the associated studies report a linear relationship between the viscosity and the particle volume fraction [60,82,147,148]. The linear relationship implies that each particle contributes to the viscosity rise independently and there is no significant interaction among them.

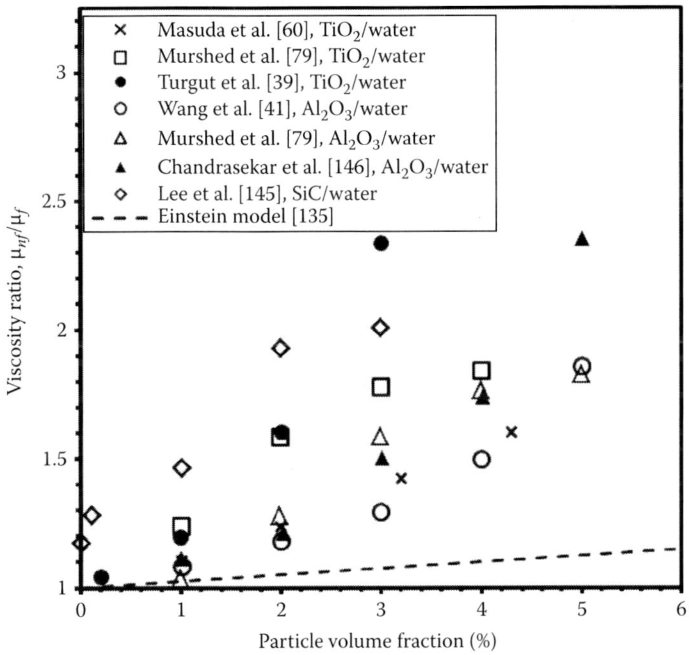

FIGURE 5.7 Experimental data of viscosity ratio for water-based nanofluids as a function of particle volume fraction. (Based on Murshed, S.M.S. et al., *Appl. Therm. Eng.*, 28(17–18), 2109, 2008.)

There are also studies that indicate nonlinearities in the volume fraction dependence of the viscosity [39,79,145,149–151]. These nonlinearities can be attributed to the interactions between the nanoparticles. In addition, the aggregation behavior of the nanoparticles might induce further nonlinearities, since aggregation becomes more pronounced with increasing particle volume fraction.

The second parameter that might affect the viscosity is the temperature. Viscosity of liquids generally decreases with increasing temperature due to the reduction in the force interactions between the neighboring liquid molecules [152]. The more important parameter here is the behavior of the viscosity ratio, μ_{nf}/μ_f, as a function of temperature. There is both qualitative and quantitative discrepancy among the data available in the literature regarding this parameter. In most studies, the viscosity variation with temperature is similar to that of the base liquid, and therefore, the viscosity ratio is approximately constant with temperature [80,139,153–155]. In some of these studies, temperature dependence has been observed beyond a certain volume fraction [139,155]. For example, Nguyen et al. [139] considered the viscosity of water-based Al_2O_3 and CuO nanofluids, and observed a significant decrease in viscosity ratio with increasing temperature for particle volume fractions higher than 7%. More interestingly, in their study [139], the rate of decrease of the viscosity ratio diminished with increasing temperature, and the ratio started to increase for $T > 45°C$. The fact that the viscosity variations with temperature are observed mostly for high volume fractions imply that the underlying mechanisms might be related to the aggregation behavior of the nanofluid and its temperature dependence. Nguyen et al. [140] argued that the effectiveness of some surfactants might diminish with increasing temperature under certain circumstances, which results in aggregation of the particles, which in turn increases the viscosity. It should also be noted that there are also a few studies which have shown the opposite temperature dependency, that is, increasing viscosity ratio with increasing temperature [82,140].

The third important parameter is the particle size. For conventional suspensions with particle sizes on the order of microns or millimeters, the classical theories predict no particle size dependence for the viscosity (see Equation 5.18 [134]). However, aggregation of nanoparticles in nanofluids has been observed to result in particle size dependency. Chen et al. [135] indicate that the volume fraction relevant to the viscosity of a nanofluid is the *effective* volume fraction which is determined by considering the outer boundaries of the aggregates as the boundaries of effective particles. One can modify the Krieger and Dougherty [138] formula (Equation 5.21) accordingly for nanofluids as [135]

$$\frac{\eta_{nf}}{\eta_f} = \left(1 - \frac{\phi_{eff}}{\phi_m}\right)^{-2.5\phi_m}, \tag{5.22}$$

where ϕ_{eff} is the effective volume fraction, which can be calculated from [135]

$$\phi_{eff} = \frac{\phi}{\phi_{m,a}}, \tag{5.23}$$

where $\phi_{m,a}$ is the maximum packing fraction of the aggregates, which can be determined from the following expression [135].

$$\phi_{m,a} = n_a \left(\frac{d_p}{d_a}\right)^3, \tag{5.24}$$

where

n_a is the number of particles in an aggregate
d_p is the particle diameter
d_a is the aggregate diameter

Therefore, effective particle volume fraction is a function of the diameter of the aggregate, diameter of the nanoparticle, and the number of particles per aggregate. According to this model, the viscosity of a nanofluid decreases with increasing particle size. There are many studies in the literature that are in agreement with this prediction [141,148,156–158]. Some researchers reported more pronounced particle size dependence at higher volume fractions [139,140]. This second observation is also in agreement with the model since aggregation effects become more pronounced at higher particle volume fractions.

On the other hand, there are also studies that report increasing viscosity with increasing particle size [139,140,159]. These results might be due to the changes in the aggregation behavior as a function of particle size. It should be noted that for well dispersed particles, that is, no aggregation, Equation 5.22 does not predict any particle size dependence, and for well dispersed particles in the dilute limit, Equation 5.22 converges to Einstein's expression for viscosity [134] (Equation 5.18).

In addition to the particle size, particle shape might also affect the viscosity of nanofluids. Timofeeva et al. [71] considered Al_2O_3/water nanofluids with particles of different shapes, and showed that elongated particles result in higher viscosity. Since the volume fraction threshold for percolation is smaller for particles with higher aspect ratios, this result is not surprising. Especially carbon nanotube-based nanofluids have the disadvantage of having large increases in viscosity which might reduce their impact on the thermal efficiency enhancement [111,142].

In addition to the effects of these nanofluid parameters on viscosity, there have also been research activities regarding the rheological behavior of the nanofluids. Many studies in the literature reported Newtonian behavior [135,145,153–155,160–167] for nanofluids, especially for low particle volume fractions, but there are also studies that observed a non-Newtonian behavior through shear thinning [80,142,143,150,159,162,165,168–171]. Yu et al. [80] measured the viscosity of ZnO/ethylene glycol nanofluids and Newtonian behavior has been observed up to 2 vol.%, whereas shear thinning behavior appeared at higher concentrations. A similar trend was reported by Phuoc and Massoudi [162] for Fe_2O_3/water nanofluids. Similar to the mechanism for the nonlinearity in particle volume fraction dependence, the non-Newtonian behavior observed is also attributed to the aggregation effects [80]. Most studies report that the non-Newtonian behavior increases with increasing particle volume fraction [80,162], which support this explanation. Duan et al. [143] measured the viscosity of Al_2O_3/water nanofluids and showed that by breaking the aggregates into smaller pieces through ultrasonic vibration, the nanofluid which was previously non-Newtonian (shear thinning) became a Newtonian fluid [143]. Namburu et al. [141] investigated SiO_2/water-ethylene glycol nanofluids and showed that the rheological behavior becomes non-Newtonian for temperatures below −10°C.

In addition to the non-Newtonian behavior and nonlinear dependence on particle volume fraction, another interesting phenomenon is the hysteresis behavior of the nanofluid viscosity. Nguyen et al. [140] measured the viscosity of Al_2O_3/water nanofluids as a function of temperature and observed a decrease in the viscosity up to about 65°C. Beyond that temperature, the viscosity starts to increase dramatically, and upon cooling, the viscosity remains relatively high when compared to the data collected during the heating cycle. This behavior might be explained by temperature dependent particle aggregation [140]. An increase in aggregation is known to increase the viscosity [135,143], and cooling down of the nanofluid is not expected to reverse the aggregation process.

5.6.3 Theoretical Models for Nanofluids

Since the classical theoretical models severely underestimate the viscosity of the nanofluids, many theoretical models and correlations have recently been developed to estimate the nanofluid viscosity. Table 5.3 lists the mathematical expressions for some of these models and correlations [124,146,172].

There have been many reviews of research findings related to nanofluid viscosity. These reviews discuss the parameters that affect the viscosity, proposed models, and correlations in more detail [85,94,135,146,172,176].

In some cases, the samples are required to be (further) diluted for the zeta potential measurements. As the zeta potential depends on the environment (e.g., pH value, ionic strength, type of the ions in the suspension, etc.) [29], it is important to use the right base liquid for the dilution, with a pH value and related properties exactly matching those of the suspension.

6.3.1.2 Particle Size Distribution (PSD) Measurements

Through the PSD measurements, one can obtain an average particle size for the particles dispersed within the base fluid. There are different statistical approaches for performing PSD analyses, the most common one being the dynamic light scattering (DLS, also known as photon correlation spectroscopy) technique.

The DLS measurements require small sample volumes (around 1–2 mL) similar to those for zeta potential measurements [31]. Such small sample volumes significantly reduce the material losses within the study period.

In some cases, the samples are required to be (further) diluted for DLS measurements; they should be carefully prepared and stored for the satisfactory sensitivity of the results. The samples should be free from dust and other external material exposure, and also the sample cups should not contain any scratches and any other defects in order to achieve accurate results.

The particle size distributions may be calculated in terms of (1) *weighted distributions based on the weight of individual particles*, (2) *number weighted distributions, a counting technique where each particle is given equal weighting irrespective of its size*, (3) *volume weighted distributions based on the volume of individual particles* (*static light scattering*), and (4) *intensity weighted distributions based on the intensity of the light scattered by the particle* (*DLS*). As a result, different techniques based on different statistical approaches can produce highly different particle size distributions [33].

This difference between different averaging methods may be exemplified using the distributions in Figure 6.2 which presents results of the same measurements belonging to the same Al_2O_3–H_2O nanofluid sample with an initial particle size of 30 ± 10 nm (1 year after the preparation) [34]. The distribution at the top is based on nanoparticle weights, while the one at the bottom is based on the number of nanoparticles.

The DLS analysis also provides a parameter called the "poly-dispersity index" which varies from 0 to 1. It is desired to have a value close to zero indicating a rather mono-dispersed sample.

6.3.1.3 Nanoparticle Dispersity Analyses

Optical microscopy may be used to gain a lot of information on the size, shape, morphology, and aggregation of particles. Little time is required for the sample preparation, and with a preliminary examination by optical microscopy, more sophisticated and quantitative characterization techniques may be planned [16]. Higher resolution is necessary for the investigation of dispersions of nanometer sized particles. In this case, electron microscopy techniques, namely, transmission electron microscopy (TEM), scanning electron microscopy (SEM), field emission scanning electron microscopy (FESEM), etc., can be used.

A sample TEM micrograph is provided in Figure 6.3. At the bottom left corner, partial nanoparticle overlaps may be observed indicating the presence of nano-clusters [35].

6.3.1.4 UV-Visible Spectroscopy Measurements

The intensity of light changes when it passes through a fluid due to scattering and absorption [36]. UV-visible spectroscopy uses this principle and through the calculation of the light absorbency ratio, one can examine the average particle size of the nanoparticles dispersed within nanofluids [37].

6.3.1.5 Sedimentation Balance Method

In this method, a tray of a sedimentation balance is called the into a nanofluid sample, right after its preparation [24,38] and the weight of the settled nanoparticles after a certain period of time is

Therefore, effective particle volume fraction is a function of the diameter of the aggregate, diameter of the nanoparticle, and the number of particles per aggregate. According to this model, the viscosity of a nanofluid decreases with increasing particle size. There are many studies in the literature that are in agreement with this prediction [141,148,156–158]. Some researchers reported more pronounced particle size dependence at higher volume fractions [139,140]. This second observation is also in agreement with the model since aggregation effects become more pronounced at higher particle volume fractions.

On the other hand, there are also studies that report increasing viscosity with increasing particle size [139,140,159]. These results might be due to the changes in the aggregation behavior as a function of particle size. It should be noted that for well dispersed particles, that is, no aggregation, Equation 5.22 does not predict any particle size dependence, and for well dispersed particles in the dilute limit, Equation 5.22 converges to Einstein's expression for viscosity [134] (Equation 5.18).

In addition to the particle size, particle shape might also affect the viscosity of nanofluids. Timofeeva et al. [71] considered Al_2O_3/water nanofluids with particles of different shapes, and showed that elongated particles result in higher viscosity. Since the volume fraction threshold for percolation is smaller for particles with higher aspect ratios, this result is not surprising. Especially carbon nanotube-based nanofluids have the disadvantage of having large increases in viscosity which might reduce their impact on the thermal efficiency enhancement [111,142].

In addition to the effects of these nanofluid parameters on viscosity, there have also been research activities regarding the rheological behavior of the nanofluids. Many studies in the literature reported Newtonian behavior [135,145,153–155,160–167] for nanofluids, especially for low particle volume fractions, but there are also studies that observed a non-Newtonian behavior through shear thinning [80,142,143,150,159,162,165,168–171]. Yu et al. [80] measured the viscosity of ZnO/ethylene glycol nanofluids and Newtonian behavior has been observed up to 2 vol.%, whereas shear thinning behavior appeared at higher concentrations. A similar trend was reported by Phuoc and Massoudi [162] for Fe_2O_3/water nanofluids. Similar to the mechanism for the nonlinearity in particle volume fraction dependence, the non-Newtonian behavior observed is also attributed to the aggregation effects [80]. Most studies report that the non-Newtonian behavior increases with increasing particle volume fraction [80,162], which support this explanation. Duan et al. [143] measured the viscosity of Al_2O_3/water nanofluids and showed that by breaking the aggregates into smaller pieces through ultrasonic vibration, the nanofluid which was previously non-Newtonian (shear thinning) became a Newtonian fluid [143]. Namburu et al. [141] investigated SiO_2/water-ethylene glycol nanofluids and showed that the rheological behavior becomes non-Newtonian for temperatures below −10°C.

In addition to the non-Newtonian behavior and nonlinear dependence on particle volume fraction, another interesting phenomenon is the hysteresis behavior of the nanofluid viscosity. Nguyen et al. [140] measured the viscosity of Al_2O_3/water nanofluids as a function of temperature and observed a decrease in the viscosity up to about 65°C. Beyond that temperature, the viscosity starts to increase dramatically, and upon cooling, the viscosity remains relatively high when compared to the data collected during the heating cycle. This behavior might be explained by temperature dependent particle aggregation [140]. An increase in aggregation is known to increase the viscosity [135,143], and cooling down of the nanofluid is not expected to reverse the aggregation process.

5.6.3 Theoretical Models for Nanofluids

Since the classical theoretical models severely underestimate the viscosity of the nanofluids, many theoretical models and correlations have recently been developed to estimate the nanofluid viscosity. Table 5.3 lists the mathematical expressions for some of these models and correlations [124,146,172].

There have been many reviews of research findings related to nanofluid viscosity. These reviews discuss the parameters that affect the viscosity, proposed models, and correlations in more detail [85,94,135,146,172,176].

TABLE 5.3
Theoretical Models and Correlations of Nanofluid Viscosity

TiO_2/water nanofluids	Tseng and Lin [173] $\mu_{nf} = 13.47e^{35.98\phi}\mu_f$
TiO_2/ethylene glycol nanofluids	Chen et al. [154] $\mu_{nf} = (1 + 10.6\phi + (10.6\phi)^2)\mu_f$
Al_2O_3/water nanofluids	Maïga et al. [174] $\mu_{nf} = (1 + 7.3\phi + 123\phi^2)\mu_f$ Nguyen et al. [139] $\mu_{nf} = 0.904e^{0.1482\phi}\mu_f$
CuO-based nanofluids	Koo and Kleinstreuer [175] $\mu_{nf} = \mu_f + 5\times10^4\beta\rho_f\phi\sqrt{\frac{\kappa_B T}{2\rho_p r_p}}\begin{bmatrix}(-134.63+1722.3\phi)\\+(0.4705-6.04\phi)T\end{bmatrix}$ $\beta = \begin{cases}0.0137(100\phi)^{-0.8229} & \text{for } \phi < 0.01\\ 0.0011(100\phi)^{-0.7272} & \text{for } \phi > 0.01\end{cases}$ κ_B is Boltzmann constant, r_p is nanoparticle radius.
CuO/water nanofluids	Kulkarni et al. [168] $\mu_{nf} = \exp\begin{bmatrix}-(2.8751+53.548\phi-107.12\phi^2)+\\(1078.3+15857\phi+20587\phi^2)(1/T)\end{bmatrix}$
All nanofluids	Chen et al. [135], based on Krieger and Dougherty [138] $\eta_{nf} = \left(1-\phi_{eff}/\phi_m\right)^{-2.5\phi_m}\eta_f,\ \phi_{eff} = \phi\left(d_a/d_p\right)^{3-D},$ where ϕ_m is the maximum particle packing fraction, D is the fractal index of the aggregate, ≈ 1.8–2.5, and d_a is the aggregate diameter [135].

Sources: Based on Özerinç, S., *Heat Transfer Enhancement with Nanofluids*, Middle East Technical University, Ankara, Turkey, 2010; Mahbubul, I.M. et al., *Int. J. Heat Mass Transfer*, 55(4), 874, 2012; Elçioğlu, E.B. et al., *J. Therm. Sci. Technol.*, 34(1), 137, 2014.

5.7 COMBINED EFFECT OF ENHANCED THERMAL CONDUCTIVITY AND INCREASED VISCOSITY ON HEAT TRANSFER

Since thermal conductivity enhancement is advantageous, and viscosity increase is undesirable for heat transfer applications, the overall impact of these two parameters on the heat transfer performance is critical. Prasher et al. [153] considered the fully developed forced convection inside a circular tube for a nanofluid with the thermophysical properties:

$$k_{nf} = k_f\left(1+K\phi\right), \tag{5.25}$$

where K is the proportionality constant for the thermal conductivity enhancement, and,

$$\mu_{nf} = \mu_f\left(1+M\phi\right), \tag{5.26}$$

where M is the proportionality constant for the viscosity increase. Prasher et al. [153] assumed that the Nusselt number (Nu) is the same for the nanofluid and the pure base liquid flow, that is, $Nu_{nf} = Nu_f$, and showed that for small particle volume fractions, the heat transfer and pressure drop for the case of the nanofluid and the pure liquid are the same if M is four times larger than K, that is [153],

$$\Delta p_{nf} \approx \Delta p_f \text{ and } h_{nf} \approx h_f, \quad \text{if } M = 4K, \text{ assuming } Nu_{nf} = Nu_f. \tag{5.27}$$

This simple analysis implies that if the viscosity increase with respect to particle volume fraction is greater than four times that of the thermal conductivity, then the application of the nanofluid for forced convection heat transfer is not feasible [153]; pressure drop will become higher and heat transfer coefficient will be reduced. It should be noted that for obtaining Equation 5.27, $Nu_{nf} = Nu_f$ is assumed [153]; whereas experiments of forced convection heat transfer with nanofluids show that Nu_{nf} is greater than Nu_f [177,178]. Therefore, depending on the Nu_{nf} enhancement, the factor that relates M to K might become higher [153]. It should also be noted that the relationship between the thermal conductivity enhancement and viscosity increase required for making the nanofluid usage feasible might also change based on the specific heat transfer application.

5.8 CONCLUSIONS

In this chapter, the current understanding about the thermophysical properties of nanofluids is summarized. The findings suggest that the enhanced thermal transport due to the nanoparticle clustering is the major mechanism responsible for the thermal conductivity enhancement of nanofluids. This conclusion implies both theoretical and practical challenges regarding the application of nanofluids in engineering systems. Larger clusters of nanoparticles are thermally advantageous since the high conductivity paths become longer, but on the other hand, very large clusters might behave similar to micron-sized particles and might sediment or clog channels [13]. Furthermore, the clustering process increases the viscosity of the nanofluids, which is not desirable. Therefore, there is a need for studies that correlate the stability, thermal conductivity, and viscosity of the nanofluids in terms of their clustering behavior.

A key issue regarding the nanofluid characterization is the detailed evaluation of the clustering characteristics of the nanofluid. Commonly used techniques for characterization include scanning electron microscopy, transmission electron microscopy, and dynamic light scattering. Although electron microscopy is a powerful technique, it usually requires the drying of the liquid matrix unless the vapor pressure of the liquid is very small [132]. The clustering parameters might change during this process. Dynamic light scattering is limited to particle volume fractions below 1%, therefore it does not provide useful information about the high volume fraction cases, which are more interesting for investigating clustering [132]. Another option is the freezing of the nanofluids [112]. Although some useful information can be obtained from these experiments, the phase change of the matrix might result in the rearrangement of the nanoparticles [112]. Therefore there is a need for new methodologies that can evaluate the microstructure of the nanofluids without introducing significant artifacts. In the investigations of the clustering behavior, it should be noted that the clustering characteristics are a function of base liquid and particle type, particle size and shape, and temperature. Furthermore, it is known that the sample preparation details such as sonication time, shelf time, surfactant type, and its quantity also affect the clustering process. Therefore, care should be taken during the systematic studies for the determination of the optimum clustering behavior.

The viscosity of nanofluids is very critical for the practical applications of nanofluids in heat transfer systems. Similar to the thermal conductivity case, the research findings regarding the viscosity suggest further experiments and modeling for understanding the effect of clustering behavior on the viscosity. The main mechanism that increases the thermal conductivity, nanoparticle clustering, has also been observed to increase the viscosity; therefore, an optimized compromise between these two effects is necessary for the feasible application of nanofluids in heat transfer processes.

REFERENCES

1. S.U.S. Choi, *Enhancing Thermal Conductivity of Fluids with Nanoparticles*, ASME Fluids Engineering Division, Washington, DC, 1995, pp. 99–105.
2. R. Taylor, S. Coulombe, T. Otanicar, P. Phelan, A. Gunawan, W. Lv et al., Small particles, big impacts: A review of the diverse applications of nanofluids, *Journal of Applied Physics*, 113 (1), 011301 (2013).
3. A.S. Ahuja, Augmentation of heat transport in laminar flow of polystyrene suspensions. I. Experiments and results, *Journal of Applied Physics*, 46 (8), 3408–3416 (1975).
4. J. Happel, Viscous flow in multiparticle systems: Slow motion of fluids relative to beds of spherical particles, *AIChE Journal*, 4 (2), 197–201 (1958).
5. R.L. Hamilton, O.K. Crosser, Thermal conductivity of heterogeneous two-component systems, *Industrial & Engineering Chemistry Fundamentals*, 1 (3), 187–191 (1962).
6. H. Gleiter, Nanocrystalline materials, *Progress in Materials Science*, 33 (4), 223–315 (1989).
7. A.K. Gupta, M. Gupta, Synthesis and surface engineering of iron oxide nanoparticles for biomedical applications, *Biomaterials*, 26 (18), 3995–4021 (2005).
8. A.-H. Lu, E.L. Salabas, F. Schüth, Magnetic nanoparticles: Synthesis, protection, functionalization, and application, *Angewandte Chemie—International Edition*, 46 (8), 1222–1244 (2007).
9. J.A. Eastman, U.S. Choi, S. Li, L.J. Thompson, S. Lee, Enhanced thermal conductivity through the development of nanofluids, *MRS Symposium Proceedings*, 457, 3–11 (1997).
10. W. Yu, H. Xie, A review on nanofluids: Preparation, stability mechanisms, and applications, *Journal of Nanomaterials*, 2012, 435873 (2012).
11. W. Yu, D.M. France, J.L. Routbort, S.U.S. Choi, Review and comparison of nanofluid thermal conductivity and heat transfer enhancements, *Heat Transfer Engineering*, 29 (5), 432–460 (2008).
12. C.G. Granqvist, R.A. Buhrman, Ultrafine metal particles, *Journal of Applied Physics*, 47 (5), 2200–2219 (1976).
13. J.A. Eastman, S.R. Phillpot, S.U.S. Choi, P. Keblinski, Thermal transport in nanofluids, *Annual Review of Materials Research*, 34 (1), 219–246 (2004).
14. S. Yatsuya, Y. Tsukasaki, K. Mihama, R. Uyeda, Preparation of extremely fine particles by vacuum evaporation onto a running oil substrate, *Journal of Crystal Growth*, 45, 490–494 (1978).
15. M. Wagener, B.S. Murty, B. Günther, Preparation of metal nanosuspensions by high-pressure DC-sputtering on running liquids, *MRS Symposium Proceedings*, 457, 149 (1996).
16. M. Wagener, B. Günther, Sputtering on liquids—A versatile process for the production of magnetic suspensions, *Journal of Magnetism and Magnetic Materials*, 201 (1–3), 41–44 (1999).
17. J.A. Eastman, S.U.S. Choi, S. Li, W. Yu, L.J. Thompson, Anomalously increased effective thermal conductivities of ethylene glycol-based nanofluids containing copper nanoparticles, *Applied Physics Letters*, 78 (6), 718–720 (2001).
18. Z. Haddad, C. Abid, H.F. Oztop, A. Mataoui, A review on how the researchers prepare their nanofluids, *International Journal of Thermal Sciences*, 76, 168–189 (2014).
19. S. Lee, S.U.S. Choi, S. Li, J.A. Eastman, Measuring thermal conductivity of fluids containing oxide nanoparticles, *Journal of Heat Transfer*, 121 (2), 280–289 (1999).
20. Y. Xuan, Q. Li, Heat transfer enhancement of nanofluids, *International Journal of Heat and Fluid Flow*, 21 (1), 58–64 (2000).
21. E.V. Timofeeva, A.N. Gavrilov, J.M. McCloskey, Y.V. Tolmachev, S. Sprunt, L.M. Lopatina et al., Thermal conductivity and particle agglomeration in alumina nanofluids: Experiment and theory, *Physical Review E—Statistical, Nonlinear, and Soft Matter Physics*, 76 (6), 061203 (2007).
22. X.F. Li, D.S. Zhu, X.J. Wang, N. Wang, J.W. Gao, H. Li, Thermal conductivity enhancement dependent pH and chemical surfactant for Cu–H_2O nanofluids, *Thermochimica Acta*, 469 (1–2), 98–103 (2008).
23. Y. Hwang, J.-K. Lee, J.-K. Lee, Y.-M. Jeong, S.-I. Cheong, Y.-C. Ahn et al., Production and dispersion stability of nanoparticles in nanofluids, *Powder Technology*, 186 (2), 145–153 (2008).
24. H. Xie, H. Lee, W. Youn, M. Choi, Nanofluids containing multiwalled carbon nanotubes and their enhanced thermal conductivities, *Journal of Applied Physics*, 94 (8), 4967–4971 (2003).
25. Y.J. Hwang, Y.C. Ahn, H.S. Shin, C.G. Lee, G.T. Kim, H.S. Park et al., Investigation on characteristics of thermal conductivity enhancement of nanofluids, *Current Applied Physics*, 6 (6), 1068–1071 (2006).
26. S.M.S. Murshed, K.C. Leong, C. Yang, Enhanced thermal conductivity of TiO_2-water based nanofluids, *International Journal of Thermal Sciences*, 44 (4), 367–373 (2005).

27. A. Ghadimi, R. Saidur, H.S.C. Metselaar, A review of nanofluid stability properties and characterization in stationary conditions, *International Journal of Heat and Mass Transfer*, 54 (17–18), 4051–4068 (2011).
28. M.J. Pastoriza-Gallego, C. Casanova, R. Paramo, B. Barbes, J.L. Legido, M.M. Pineiro, A study on stability and thermophysical properties (density and viscosity) of Al_2O_3 in water nanofluid, *Journal of Applied Physics*, 106 (6), 064301 (2009).
29. J. Buongiorno, Convective transport in nanofluids, *Journal of Heat Transfer*, 128 (3), 240–250 (2006).
30. J. Avsec, M. Oblak, The calculation of thermal conductivity, viscosity and thermodynamic properties for nanofluids on the basis of statistical nanomechanics, *International Journal of Heat and Mass Transfer*, 50 (21–22), 4331–4341 (2007).
31. S.-Q. Zhou, R. Ni, Measurement of the specific heat capacity of water-based Al_2O_3 nanofluid, *Applied Physics Letters*, 92 (9), 093123 (2008).
32. Y. Ding, H. Alias, D. Wen, R.A. Williams, Heat transfer of aqueous suspensions of carbon nanotubes (CNT nanofluids), *International Journal of Heat and Mass Transfer*, 49 (1–2), 240–250 (2006).
33. H.T. Zhu, C.Y. Zhang, Y.M. Tang, J.X. Wang, Novel synthesis and thermal conductivity of CuO nanofluid, *Journal of Physical Chemistry C*, 111 (4), 1646–1650 (2007).
34. T.-K. Hong, H.-S. Yang, C.J. Choi, Study of the enhanced thermal conductivity of Fe nanofluids, *Journal of Applied Physics*, 97 (6), 064311 (2005).
35. M. Beck, Y. Yuan, P. Warrier, A. Teja, The effect of particle size on the thermal conductivity of alumina nanofluids, *Journal of Nanoparticle Research*, 11 (5), 1129–1136 (2009).
36. X. Zhang, H. Gu, M. Fujii, Effective thermal conductivity and thermal diffusivity of nanofluids containing spherical and cylindrical nanoparticles, *Journal of Applied Physics*, 100 (4), 044325 (2006).
37. S.K. Das, N. Putra, P. Thiesen, W. Roetzel, Temperature dependence of thermal conductivity enhancement for nanofluids, *Journal of Heat Transfer*, 125 (4), 567–574 (2003).
38. D.-W. Oh, A. Jain, J.K. Eaton, K.E. Goodson, J.S. Lee, Thermal conductivity measurement and sedimentation detection of aluminum oxide nanofluids by using the 3ω method, *International Journal of Heat and Fluid Flow*, 29 (5), 1456–1461 (2008).
39. A. Turgut, I. Tavman, M. Chirtoc, H. Schuchmann, C. Sauter, S. Tavman, Thermal conductivity and viscosity measurements of water-based TiO_2 nanofluids, *International Journal of Thermophysics*, 30 (4), 1213–1226 (2009).
40. B. Barbés, R. Páramo, E. Blanco, M.J. Pastoriza-Gallego, M.M. Piñeiro, J.L. Legido et al., Thermal conductivity and specific heat capacity measurements of Al_2O_3 nanofluids, *Journal of Thermal Analysis and Calorimetry*, 111 (2), 1615–1625 (2013).
41. X. Wang, S.U.S. Choi, X. Xu, Thermal conductivity of nanoparticle-fluid mixture, *Journal of Thermophysics and Heat Transfer*, 13 (4), 474–480 (1999).
42. S.A. Putnam, D.G. Cahill, P.V. Braun, Z. Ge, R.G. Shimmin, Thermal conductivity of nanoparticle suspensions, *Journal of Applied Physics*, 99 (8), 084308 (2006).
43. W. Czarnetzki, W. Roetzel, Temperature oscillation techniques for simultaneous measurement of thermal diffusivity and conductivity, *International Journal of Thermophysics*, 16 (2), 413–422 (1995).
44. A.J. Schmidt, M. Chiesa, D.H. Torchinsky, J.A. Johnson, K.A. Nelson, G. Chen, Thermal conductivity of nanoparticle suspensions in insulating media measured with a transient optical grating and a hotwire, *Journal of Applied Physics*, 103 (8), 083529 (2008).
45. H. Eichler, G. Salje, H. Stahl, Thermal diffusion measurements using spatially periodic temperature distributions induced by laser light, *Journal of Applied Physics*, 44 (12), 5383–5388 (1973).
46. J. Buongiorno, D.C. Venerus, N. Prabhat, T. McKrell, J. Townsend, R. Christianson et al., A benchmark study on the thermal conductivity of nanofluids, *Journal of Applied Physics*, 106 (9), 094312 (2009).
47. P. Keblinski, S.R. Phillpot, S.U.S. Choi, J.A. Eastman, Mechanisms of heat flow in suspensions of nano-sized particles (nanofluids), *International Journal of Heat and Mass Transfer*, 45 (4), 855–863 (2002).
48. J.A. Eastman, U.S. Choi, S. Li, G. Soyez, L.J. Thompson, R.J. DiMelfi, Novel thermal properties of nanostructured materials, *Materials Science Forum*, 312, 629–634 (1999).
49. K.S. Hong, T.-K. Hong, H.-S. Yang, Thermal conductivity of Fe nanofluids depending on the cluster size of nanoparticles, *Applied Physics Letters*, 88 (3), 031901 (2006).
50. J. Eapen, W.C. Williams, J. Buongiorno, L. Hu, S. Yip, R. Rusconi et al., Mean-field versus microconvection effects in nanofluid thermal conduction, *Physical Review Letters*, 99 (9), 095901 (2007).
51. J. Eapen, R. Rusconi, R. Piazza, S. Yip, The classical nature of thermal conduction in nanofluids, *Journal of Heat Transfer*, 132 (10), 102402 (2010).

105. H. Babaei, P. Keblinski, J.M. Khodadadi, A proof for insignificant effect of Brownian motion-induced micro-convection on thermal conductivity of nanofluids by utilizing molecular dynamics simulations, *Journal of Applied Physics*, 113 (8), 084302 (2013).
106. W. Evans, J. Fish, P. Keblinski, Role of Brownian motion hydrodynamics on nanofluid thermal conductivity, *Applied Physics Letters*, 88 (9), 093116 (2006).
107. M. Vladkov, J.-L. Barrat, Modeling transient absorption and thermal conductivity in a simple nanofluid, *Nano Letters*, 6 (6), 1224–1228 (2006).
108. R. Rusconi, E. Rodari, R. Piazza, Optical measurements of the thermal properties of nanofluids, *Applied Physics Letters*, 89 (26), 261916 (2006).
109. D.C. Venerus, M.S. Kabadi, S. Lee, V. Perez-Luna, Study of thermal transport in nanoparticle suspensions using forced Rayleigh scattering, *Journal of Applied Physics*, 100 (9), 094310 (2006).
110. X. Zhang, H. Gu, M. Fujii, Experimental study on the effective thermal conductivity and thermal diffusivity of nanofluids, *International Journal of Thermophysics*, 27 (2), 569–580 (2006).
111. T.X. Phuoc, M. Massoudi, R.-H. Chen, Viscosity and thermal conductivity of nanofluids containing multi-walled carbon nanotubes stabilized by chitosan, *International Journal of Thermal Sciences*, 50 (1), 12–18 (2011).
112. J.W. Gao, R.T. Zheng, H. Ohtani, D.S. Zhu, G. Chen, Experimental investigation of heat conduction mechanisms in nanofluids: Clue on clustering, *Nano Letters*, 9 (12), 4128–4132 (2009).
113. C.-J. Yu, A.G. Richter, A. Datta, M.K. Durbin, P. Dutta, Observation of molecular layering in thin liquid films using x-ray reflectivity, *Physical Review Letters*, 82 (11), 2326 (1999).
114. W. Yu, S.U.S. Choi, The role of interfacial layers in the enhanced thermal conductivity of nanofluids: A renovated Maxwell model, *Journal of Nanoparticle Research*, 5 (1), 167–171 (2003).
115. L.M. Schwartz, E.J. Garboczi, D.P. Bentz, Interfacial transport in porous media: Application to DC electrical conductivity of mortars, *Journal of Applied Physics*, 78 (10), 5898–5908 (1995).
116. W. Yu, S.U.S. Choi, The role of interfacial layers in the enhanced thermal conductivity of nanofluids: A renovated Hamilton–Crosser model, *Journal of Nanoparticle Research*, 6 (4), 355–361 (2004).
117. B.-X. Wang, L.-P. Zhou, X.-F. Peng, A fractal model for predicting the effective thermal conductivity of liquid with suspension of nanoparticles, *International Journal of Heat and Mass Transfer*, 46 (14), 2665–2672 (2003).
118. H. Xie, M. Fujii, X. Zhang, Effect of interfacial nanolayer on the effective thermal conductivity of nanoparticle-fluid mixture, *International Journal of Heat and Mass Transfer*, 48 (14), 2926–2932 (2005).
119. Q. Xue, W.-M. Xu, A model of thermal conductivity of nanofluids with interfacial shells, *Materials Chemistry and Physics*, 90 (2–3), 298–301 (2005).
120. Y.-H. Li, W. Qu, J.-C. Feng, Temperature dependence of thermal conductivity of nanofluids, *Chinese Physics Letters*, 25 (9), 3319–3322 (2008).
121. C. Sitprasert, P. Dechaumphai, V. Juntasaro, A thermal conductivity model for nanofluids including effect of the temperature-dependent interfacial layer, *Journal of Nanoparticle Research*, 11 (6), 1465–1476 (2009).
122. L. Xue, P. Keblinski, S.R. Phillpot, S.U.-S. Choi, J.A. Eastman, Effect of liquid layering at the liquid-solid interface on thermal transport, *International Journal of Heat and Mass Transfer*, 47 (19–20), 4277–4284 (2004).
123. O.M. Wilson, X. Hu, D.G. Cahill, P.V. Braun, Colloidal metal particles as probes of nanoscale thermal transport in fluids, *Physical Review B*, 66 (22), 224301 (2002).
124. S. Özerinç, *Heat Transfer Enhancement with Nanofluids*, Middle East Technical University, Ankara, Turkey, 2010.
125. H. Xie, J. Wang, T. Xi, Y. Liu, F. Ai, Q. Wu, Thermal conductivity enhancement of suspensions containing nanosized alumina particles, *Journal of Applied Physics*, 91 (7), 4568–4572 (2002).
126. H. Xie, J. Wang, T. Xi, Y. Liu, Thermal conductivity of suspensions containing nanosized SiC particles, *International Journal of Thermophysics*, 23 (2), 571–580 (2002).
127. M.J. Assael, I.N. Metaxa, J. Arvanitidis, D. Christofilos, C. Lioutas, Thermal conductivity enhancement in aqueous suspensions of carbon multi-walled and double-walled nanotubes in the presence of two different dispersants. In *Proceedings of the 27th International Thermal Conductivity Conference*, H. Wang, W.D. Porter (eds.), DEStech Publications, Lancaster, PA, 2005.
128. R. Zheng, J. Gao, J. Wang, S.-P. Feng, H. Ohtani, J. Wang et al., Thermal percolation in stable graphite suspensions, *Nano Letters*, 12 (1), 188–192 (2012).
129. L. Ma, J. Wang, A.M. Marconnet, A.C. Barbati, G.H. McKinley, W. Liu et al., Viscosity and thermal conductivity of stable graphite suspensions near percolation, *Nano Letters*, 15 (1), 127–133 (2014).

130. D. Lee, J.-W. Kim, B.G. Kim, A new parameter to control heat transport in nanofluids: Surface charge state of the particle in suspension, *Journal of Physical Chemistry B*, 110 (9), 4323–4328 (2006).
131. R. Prasher, P.E. Phelan, P. Bhattacharya, Effect of aggregation kinetics on the thermal conductivity of nanoscale colloidal solutions (nanofluid), *Nano Letters*, 6 (7), 1529–1534 (2006).
132. P. Keblinski, R. Prasher, J. Eapen, Thermal conductance of nanofluids: Is the controversy over?, *Journal of Nanoparticle Research*, 10 (7), 1089–1097 (2008).
133. J. Mewis, N.J. Wagner, *Colloidal Suspension Rheology*, Cambridge University Press, Cambridge, U.K., 2012.
134. A. Einstein, A new determination of the molecular dimensions, *Annalen Der Physik*, 324 (2), 289–306 (1906).
135. H. Chen, Y. Ding, C. Tan, Rheological behaviour of nanofluids, *New Journal of Physics*, 9 (10), 367 (2007).
136. H.C. Brinkman, The viscosity of concentrated suspensions and solutions, *The Journal of Chemical Physics*, 20 (4), 571 (1952).
137. G.K. Batchelor, The effect of Brownian motion on the bulk stress in a suspension of spherical particles, *Journal of Fluid Mechanics Digital Archive*, 83 (01), 97–117 (1977).
138. I.M. Krieger, T.J. Dougherty, A mechanism for non-Newtonian flow in suspensions of rigid spheres, *Journal of Rheology*, 3 (1), 137–152 (1959).
139. C.T. Nguyen, F. Desgranges, G. Roy, N. Galanis, T. Maré, S. Boucher et al., Temperature and particle-size dependent viscosity data for water-based nanofluids—Hysteresis phenomenon, *International Journal of Heat and Fluid Flow*, 28 (6), 1492–1506 (2007).
140. C.T. Nguyen, F. Desgranges, N. Galanis, G. Roy, T. Maré, S. Boucher et al., Viscosity data for Al_2O_3–water nanofluid-hysteresis: Is heat transfer enhancement using nanofluids reliable?, *International Journal of Thermal Sciences*, 47 (2), 103–111 (2008).
141. P.K. Namburu, D.P. Kulkarni, A. Dandekar, D.K. Das, Experimental investigation of viscosity and specific heat of silicon dioxide nanofluids, *Micro and Nano Letters*, 2 (3), 67–71 (2007).
142. P. Garg, J.L. Alvarado, C. Marsh, T.A. Carlson, D.A. Kessler, K. Annamalai, An experimental study on the effect of ultrasonication on viscosity and heat transfer performance of multi-wall carbon nanotube-based aqueous nanofluids, *International Journal of Heat and Mass Transfer*, 52 (21–22), 5090–5101 (2009).
143. F. Duan, D. Kwek, A. Crivoi, Viscosity affected by nanoparticle aggregation in Al_2O_3-water nanofluids, *Nanoscale Research Letters*, 6 (1), 1–5 (2011).
144. S.W. Lee, S.D. Park, S. Kang, I.C. Bang, J.H. Kim, Investigation of viscosity and thermal conductivity of SiC nanofluids for heat transfer applications, *International Journal of Heat and Mass Transfer*, 54 (1–3), 433–438 (2011).
145. M. Chandrasekar, S. Suresh, A. Chandra Bose, Experimental investigations and theoretical determination of thermal conductivity and viscosity of Al_2O_3/water nanofluid, *Experimental Thermal and Fluid Science*, 34 (2), 210–216 (2010).
146. I.M. Mahbubul, R. Saidur, M.A. Amalina, Latest developments on the viscosity of nanofluids, *International Journal of Heat and Mass Transfer*, 55 (4), 874–885 (2012).
147. B.C. Pak, Y.I. Cho, Hydrodynamic and heat transfer study of dispersed fluids with submicron metallic oxide particles, *Experimental Heat Transfer*, 11 (2), 151 (1998).
148. K.B. Anoop, T. Sundararajan, S.K. Das, Effect of particle size on the convective heat transfer in nanofluid in the developing region, *International Journal of Heat and Mass Transfer*, 52 (9–10), 2189–2195 (2009).
149. K.B. Anoop, S. Kabelac, T. Sundararajan, S.K. Das, Rheological and flow characteristics of nanofluids: Influence of electroviscous effects and particle agglomeration, *Journal of Applied Physics*, 106 (3), 034909 (2009).
150. M. Kole, T.K. Dey, Viscosity of alumina nanoparticles dispersed in car engine coolant, *Experimental Thermal and Fluid Science*, 34 (6), 677–683 (2010).
151. L. Chen, H. Xie, Y. Li, W. Yu, Nanofluids containing carbon nanotubes treated by mechanochemical reaction, *Thermochimica Acta*, 477 (1–2), 21–24 (2008).
152. V. Ghosh, *Viscosity of Liquids: Theory, Estimation, Experiment, and Data*, Springer Science & Business Media, Dordrecht, the Netherlands, 2007
153. R. Prasher, D. Song, J. Wang, P. Phelan, Measurements of nanofluid viscosity and its implications for thermal applications, *Applied Physics Letters*, 89 (13), 133108 (2006).
154. H. Chen, Y. Ding, Y. He, C. Tan, Rheological behaviour of ethylene glycol based titania nanofluids, *Chemical Physics Letters*, 444 (4–6), 333–337 (2007).
155. P.K. Namburu, D.P. Kulkarni, D. Misra, D.K. Das, Viscosity of copper oxide nanoparticles dispersed in ethylene glycol and water mixture, *Experimental Thermal and Fluid Science*, 32 (2), 397–402 (2007).

156. W.-Q. Lu, Q.-M. Fan, Study for the particle's scale effect on some thermophysical properties of nanofluids by a simplified molecular dynamics method, *Engineering Analysis with Boundary Elements*, 32 (4), 282–289 (2008).
157. J. Chevalier, O. Tillement, F. Ayela, Rheological properties of nanofluids flowing through microchannels, *Applied Physics Letters*, 91 (23), 233103 (2007).
158. M.J. Pastoriza-Gallego, C. Casanova, J.L. Legido, M.M. Piñeiro, CuO in water nanofluid: Influence of particle size and polydispersity on volumetric behaviour and viscosity, *Fluid Phase Equilibria*, 300 (1–2), 188–196 (2011).
159. Y. He, Y. Jin, H. Chen, Y. Ding, D. Cang, H. Lu, Heat transfer and flow behaviour of aqueous suspensions of TiO_2 nanoparticles (nanofluids) flowing upward through a vertical pipe, *International Journal of Heat and Mass Transfer*, 50 (11–12), 2272–2281 (2007).
160. N. Putra, W. Roetzel, S.K. Das, Natural convection of nano-fluids, *Heat and Mass Transfer/Waerme-Und Stoffuebertragung*, 39 (8–9), 775–784 (2003).
161. Y. Yang, Z.G. Zhang, E.A. Grulke, W.B. Anderson, G. Wu, Heat transfer properties of nanoparticle-in-fluid dispersions (nanofluids) in laminar flow, *International Journal of Heat and Mass Transfer*, 48 (6), 1107–1116 (2005).
162. T.X. Phuoc, M. Massoudi, Experimental observations of the effects of shear rates and particle concentration on the viscosity of Fe_2O_3-deionized water nanofluids, *International Journal of Thermal Sciences*, 48 (7), 1294–1301 (2009).
163. D.P. Kulkarni, D.K. Das, R.S. Vajjha, Application of nanofluids in heating buildings and reducing pollution, *Applied Energy*, 86 (12), 2566–2573 (2009).
164. F. Boschini, A. Rulmont, R. Cloots, R. Moreno, Rheological behaviour of $BaZrO_3$ suspensions in non-aqueous media, *Ceramics International*, 35 (3), 1007–1013 (2009).
165. H. Chen, S. Witharana, Y. Jin, C. Kim, Y. Ding, Predicting thermal conductivity of liquid suspensions of nanoparticles (nanofluids) based on rheology, *Particuology*, 7 (2), 151–157 (2009).
166. H. Zhu, C. Li, D. Wu, C. Zhang, Y. Yin, Preparation, characterization, viscosity and thermal conductivity of $CaCO_3$ aqueous nanofluids, *Science China Technological Sciences*, 53 (2), 360–368 (2010).
167. G.H. Ko, K. Heo, K. Lee, D.S. Kim, C. Kim, Y. Sohn et al., An experimental study on the pressure drop of nanofluids containing carbon nanotubes in a horizontal tube, *International Journal of Heat and Mass Transfer*, 50 (23–24), 4749–4753 (2007).
168. D.P. Kulkarni, D.K. Das, G.A. Chukwu, Temperature dependent rheological property of copper oxide nanoparticles suspension (nanofluid), *Journal of Nanoscience and Nanotechnology*, 6, 1150–1154 (2006).
169. H. Chen, W. Yang, Y. He, Y. Ding, L. Zhang, C. Tan et al., Heat transfer and flow behaviour of aqueous suspensions of titanate nanotubes (nanofluids), *Powder Technology*, 183 (1), 63–72 (2008).
170. H. Chen, Y. Ding, A. Lapkin, X. Fan, Rheological behaviour of ethylene glycol-titanate nanotube nanofluids, *Journal of Nanoparticle Research*, 11 (6), 1513–1520 (2009).
171. M. Hojjat, S.G. Etemad, R. Bagheri, J. Thibault, Rheological characteristics of non-Newtonian nanofluids: Experimental investigation, *International Communications in Heat and Mass Transfer*, 38 (2), 144–148 (2011).
172. E.B. Elçioğlu, A. Güvenç Yazıcıoğlu, S. Kakaç, A comparative evaluation on nanofluid viscosity, *Journal of Thermal Science and Technology*, 34 (1), 137–151 (2014).
173. W.J. Tseng, K.-C. Lin, Rheology and colloidal structure of aqueous TiO_2 nanoparticle suspensions, *Materials Science and Engineering A*, 355 (1–2), 186–192 (2003).
174. S.E.B. Maïga, C.T. Nguyen, N. Galanis, G. Roy, Heat transfer behaviours of nanofluids in a uniformly heated tube, *Superlattices and Microstructures*, 35 (3–6), 543–557 (2004).
175. J. Koo, C. Kleinstreuer, Laminar nanofluid flow in microheat-sinks, *International Journal of Heat and Mass Transfer*, 48(13), 2652–2661 (2005).
176. Y. Li, J. Zhou, S. Tung, E. Schneider, S. Xi, A review on development of nanofluid preparation and characterization, *Powder Technology*, 196 (2), 89–101 (2009).
177. S. Kakaç, A. Pramuanjaroenkij, Review of convective heat transfer enhancement with nanofluids, *International Journal of Heat and Mass Transfer*, 52 (13–14), 3187–3196 (2009).
178. S. Özerinç, A. Güvenç Yazıcıoğlu, S. Kakaç, Numerical analysis of laminar forced convection with temperature-dependent thermal conductivity of nanofluids and thermal dispersion, *International Journal of Thermal Sciences*, 62, 138–148 (2012).

6 Stability of Nanofluids
Fundamentals, State-of-the-Art, and Potential Applications

Eskişehir B. Elçioğlu, E. Şimşek, and Tuba Okutucu-Özyurt

CONTENTS

6.1 INTRODUCTION

Nanofluids as defined by Choi and Eastman [1] are an "innovative new class of heat transfer fluids," which "can be engineered by suspending metallic nanoparticles in conventional heat transfer fluids." Generally speaking, nanofluids are colloidal dispersions of differently shaped (e.g., spherical, cylindrical, rod-like, etc.) nanoparticles (e.g., Al, Al_2O_3, Cu, CuO, Au, Ag, SiO_2, TiO_2, and carbon nanotubes [CNTs]) within base fluids (conventional coolants such as deionized water, engine oil, acetone, ethylene glycol) [2]. Nanofluids have gained significant attention in heat transfer research since 1990s, mainly due to their improved thermophysical properties, such as thermal conductivity, and specific heat. It is critical to sustain the improved characteristics, in order to utilize nanofluids reliably in all potential applications, including heat transfer enhancement.

While investigating possible heat transfer enhancement mechanisms, researchers mainly focused on the estimation of nanofluid thermal conductivity. The shape and material of the nanoparticles, the type of the base fluid, and the volumetric concentration are some parameters affecting the nanofluid thermal conductivity that need to be addressed adequately in the developed models.

In addition to the aforementioned parameters, the type and concentration of the additives as well as the nanofluid pH also have significant impact on nanofluid thermal conductivity [3].

The Maxwell equation is one of the most frequently used models, which is valid for dilute or spherical particles. The model only takes into account the nanofluid volume concentration and the thermal conductivities of the nanoparticle and the base fluid, and does not consider the interaction of the particles [4,5]. Other classical models such as Hamilton–Crosser and Bruggeman also do not account for the nanoparticle-fluid interfacial layer, or the particle size and distribution in the nanofluid [6].

Theoretically, increasing the nanoparticle concentration (φ) causes the thermal conductivity and the heat transfer coefficient to increase. Increased concentration results in a decrease in the distance between the dispersed nanoparticles in the base fluid. Eventually, the nanoparticles tend to have particle–particle interactions, and form nanoparticle aggregates [7]. Figure 6.1 illustrates the interaction and aggregation behavior.

The particle–particle interaction has been elaborated by Chen et al. [7]. For an aggregated nanoparticle suspension, the effective volume fraction of nanoparticle aggregates is given by

$$\varphi_a = \frac{\varphi}{\varphi_{ma}}, \quad n_a = \varphi_{ma}\left(\frac{a_a}{a}\right)^3 \tag{6.1}$$

where φ_{ma}, n_a, a_a, and a stand for the maximum packing fraction of nanoparticle aggregates, the number of particles contained in an aggregate, the aggregate radius, and the primary particle radius, respectively [7].

The formation of nanoparticle aggregates hinders the practical usage of nanofluids in heat transfer applications. In an aggregated nanoparticle suspension, the weights of the particles increase considerably, which in turn increase the tendency of nanoparticle settling (sedimentation) due to the gravitational forces. Such settlements may cause the undesired clogging of flow channels, and the erosion of the channel walls. Moreover, as the aggregates start touching each other, an increase in the viscosity observed even at volume fractions as low as 0.2% [8]. Besides, the hydraulic diameter

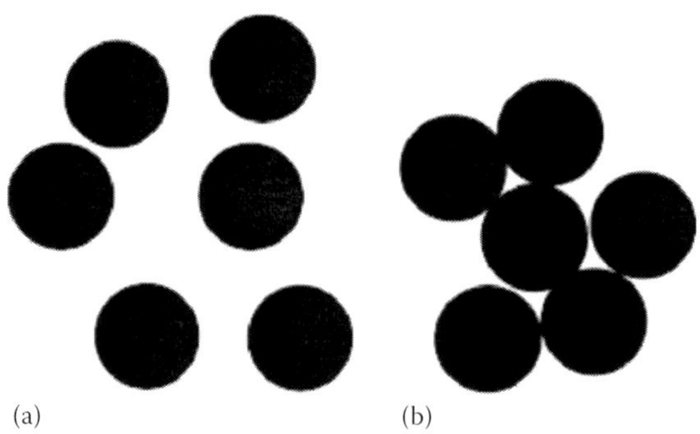

FIGURE 6.1 Schematic description of nanoparticles inside a nanofluid: (a) nanoparticle interaction, (b) nanoparticle aggregation.

of the channel may decrease for such a condition. As a result, it will no longer be possible to perform thermal or hydrodynamic calculations (e.g., heat transfer coefficient, pressure drop, and pumping power) with high accuracy.

The lack of long-term stability in nanofluids, deposition on walls, erosion, and agglomeration may cause operational problems [9]. Such problems may occur especially during heating of the nanofluid. Koyuncuoğlu et al. [10] used a high-speed camera attached to an optical microscope to illustrate the adherence of silver nanoparticles coated with a fluorescent silica-shell in DI-water. Initially, well-dispersed nanoparticles flowing through microchannels start accumulating on channel walls during heating.

Another critical parameter for the commercial usage of nanofluids is the viscosity, which is directly related to the pumping power. Although the nanoparticle addition improves the heat transfer performance of the base fluid, it also increases the pressure drop, and hence, the pumping power requirement. The pumping power is a major limitation that should be considered carefully for the feasible usage of nanofluids in a system. Extensive research was done on the pressure drop in nanofluid flow mostly through straight tubes. Xuan and Li [11] experimentally investigated the friction factor for Cu-water nanofluids in turbulent flow to evaluate the pressure drop. Different volume fractions (1%, 1.2%, 1.5%, and 2.0%) of the Cu-water nanofluid flowing through a straight brass tube have been investigated. The experiments showed that the dilute nanofluids have nearly the same friction factor as that for water at the same Reynolds numbers.

Although a nanofluid is a two-phase mixture, single-phase correlations are widely used to estimate the friction factor [11]. It should also be noted that in real systems, in addition to the straight tube sections, components such as elbows, T-joints, etc., exist, and must be considered. Routbort et al. [12] experimentally investigated the effects of elbows, pipes, and expansions on the pressure drop in a flow system. In the experiments, alumina-in-water at 2%, 4%, and 8%, and SiC-in-50/50 ethylene glycol/water mixture at 2.2% volumetric concentrations have been used as the nanofluids. The results were parallel to those obtained by Xuan and Li [11]. The pumping power has been calculated by considering a single-phase liquid with nanofluid properties. The authors observed that alumina-in-water nanofluid was not a good option for commercial usage due to the high pumping power requirement. However, the SiC nanofluid was found promising when the temperature and particle size, which directly affect the nanofluid viscosity, are altered (increased in this case) such that the pumping power decreases.

To assess whether the nanofluid use is effective or not, the Mouromtseff number, *Mo*, has been used. The number is defined as $Mo = (\rho^a k^b c_p^c)/\mu^d$ with powers *a*, *b*, *c*, and *d* to be determined based on the heat transfer mode and the empirical heat transfer correlation used [13]. The Mouromtseff number and its importance are discussed in Section 6.5.1, in detail.

The thermophysical properties touched on earlier are only meaningful for stable nanofluid suspensions. The classical paradigm of *stability* refers to a rather controllable and predictable condition of materials. Stability of nanofluids is also a complicated condition, as it directly influences the thermophysical properties and heat transfer performance. Preserving the improvements in the thermophysical properties of nanofluids is essential. Inevitable deterioration of dispersion stability can easily cut these improvements off and turn the use of nanofluids from advantageous to indeterminate.

As the size of solid particles shrink to the nanometer range, their behavior and properties tend to differ significantly from their bulk counterparts. This is mainly due to the increase in the fraction of atoms located at the particle surface (activation of the particle surface), and the increased surface area of the particles [14]. Yokoyama et al. [14] studied the relation between the nanoparticle size and the number ratio of surface atoms to the total. They reported that the number ratio of surface atoms to the total increases significantly, as the particle size decreases. Likewise, the surface area-to-volume ratio of nanoparticles is considerably greater compared to that of larger particles. As heat transfer occurs at the surface of nanoparticles [1], one can comprehend the reason for the improved thermal characteristics of nanofluids compared to those of base fluids, even though the nanoparticles constitute the smaller portion of a nanofluid (typically 1%–5% volume fraction).

In some cases, the samples are required to be (further) diluted for the zeta potential measurements. As the zeta potential depends on the environment (e.g., pH value, ionic strength, type of the ions in the suspension, etc.) [29], it is important to use the right base liquid for the dilution, with a pH value and related properties exactly matching those of the suspension.

6.3.1.2 Particle Size Distribution (PSD) Measurements

Through the PSD measurements, one can obtain an average particle size for the particles dispersed within the base fluid. There are different statistical approaches for performing PSD analyses, the most common one being the dynamic light scattering (DLS, also known as photon correlation spectroscopy) technique.

The DLS measurements require small sample volumes (around 1–2 mL) similar to those for zeta potential measurements [31]. Such small sample volumes significantly reduce the material losses within the study period.

In some cases, the samples are required to be (further) diluted for DLS measurements; they should be carefully prepared and stored for the satisfactory sensitivity of the results. The samples should be free from dust and other external material exposure, and also the sample cups should not contain any scratches and any other defects in order to achieve accurate results.

The particle size distributions may be calculated in terms of (1) *weighted distributions based on the weight of individual particles*, (2) *number weighted distributions, a counting technique where each particle is given equal weighting irrespective of its size*, (3) *volume weighted distributions based on the volume of individual particles* (*static light scattering*), and (4) *intensity weighted distributions based on the intensity of the light scattered by the particle* (*DLS*). As a result, different techniques based on different statistical approaches can produce highly different particle size distributions [33].

This difference between different averaging methods may be exemplified using the distributions in Figure 6.2 which presents results of the same measurements belonging to the same Al_2O_3–H_2O nanofluid sample with an initial particle size of 30 ± 10 nm (1 year after the preparation) [34]. The distribution at the top is based on nanoparticle weights, while the one at the bottom is based on the number of nanoparticles.

The DLS analysis also provides a parameter called the “poly-dispersity index” which varies from 0 to 1. It is desired to have a value close to zero indicating a rather mono-dispersed sample.

6.3.1.3 Nanoparticle Dispersity Analyses

Optical microscopy may be used to gain a lot of information on the size, shape, morphology, and aggregation of particles. Little time is required for the sample preparation, and with a preliminary examination by optical microscopy, more sophisticated and quantitative characterization techniques may be planned [16]. Higher resolution is necessary for the investigation of dispersions of nanometer sized particles. In this case, electron microscopy techniques, namely, transmission electron microscopy (TEM), scanning electron microscopy (SEM), field emission scanning electron microscopy (FESEM), etc., can be used.

A sample TEM micrograph is provided in Figure 6.3. At the bottom left corner, partial nanoparticle overlaps may be observed indicating the presence of nano-clusters [35].

6.3.1.4 UV-Visible Spectroscopy Measurements

The intensity of light changes when it passes through a fluid due to scattering and absorption [36]. UV-visible spectroscopy uses this principle and through the calculation of the light absorbency ratio, one can examine the average particle size of the nanoparticles dispersed within nanofluids [37].

6.3.1.5 Sedimentation Balance Method

In this method, a tray of a sedimentation balance is called the into a nanofluid sample, right after its preparation [24,38] and the weight of the settled nanoparticles after a certain period of time is

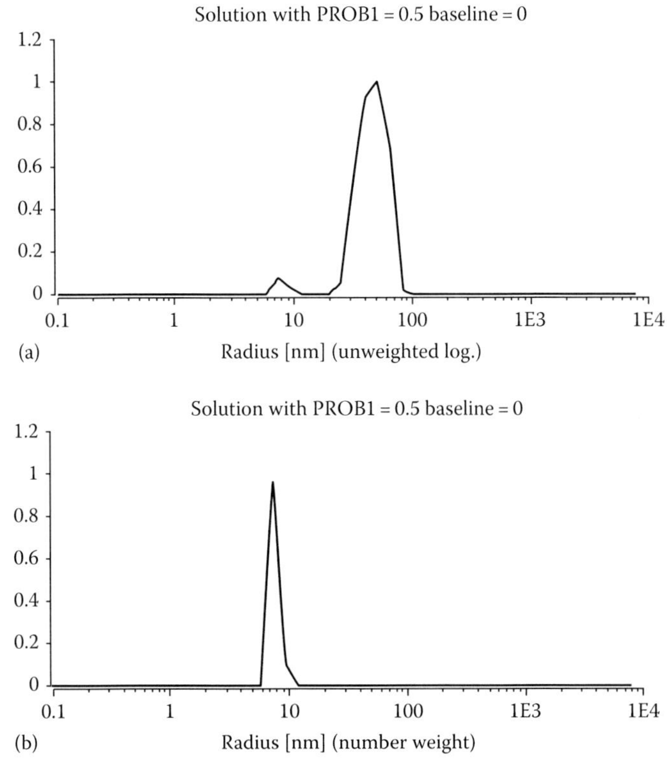

FIGURE 6.2 Demonstration of the differences between the particle size distribution measurements based on (a) nanoparticle weight and (b) number of nanoparticles. (From Elçioğlu, E.B., Experimental investigation on the dispersion stability of nanofluids, ME 552-Micro- and Nanoscale Heat Transfer Term Project, Department of Mechanical Engineering, Middle East Technical University, Ankara, Turkey, January 2014.)

measured. The essential components of a sedimentation balance setup are a vibration free table, a high precision weighing balance, temperature and humidity controllers, and a computer processor [38].

Through this approach, the fraction of nanoparticles can be calculated by disregarding the fraction of sediment nanoparticles as

$$F_s = \frac{W_T - W}{W_T} \tag{6.4}$$

In Equation 6.4, the F_s, W_T, and W stand, respectively, for the suspension fraction of nanoparticles at a given time, the total weight of nanoparticles, and the weight of sediment nanoparticles at a given time [38].

Shelf lifetime (or stability) of distilled water, and ethylene glycol based clay, Al_2O_3, and CeO_2 nanofluids was studied by Haghighi et al. [39] via the sedimentation balance method. The results indicated a reasonably good stability of distilled water based Al_2O_3 nanofluids.

6.3.1.6 3ω Method

The 3ω method has been a widely used thermal conductivity measurement technique. By changing the orientation of the 3ω device, Oh et al. [40] examined nanofluid stability by detecting the spatial variation (caused by the nanoparticle aggregation) of the thermal conductivities of Al_2O_3-EG and Al_2O_3-deionized water nanofluids.

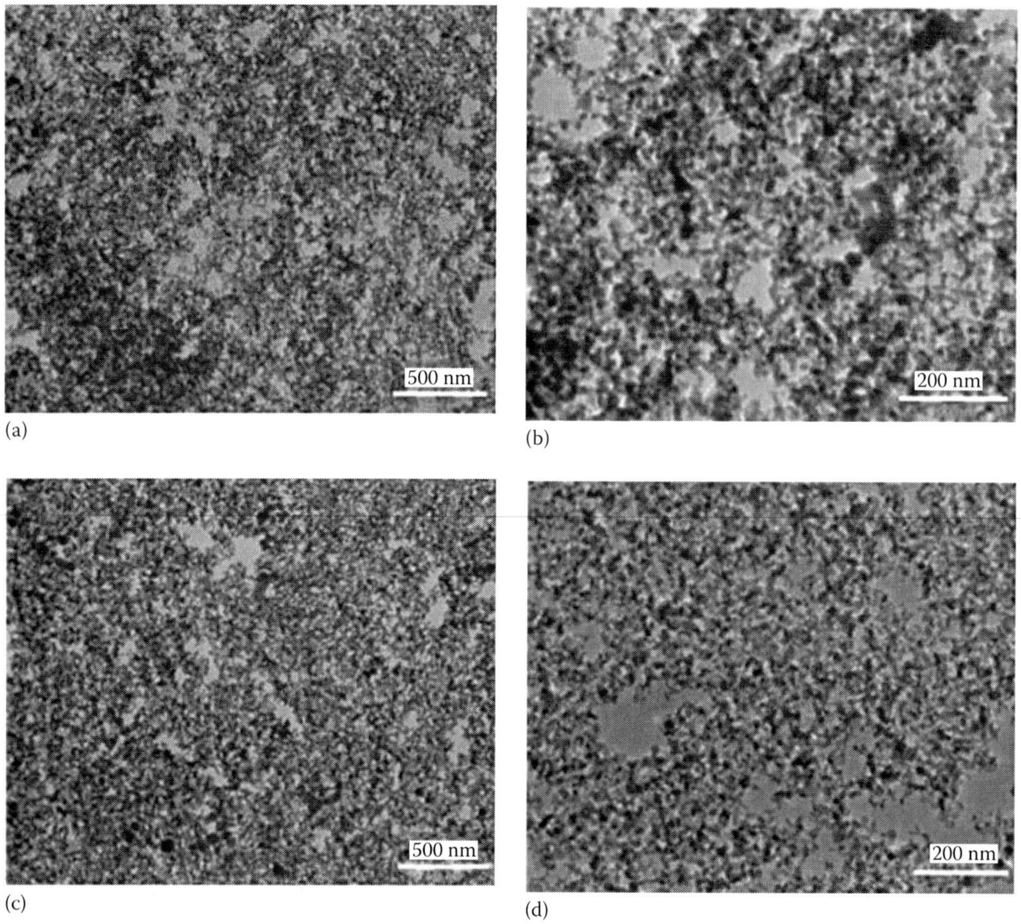

FIGURE 6.3 TEM images of Al_2O_3-water nanofluids ultrasonicated for 3 h at (a, b) 25% and (c, d) 50% ultrasonication amplitudes with different magnifications. (Reprinted from *Ultrason. Sonochem.*, 26, Mahbubul, I.M., Saidur, R., Amalina, M.A., Elcioglu, E.B., and Okutucu-Ozyurt, T., Effective ultrasonication process for better colloidal dispersion of nanofluid, 361–369. Copyright 2015, with permission from Elsevier.)

6.3.1.7 Sediment Photography Capturing

As frequently touched on in the *Literature Review* (Section 6.3.2), this method is a widely utilized one as it has the ease of applicability (performed through visual inspections only) and does not require any quantitative measurements and analyses. As a result of gravity, large and heavy nanoparticle clusters get more prone to settling down compared to the primary nanoparticles. Sedimentation stands as a critical concern for nanofluid flow through channels, not only because the sediments can cause damages on the channel walls (through erosion, for example), but they also cause the available nanoparticle concentration to decrease, which theoretically reduces the potential improvements in thermophysical properties.

6.3.2 Literature Review

Alberola et al. [41] studied the vol.% characterization of 0.5 vol.% clay mineral halloysite (Hal: $Al_2Si_2O_5(OH)_4{\cdot}nH_2O$ with n = 0–2) nanotube-water suspensions using SEM, TEM, WAXS (wide angle x-ray scattering), FTIR (Fourier transform infrared) and TGA (thermo-gravimetric) analyses.

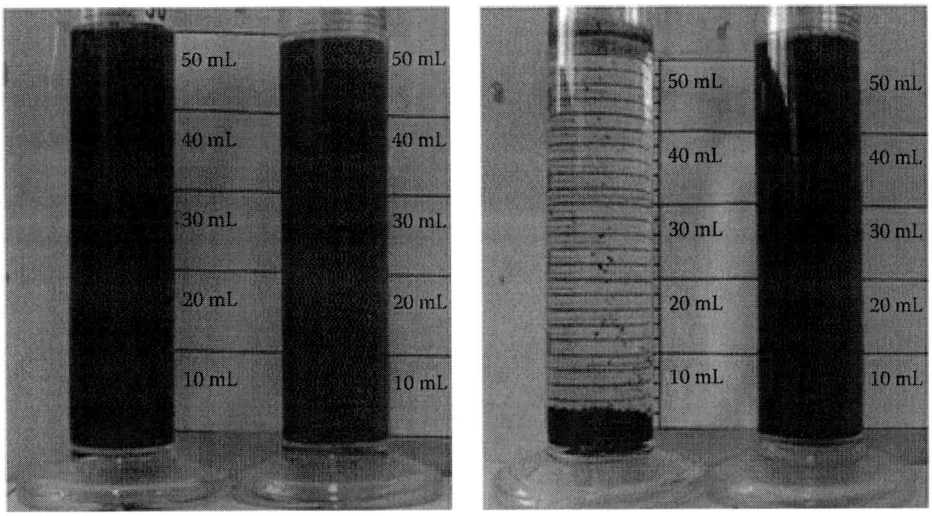

(a) (b)

FIGURE 6.4 Dispersity of MWCNTs (MWCNT CM-95) and oxidized MWCNTs (OMWCNT CM-95) after sonicator dispersion (a) and after 24 h (b). (Reprinted from *Renew. Energ.*, 65, Park, S.S. and Kim, N.J., A study on the characteristics of carbon nanofluid for heat transfer enhancement of heat pipe, 123–129. Copyright 2014, with permission from Elsevier.)

The Hal nanotubes were of diameters smaller than 100 nm and lengths ranging from about 500 nm to over 1.2 μm. The nanofluids were prepared via the two-step method and the dispersion of dry powders was obtained using an ultrasonic probe for 2 min. The highest zeta potential (–39 mV) was obtained at pH = 12.

Park and Kim [42] formulated water-based multiwalled CNT (MWCNT) nanofluids through a chemical reformation process. The investigated nanofluids were produced via the two-step method and ultrasonication for 2 h was applied. The diameters and lengths of the studied nanofluids (CM-95 and CM-100) were 10–15 nm by 10–15 nm and 10–20 nm by ~200 nm, respectively. After 24 h from the preparation, the nanotubes within the mechanically agitated nanofluids were accumulated at the bottom of the sample containers. Figure 6.4 shows the difference between the dispersities of MWCNTs and oxidized MWCNTs.

It is evident from the figure that the nanofluid containing oxidized MWCNTs preserves its dispersity with time. The viscosity increase of the oxidized MWCNT nanofluid was lower than that for the non-oxidized one, and the increase in the thermal conductivity of the oxidized MWCNT sample was satisfactory.

Chen and Xie [43] studied the stability of water based MWCNT (of 30–50 nm diameter and 20 μm length) nanofluids with and without cationic gemini surfactant, through the UV-visible absorption. In the nanofluids without the surfactant, aggregates were present, whereas the zeta potential of the sample with surfactant was greater than 50 mV, indicating a high stability. Also, the samples containing gemini surfactant with longer spacer chain length revealed poor stability. Chen et al. [44] also studied the stability of distilled water, ethylene glycol, and glycerol based MWCNT nanofluids, with the aim of surfactant-free dispersions. The MWCNT's were treated with mechanochemical reaction method to obtain improved dispersibility. The results revealed that the pristine carbon nanotubes were prone to form aggregates, whereas the treated ones remained stable in the suspension without any visible precipitation for months.

Stability of water based SWCNH (single walled carbon nanohorn, with an equivalent diameter of 100 nm), TiO_2 (21 nm) and CuO (30–50 nm) nanofluids was investigated by Fedele et al. [45].

The effect of the dispersion method (sonication, ball-milling, and high-pressure homogenization) on stability was investigated. SDS (sodium *n*-dodecyl sulfate) and polyethylene glycol (PEG) were used as dispersants for SWCNH and TiO_2 nanofluids, respectively. The results revealed that high pressure homogenization provided better stability compared to the other two methods. Hwang et al. [46] compared the effectiveness of different ultrasonication methods (in two-step method) and the magnetron sputtering (in one-step method) on the stability of carbon black and Ag nanofluids. The high pressure homogenization was the most effective one among the two-step methods, and extremely stable nanofluids were obtained through the use of magnetron sputtering.

Stability of deionized water, ethylene glycol and oil based MWCNT (of 10–30 nm diameter and 10–50 μm length) nanofluids were studied by Hwang et al. [47] together with fullerene (C_{60} and C_{60} and C_{70}, 10 nm diameter), CuO (35.4 nm diameter) and SiO_2 (12 nm diameter) nanofluids through UV-vis spectrophotometry. An ultrasonic disruptor was used at the preparation stage. Addition of the surfactant SDS improved the stability of nanoparticles in water.

Mao et al. [48] investigated the suspension stability of 0.5 wt.% Al_2O_3-water nanofluid under different ultrasonic vibration times, pH values, and dispersant (sodium dodecyl benzene sulfonate, SDBS) concentrations. The diameter of the nanoparticles was reported as 10 nm. They reported improvements in the stability for ultrasonications longer than 0.5 h and at relatively high SDBS concentrations. On the other hand, for SDBS concentrations higher than 0.5 wt.%, the adsorption of the nanoparticle surfaces were reported to be oversaturated, and the stability of the suspension got deteriorated. Figure 6.5 shows images of studied samples. The most satisfactory dispersity of Al_2O_3 (0.5 wt.%) nanofluid was obtained for an SDBS mass fraction of 0.5%, a base fluid pH value of 7, and an ultrasonic vibration time of 1 h.

Nine et al. [49] experimentally investigated the stability of nanofluids (1 wt.%) in a single loop oscillating heat pipe. The prepared nanofluid samples contained Cu and Ag (with mean diameters < 100 nm), and the base fluid was deionized water. No dispersants were added to the suspensions. Changes in the color, deposition, dispersity, tendency of oxidation, disintegration, agglomeration and thermal conductivity of the metallic nanofluids experiencing heat transfer were observed. The authors stressed that the change in color might be due to (1) *the deposition of nanoparticles on heat transfer surface lowering the overall concentration yielding a lighter color*, and (2) *the chemical instability of the nanofluids*. After the heat transfer experiments, the Cu particles were irregular in shape and smaller in size, and the Ag particles were in agglomerated state but still smaller than their original size. The dispersion quality depended on particle size, concentration and purity, particle adhesive property, and pH of the suspension [49].

Ijam et al. [50] studied the stability and some of the thermophysical properties of graphene oxide nanosheet (GON) in water. The nanofluids were prepared via the two-step method. Thermal conductivities of all samples remained the same for 14 days, indicating a high stability. No sedimentation was observed for the samples relaxed for 5 months, as shown in Figure 6.6. The results were also confirmed through UV-visible spectroscopy.

Ponmani et al. [51] investigated the stability of deionized water and XG (dispersant) based ZnO (diameter < 50 nm) and CuO nanofluids using SEM and DLS measurements. The samples were prepared via the two-step method. The presence of XG as the dispersant resulted in a better dispersion by forming a gel-like structure around the nanoparticles. For 0.5 wt.% nanofluids, significant agglomeration was observed for both nanofluids after the first 3 days, and the agglomeration got more pronounced for higher nanoparticle concentrations. The CuO nanofluid with XG was more stable than the ZnO nanofluid in the same base system.

Hordy et al. [52] studied the stability of MWCNTs (with an average diameter of 30 nm and a length of 4 μm) in water/EG/PG/Therminol® VP-1 heat transfer fluid in the long term, and under different heating conditions. The surfaces of MWCNTs were functionalized with low pressure capacitively-coupled radio frequency glow discharge plasma sustained in an argon/oxygen/ethane

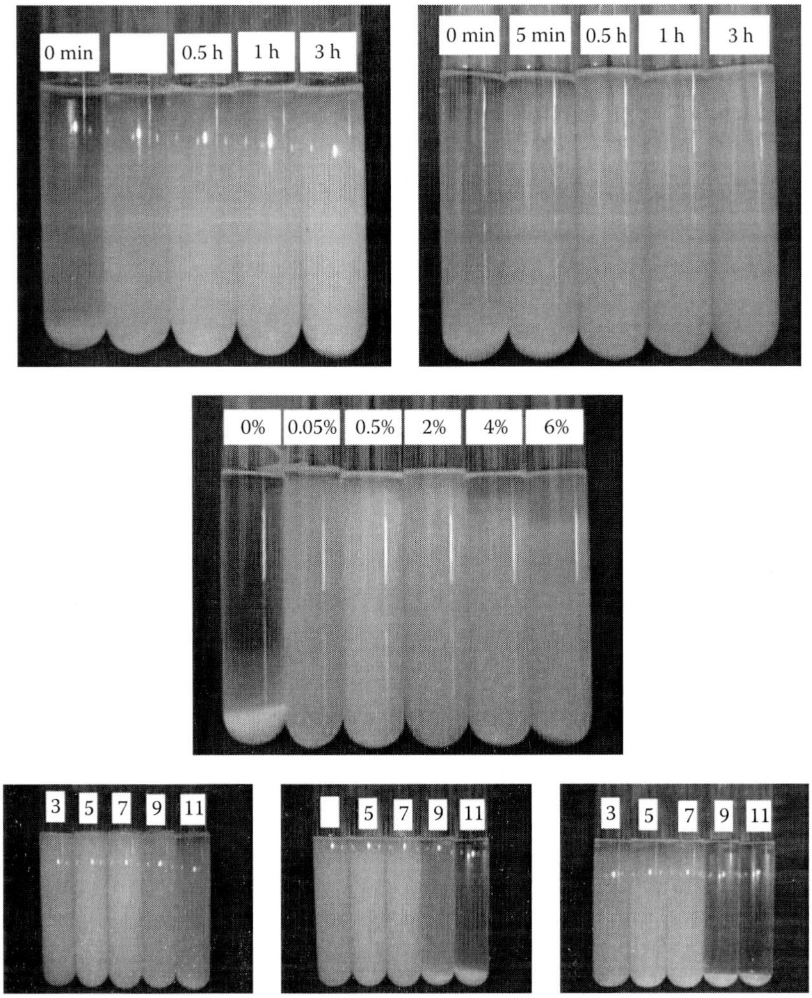

FIGURE 6.5 Top to bottom: effects on the ultrasonic vibration (without dispersant (left) and with dispersant (right)), dispersant concentration, and pH value (1 h, 2 h, and 2 days relaxation—from left to right) on nanofluid stability. (Reprinted from *Int. J. Adv. Manuf. Tech.*, 71(9–12), Mao, C., Zou, H., Zhou, X., Huang, Y., Gan, H., and Zhou, Z., Analysis of suspension stability for nanofluid applied in minimum quantity lubricant grinding, 2073–2081. Copyright 2014, with permission from Elsevier.)

FIGURE 6.6 The samples right after preparation (left) and the same samples after 5 months. (Reprinted from *J. Mater. Sci.*, 49(17), Ijam, A., Moradi Golsheikh, A.M., Saidur, R., and Ganesan, P., A glycerol–water-based nanofluid containing graphene oxide nanosheets, 5934–5944. Copyright 2014, with permission from Elsevier.)

gas mixture to make them dispersible in water. To study the long-term stability, the samples were stored at room temperature and were not treated with agitation.

Mahbubul et al. [35] investigated the optimum ultrasonication conditions for 0.5 vol.% Al_2O_3-water nanofluid prepared via the two-step method. The samples were homogenized using an ultrasonic horn for 1–5 h at 25% and 50% ultrasonication amplitude. The initial particle size was provided as 13 nm, and 0.5 vol.% sample was selected for further stability investigation as it yielded better dispersion characteristics among other concentrations. The dispersion characteristics of the samples were examined through TEM, PSD, and zeta potential analyses. Better particle dispersion, smaller aggregate size, and higher zeta potentials were obtained for 3 and 5 h long ultrasonications and at 50% and 25% sonicator power amplitudes, respectively. In another study, Mahbubul et al. [30] investigated the effect of discrete ultrasonication perioas, for example, 2 min ON and 2 min OFF. After 30 days from the preparation, sedimentation was evident for the sample prepared without ultrasonication, while it was negligible for the ones treated with ultrasonication.

Elçioğlu and Okutucu-Özyurt [31] studied the dispersion stability of 1-year-old Al_2O_3-H_2O nanofluids of 30 ± 10 nm initial particle size through zeta potential and PSD measurements. They measured the PSD of the 0.5, 0.25, and 0.125 vol.% samples in a weekly manner. The average particle sizes of the 1-year-old Al_2O_3-H_2O samples varied between 70–90, 60–85, and 55–75 nm for no ultrasonication and after 45 min, 60 min, and 2 h ultrasonications, for 0.125, 0.25, and 0.5 vol.% samples, respectively, over 3–4 weeks. The results revealed a decrease in the particle sizes for all samples examined. On the other hand, the initial particle size of 30 ± 10 nm could not be re-established with the ultrasonications tried. As the samples did not show any nanoparticle sedimentation, it can be concluded that the sedimentation photograph capturing alone may not be sufficient to conclude about the PSD of nanofluids.

Ilyas et al. [53] investigated the settling characteristics of Al_2O_3 (with a diameters of 40 and 50 nm) based in EG-water binary fluid nanofluids. Nanoparticle weight concentrations of 0.1%, 0.3%, and 0.5% were considered with different proportions of EG-water mixtures containing no surfactants or stabilizers. The sedimentation behavior of the samples with and without ultrasonication was examined through the measurement of the sedimentation height via visualization method in batch sedimentation. The authors measured the *sediment ratio* defined as the ratio of the sediment volume to the total volume. The sedimentation height increased with time due to the gravity and agglomeration effects. Also, the sediment height was smaller when the concentration of EG was higher. The use of ultrasonic agitation provided better stability.

Subramaniyan et al. [54] studied the stability of TiO_2-water nanofluids via the measurement of interfacial layer thickness and attenuation of transmitted light intensity. This technique was expressed as economic, reliable but time consuming when compared to the zeta potential measurements.

Kazemi-Beydokhti et al. [55] examined the effects of the size, volume fraction, and material of nanoparticles, the type of the base fluid, the sonication, and the elapsed time on the enhanced thermal conductivity and stability. Experiments have been performed with nanofluids synthesized with different methods. No additives or stabilizers have been used during the preparation. The stability of the nanofluids has been analyzed using UV-visible scanning spectrophotometer, and visual inspection. There was an optimum sonication time, for which maximum thermal conductivity and minimum hydrodynamic cluster size was observed. As the time elapsed, a decrease in the thermal conductivity has been observed due to particle clustering and agglomeration.

Pang and Kang [56] studied the dispersion stability of 0.1 vol.% Al_2O_3 (with a primary size of 40–50 nm) nanoparticles in H_2O/CH_3OH binary mixture through zeta potential measurements, visualization, and Tyndall effect (light scattering by the particles in the suspension). The samples were prepared via the two-step method (with ultrasonication) and they contained 0.05 wt.% gum Arabic (GA) as a staric stabilizer. The dispersity of the nanoparticles within the nanofluid was examined by taking photographs of the Tyndall effect. The concentration of particles was maximum right after preparation and it decreased within 1 week due to the weak stability. The addition of NaCl resulted in a poor stability [56].

Huang and Wang [57] studied the dispersion behavior of Al_2O_3-water and Cu-water nanofluids through the absorbency sediment photography capturing and zeta potential measurements. The average particle sizes were 15–50 nm and 25–60 nm for Al_2O_3 and Cu nanoparticles, respectively. The nanofluid samples contained SDBS as a surfactant. The absorbency tests were performed on 0.1 wt.% Al_2O_3-water and 0.1 wt.% Cu-water samples. The presence of SDBS resulted in better dispersion. The optimum pH values for the studied nanofluids were reported as 8.0 and 9.5 for Al_2O_3 and Cu, respectively.

Wang et al. [58] investigated the dispersion stability of 0.05 by mass % Al_2O_3 (plate shaped)-water and 0.01 by mass % TiO_2 (spherical)-water nanofluids through the measurements of zeta potential and light absorbency. The mean particle diameters were estimated to be 25–50 nm. As a surfactant, SDS was used. The optimum pH values providing the highest zeta potential and absorbency were provided as $pH_{Al_2O_3} \approx 6.0$, and $pH_{TiO_2} \approx 9.5$.

Thrush [59] investigated the dispersion stability of surface functionalized Al_2O_3-Ergon P70N (a light paraffinic oil) nanofluids containing oleic acid or PIBSA (polyisobutylene succinic anhydride) as a surfactant. Better dispersion was obtained with the use of ultrasonic disruptor compared to the ultrasonic bath, as smaller average particle sizes were achievable at lower specific energies with the disruptor. The addition of the surfactant before the ultrasonication resulted in better dispersion and smaller average particle size. The most stable sample had a 28:1 ratio between the oleic acid and the nanoparticles (14 wt.% oleic acid at 0.5 wt.% alumina). The dependence of stability on oleic acid concentration is presented by the sedimentation photographs in Figure 6.7.

In addition, after 4 days of relaxation, the samples containing the surfactant PIBSA showed better stability (less sedimentation) compared to the oleic acid stabilized nanofluids.

Sahooli and Sabbaghi [60] studied the stability of CuO-EG nanofluids prepared via the two-step method and containing PVP (polyvinyl pyrrolidone) as a surfactant, through the measurements of zeta potential and absorbency. The average diameter of CuO nanopowders was provided as 4 nm. The sedimentation experiments were performed on the nanofluids mechanically stirred or ultrasonicated, which were allowed to sediment for a few days. The sedimentation photographs showed that the stability of the sample with surfactant could last about 1 month in the stationary state without any sedimentation, while that with no surfactant had poor stability, in comparison. The optimum pH values for the samples with and without PVP as 9.5 and 10.5, respectively and the optimum PVP concentration was 0.05 wt.% [60].

Li et al. [61] worked on a water based SiO_2-coated graphene nanofluid to investigate the effect of coating on thermal conductivity and stability. Three samples were experimented: graphene-SiO_2, graphene, and graphene with SDBS as the surfactant. Firstly, the effects of the surfactant and the coating on the stability were assessed by sediment photography capturing, and UV-visible absorbency methods.

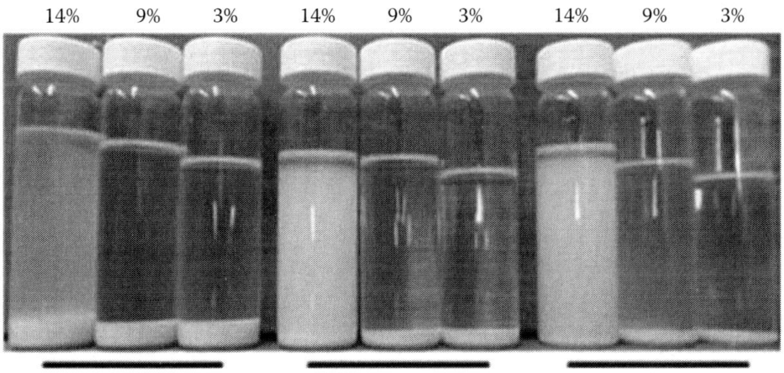

FIGURE 6.7 Sedimentation photographs showing the dispersion behavior of different sample concentrations, 4 days after a 120 W, 150 s lasting disruptor ultrasonication. (From Thrush, S.J., MS thesis, Oakland University, Rochester, MI, 2012.)

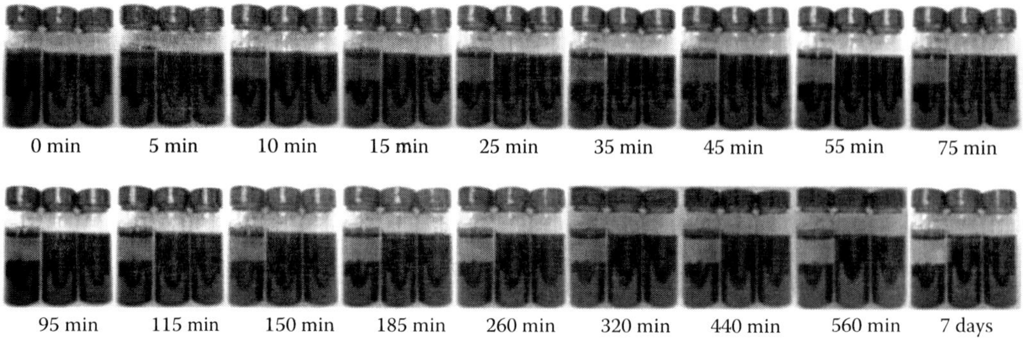

FIGURE 6.8 Sediment images of graphene, SiO_2 coated graphene, and graphene with SDBS as surfactant. (Reprinted from *Thermochim. Acta*, 595, Li, X., Chen, Y., Mo, S., Jia, L., and Shao, X., Effect of surface modification on the stability and thermal conductivity of water-based SiO_2-coated graphene nanofluid, 6–10. Copyright 2014, with permission from Elsevier.)

In Figure 6.8 sediment images of the three samples (from left to right: graphene, SiO_2 coated graphene, and graphene with SDBS as surfactant) are shown. Graphene without any surfactant and coating settled very quickly; the color change from the bottom to the top was very clear after 10 min. However, in the sample with SiO_2 coating, or that with added SDBS, no settlement was observed in the photographs. The UV-visible absorbency results are presented in Figure 6.9. All three samples have a common sharp decrease in the absorbance during the first 2 days, and then the trends deviated from each other.

The results obtained from the sediment images and the UV-visible absorbency indicated that, the inorganic coating or the addition of surfactant had similar effects on the stability; both can be used for the stability improvement.

Elango et al. [62] studied the usage of nanofluids for heat transfer enhancement in a solar still which functions to distilled water using solar energy. Different metallic nanoparticles (zinc oxide,

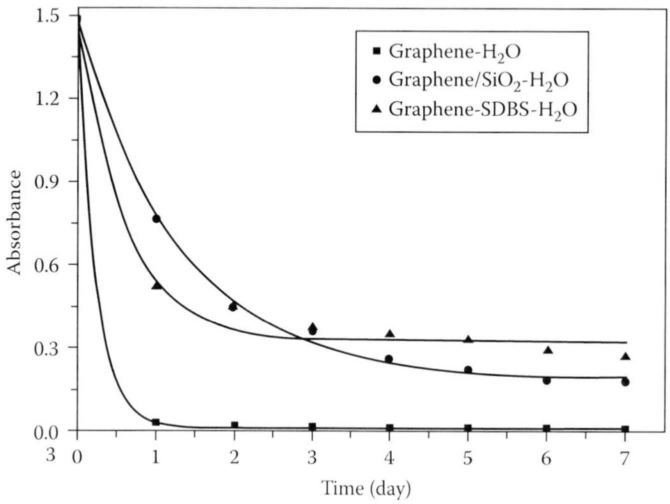

FIGURE 6.9 UV-visible absorbency of graphene-water, graphene/SiO_2, and graphene-water nanofluid with SDBS. (Reprinted from *Thermochim. Acta*, 595, Li, X., Chen, Y., Mo, S., Jia, L., and Shao, X., Effect of surface modification on the stability and thermal conductivity of water-based SiO_2-coated graphene nanofluid, 6–10. Copyright 2014, with permission from Elsevier.)

iron oxide, aluminum oxide, and tin oxide) were dissolved in water. Stability analysis has been made using zeta potential and particle size distribution analyses. The stabilities of water-based metal-oxide nanofluids of 0.05% and 0.1% concentrations have been compared. It has been observed that the sample with 0.1% concentration was comparatively more stable. The study emphasized that the performance of a solar still is directly affected by the stability as well as the thermal conductivity of the working fluid, and, naturally, the climatic conditions.

Leong et al. [63] studied stability and thermal conductivity of MWCNTs with and without surfactant. Ten samples have been used in the experiments. CNTs were dissolved in ethylene glycol or ethylene glycol-water mixtures with different concentrations. A surfactant has been added to five of the samples whereas the other five were without any surfactant. Stability analyses were made by zeta potential measurements, and observation. Images of the samples were taken for 4 weeks, and coagulation was observed in the 0.01 and 0.05 wt.% samples (with no surfactant). No settlement occurred in the nanofluid with the surfactant. Nanofluids with the surfactant showed comparatively higher absolute zeta potential values, hence higher stability. A comparison among the surfactant added nanofluids indicated that carbon nanotubes dissolved in ethylene glycol-water mixture with hexadecyltrimethylammonium bromide as the surfactant had the highest absolute zeta potential.

Allouni et al. [64] investigated the colloidal stability of TiO_2 nanoparticles in cell culture medium through the sedimentation, DLS, and zeta potential measurements, and concluded that the particle concentration, ionic strength, and the presence of proteins had important impacts on the colloidal stability of TiO_2 nanoparticles in a cell culture medium (e.g., RPMI).

Chang et al. [65] aimed to investigate the properties and stability of Co/Ag nanocompound particles dispersed in deionized water. The morphology and size distribution of nanoparticles were investigated through SEM and particle size analyzer. The nanoparticles had a good dispersion even without the presence of a surfactant.

Safaei-Naeini et al. investigated the stability of distilled H_2O based TiO_2 nanofluids by observing their sedimentation time and absorbency [66]. With increased nanoparticle concentration, stability of the samples decreased.

Zhu et al. [67] studied Al_2O_3-H_2O nanofluids (with a median of particle size distribution of 15–50 nm), and reported a strong dependence of thermal conductivity on the dispersion pH value. The optimum pH corresponding to the 0.05 wt.% sample containing SDBS as a surfactant was 8, which corresponded to a zeta potential as high as −40 mV. The optimum pH for SiO_2-water nanofluids was reported by Mondragon et al. [68] as greater than 7, which provided a stable condition that could be sustained for at least 48 h.

Raykar and Singh [69] conducted experiments with ZnO nanoparticles in water with acetylacetone (acac) ligand as a stabilizer. The size of the nanoparticle cluster decreased with the addition of acac as it broke the ZnO cluster down. The samples in the mentioned study were stable for up to 9 months to 1 year.

Yu et al. [70] investigated the stability of Cu-EG nanofluids prepared by two-step method with the addition of PVP as a surfactant. PVP improved the stability significantly, and the prepared nanofluids showed no settlement. On the other hand, the authors mentioned the drawbacks of adding surfactant as the increment in viscosity and reduction in thermal conductivity.

Rashmi et al. [71] studied the stability of MWCNT-distilled H_2O nanofluids prepared by two-step method with and without GA. Ultrasonication times longer than the optimum and the amount of GA higher than the optimum resulted in a decrease in stability.

Abbasi et al. [72] studied the stability of DI-water based hybrid MWCNT and Al_2O_3 nanofluids with GA as the surfactant. All functionalized MWCNTs remained well-dispersed within the nanofluids for several weeks. The functionalization did not cause any difference in the thermal conductivities of the nanofluids. Wusiman et al. [73] investigated the stability of MWCNT-distilled H_2O nanofluids for different type and amount of surfactants (SDS and SDBS). The addition of surfactants decreased the thermal conductivity, with less decrease caused by SDBS.

Lee et al. [74] used DLS and hydrometer together to analyze the dispersion stability of five nanofluid samples (water based Ag, Al_2O_3, SiO_2, and EG based ZnO, Al_2O_3). According to the measurements taken for 31 days, the volume fraction, the particle size (except for the EG based ZnO nanofluid), and the thermal conductivity of all nanofluids decreased with time.

6.3.3 Ways to Improve Nanofluid Stability

Sustaining the stability of nanofluids for reliable applications is crucial. Several methods have been found in the literature to maintain the stability. These methods include (1) *addition of surfactants*, (2) *adjustment of the pH value of the dispersion*, (3) *nanoparticle surface modifications*, and (4) *mechanical agitation* (*ultrasonication*).

Surface active agents (abbreviated as *surfactants*) are organic molecules that can be adsorbed at interfaces when dissolved in low concentrations [75]. Surfactants are used to prevent the nanoparticle aggregation. Some of the most commonly used ones in nanofluids literature are GA, sodium octanoate (SOCT), salt, oleic acid, sodium dodecyl benzene sulfonate (SDBS), and polyvinyl pyrrolidone (PVP) [36]. In nanofluids, an interface is formed between the nanoparticles and the base fluid. By the addition of a surfactant, an increase is observed in the contact of the two materials causing an increase in the wettability. The reason for this increase is the surfactant located at the interface creating some degree of continuity between the two phases [76]. The effective use of surfactants depends on many factors, such as the working temperature, type of the surfactants, nanoparticles and the base fluid, and their concentration in the mixture. The addition of surfactants is a purely chemical process, and the surfactant behavior in the nanofluid should be carefully monitored. Thrush [59] examined the appropriate step for adding the surfactant in the two-step method, and reported that the addition before and after ultrasonication resulted in different dispersions. The condition of surfactants should also be studied carefully, in relation with temperature changes.

The *pH values* of nanofluids can be adjusted to improve the stability as the zeta potential strongly depends on pH. To maximize the zeta potential, which is required for a mono-dispersed and sediment-free suspension, the pH value of the suspensions should be modified using appropriate amounts of acidic or alkaline chemical agents such as HCl and NaOH. There exists an optimum pH value for a stable nanofluid which can be achieved by making the addition of these agents in trace amounts.

The *nanoparticle surface chemistry* directly influences the dispersion. The surfaces of nanoparticles can be modified to have appropriate dispersions, since the surface properties make nanoparticles dispersible/compatible with the host fluid. Schrand et al. [77] demonstrated the effect of MWCNT surface properties on their dispersion within water. The hydrophobic character of MWCNT surfaces prevented their dispersion within water and needed intensive ultrasonication even for a non-sufficient yet acceptable dispersion. The researchers reported that the ultrasonication could not yield long-term stability as the MWCNTs were only temporarily suspended in water. Kamiya and Iijima [78] reviewed the surface modifications of nanoparticles to have an improved dispersion within liquid media.

Ultrasonication has been one of the most commonly used methods to end up with a stable suspension. Taurozzi et al. [79] provided guidelines for the preparation of nanoparticle dispersions in liquid media by the application of ultrasonic energy. They classified the ultrasonication methods into two sub-groups: (1) *direct ultrasonication* (*via an ultrasound probe or horn*), (2) *indirect ultrasonication* (*via a sonication bath*). They recommended the use of direct sonication for the dispersion of dry powders as it provides higher effective ultrasonic energy output. The indirect sonication could be used for resuspending a preprocessed dispersion with direct ultrasonication or dispersion subjected to unintended modifications and damages under direct sonication. Taurozzi et al. [79] stated that sonication, as a highly system-specific procedure could "result in either breakdown of the nanoparticle clusters or further agglomeration, as well as other effects including chemical reactions."

6.4 CHANGE IN THERMOPHYSICAL PROPERTIES DUE TO THE STABILITY DETERIORATION

The thermophysical properties of nanofluids are defined based on their colloidal and morphological characteristics. The properties depend on parameters, including, but not limited to, the fraction, diameter and shape of nanoparticles, dispersion temperature, etc. The dependence of nanofluid thermophysical properties on the aforementioned parameters have been validated through experiments.

The potential improvements in thermal conductivity led to extensive research on nanofluids. The increase in nanofluid thermal conductivity compared to that of pure fluids has been attributed in [80] to the following mechanisms: (1) *nanoparticle-liquid interaction in the form of interfacial thermal resistance*, (2) *formation of condensed nano-layers around the particles*, (3) *the particle size effects*, (4) *nanoparticle agglomeration*, (5) *the micro-convection due to Brownian motion of nanoparticles in the liquid*, (6) *surface plasmon resonance*, and (7) *near field radiation*. Since nanoparticle aggregation is an inevitable outcome in the long term, its effects on thermophysical properties are of critical importance.

By presenting the behavior illustrated in Figure 6.10, Wen et al. [81] mentioned an optimum aggregation level which could provide a maximum thermal conductivity that is far beyond the predictions for homogeneous dispersions.

Prasher et al. [82] investigated the role of aggregation on the thermal conductivity of nanofluids. The authors presented a three-level homogenization theory for the evaluation of the effective thermal conductivity of nanofluids involving fractal clusters. The authors considered the average radius of gyration of the aggregates, and the fractal and chemical dimensions of the aggregates, and demonstrated that the presence of these fractal aggregates could lead to enhancement in thermal conductivity more than that estimated by the homogenization theories of well-dispersed composites.

A schematic of the modeling of Prasher et al. [82] is shown in Figure 6.11. The modeling was based on the consideration of the linear chains spanning the whole cluster (aggregate), and side chains. Due to the conductivity of the linear chains, the authors reported that the thermal conductivity of a clustered nanofluid was enhanced more compared to a well-dispersed suspension. On the other hand, as time elapses, the nanoparticle aggregates get heavier, and sedimentation occurs. Nanoparticle sedimentation reduces the effective thermal conductivity of nanofluids due to decreased nanoparticle concentration dispersed within the nanofluid, as also stated in [53]. A decrease in the thermal conductivity as a result of nanoparticle aggregation has also been reported by Sharma et al. [83] for silver-ethylene glycol nanofluids.

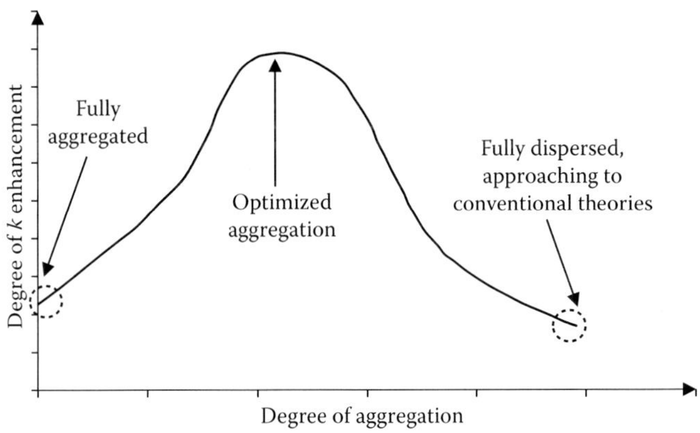

FIGURE 6.10 Effective thermal conductivity enhancement of nanofluids with nanoparticle aggregation. (Reprinted from *Particuology*, 7(2), Wen, D., Lin, G., Vafaei, S., and Zhang, K., Review of nanofluids for heat transfer applications, 141–150. Copyright 2009, with permission from Elsevier.)

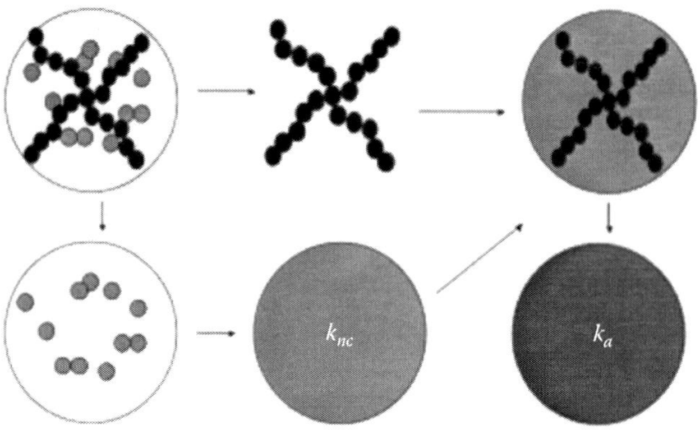

FIGURE 6.11 The schematic of an aggregate consisting of linear and side chains. (Reprinted from *Appl. Phys. Lett.*, 89, Prasher, R., Evans, W., Meakin, P., Fish, J., Phelan, P., and Keblinski, P., Effect of aggregation on thermal conduction in colloidal nanofluids, 1–3. Copyright 2006, with permission from Elsevier.)

Sadri et al. [84] studied the stability, thermal conductivity and viscosity of MWCNT-distilled water nanofluids containing GA, SDBS, or SDS as surfactants. The nanotube outer diameter and length were provided as 20–30 nm and 10–30 μm, respectively. The addition surfactants lowered the thermal conductivity of the base fluid, and addition of 0.25 wt.% GA yielded the highest thermal conductivity for 0.5 wt.% MWCNT-water nanofluids. Thermal conductivity enhancements were observed with increasing ultrasonication durations from 2 to 40 min. The viscosity of the suspensions increased with 2–7 min-ultrasonication, and it decreased for longer sonication periods. The results indicated that the ultrasonication broke down the clusters and aggregates, but longer ultrasonications resulted in fragmentation of MWCNTs into smaller particles (shortening of the mean length), yielding lower viscosities.

As stated by Timofeeva in [80] a nanofluid can be described as a three-phase system consisting of the nanoparticles, base fluid, and the interface. The author mentions the establishment of dispersion or agglomeration equilibrium based on the "particle concentration and the magnitude of the particle–particle interactions." An increase in agglomeration could increase the thermal conductivity. However, nanoparticle clustering results in an unwanted increase in viscosity and sedimentation. The thermal conductivity rise is an apparent advantage for heat transfer applications. On the other hand, the dramatic increase in viscosity is a critical drawback causing the pumping power requirement to increase. Hence, there is a need to have criteria for the flow conditions to determine whether the use of nanofluid is feasible or not. The feasibility criteria are detailed under Section 6.5.

Yu et al. [85] studied the thermal conductivity of liquid paraffin based nanofluids containing copper nanoparticles. The results showed that the thermal conductivity did not change with time for the samples containing dispersant, whereas it did vary with time for the samples without dispersant. This indicates that for unstable nanofluids, the thermophysical properties may change with time, while for stable nanofluids, they are maintained for a certain period of time.

6.5 NANOFLUID FEASIBILITY CRITERIA FOR LAMINAR AND TURBULENT FLOW CONDITIONS

For forced convection heat transfer applications, the maximum allowable pressure drop and the pumping power are equally important as the design of the heat transfer unit (area) [87]. Hence, in addition to the thermal issues, the effective operation of the system should also be considered.

For the implementation of nanofluids in a variety of heat transfer practices, development of a fundamental understanding on the essential principles of heat transfer and wall friction are of critical importance [86]. The fluid viscosity is directly proportional to the friction factor [87], hence, the pressure drop, and the pumping power required to circulate the working fluid through the system. Then, viscosity as a transport property stands as an important parameter in the criteria for feasible nanofluid usage in heat transfer applications.

6.5.1 Nanofluid Figure of Merit in Different Flow Conditions

The relatively high thermal conductivities of nanofluids made them promising working fluid candidates in forced convection heat transfer applications. On the other hand, their high viscosity and tendency of easily transforming into a non-Newtonian fluid is a serious penalty, as it directly increases the pumping power requirement.

The efficiency criteria of nanofluid use in convection heat transfer applications are available in literature for fully developed laminar and turbulent flow conditions. For the laminar flow in circular tubes, the criteria are given in [88,89] as

$$\frac{\mu_{nf}}{\mu_{bf}} \approx 1 + C_{\mu}\varphi, \quad \frac{k_{nf}}{k_{bf}} \approx 1 + C_{k}\varphi, \quad \frac{C_{\mu}}{C_{k}} < 4 \tag{6.5}$$

where

- the subscripts nf and bf stand for nanofluid and base fluid
- k and μ denote the thermal conductivity and viscosity
- C_{μ} and C_{k} are viscosity and thermal conductivity enhancement coefficients, respectively

For turbulent flow, the performance criterion depends not only on the thermal conductivity, but also on the density, constant pressure specific heat, and viscosity of the nanofluid. The dependence is given by the Mouromtseff number, which is briefly defined in Section 6.1. The constant powers appearing in the Mouromtseff number are determined empirically, and given in [88] as

$$Mo = \frac{\rho^{0.8} k^{0.67} c_p^{\,0.33}}{\mu^{0.47}}. \tag{6.6}$$

The liquid with greater Mouromtseff value provides higher heat transfer rate [88].

As may be inferred from the previous relations, the increase in the viscosity of nanofluids results in significant deterioration of their efficient use in both laminar and turbulent flow conditions. Establishing a mono-dispersed condition comes to the fore in here, as well, in order to preserve the thermal conductivity enhancement without significant rise in the viscosity.

6.5.2 Heat Transfer Enhancement of Nanofluids in Different Flow Conditions

Yu et al. [5] reviewed the literature for the heat transfer enhancement with nanofluids in different flow regimes. In laminar flow, heat transfer augments as the particle volume concentration increases. At low volumetric concentrations, the heat transfer enhancement is nearly independent from *Re*. At higher volumetric concentrations, on the other hand, a greater *Re* leads to enhanced heat transfer. Similarly, in turbulent flow, as the volumetric concentration increased an enhancement in heat transfer occurred, and the *Re* did not affect the enhancement. Despite the extensive research in the laminar and turbulent flow regimes of nanofluids, there is little work in pool boiling. According to the limited number of studies, it has been observed that the nanofluid use did not improve heat transfer in pool boiling mainly due to the deposition of nanoparticles on heated surfaces. It can be concluded that laminar or turbulent nanofluid flows seem to have less stability problems. However, stability is an issue in pool boiling of nanofluids.

6.6 EXAMPLE APPLICATION AREAS OF STABLE NANOFLUIDS

In the last two decades, extensive research has been conducted in laboratories to understand the heat transfer performance of nanofluids. More recently, nanofluids gained attention from the industry, and its application areas have been enlarged. A review of nanofluid applications has been provided recently by Wong and De Leon (Chapter 1 in [90]). To have an effective usage of nanofluids in these application areas, the stability of the nanofluids should be maintained. Nanofluids have been reported as promising mainly in the following heat transfer applications: refrigeration systems (nanofluids used in those systems are called *nanorefrigerants*) [91–97], electronics industry [98,99], automotive industry [5], quenching heat treatment [100–104], metal processing [105–108], and thermal storage [109–111]. Their use as cancer therapeutics in biomedical applications [112] has also been studied.

6.7 CONCLUSIONS

As nanofluid research is highly interdisciplinary, investigating the nanofluid stability requires experimental capabilities and theoretical understanding in many scientific areas including, but not limited to, engineering, materials science, and chemistry. As also emphasized by Li et al. [113], it is essential to optimize nanofluid properties including the long-term stability for practical applications. Investigating nanofluid stability is of critical importance in order to provide a direction for prospective nanofluid experiments, as well as to ensure the safety, reliability and repeatability of the applications. In so doing, this chapter touched upon the following:

1. The state-of-the-art, the measurement instruments and techniques of nanofluid stability
2. The fundamentals of nanofluid stability to establish the link between the theory and experiments
3. The impacts of stability change on the nanofluid thermophysical properties
4. Feasibility criteria for laminar and turbulent nanofluid flows
5. Cutting-edge research in nanofluids field by highlighting the applications

The main conclusions drawn from the studies in the literature on nanofluid stability may be summarized as follows:

- Stability of nanofluids must be attained and maintained in all applications.
- Zeta potential measurements, light scattering techniques, microscopy techniques, UV-visible spectrophotometer measurements, and sediment photography capturing methods are the main tools for stability assessment.
- Stability may be improved by means of surfactant addition, pH adjustment, nanoparticle surface modifications, and ultrasonication.
- The addition of surfactants increases stability with the cost of decreased thermal conductivity.
- Although particle aggregation deteriorates stability, it is shown to sometimes improve thermal conductivity. There is an optimum aggregation level providing a maximum thermal conductivity beyond the predictions for homogeneous dispersions.
- Stable nanofluid concentration should be adjusted considering the undesired increase in viscosity to avoid high pumping powers.
- For the long-term usage of nanofluids as coolants/refrigerants and in electronics, automotive, heat treatment, metal processing, thermal storage, and biomedical applications, their stability must be established.
- In nanofluid applications, flow and/or oscillations are preferred rather that stationary systems in terms of stability.

To conclude this chapter, the stability of nanofluids deserves much more attention, and further research is required in terms of the fundamentals and applications. It is beyond any doubt that caution should be taken before using nanofluids in industrial applications, and standards must be established on the preparation and maintenance of nanofluids for stable usage.

REFERENCES

1. S. U. S. Choi and J. A. Eastman, *Enhancing Thermal Conductivity of Fluids with Nanoparticles*, ASME International Mechanical Engineering Congress and Exposition, San Francisco, CA, November 12–17, 1995.
2. A. Sergis and Y. Hardalupas, Anomalous heat transfer modes of nanofluids: A review based on statistical analysis, *Nanoscale Research Letters*, 6, 391, 1–37 (2011).
3. S. Ozerinc, S. Kakac, and A. Guvenc Yazicioglu, Enhanced thermal conductivity of nanofluids: A state-of-the-art review, *Microfluidics and Nanofluidics*, 8(2), 145–170 (2010).
4. W. Yu and S. U. S. Choi, The role of interfacial layers in the enhanced thermal conductivity of nanofluids: A renovated Maxwell model, *Journal of Nanoparticle Research*, 5(1–2), 167–171 (2003).
5. W. Yu, D. M. France, J. L. Routbort, and S. U. S. Choi, Review and comparison of nanofluid thermal conductivity and heat transfer enhancements, *Heat Transfer Engineering*, 29(5), 432–460 (2008).
6. S. M. S. Murshed, K. C. Leong, and C. Yang, Investigations of thermal conductivity and viscosity of nanofluids, *International Journal of Thermal Sciences*, 47(5), 560–568 (2007).
7. H. Chen, Y. Ding and C. Tan, Rheological behaviour of nanofluids, *New Journal of Physics*, 9, 367, 1–24 (2007).
8. K. Kwak and C. Kim, Viscosity and thermal conductivity of copper oxide nanofluid dispersed in ethylene glycol, *Korea-Australia Rheology Journal*, 17, 35–40 (2005).
9. H. Setia, R. Gupta, and R. K. Wanchoo, Stability of nanofluids, *Materials Science Forum*, 757, 139–149 (2013).
10. A. Koyuncuoğlu, R. Jafari, M. Redmond, T. Okutucu-Özyurt, and H. Külah, MEMS Tabanlı Mikrokanallı Isı Kuyularında Su ve Nanoakışkanlar ile Isı Transferi İyileştirme Deneyleri (Heat transfer enhancement experiments in MEMS based microchannel heat sinks with water and nanofluids), *Proceedings of ULIBTK'13, 19. Turkish Thermal Sciences and Techniques Congress*, Samsun, Turkey, September 9–12, 2013.
11. Y. Xuan and Q. Li, Investigation on convective heat transfer and flow features of nanofluids, *Journal of Heat Transfer*, 125(1), 151–155 (2003).
12. J. L. Routbort, D. Singh, E. V. Timofeeva, W. Yu, and D. M. France, Pumping power of nanofluids in a flowing system, *Journal of Nanoparticle Research*, 13(3), 931–937 (2011).
13. A. E. Bergles and A. Bar-Cohen, Direct Liquid Cooling Of Microelectronic Components, in *Advances Thermal Modeling of Electronic Components and Systems*, Vol. 2 (Eds.: A. Bar-Cohen and A. D. Kraus), ASME Press, New York, 1990.
14. T. Yokoyama, H. Masuda, M. Suzuki, K. Ehara, K. Nogi, M. Fuji, T. Fukui et al., Basic properties and measuring methods of nanoparticles, in *Nanoparticle Technology Handbook* (Eds.: M. Hosokawa, K. Nogi, M. Naito, and T. Yokoyama), 1st edn., pp. 5–48, Elsevier, Oxford, U.K., 2007.
15. H-T. Zhu, S-Q. Liu, L. Xu and C-Y. Zhang, Preparation, characterization and thermal properties of nanofluids, in *Leading Edge Nanotechnology Research Developments* (Ed.: D. M. Sabatini), pp. 5–38, Nova Science Publishers, Inc., New York, 2007.
16. E. Kissa, *Dispersions Characterization, Testing and Measurement*, Markel Dekker Inc., New York, 1999.
17. G. T. Barnes and I. R. Gentle, *Interfacial Science: An Introduction*, 2nd edn., Oxford University Press, New York, 2011.
18. S. K. Das, S. U. S. Choi, W. Yu, and T. Pradeep, *Nanofluids Science and Technology*, John Wiley & Sons, Inc., Hoboken, NJ, 2008.
19. A. V. Delgado, F. González-Caballero, R. J. Hunter, L. K. Koopal, and J. Lyklema, Measurement and interpretation of electrokinetic phenomena, *Journal of Colloid and Interface Science*, 309(2), 194–224 (2007).
20. J. Kim and D. F. Lawler, Characteristics of zeta potential distribution in silica particles, *Bulletin of the Korean Chemical Society*, 26, 1083–1089 (2005).
21. B. Salopek, D. Krasić, and S. Filipović, Measurement and application of zeta-potential, *Rudarsko-Geoloiko-Naftni Zbornik*, 4, 147–151 (1992).

22. A. Turgut, I. Tavman, L. Cetin, M. Chirtoc, and O. Fudym, Preparation and characterization of nanofluids containing alumina nanoparticles, ICHMT Digital Library Online, Volume 0, 2011 Issue—TMNN-2010, *Proceedings of the International Symposium on Thermal and Materials Nanoscience and Nanotechnology*, Antalya, Turkey, May 29–June 3, 2011.
23. D. J. Shaw, *Introduction to Colloid and Surface Chemistry*, 4th edn., Butterworth-Heinemann Ltd, Oxford, U.K., 1992.
24. A. Ghadimi, R. Saidur, and H. S. C. Metselaar, A review of nanofluid stability properties and characterization in stationary conditions, *International Journal of Heat and Mass Transfer*, 54(17–18), 4051–4068 (2011).
25. R. Prasher, P. Bhattacharya, and P. E. Phelan, Thermal conductivity of nanoscale colloidal solutions (nanofluids), *Physical Review Letters*, 94(2), 025901, 1–4 (2005).
26. W. Evans, J. Fish, and P. Keblinski, Role of Brownian motion hydrodynamics on nanofluid thermal conductivity, *Applied Physics Letters*, 88(9), 093116, 1–3 (2006).
27. C. Koral, H. Altan, H. A. Vural, B. Ortaç, and T. Okutucu, Terahertz time domain spectroscopy investigation of nanofluids, *Proceedings of the Eighth Nanoscience and Nanotechnology Congress, Eighth Nanoscience and Nanotechnology Congress*, Ankara, Turkey, June 25–29, 2012.
28. W. C. Williams, I. C. Bang, E. Forrest, L. W. Hu, and J. Buongiorno, Preparation and characterization of various nanofluids, *NSTI-Nanotech*, 2, 408–411 (2006).
29. R. Xu, Progress in nanoparticles characterization: Sizing and zeta potential measurement, *Particuology*, 6(2), 112–115 (2008).
30. I. M. Mahbubul, T. H. Chong, S. S. Khaleduzzaman, I. M. Shahrul, R. Saidur, B. D. Long, and M. A. Amalina, Effect of ultrasonication duration on colloidal structure and viscosity of alumina–water nanofluid, *Industrial & Engineering Chemistry Research*, 53(16) 6677–6684 (2014).
31. E. B. Elçioğlu and T. Okutucu-Özyurt, An experimental study on the dispersion stability of alumina-water nanofluids via particle size distribution and zeta potential measurements, *Proceedings of THERMAM 2014 and Third Rostocker Symposium on Thermophysical Properties for Technical Thermodynamics*, Izmir, Turkey, June 12–15, 2014.
32. B. B. Weiner, W. W. Tscharnuter, and D. Fairhurst, Zeta potential: A new approach, *Canadian Mineral Analysts Meeting*, Manitoba, Winnipeg, Canada, September 8–12, 1993.
33. A Basic Guide to Particle Characterization, http://golik.co.il/Data/ABasicGuidtoParticleCharacterization(2)_1962085150.pdf, accessed March 2, 2015.
34. E. B. Elçioğlu, Experimental investigation on the dispersion stability of nanofluids, ME 552-Micro- and Nanoscale Heat Transfer Term Project, Department of Mechanical Engineering, Middle East Technical University, Ankara, Turkey, January 2014.
35. I. M. Mahbubul, R. Saidur, M. A. Amalina, E. B. Elcioglu, and T. Okutucu-Ozyurt, Effective ultrasonication process for better colloidal dispersion of nanofluid, *Ultrasonics Sonochemistry*, 26, 361–369 (2015).
36. E. Şimşek, E. B. Elçioğlu, and T. Okutucu-Özyurt, Stability of nanofluids: A critical review, *International Symposium on Convective Heat and Mass Transfer (CONV-14)*, Kuşadası, Turkey, June 8–13, 2014.
37. M. Mehrali, E. Sadeghinezhad, S. T. Latibari, S. N. Kazi, M. Mehrali, M. N. B. M. Zubir, and H. S. C. Metselaar, Investigation of thermal conductivity and rheological properties of nanofluids containing graphene nanoplatelets, *Nanoscale Research Letters*, 9, 15, 1–12 (2014).
38. M. Behi and S. A. Mirmohammadi, Investigation on thermal conductivity, viscosity and stability of nanofluids, MS Thesis, Royal Institute of Technology (KTH), Stockholm, Sweden, 2012.
39. E. B. Haghighi, N. Nikkam, M. Saleemi, M. Behi, S. A. Mirmohammadi, H. Poth, R. Khodabandeh, M. S. Toprak, M. Muhammed, and B. Palm, Shelf stability of nanofluids and its effect on thermal conductivity and viscosity, *Measurement Science and Technology*, 24, 105301, 1–11 (2013).
40. D-W. Oh, A. Jain, J. K. Eaton, K. E. Goodson, and J. S. Lee, Thermal conductivity measurement and sedimentation detection of aluminum oxide nanofluids by using the 3ω method, *International Journal of Heat and Fluid Flow*, 29(5), 1456–1461 (2008).
41. J. A. Alberola, R. Mondragón, J. E. Juliá, L. Hernández, and L. Cabedo, Characterization of halloysite-water nanofluid for heat transfer applications, *Applied Clay Science*, 99, 54–61 (2014).
42. S. S. Park and N. J. Kim, A study on the characteristics of carbon nanofluid for heat transfer enhancement of heat pipe, *Renewable Energy*, 65, 123–129 (2014).
43. L. Chen and H. Xie, Properties of carbon nanotube nanofluids stabilized by cationic gemini surfactant, *Thermochimica Acta*, 506(1–2), 62–66 (2010).
44. L. Chen, H. Xie, Y. Li, and W. Yu, Nanofluids containing carbon nanotubes treated by mechanochemical reaction, *Thermochimica Acta*, 477(1–2), 21–24 (2008).

45. L. Fedele, L. Colla, S. Bobbo, S. Barison, and F. Agresti, Experimental stability analysis of different water based nanofluids, *Nanoscale Research Letters*, 6, 30, 1–8 (2011).
46. Y. Hwang, J.-K. Lee, J.-K. Lee, Y.-M. Jeong, S.-I. Cheong, Y.-C. Ahn, and S. H. Kim, Production and dispersion stability of nanoparticles in nanofluids, *Powder Technology*, 186(2), 145–153 (2008).
47. Y. Hwang, J. K. Lee, C. H. Lee, Y. M. Jung, S. I. Cheong, C. G. Lee, B. C. Ku, and S. P. Jang, Stability and thermal conductivity characteristics of nanofluids, *Thermochimica Acta*, 455(1–2), 70–74 (2007).
48. C. Mao, H. Zou, X. Zhou, Y. Huang, H. Gan, and Z. Zhou, Analysis of suspension stability for nanofluid applied in minimum quantity lubricant grinding, *The International Journal of Advanced Manufacturing Technology*, 71(9–12), 2073–2081 (2014).
49. M. J. Nine, H. Chung, M. R. Tanshen, N. A. B. Abu Osman, and H. Jeong, Is metal nanofluid reliable as heat carrier?, *Journal of Hazardous Materials*, 273, 183–191 (2014).
50. A. Ijam, A. M. Moradi Golsheikh, R. Saidur, and P. Ganesan, A glycerol–water-based nanofluid containing graphene oxide nanosheets, *Journal of Materials Science*, 49(17), 5934–5944 (2014).
51. S. Ponmani, J. K. M. William, R. Samuel, R. Nagarajanc, and J. S. Sangwai, Formation and characterization of thermal and electrical properties of CuO and ZnO nanofluids in xanthan gum, *Colloids and Surfaces A: Physicochemical and Engineering Aspects*, 443, 37–43 (2014).
52. N. Hordy, D. Rabilloud, J.-L. Meunier, and S. Coulombe, High temperature and long-term stability of carbon nanotube nanofluids for direct absorption solar thermal collectors, *Solar Energy*, 105, 82–90 (2014).
53. S. U. Ilyas, R. Pendyala, and N. Marneni, Settling characteristics of alumina nanoparticles in ethanol-water mixtures, *Applied Mechanics and Materials*, 372, 143–148 (2013).
54. A. L. Subramaniyan, J. Sabareswaran, S. Lakshmipriya, M. Kotaisamy, and R. Ilangovan, Alternative characterisation techniques for stability of nanofluids, *International Journal of ChemTech Research*, 6, 2080–2082 (2014).
55. A. Kazemi-Beydokhti, S. Z. Heris, N. Moghadam, M. Shariati-Niasar, and A. A. Hamidi, Experimental investigation of parameters affecting nanofluid effective thermal conductivity, *Chemical Engineering Communications*, 201(5), 593–611 (2014).
56. C. Pang and Y. T. Kang, Stability and thermal conductivity characteristics of nanofluids (H_2O/CH_3OH + NaCl + Al_2O_3 nanoparticles) for CO_2 absorption application, *International Refrigeration and Air Conditioning Conference*, West Lafayette, IN, July 16–19, 2012.
57. J. Huang and X. Wang, Influence of pH on the stability characteristics of nanofluids, *Photonics and Optoelectronics*, Wuhan, People's Republic of China, August 14–16, 2009.
58. X.-J. Wang, H. Li, X.-F. Li, Z.-F. Wang, and F. Lin, Stability of TiO_2 and Al_2O_3 nanofluids, *Chinese Physics Letters*, 28(8), 086601, 1–4 (2011).
59. S. J. Thrush, Investigation of dispersion stability, and tribological performance of oil-based aluminum oxide nanofluids, MS thesis, Oakland University, Rochester, MI, 2012.
60. M. Sahooli and S. Sabbaghi, CuO nanofluids: The synthesis and investigation of stability and thermal conductivity, *Journal of Nanofluids*, 1, 155–160 (2012).
61. X. Li, Y. Chen, S. Mo, L. Jia, and X. Shao, Effect of surface modification on the stability and thermal conductivity of water-based SiO_2-coated graphene nanofluid, *Thermochimica Acta*, 595, 6–10 (2014).
62. T. Elango, A. Kannan, and K. Kalidasa Murugavel, Performance study on single basin single slope solar still with different water nanofluids, *Desalination*, 360, 45–51 (2015).
63. K. Y. Leong, N. Mohd Hanafi, R. Mohd Sohaimi, and N. H. Amer, The effect of surfactant on stability and thermal conductivity of carbon nanotube based nanofluids, *Thermal Science OnLine-First*, 1(00), 1–12 (2014).
64. Z. E. Allouni, M. R. Cimpan, P. J. Høl, T. Skodvin, and N. R. Gjerdet, Agglomeration and sedimentation of TiO_2 nanoparticles in cell culture medium, *Colloids and Surfaces B: Biointerfaces*, 68(1), 83–87 (2009).
65. H. Chang, M.-J. Kao, C.-S. Jwo, C.-G. Kuo, Y.-H. Yeh, and W.-C. Tzeng, Preparation of Co/Ag nanocompound fluid using ASNSS with aid of ultrasonic orthogonal vibration, *Journal of Alloys and Compounds*, 504, S376–S379 (2010).
66. Y. Safaei-Naeini, M. Aminzare, F. Golestani-Fard, F. Khorasanizadeh, and E. Salahi, Suspension stability of titania nanoparticles studied by uv-vis spectroscopy method, *Iranian Journal of Materials Science & Engineering*, 9, 62–68 (2012).
67. D. Zhu, X. Li, N. Wang, X. Wang, J. Gao, and H. Li, Dispersion behavior and thermal conductivity characteristics of Al_2O_3–H_2O nanofluids, *Current Applied Physics*, 9(1), 131–139 (2009).
68. R. Mondragon, J. E. Julia, A. Barba, and J. C. Jarque, Characterization of silica–water nanofluids dispersed with an ultrasound probe: A study of their physical properties and stability, *Powder Technology*, 224, 138–146 (2012).

69. V. S. Raykar and A. K. Singh, Thermal and rheological behavior of acetylacetone stabilized ZnO nanofluids, *Thermochimica Acta*, 502(1–2), 60–65 (2010).
70. W. Yu, H. Xie, L. Chen, and Y. Li, Investigation on the thermal transport properties of ethylene glycol-based nanofluids containing copper nanoparticles, *Powder Technology*, 197(3), 218–221 (2010).
71. W. Rashmi, A. F. Ismail, I. Sopyan, A. T. Jameel, F. Yusof, M. Khalid, and N. M. Mubarak, Stability and thermal conductivity enhancement of carbon nanotube nanofluid using gum arabic, *Journal of Experimental Nanoscience*, 6(6), 567–579 (2011).
72. S. M. Abbasi, A. Rashidi, A. Nemati, and K. Arzani, The effect of functionalisation method on the stability and the thermal conductivity of nanofluid hybrids of carbon nanotubes/gamma alumina, *Ceramics International*, 39(4), 3885–3891 (2013).
73. K. Wusiman, H. Jeong, K. Tulugan, H. Afrianto, and H. Chung, Thermal performance of multi-walled carbon nanotubes (MWCNTs) in aqueous suspensions with surfactants SDBS and SDS, *International Communications in Heat and Mass Transfer*, 41, 28–33 (2013).
74. J. Lee, K. Han, and J. Koo, A novel method to evaluate dispersion stability of nanofluids, *International Journal of Heat and Mass Transfer*, 70, 421–429 (2014).
75. J. Eastoe, Surfactant aggregation and adsorption at interfaces, in *Colloid Science Principles, Methods and Applications* (Ed.: T. Cosgrove), 2nd edn., pp. 61–88, John Wiley & Sons Ltd, West Sussex, U.K., 2010.
76. W. Yu and H. Xie, A review on nanofluids: Preparation, stability mechanisms, and applications, *Journal of Nanomaterials*, Article ID: 435873, 1–17 (2012).
77. A. M. Schrand, J. J. Schlager, L. Dai and S. M. Hussain, Preparation of cells for Assessing ultrastructural localization of nanoparticles with transmission electron microscopy, *Nature Protocols*, 5(4), 744–757 (2010).
78. H. Kamiya and M. Iijima, Surface modification and characterization for dispersion stability of inorganic nanometer-scaled particles in liquid media, *Science and Technology of Advanced Materials*, 11, 044304, 1–7 (2010).
79. J. S. Taurozzi, V. A. Hackley, and M. R. Wiesner, Preparation of nanoparticle dispersions from powdered material using ultrasonic disruption, Version 1.1, NIST Special Publication 1200-2, NanoEHS Protocols, June 2012.
80. E. V. Timofeeva, Nanofluids for heat transfer —Potential and engineering strategies, in *Two Phase Flow, Phase Change and Numerical Modeling* (Ed.: A. Ahsan), pp. 435–450, InTech, Rijeka, Croatia, 2011.
81. D. Wen, G. Lin, S. Vafaei, and K. Zhang, Review of nanofluids for heat transfer applications, *Particuology*, 7(2), 141–150 (2009).
82. R. Prasher, W. Evans, P. Meakin, J. Fish, P. Phelan, and P. Keblinski, Effect of aggregation on thermal conduction in colloidal nanofluids, *Applied Physics Letters*, 89, 143119, 1–3 (2006).
83. P. Sharma, I-H. Baek, T. Cho, S. Park, and K. B. Lee, Enhancement of thermal conductivity of ethylene glycol based silver nanofluids, *Powder Technology*, 208(1), 7–19 (2011).
84. R. Sadri, G. Ahmadi, H. Togun, M. Dahari, S. N. Kazi, E. Sadeghinezhad, and N. Zubir, An experimental study on thermal conductivity and viscosity of nanofluids containing carbon nanotubes, *Nanoscale Research Letters*, 9(1), 151 (2014).
85. W. Yu, H. Xie, and X. Wang, Enhanced thermal conductivity of liquid paraffin based nanofluids containing copper nanoparticles, *Journal of Dispersion Science and Technology*, 32(7), 948–951 (2011).
86. S. Kakaç and A. Pramuanjaroenkij, Review of convective heat transfer enhancement with nanofluids, *International Journal of Heat and Mass Transfer*, 52(13–14), 3187–3196 (2009).
87. S. Kakaç and H. Liu, *Heat Exchangers Selection, Rating, and Thermal Design*, 2nd edn., CRC Press, Boca Raton, FL, 2002.
88. E. V. Timofeeva, J. L. Routbort, and D. Singh, Particle shape effects on thermophysical properties of alumina nanofluids, *Journal of Applied Physics*, 106(1), 014304, 1–10 (2009).
89. R. Prasher, D. Song, J. Wang, and P. Phelan, Measurements of nanofluid viscosity and its implications for thermal applications, *Applied Physics Letters*, 89, 133108, 1–3 (2006).
90. R. Saidur, S. N. Kazi, M. S. Hossain, M. M. Rahman, and H. A. Mohammed, A review on the performance of nanoparticles suspended with refrigerants and lubricating oils in refrigeration systems, *Renewable and Sustainable Energy Reviews*, 15(1), 310–323 (2011).
91. K. V. Wong and O. De Leon, Review of nanofluid applications, in *Nanoparticle Heat Transfer and Fluid Flow* (Eds.: W. J. Minkowycz, E. M. Sparrow, and J. P. Abraham), pp. 1–24, CRC Press, Boca Raton, FL, 2013.

92. M. Abbas, R. G. Walvekar, M. T. Hajibeigy, and F. S. Javadi, Efficient air-condition unit by using nano-refrigerant, *EURECA 2013*, Selangor Darul Ehsan, Malaysia, July 1–2, 2013.
93. I. M. Mahbubul, S. A. Fadhilah, R. Saidur, K. Y. Leong, and M. A. Amalina, Thermophysical properties and heat transfer performance of Al_2O_3/R-134a nanorefrigerants, *International Journal of Heat and Mass Transfer*, 57(1), 100–108 (2013).
94. T. Coumaressin and K. Palaniradja, Performance analysis of a refrigeration system using nano fluid, *International Journal of Advanced Mechanical Engineering*, 4, 459–470 (2014).
95. R. R. Kumar, K. Sridhar, and M. Narasimha, Heat transfer enhancement in domestic refrigerator using R600a/mineral oil/nano-Al_2O_3 as working fluid, *International Journal of Computational Engineering Research*, 03(4), 42–50 (2013).
96. S. Bi, K. Guo, Z. Liu, and J. Wu, Performance of a domestic refrigerator using TiO_2-R600a nano-refrigerant as working fluid, *Energy Conversion and Management*, 52(1), 733–737 (2011).
97. S. A. Fadhilah, R. S. Marhamah, and A. H. Izzat, Copper oxide nanoparticles for advanced refrigerant thermophysical properties: Mathematical modeling, *Journal of Nanoparticles*, Article ID: 890751, 1–5 (2014).
98. C. T. Nguyen, G. Roy, C. Gauthier, and N. Galanis, Heat transfer enhancement using Al_2O_3–water nanofluid for an electronic liquid cooling system, *Applied Thermal Engineering*, 27(8–9), 1501–1506 (2007).
99. H. B. Ma, C. Wilson, B. Borgmeyer, K. Park, Q. Yu, S. U. S. Choi, and M. Tirumala, Effect of nanofluid on the heat transport capability in an oscillating heat pipe, *Applied Physics Letters*, 88, 143116, 1–3 (2006).
100. N. Kosseifi, E. Hachem, L. Silva, S. Boyer, E. Massoni, and T. Coupez, Numerical simulation of boiling during the quenching process, *10e Colloque National En Calcul Des Structures*, Giens, France, May 2011.
101. G. Ramesh and N. K. Prabhu, Review of thermo-physical properties, wetting and heat transfer characteristics of nanofluids and their applicability in industrial quench heat treatment, *Nanoscale Research Letters*, 6(1):334, 1–15 (2011).
102. S-Y. Chun, I. C. Bang, Y-J. Choo, and C-H. Song, Heat transfer characteristics of Si and SiC nanofluids during a rapid quenching and nanoparticles deposition effects, *International Journal of Heat and Mass Transfer*, 54(5–6), 1217–1223 (2011).
103. H. H. Khoshmehr, A. Saboonchi, and M. B. Shafii, The quenching of silver rod in boiling carbon nano tube-water nanofluid, *International Journal of Thermal Sciences*, 75, 95–104 (2014).
104. K. Babu and T. S. Psasanna Kumar, Effect of CNT concentration and agitation on surface heat flux during quenching in CNT nanofluids, *International Journal of Heat and Mass Transfer*, 54(1–3), 106–117 (2011).
105. A. Goyal, J. S. Tiwana, and A. Pal, A review study on minimum quantity lubrication in machining, *International Journal of Recent Development in Engineering and Technology*, 2(5), 37–40 (2014).
106. J. S. Nam, P-H. Lee, and S. W. Lee, Experimental characterization of micro-drilling process using nanofluid minimum quantity lubrication, *International Journal of Machine Tools and Manufacture*, 51(7–8), 649–652 (2011).
107. M. M. S. Prasad and R. R. Srikant, Performance evaluation of nano graphite inclusions in cutting fluids with MQL technique in turning of AISI 1040 steel, *International Journal of Research in Engineering and Technology*, 02(11), 381–393 (2013).
108. V. Vasu and K. M. Kumar, Analysis of nanofluids as cutting fluid in grinding EN-31 Steel, *Nano-Micro Letters*, 3(4), 209–214 (2011).
109. H. N. Obaid, M. A. Habeeb, F. L. Rashid and A. Hashim, Thermal energy storage by nanofluids, *Journal of Energy Technologies and Policy*, 3, 34–36 (2013).
110. S. Wu, D. Zhu, X. Li, H. Li, and J. Lei, Thermal energy storage behavior of Al_2O_3–H_2O nanofluid, *Thermochimica Acta*, 483(1–2), 73–77 (2009).
111. X. J. Wang, X. F. Li, Y. H. Xu, and D. S. Zhu, Thermal energy storage characteristics of Cu–H_2O nanofluid, *Energy*, 78, 212–217 (2014).
112. K. V. Wong and O. De Leon, Applications of nanofluids: Current and future, *Advances in Mechanical Engineering*, Article ID: 519659, 1–11 (2010).
113. Y. Li, J. Zhou, S. Tung, E. Schneider, and S. Xi, A review on development of nanofluid preparation and characterization, *Powder Technology*, 196(2), 89–101 (2009).

7 Numerical Approaches for Convective Heat Transfer with Nanofluids

Ismail Ozan Sert, Nilay Sezer-Uzol, and Sadık Kakaç

CONTENTS

7.1 INTRODUCTION

Nanofluids which constitute base fluids and metallic or nonmetallic particles having sizes less than 100 nm are known to have better thermal conductivity than conventional fluids. Such heat transfer enhancement characteristics of nanofluids can be a very important factor in many engineering applications. Nanofluids were first introduced by Choi [1]. There have been many experimental, theoretical, and numerical studies in the literature. Since then, the heat transfer characteristics of nanofluids such as thermal conductivity, convection heat transfer coefficient, viscosity, heat transfer enhancement mechanisms, and models for these physical characteristics are being investigated. Even with low concentration of nanoparticle suspension, anomalously high thermal conductivities are obtained. Hence, many experimental and theoretical studies are conducted to investigate the mechanisms of this anomalous heat transfer enhancement. In many experimental studies in the literature, it is observed that the enhancement of the heat transfer coefficient is higher than the thermal conductivity [2]. Then, these additional and complex enhancement mechanisms are also studied and explained theoretically and experimentally by many scientists. Additionally, many numerical simulations are performed to understand the physical mechanisms of the heat transfer enhancement with nanofluids.

For Computational Fluid Dynamics (CFD) simulation of nanofluids, there are two different numerical approaches: the single-phase modeling and the two-phase modeling as shown in Figure 7.1, depending on the mathematical definition of nanofluids according to various physical and modeling assumptions.

The first approach is the single-phase model that assumes that the fluid phase and the particle phase are in thermal equilibrium and there is no slip velocity between each phase. On the other hand, the two-phase flow approach considers the fluid and nanoparticles individually and the interactions between each phase are accomplished by introducing the relative terms into the

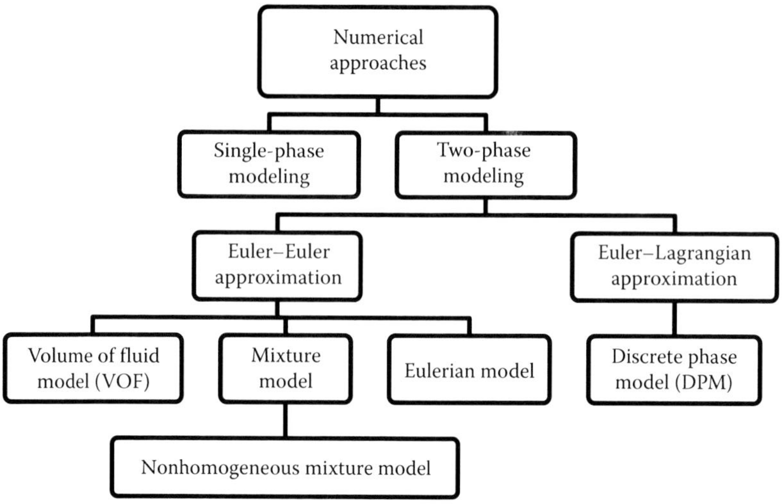

FIGURE 7.1 Approaches for the numerical simulation of nanofluids.

governing equations. Two-phase flow models are defined in several forms for different flow regimes such as Euler–Euler approximation (i.e., volume of fluids, mixture, and Eulerian models) or Euler–Lagrangian approximation (i.e., discrete phase model).

The two-phase modeling is defined by two main mathematical approaches as Euler–Euler and Euler–Lagrangian approximations. The main difference between these approximations is the way they simulate the secondary phase (particles). Both approximations treat the fluid phase as a continuum medium and the Navier–Stokes equations are solved for numerical analysis. On the other hand, while the Euler–Euler approximation treats the secondary phase (particle) as continuum medium like the primary phase (fluid), the Euler–Lagrangian approximation solves the particle tracking equations in Lagrangian form for the secondary phase, which is coupled with the Navier–Stokes equation of the primary phase.

There are many studies in the literature, which modeled the nanofluids following the single-phase approach assuming the particles and the fluid as a single medium with a same velocity [3–6]. However, the single-phase results are mostly based on speculation and they fell short of the expectations. In other words, the single-phase models are not able to simulate this anomalous heat transfer enhancement. Several studies in the literature have interpreted this difference between the experiments and numerical analyses as the thermal dispersion effect [3,7]. In order to compensate for this difference, another term resulting from the thermal dispersion is proposed to be added to the thermal conductivity of the nanofluid.

The main physical mechanisms, which play a role on the heat transfer enhancement with nanofluids, are discussed by many scientists. Buongiorno [8] presented a theoretical benchmark study on nanofluid flow analysis. He carried out order of magnitude analysis to investigate the most useful set of equations to model nanofluid flows. He analyzed different slip mechanisms that may lead to enhancement in heat transport such as inertia of the nanoparticles, Brownian diffusion, thermophoresis, diffusiophoresis, Magnus effect, fluid drainage, and gravity. In this study, water around nanoparticles can be regarded as continuum since Knudsen number is relatively small ($Kn < 0.3$) although the size of the nanoparticles is small as 1 nm. Therefore, water flow around nanoparticles is continuous and solution by Navier–Stokes equations for momentum transport is reasonable. As a result of his order of magnitude analyses, he concluded that Brownian diffusion and thermophoresis are relatively important slip mechanisms but still have a negligible impact on heat transport. The main effect of these mechanisms is local change in thermo-physical properties caused by particle migration due to diffusion. Heat transfer enhancement arises only via change in physical properties of the fluid. According to Buongiorno, the reason behind the incompatible experimental data with the correlations for the pure fluid is nonuniform nanoparticle distribution along the channel which is caused by particle migration due to the mentioned slip mechanisms. Regarding these theoretical outcomings, he suggested a two-phase flow model for nanofluids, which is a two-fluid four equation nonhomogenous equilibrium model. This model constitutes of nanofluid (mixture) continuity, momentum and energy equations, and nanoparticles diffusion equation. By this model, local change of particle volume fraction and physical properties can be included and their effect on heat transfer can be investigated.

Hwang et al. [9] performed an experimental study in hydrodynamically fully developed and thermally developing laminar flow regime with Al_2O_3/water nanofluid. They stated that the Jang and Choi model [10] can predict both the experimental results and the results presented by Lee et al. [11] with 1% error. The enhancement in convective heat transfer coefficient of water-based Al_2O_3 nanofluids according to the Reynolds number is found to be about constant in fully developed laminar flow region. Enhancement of heat transfer coefficient (h_{enh} = 8%) is obtained as much higher than the enhancement of effective conductivity (k_{enh} = 1.44%) at the same volume fraction (0.3%). They also investigated pressure drop characteristics in nanofluid flows and revealed that friction factor correlation for the single-phase flow can be extended to water-based Al_2O_3 nanofluids. In addition, they performed an order of magnitude analysis similar to Buongiorno [8] with original experimental parameters and found that the energy transfer by nanoparticle dispersion (thermal dispersion) is negligible. Thermophoresis and Brownian diffusion are major effects on particle migration, and the effect of viscosity gradient and non-uniform shear rate on particle migration can be negligible. They explained the reason of excessive enhancement as flattening of the velocity profile due to particle migration that causes the thermal conductivity enhancement.

These studies indicated that the main mechanisms that lead the additional enhancement on the heat transfer are Brownian and thermophoresis diffusions. In order to include these main physical mechanisms into the mathematical calculations, there are several theoretical and experimental thermal conductivity models, which are interpreted for the single-phase model approach. However, the effect of the new conductivity models on the single-phase heat transfer is not satisfactory. Thus, instead of including the effect of Brownian and thermophoresis diffusions as an extra term into the thermal conductivity, the two-phase model approach is employed to simulate the contributions of these physical mechanisms to the heat transfer enhancement analysis.

Over the past decade, many numerical studies have been performed to simulate heat transfer with nanofluids by using different numerical approaches such as single-phase models and Eulerian or Lagrangian two-phase models. Some of these studies are reviewed in the following corresponding sections of the numerical approaches. For completeness, the mathematical representations of the numerical approaches described in Figure 7.1 are given in Sections 7.2 and 7.3, and especially, the two-phase modeling is presented in detail.

7.2 SINGLE-PHASE MODELING

The single-phase modeling assumes that the phases are in thermal equilibrium and the relative velocity between phases is zero. The effect of the nanoparticle suspension in the base fluid is interpreted with the thermo-physical properties (i.e., nanofluid viscosity, thermal conductivity, density, etc.). The general form of the governing equations for the single-phase modeling is defined by the following equations [12]:

Conservation of mass (continuity equation):

$$\nabla \cdot \left(\rho \vec{\upsilon}\right) = 0 \tag{7.1}$$

Fard et al. [25] compared the single-phase and two-phase Eulerian models to predict the laminar heat transfer in a tube with the constant wall temperature boundary condition. In this study, the Ranz-Marshall model [26] was used in the energy equation to evaluate the interphase heat transfer. This correlation that is employed to calculate the Nusselt number of the particles is valid for $0 < Re < 200$ and $0 < Pr < 250$. The Ishii and Zuber drag law [27] was employed to evaluate the particle drag. The conductivity of the nanofluid was calculated by the Yu and Choi correlation [28]. The CFD results obtained by using the commercial software, CFX, were compared with the experimental data of the previous study which was conducted by Heris et al. [29,30].

There are three different nanofluids (Cu/water, CuO/water and Al_2O_3/water) available in the experimental study of Heris, and all were investigated numerically by Fard et al. [25]. In Figures 7.8 and 7.9, the heat transfer coefficients of Cu/water nanofluid with the 0.2% and 2% volume fractions versus the Peclet number are compared with the experimental results. It is observed that the heat transfer coefficient increases with the volume fraction and the Peclet number. The relative error between the single-phase results and the experimental data is 16.8%, whereas the error is 8.07% for the two-phase results for the 0.2% volume fraction case. The relative errors with the 2% volume fraction are 17.6% and 11.08% for the single-phase model and the two-phase model, respectively. Hence, it can be said that the two-phase model results are in better agreement with the experiments. They concluded that the CFD results underpredict the experimental results for all cases. This underprediction is explained by other mechanisms, which are disregarded at the CFD simulations such as the particle migration, clustering, and dispersion, etc.

Another numerical study was performed by Akbari et al. [16]. They performed a comparative analysis of the single- and two-phase models with the CFD simulations of nanofluid convective heat transfer. In this study, the laminar mixed convection at the Reynolds numbers of 1050 and 1600 at the volume fractions less than 2% was analyzed. The temperature-dependent properties and the Schiller and Naumann correlation [20] for the particle drag were used. The CFD simulations were performed by using the single-phase modeling and by using the different types of two-phase models, that is, the VOF model, the mixture model, and the Eulerian model. The single-phase and two-phase models predict almost the same hydrodynamic fields but very different thermal fields.

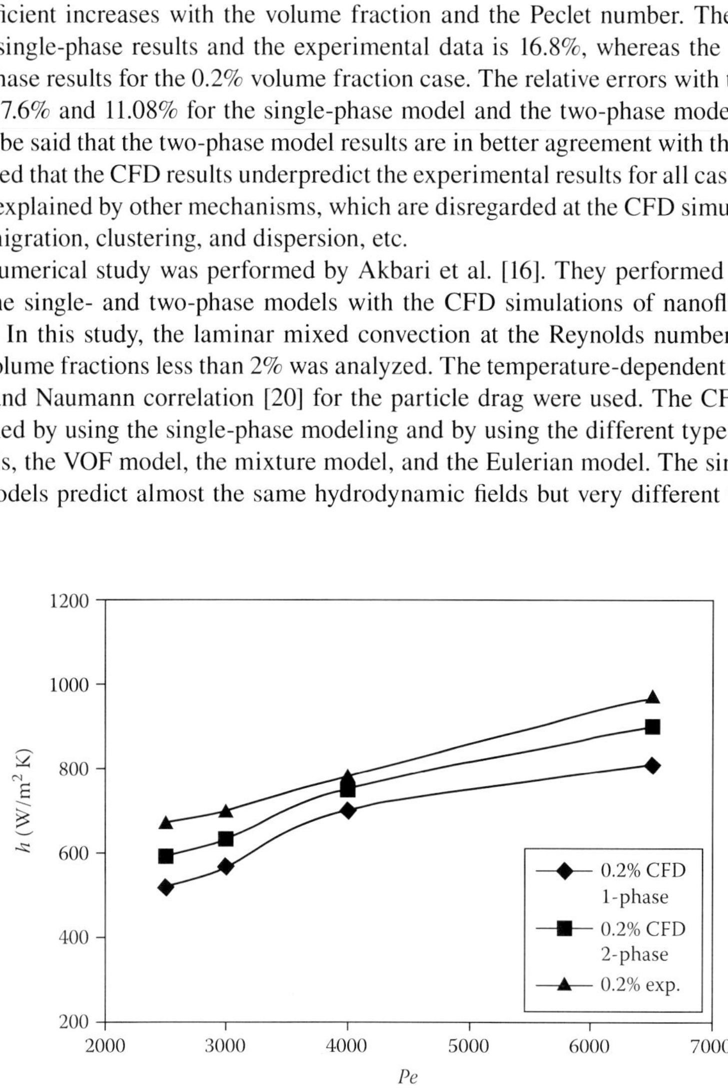

FIGURE 7.8 Comparison between Eulerian model predictions and experimental data for Cu/water nanofluid with 0.2% volume fraction. (Reprinted from *Commun. Heat Mass Transfer*, 37, Fard, M.H., Esfahany, M.N., and Talaie, M.R., Numerical study of convective heat transfer of nanofluids in a circular tube two-phase model versus single-phase modelInt, 91–97. Copyright 2010, with permission from Elsevier.)

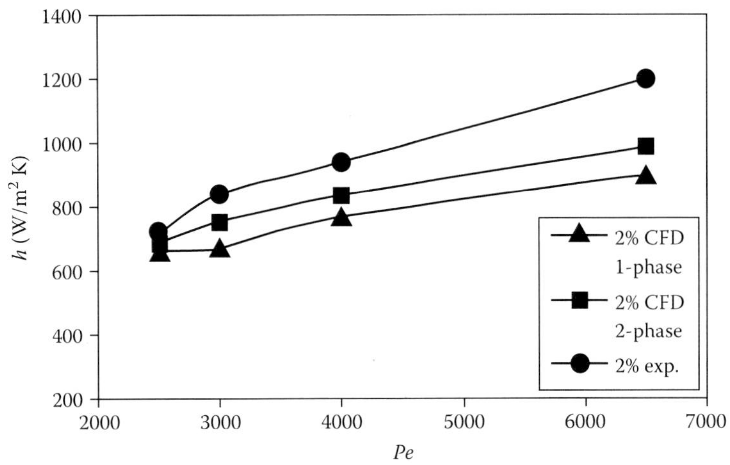

FIGURE 7.9 Comparison between Eulerian model predictions and experimental data for Cu/water nanofluid with 2% volume fraction. (Reprinted from *Commun. Heat Mass Transfer*, 37, Fard, M.H., Esfahany, M.N., and Talaie, M.R., Numerical study of convective heat transfer of nanofluids in a circular tube two-phase model versus single-phase modelInt, 91–97. Copyright 2010, with permission from Elsevier.)

Although, the two-phase models give closer results to the experimental data than the single-phase model, surprisingly the predictions by the three different two-phase models are essentially the same.

Despite the enhancement of the heat transfer coefficients by employing the Eulerian two-phase model, there is still a significant gap between the numerical and experimental results. This difference can be caused by the momentum interaction force between the phases. This force is defined as an extra term in the momentum equation that depends on the relative velocities similar to the mixture model. It has been shown by several studies in the literature that the relative velocities between the phases are very small [8,17,19].

As a summary, the physical mechanisms occur between the particles and fluid. The interactions must be well defined for nanoscale and also the implementation of these mechanisms into the numerical calculations should be achieved with more appropriate correlations and approximations.

7.3.2 Euler–Lagrangian Approximation

In the Euler–Lagrangian approximation, the fluid phase is modeled as a continuum medium by solving the Navier–Stokes equations and the particle phase is modeled by the particle motion theory in the Lagrangian reference frame.

The DPM is an Euler–Lagrangian approximation, which is used for the numerical simulations of disperse-phase flows in the literature such as spray dryers, fuel combustion or nanofluid flows, etc. This model is more appropriate for the cases with the low volume fraction of the secondary phase.

7.3.2.1 Discrete Phase Model (DPM)

In the DPM, the governing equations for the fluid phase are similar as the single-phase approach, except that there are the source term and sink term, which represent the integrated effects of the momentum and energy exchange with the base fluid and particles.

The governing equations for the continuum phase (fluid) in the DPM model are [15,31,32]

Continuity equation:

$$\nabla \cdot (\rho \vec{\upsilon}) = 0 \tag{7.1}$$

Momentum equation:

$$\nabla \cdot \left(\rho \vec{\upsilon} \vec{\upsilon}\right) = -\nabla p + \mu \nabla^2 \vec{\upsilon} + S_m \tag{7.35}$$

Energy equation:

$$\nabla \cdot \left(\rho \vec{\upsilon} c_p T\right) = \nabla \cdot \left(k \cdot \nabla T\right) + S_e \tag{7.36}$$

where S_m and S_e are the exchange source/sink terms and the viscous dissipation is neglected in the energy equation.

On the other hand, the secondary phase (particles) is modeled by the particle motion equation in the Lagrangian form for the spherical particles [33] as

Particle motion equation:

$$\frac{du_p}{dt} = F_D(\vec{\upsilon} - \vec{\upsilon}_p) + \frac{\vec{g}(\rho_p - \rho)}{\rho_p} + \vec{F} \tag{7.37}$$

where

$F_D(\vec{\upsilon} - \vec{\upsilon}_q)$ is the drag force per unit mass

F_D is the drag coefficient which can be specified by the flow characteristics

In nanofluid flows, $\text{Re}_d \sim 0.01$ and the Stokes resistance law is valid for $\text{Re}_d < 0.1$. Hence, the drag coefficient is calculated as [31,34]

$$\text{Re}_d = \frac{\rho d_p \left|\vec{\upsilon}_p - \vec{\upsilon}\right|}{\mu} \tag{7.38}$$

$$F_D = \frac{18\mu}{d_p^2 \rho_p C_c} \tag{7.39}$$

where, μ is viscosity of fluid, and d_p and ρ_p are the particle diameter and density, respectively. C_c is the Cunningham correction factor which is necessary to employ the Stokes' resistance law to sub-micron particles given as [34]

$$C_c = 1 + \frac{2\lambda}{d_p}\left(1.257 + 0.4e^{(1.1d/2\lambda)}\right) \tag{7.40}$$

where λ is the mean free path of the particles.

In the particle motion equation, F is an additional force term that can include the important mechanisms such as thermophoretic force and Brownian force individually. The thermophoretic force, which is caused by the temperature gradients, can be defined as a constant, polynomial, or an algebraic function according to the flow characteristics. Also, another important term, the Brownian force, which is caused by the random motion of the sub-micron particles, can be included as the additional force term.

In the Euler–Euler models, these additional force effects are introduced into numerical analysis by the definition of nanofluid physical properties (i.e., thermal conductivity, viscosity, etc.) or by the exchange parameters. Unlike the Euler–Euler approaches, in the DPM the effect of the additional physical mechanism can be inserted into the calculations as source terms in the equations directly.

After solving the particle motion equation, the momentum exchange between the particles and the base fluid is computed by examining the change in the momentum of a particle as it passes through each control volume in the computational domain. This momentum change [15,31] is computed by

$$S_m = \sum_{np} \frac{m_p}{\delta V} \frac{d\vec{\upsilon}_p}{dt} \tag{7.41}$$

where

δV is the cell volume

np is the number of particles within the cell volume

The energy change for the spherical particles can be evaluated with the same methodology as for the momentum change:

$$\rho_p c_{p,p} \frac{dT_p}{dt} = \frac{6h}{d_p}\left(T - T_p\right) \tag{7.42}$$

where h is the heat transfer coefficient and calculated from the Ranz and Marshall correlation [26] as

$$Nu = \frac{hd_p}{k_f} = 2.0 + 0.6 \cdot \mathrm{Re}_d^{1/2} \cdot \mathrm{Pr}^{1/3} \tag{7.43}$$

which is valid for $1 < \mathrm{Re}_d \cdot \mathrm{Pr}^{2/3} < 5.10^4$, where Re_d is defined in Equation 7.36.

The heat transfer from the continuous phase to the discrete phase is computed by examining the change in the thermal energy of a particle as it passes through each control volume and can be computed by

$$S_e = \sum_{np} \frac{m_p}{\delta V} c_p \frac{dT_p}{dt} \tag{7.44}$$

Since the Ranz and Marshall correlation is valid for the range given earlier, there is also an approximation of the DPM model applied to the nanoparticles, which is slightly outside the lower limit, being around 0.5 instead of 1 [31]. Instead of employing the Ranz and Marshall correlation, more appropriate correlations can be developed in future for the particle Nusselt number calculations. The Ranz and Marshall correlation is also used in the Eulerian two-phase model in the study of the Fard et al. [25].

Bianco et al. [31] investigated the developing laminar forced convection flow of a water–Al_2O_3 nanofluid in a circular tube, subjected to a constant and uniform heat flux at the wall, by employing the DPM. They compared the single-phase and two-phase model results with the constant and temperature-dependent physical properties. The effect of the nanoparticle volume fraction and the numerical calculation of the average heat transfer coefficients as a function of Reynolds number are shown in Figure 7.10. Also, some values are given in Table 7.1 for Re = 250 with the temperature dependent physical properties.

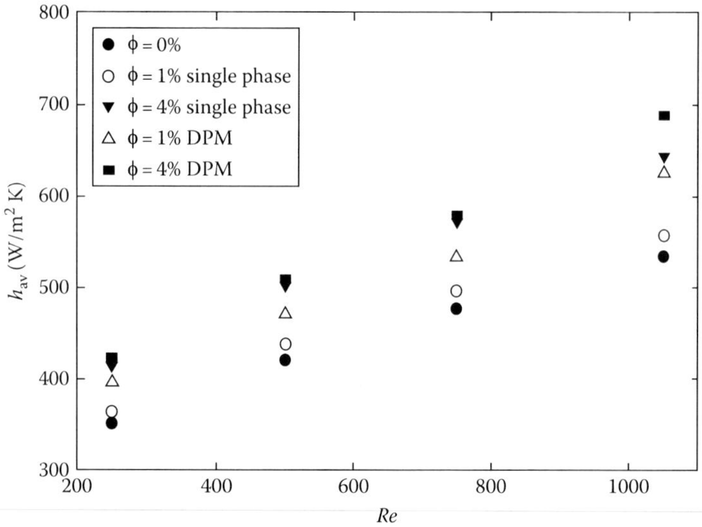

FIGURE 7.10 The average heat transfer coefficients at different Reynolds numbers with temperature-dependent physical properties. (Reprinted from *Appl. Therm. Eng.*, 29, Bianco, V., Chiacchio, F., Manca, S., and Nardini, S., Numerical investigation of nanofluids forced convection in circular tubes, 3632–3642. Copyright 2009, with permission from Elsevier.)

TABLE 7.1
Heat Transfer Coefficients and Nusselt Numbers for Single- and Two-Phase Models for Re = 250, q = 5000 W/m²

Model	h_{ave}	h_{nf}/h_f	Nu_{av}	Nu_{nf}/Nu_f
Water	375.5	1	6.29	1.000
Single-phase 1%	397.7	1.059	6.48	1.030
Two-phase 1%	420.7	1.12	6.86	1.090
Single-phase 4%	443.6	1.181	7.23	1.150
Two-phase 4%	445.8	1.187	7.27	1.155

Source: Reprinted from *Appl. Therm. Eng.*, 29, Bianco, V., Chiacchio, F., Manca, S., and Nardini, S., Numerical investigation of nanofluids forced convection in circular tubes, 3632–3642. Copyright 2009, with permission from Elsevier.

The numerical results of Bianco et al. [31] were also compared with the experimental study of Wen and Ding [21] for Re = 1050. In Table 7.2, the local Nusselt number and its ratio with the local Nusselt number for the base fluid are given at z/D = 26 and 63 and for all the investigated models. It can be observed that the results of the DPM with the variable physical properties are in better agreement with the experimental results compared to the other models. Thus, the DPM model has a promise to model the nanofluid heat transfer characteristics by mathematically modeling the every single additional force between the particles and the base fluid individually.

He et al. [32] also performed CFD simulations by using both the single-phase model and the combined Euler–Lagrange DPM model for the convective heat transfer of a laminar flow of TiO_2 nanofluids in a tube under constant wall heat flux boundary condition. They compared both numerical

TABLE 7.2
Local Nusselt Number Comparison

	Wen and Ding		Single-Phase		Two-Phase	
x/L	*Nu*	Nu_{nf}/Nu_f	*Nu*	Nu_{nf}/Nu_f	*Nu*	Nu_{nf}/Nu_f
26	12.3	1.15	9.65	1.03	10.87	1.16
63	8	1.13	7.44	1.04	8.35	1.17

Source: Reprinted from *Appl. Therm. Eng.*, 29, Bianco, V., Chiacchio, F., Manca, S., and Nardini, S., Numerical investigation of nanofluids forced convection in circular tubes, 3632–3642. Copyright 2009, with permission from Elsevier.

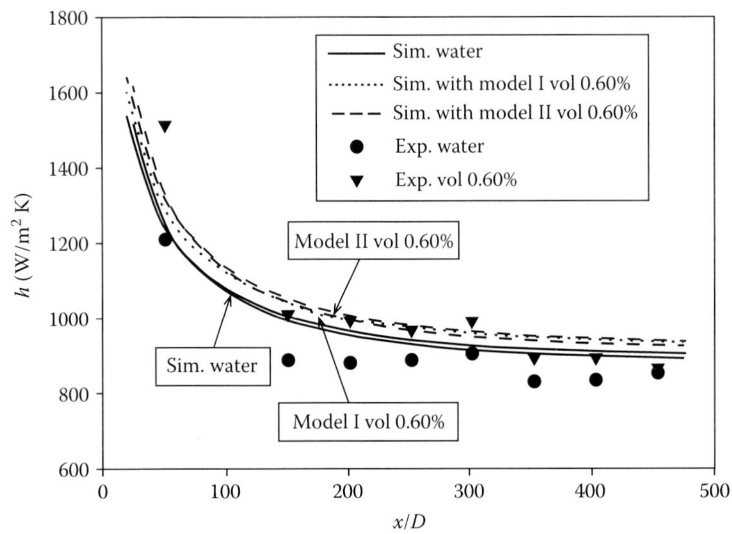

FIGURE 7.11 Comparison of the heat transfer coefficients along the tube with Re = 900. (Reprinted from *Appl. Therm. Eng.*, 29, He, Y., Men, Y., Zhao, Y., Lu, H., and Ding, Y., Numerical investigation into the convective heat transfer of TiO_2 nanofluids flowing through a straight tube under the laminar flow conditions, 1965–1972. Copyright 2009, with permission from Elsevier.)

simulations with the experimental results as shown in Figure 7.11. It can be seen that the heat transfer coefficient calculated with the two-phase model is slightly higher than the one with the single-phase model, because of the movement of nanoparticles and the interactions between the particles and the base fluid. The similarity between the single-phase and two-phase calculations can be caused by the nanoparticle parameters, boundary conditions, and the definition of the additional forces.

The effects of the Brownian motion and thermophoresis on heat transfer enhancement were investigated individually by He et al. [32] by using the commercial CFD solver, Fluent, like many other studies in the literature. The two-phase numerical simulations were performed and compared with and without the effects of the Brownian motion and thermophoresis. As shown in Figures 7.12 and 7.13, it is observed that the effects of the particle migration forces have very important roles on the heat transfer enhancement, especially the effect of the Brownian motion.

The effects of Brownian motion and thermophoresis are usually introduced by the UDF into the governing equations in the CFD solvers such as Fluent. Thus, the definition and contribution of

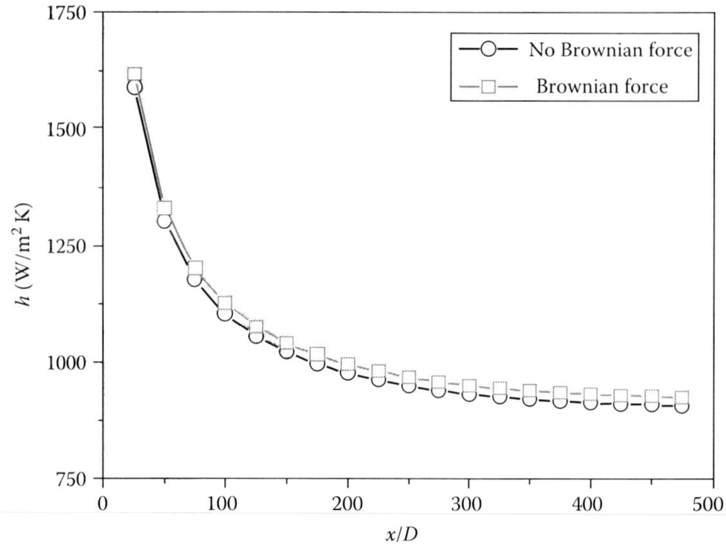

FIGURE 7.12 Effect of Brownian motion. (Reprinted from *Appl. Therm. Eng.*, 29, He, Y., Men, Y., Zhao, Y., Lu, H., and Ding, Y., Numerical investigation into the convective heat transfer of TiO_2 nanofluids flowing through a straight tube under the laminar flow conditions, 1965–1972. Copyright 2009, with permission from Elsevier.)

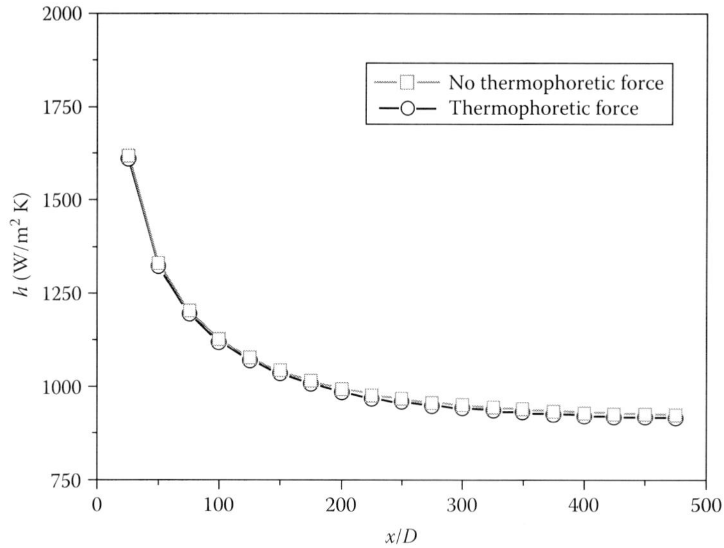

FIGURE 7.13 Effect of thermophoresis. (Reprinted from *Appl. Therm. Eng.*, 29, He, Y., Men, Y., Zhao, Y., Lu, H., and Ding, Y., Numerical investigation into the convective heat transfer of TiO_2 nanofluids flowing through a straight tube under the laminar flow conditions, 1965–1972. Copyright 2009, with permission from Elsevier.)

these forces may be different from one study to another according to definition, by the user, of the Brownian motion, thermophoretic forces or any other extra forces, which are acting on the particle fluid interactions. Therefore, the DPM numerical simulations can be updated by including the effect of the desired physical mechanisms into the Lagrangian particle motion equation, which leads to observe the effect of the each physical mechanism individually.

7.4 DISCUSSIONS

The recent experimental and numerical studies suggest the following observations for heat transfer characteristics of nanofluids:

- Heat transfer enhancement with nanofluids is a challenging problem due to the difficulties to model the physical mechanisms between the particles and fluids. The single-phase models are not capable of modeling these complex mechanisms.
- The interactions between phases are tried to be modeled with several two-phase models, which are described in this chapter. Each two-phase model has advantages and disadvantages according to the computational aspects and the ability of modeling some specific physical forces/mechanisms between each phase.
- Since the experimental data of the nanofluids have some discrepancies due to stability of nanoparticles (because of the coagulation and sedimentation, etc.), the numerical modeling of nanofluids and validation of the numerical analysis with the available experimental data is more challenging and difficult. However, it is obvious that the two-phase model analysis carried the numerical calculations at least one step forward in terms of simulating the physics of nanofluids.
- As the new aspects of thermal and hydrodynamical characteristics of nanofluids being discovered, the numerical approaches would provide better predictions in numerical calculations of nanofluids.

A short summary of advantages and disadvantages of the numerical approaches being used for nanofluids can be listed as follows:

- Despite the advantages of the mixture model such as easy implementation of physical properties and the less computational power requirements, some studies showed that the two-phase mixture model and single-phase model results are very similar [19] due to the low relative velocities between fluid and the nanoparticles. The main difference between the single-phase model and the two-phase mixture models is the effect of the drift velocities, which are found to be very small because of the decrement of the relaxation time, which is dependent on the particle size. This result is also explained by several studies in the literature [8,17,19].
- On the other hand, in nonhomogenous mixture model, which is based on the theoretical study of the Buongiorno [8], the relative velocities between phases are neglected. The extra diffusion equation is employed to include effects of the Brownian motion and thermophoretic forces. By this way, the heat transfer enhancement due to the particle migration along the radial direction of the circular channel is modeled. This migration effects caused an additional heat transfer increment compared to the single-phase and two-phase mixture models. Hence, the abnormal heat transfer enhancement is modeled. Despite all other two-phase models, which are mostly employed with the commercial CFD solver Fluent, this model is employed by the in-house code, which is developed by Heyhat et al. [24]. This model has promising results, but there is no experimental data comparison and the code is only developed for steady-state two-dimensional axisymmetric flows.
- More complex and computationally expensive Eulerian two-phase model is employed by several numerical studies in the literature. The major differences between the mixture model and Eulerian model are as follows:
  - The governing equations are solved for each phase instead of the mixture only in the Eulerian models.
  - The forces between phases are introduced by the interphase exchange parameters of momentum and energy equations in the Eulerian models.

- Since the momentum interaction force between phases depends on the relative velocities, which are very small, the heat transfer results obtained by the Eulerian model are higher than the single-phase model, but the experimental results are still underpredicted.
- The DPM, which is an Euler–Lagrangian approximation, is modeled by Lagrangian particle motion equation which includes the effects of several forces particularly. The effects of these forces are introduced as source terms into the momentum and energy equations, which are governing the fluid phase. Despite the complex computational set up due to the coupling of Euler–Lagrangian calculations, particle tracking, and time-dependent parameters, DPM is a promising approach in terms of implementing each physical mechanism as an additional force into the Lagrangian particle equation.

REFERENCES

1. S.U.S. Choi, Enhancing thermal conductivity of fluids with nanoparticles, in Siginer DA, Wang HP (eds.) *Developments and Applications of Non-Newtonian Flows*, The American Society of Mechanical Engineers, New York, FED-Vol. 231/MD-Vol. 66, pp. 99–105, 1995.
2. Q. Li and Y. Xuan, Convective heat transfer and flow characteristics of Cu–water nanofluid, *Sci. China Ser. E Technol. Sci.*, 45 (4), 408–416 (2002)
3. S. Özerinç, A.G. Yazıcıoglu, and S. Kakaç, Numerical analysis of laminar forced convection with temperature-dependent thermal conductivity of nanofluids and thermal dispersion, *Int. J. Therm. Sci.*, 62, 138–148 (2012).
4. S.Z. Heris, M.N. Esfahany, and G. Etemad, Numerical investigation of nanofluid laminar convection heat transfer through a circular tube, *Numer. Heat Transfer A*, 52(11), 1043–1058 (2007).
5. I.O. Sert, N. Sezer-Uzol, and S. Kakac, Analysis of transient laminar forced convection of nanofluids in circular channels, *ASME 2012 International Mechanical Engineering Congress and Exposition*, Houston, TX, November 9–15, 2012.
6. S.E.B. Maïga, C.T. Nguyen, N. Galanis, and G. Roy, Heat transfer behaviors of nanofluids in a uniformly heated tube, *Superlattice Microst.*, 35, 543–557 (2004).
7. Y. Xuan and W. Roetzel, Conceptions for heat transfer correlation of nanofluids, *Int. J. Heat Mass Transfer*, 43, 3701–3707 (2000).
8. J. Buongiorno, Convective transport in nanofluids, *Trans. ASME*,128, 240–250 (2006).
9. K.S. Hwang, S.P. Jang, and S.U.S. Choi, Flow and convective heat transfer characteristics of water-based Al_2O_3 nanofluids in fully developed laminar flow regime, *Int. J. Heat Mass Transfer*, 52, 193–199 (2009).
10. S.P. Jang and S.U.S. Choi, Role of Brownian motion in the enhanced thermal conductivity of nanofluids, *Appl. Phys. Lett.*, 84, 4316–4318 (2004).
11. S. Lee, S.U.S. Choi, S. Li, and J.A. Eastman, Measuring thermal conductivity of fluids containing oxide nanoparticles, *ASME J. Heat Transfer*, 121, 280–289 (1999).
12. S. Kakac, Y. Yener, and A. Pramuanjaroenkij, *Convective Heat Transfer*, CRC Press, Boca Raton, FL, 2013.
13. S. Kakac and A. Pramuanjaroenkij, Review of convective heat transfer enhancement with nanofluids, *Int. J. Heat Mass Transfer*, 52, 3187–3196 (2009).
14. B. Apacoglu, O. Kirez, S. Kakaç, and A.G. Yazicioglu, Enhancement of convective heat transfer in laminar and turbulent flows with nanofluids, *2011 VIII Minsk International Seminar*, Minsk, Belarus, September 12–15, 2011.
15. Fluent 6.2 User Manual, Fluent Incorporated, New York, 2006.
16. M. Akbari, N. Galanis, and A. Behzadmehr, Comparative analysis of single nd two-phase models for CFD studies of nanofluid heat transfer, *Int. J. Therm. Sci.*, 50, 1343–1354 (2011).
17. S. Mirmasoumi and A. Behzadmehr, Numerical study of laminar mixed convection of a nanofluid in a horizontal tube using two-phase mixture model, *Appl. Therm. Eng.*, 28, 717–727 (2008).
18. R. Lotfi, Y. Saboohi, and A.M. Rashidi, Numerical study of forced convective heat transfer of nanofluids: Comparison of different approaches, *Int. Commun. Heat Mass Transfer*, 37, 74–78 (2010).
19. I.O. Sert, N. Sezer-Uzol, A.G. Yazicioglu, and S. Kakac, Enhancement of convective heat transfer with nanofluids, single-phase and two-phase analysis, *2013 Fifth ICAPM*, Cluj-Napoca, Romania, August 25–28, 2013.

20. L. Schiller, Z. Naumann, A drag coefficient correlation, *Z. Ver. Deutsch. Ing.*, 77, 318–320 (1935).
21. D. Wen and Y. Ding, Experimental investigation into convective heat transfer of nanofluids at the entrance region under laminar flow conditions, *Int. J. Heat Mass Transfer*, 47, 5181–5188 (2004).
22. D. Kim, Y. Kwon, Y. Cho, C. Li, S. Cheong, Y. Hwang, J. Lee, D. Hong, and S. Moon, Convective heat transfer characteristics of nanofluids under laminar and turbulent flow conditions, *Curr. Appl. Phys.*, 9(2), 119–123 (2009).
23. J. Koo and C. Kleinstreuer, A new thermal conductivity model for nanofluids, *J. Nanopart. Res.*, 6(6), 577–588 (2004).
24. M.M. Heyhat and F.Kowsary, Effect of particle migration on flow and convective heat transfer of nanofluids flowing through a circular pipe, *J. Heat Transfer*, 132, 062401 (2010).
25. M.H. Fard, M.N. Esfahany, and M.R. Talaie, Numerical study of convective heat transfer of nanofluids in a circular tube two-phase model versus single-phase modelInt. *Commun. Heat Mass Transfer*, 37, 91–97 (2010).
26. W.E. Ranz and W.R. Marshall, Jr., Evaporation from drops, Part I–II, *Chem. Eng. Prog.*, 48 (3), 141–146, 173–180 (1952).
27. M. Ishii and N. Zuber, Drag coefficient and relative velocity in bubbly, droplet or particulate flows A.I.Ch.E. *71st Annual Meeting*, Miami, FL, 1978.
28. W. Yu and U.S. Choi, The role of international layers in the enhanced thermal conductivity of nanofluids, a renovated Maxwell Model, *J. Nanopart. Res.*, 5, 167–171 (2003).
29. S.Z. Heris, S. Gh. Etemad, and M.N. Esfahany, Experimental investigation of oxide nanofluid laminar flow convective heat transfer, in circular tube, *Int. Commun. Heat Mass Transfer*, 33, 529–533 (2006).
30. S.Z. Heris, M.N. Esfahany, and S. Gh. Etemad, Experimental investigation of convective heat transfer of Al_2O_3/water nanofluid in circular tube, *Int. J. Heat Fluid Flow*, 28, 203–210 (2007).
31. V. Bianco, F. Chiacchio, S. Manca, and S. Nardini, Numerical investigation of nanofluids forced convection in circular tubes, *Appl. Therm. Eng.*, 29, 3632–3642 (2009).
32. Y. He, Y. Men, Y. Zhao, H. Lu, and Y. Ding, Numerical investigation into the convective heat transfer of TiO_2 nanofluids flowing through a straight tube under the laminar flow conditions, *Appl. Therm. Eng.*, 29, 1965–1972 (2009).
33. H. Ounis, G. Ahmadi, and J.B. McLaughlin, Brownian diffusion of submicrometer particles in the viscous sublayer, *J. Colloid Interface Sci.*, 143(1), 266–277 (1991).
34. S.K. Das, S.U.S. Choi, W. Yu, and T. Pradeep, *Nanofluids Science and Technology*, John Wiley & Sons, Hoboken, NJ, 2008.

8 Heat Pipes and Nanotechnologies

Leonard L. Vasiliev and Leonid L. Vasiliev, Jr.

CONTENTS

8.1 INTRODUCTION

Energy is one of the key motifs of our century. Despite the focus being mostly on saving electrical energy, effective transfer and saving thermal energy is of comparable importance. Heat pipes and thermosyphons are effective tools to solve the program of saving thermal energy. Heat pipes and thermosyphons are very flexible systems with regard to their effective thermal control of different heat-loaded devices. A new stream in the modern heat pipe technology is related to nanofluid (NF) and nanoparticles (NPs) application. NFs are engineered by suspending ultra-fine metallic (Cu, Au) or nonmetallic (Al_2O_3, TiO_2, ZrO_2) nanometer dimension particles in traditional fluids such as water, oil, and hydrocarbons. Some theoretical and experimental investigations have revealed that NFs have remarkably higher thermal conductivity and greater heat transfer characteristics than conventional pure fluids [1–7]. In the bibliography some experimental data published with NFs in conventional heat pipes made from metal [8–10] are available. Actually, the micro and NPs as nanocoating (NC) of the heat pipe evaporators and thermosyphons are considered as a means to increase the heat transfer coefficient with working fluid evaporation and boiling on the heat-loaded wall [3,11]. A number of studies on the evaporation phenomena in grooves of heat pipes (GHP) have been carried out over the past decades [12,13]. Most researchers have focused their attention on the liquid evaporation on menisci formed in smooth grooves with extended thin films [14,15].

In 1988, an innovative method was proposed by Vasiliev et al. [16] to enhance evaporative heat transfer in grooves. In this study, the inner surface of trapezoidal GHP was coated by a thin mini-porous layer to make the dry surface act as an extended evaporation surface with high heat transfer performance. Evaporation heat transfer in minigrooves covered by a porous coating occurs on the meniscus formed inside the groove and the menisci formed inside the thin fine porous layer. This procedure not only improves capillary forces action but also extends the evaporating surface with high heat transfer performance in comparison with the smooth groove. The thermal resistance is between 1.3 and 1.4 times lower for GHP with a nanoporous layer (between 0.021 and 0.018 W/K) than for a classical GHP (between 0.025 and 0.035 W/K). The analytical model of GHP with porous coating was developed by Jinliang Wang and Ivan Catton [17]. The authors of this publication show that evaporation heat transfer in the groove covered by a thin porous layer is three to six times higher than in the groove without a porous layer. As a result of this action in the new advance design of the GHP, significant improvement of heat transfer coefficient inside grooves was achieved. Recently, a new nanotechnology was developed to produce loop polymer thermosyphons capable of long-term operation without permeation by air or working fluids through their walls. Suggested composite polymer materials with high effective thermal conductivity are used as the case material and give possibilities to design a wide range of new heat transfer equipment. Micro and nanofiber reinforcement of polymers increases its mechanical resistance and decreases the thermal expansion, whereas thermal conductivity tends to increase. Experiments on thermal conductivity $k(T)$ are most useful tools for investigating scattering processes, interaction between electrons and phonons. The discovery of high temperature superconductors has stimulated research and experimental technique of composite polymer materials based on high-molecular polymers with fibrous fillers and polymer reinforced plastics used as the envelope of the magnet and advanced polymers heat exchangers for magnets cooling. Actually, polymer heat transfer equipment is used in different devices [18–20]. Polymer–metal composites are becoming an attractive topic due to their unique surface morphology [21]. They can be made on the base of polymeric films metalized from one, or both sides with a noble metal (gold or platinum) [22,23]. Considerable efforts have been devoted to the design and fabrication of controlled organic/inorganic composites with novel properties, including optical, electrical, chemical, biological, and mechanical properties [24,25]. In these hybrid systems, phase separation occurs naturally because they are composed of two materials with totally different chemical characteristics [26]. Besides the polymer–metal composites, the carbon fiber-reinforced carbon composites, epoxy and phenolformaldehyde composites reinforced by glass and carbon wires, and polyamide composite materials with nanocarbon filaments and NPs are also the subject of interest in designing the polymer loop thermosyphons and heat pipes [27]. The envelope of heat pipes and thermosyphons have the effective thermal conductivity 10–40 times higher than pure polymer material. It is interesting to use the evaporator and condenser flat form of heat pipes and thermosyphons for cooling the heat-generating elements (electronics, LED, and MEMS). Actually, some polymer heat transfer components are used in different devices [20,28]. Unfortunately, some polymers (e.g., epoxy complex compound) can act as a source of localized heating and initiate thermal runway when it is stressed to the point of fracture by a combination of magnetic hoop stress and stress introduced by different thermal contraction between the epoxy and superconducting wire during cool down and thermal cycling. This stress could be considered as a source of crack generation and porous media inside reinforced polymers as a function of a number of thermal cycling. The effect of aging on thermal conductivity and thermal diffusivity is of interest, because many applications of polymers are expected to be in service for 20, or more, years. Linear thermal expansion is also of importance in insulation system design, since the opening of cracks and generation of porous structures around slabs, or within the foams can cause the convective transfer of heat to become prohibitively high, Luikov et al. [29].

The topic of the present chapter is related to the field of cooling technologies for air-conditioning, refrigeration, LED cooling systems, superconductive devices, and electronics package. It is based on the innovative means for thermal energy handling at better operating conditions in comparison to traditional technologies.

8.2 NANOFLUIDS FOR HEAT TRANSFER INTENSIFICATION IN MINI LOOP THERMOSYPHONS WITH A TRANSPARENT ENVELOPE

It is interesting to design a loop mini thermosyphon from transparent material (glass, plastic) and to heat NPs dispersed in water or hydrocarbons inside the thermosyphon evaporator by radiation [30]. Bubbles generation around NPs in mini volume of the evaporator (Figure 8.1) creates a mechanical pulse action (mini pump) on the working fluid.

This laser beam pulse is working as a two-phase mini pump, which initiates NF circulation inside a flat mini channel (1 mm of thickness; Figure 8.2). Such loop thermosyphons and pulsating heat pipes have a good perspective for space applications in the system of satellite thermal control [31].

Transparent flexible heat pipes and thermosyphons made from glass or plastic [32] have a real practical interest for fuel mini/microcells thermal control and photoelectronic components cooling due to different sources of the energy input and sink application. High-temperature heat transfer devices can be used in power stations as transparent (glass) pulsating heat pipe heat exchangers in the air preheater for furnaces and boilers. One of the forthcoming topics of this work is the investigation of the influence of metal (Au) and metal oxide NPs (Al_2O_3 NPs) immersed in the fluid (water) on the local bubble generation and two-phase heat transfer intensification to compare with pure water [30,33]. As it was previously shown [34,35], the heat flow generated by light-absorbing NPs (Au) initiates more intense bubble generation when short laser pulses are used as primary sources of thermal energy (Figure 8.3). A temporal scale of the photothermal conversion of energy is limited by the duration of a laser pulse and provides good thermal confinement of the heat release in NPs. The main role of such energy absorbing NPs is to generate bubbles and to heat NF in volume. The limitation of this method for vapor generation is in delivery of the energy into the point of interest: It should be optically transparent to allow optical radiation to reach NPs. NPs are considered as additional centers of nucleation due to an increased surface of liquid/solid interaction.

Do NPs in such a case stimulate the appearance of an earlier threshold of vapor generation? The second aim is to validate this hypothesis and evaluate the influence of "passive" NPs (Al_2O_3 particles are nonabsorbing energy media) as the element of heterogeneity in the fluid when the energy threshold of bubble generation decreases.

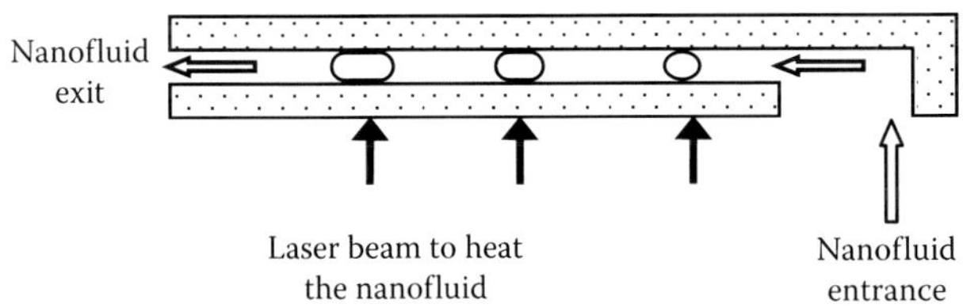

FIGURE 8.1 Flat transparent mini-evaporator with NF flow, heated by the laser beam.

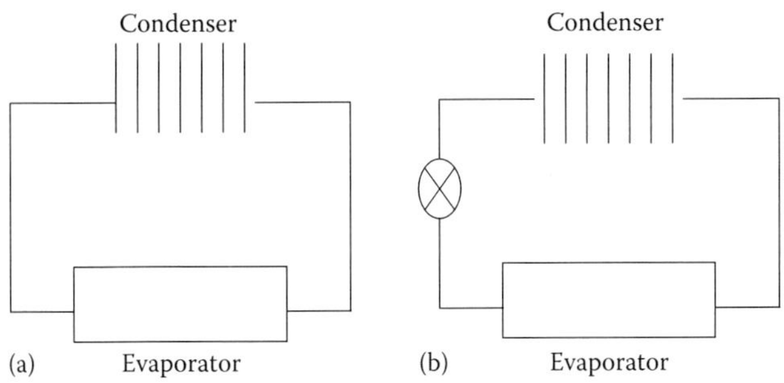

FIGURE 8.2 Schematic of a two-phase loop with NF (without mechanical pump (a) and with mechanical pump (b)).

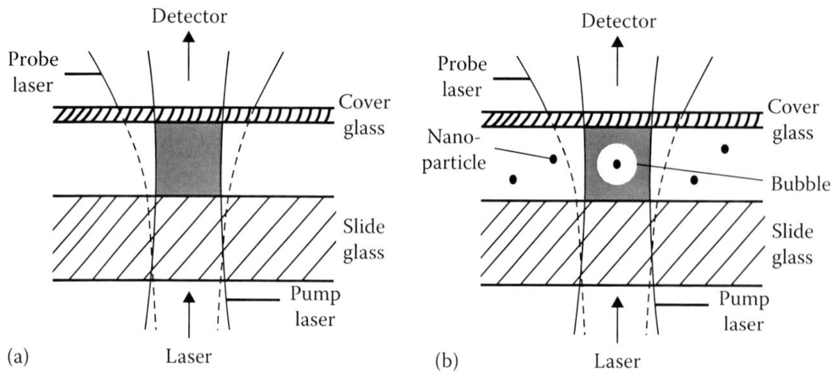

FIGURE 8.3 Visualization of bubble generation by a short laser pulse (532 nm, 10 ns) in a minichannel of the flat evaporator with NF (water + Al_2O_3 NPs): (a) control pure water; (b) NF; particle diameter $D_{Al_2O_3} < 220$ nm.

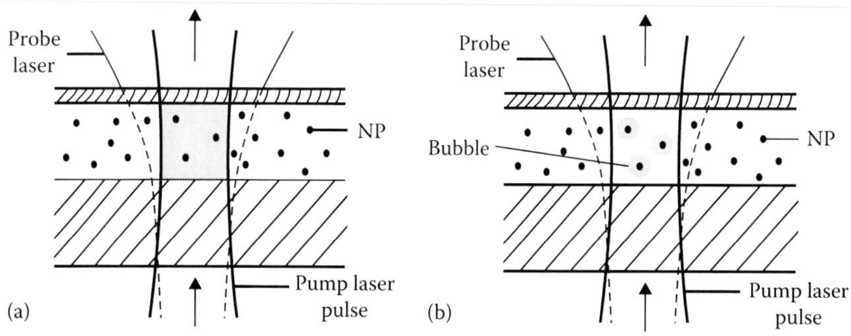

FIGURE 8.4 Experimental model: laser-induced heating of the volume of liquids with Al_2O_3 nonabsorbing NPs (a); generation of laser-induced bubbles around gold light-absorbing NPs (b); single laser pulse: 532 nm, 10 ns.

Such NPs are considered now as additional centers of nucleation due to increased surface of liquid/solid interaction (Figure 8.4). For all the cases studied, the bubble-specific photothermal signals (Figures 8.3 through 8.5), that is, PT–responses and PT–images, were detected and evaluated. The bubble-specific PT–response has a negative symmetrical profile; its front describes bubble expansion and the tail describes bubble collapse. The length of the bubble-related signal response indicates a bubble lifetime (Figure 8.5a). In the homogeneous media bubbles emerge on the entire area of the pump laser beam.

The diameter of the bubbles is much smaller than that for heated volume (cell or laser beam, Figure 8.5b). Irrespective of the medium, the bubble generation process had statistical nature with the bubble generation probability (*PRB*) from 0 to 1. The energy threshold of bubble generation for light-absorbing gold particles is less to compare with Al_2O_3 particles and pure water. Therefore, there is a possibility to organize a laser beam interaction with NPs inside mini channels (1–2 mm) of loop thermosyphon. The mini bubbles generation around the NPs is the reason of the NF flow circulation inside mini loop thermosyphon, or pulsating heat pipe, Figure 8.2. Photothermal generation and detection of vapor bubbles in the liquid was performed using the photothermal (PT) microscope [35]. Photo-induced bubbles were detected and measured by the optical method—thermal lens. The probe beam (He-Ne laser, model LGN-224, Russia, 633 nm, 0.2 mW, beam diameter at sample plane 22 μm) passes through the object, the collecting lens, then through the axial pinhole and is registered with photodetector. The output signal from the photodetector was registered with a digital oscilloscope (computer board Bordo-211 from diagnostic systems, Belarus) with 10 ns

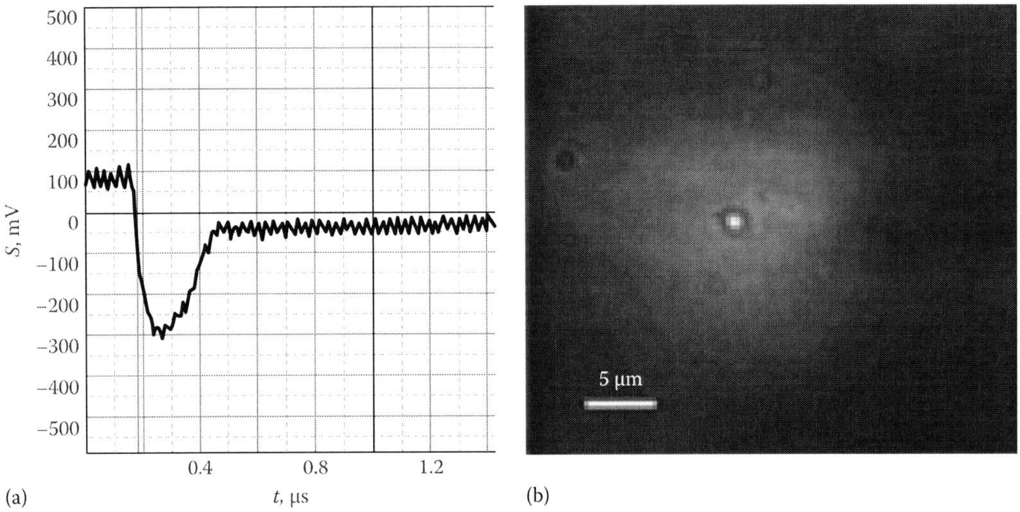

FIGURE 8.5 Photo-thermal signals obtained from an individual bubble generated in a transparent mini-evaporator: PT-response with bubble-specific shape (a) and PT-image (b). *Y* axis is for the output of the photodetector (mV).

temporal resolution and analyzed by a computer. In absence of the bubbles the amplitude of thermal lens signal is roughly proportional to the mean temperature in the probed volume. The vapor bubbles were imaged with time-gated CCD camera using the portion of pump laser pulse to form the image.

8.3 NANOCOATING IN HEAT PIPES

Actually the NC of the heat pipe and thermosyphon evaporators is considered as a means to increase the heat transfer coefficient with working fluid evaporation and boiling on the heat-loaded wall.

It is known that the heat transfer intensity with evaporation in thin liquid films greatly exceeds the heat transfer intensity of the liquid pool boiling. Though capillary grooves possess indisputable advantages, they present certain restrictions in evaporation and boiling of liquid. It is related to some special features of heat transfer in the grooves. Intensive heat transfer occurs in the thin film region that extends from the intrinsic meniscus. However, the extended thin film in the grooves with different sections constitutes only a small portion of the total surface of the grooves. A large fraction of the surface in the GHP evaporator is covered by the thick liquid film or the intrinsic meniscus, where the local heat transfer coefficient is particularly low due to the small thermal conductivity of the liquid or dry part of the groove, where only natural or mixed vapor convection exists between the solid surface and the vapor phase. Compared to the heat transfer with phase changes, the convective heat transfer coefficient in the dry area of the groove (Figure 8.6a) is insignificant and can be safely neglected. As the heat flux increases, the meniscus in the groove recedes. The dry area, a region of pronouncedly low heat transfer performance, increases as well.

The surface of trapezoidal grooves (GHP) covered by a thin layer of a sintered powder ensures an extended surface of evaporation with high heat transfer intensity [36]. In this case, the heat transfer occurred inside the porous coating of the groove (on its edge and bottom parts simultaneously; Figures 8.7a and b). On the extended porous surface the evaporation is accompanied by high heat transfer in comparison with the smooth groove. The liquid film availability inside

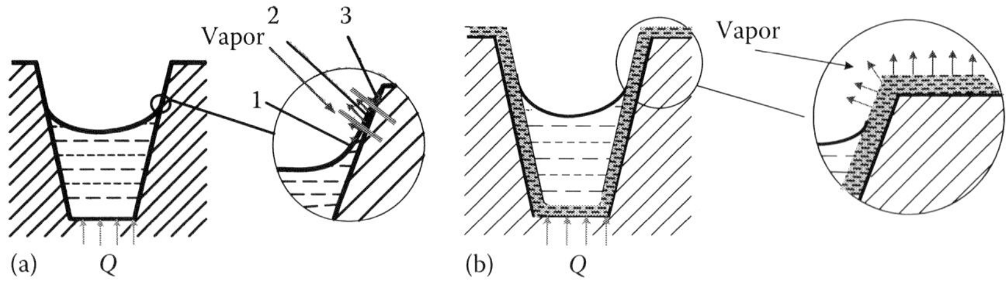

FIGURE 8.6 Heating and cooling parts of a heat pipe with smooth grooves (a); Heating and cooling parts of a grooved heat pipe with porous coating (b).

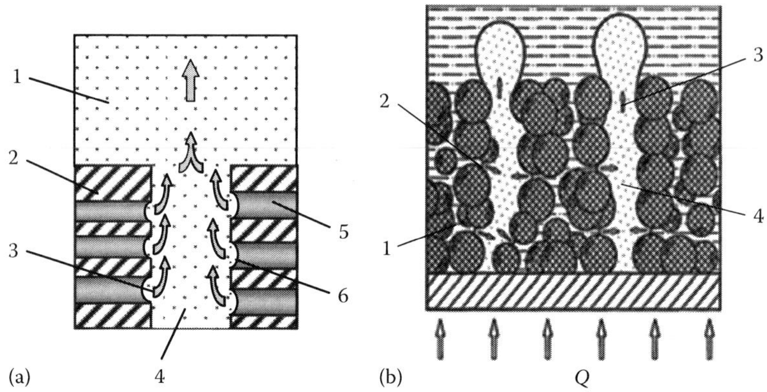

FIGURE 8.7 Model of heat transfer on the edge (a) and bottom (b) of the GHP minigroove with porous coating: (a) upper part of the groove edge: (1) vapor, (2) solid part of the wick, (3) vapor jet, (4) macropore free of liquid, (5) micropore with capillary liquid flow, (6) meniscus of liquid interface; (b) bottom part of the groove (liquid pool): (1) micropore, (2) interface meniscus of evaporation, (3) vapor bubble, (4) vapor stream.

the porous volume of the wick is ensured by capillary forces. The porous coating of the groove edge turns to be wetted uniformly. The area of the evaporation is increased manifold, Figure 8.7. The porous coating has a great potential to increase the wettability and heat transfer intensity in small-size heat transfer devices such as mini heat pipes and miniature heat exchangers. In this work the investigation of ammonia boiling and evaporation heat transfer in minigrooves inside a single horizontal heat loaded tube (smooth and with porous coating) was realized. The ammonia—a long-term alternative refrigerant—is used in compact heat exchangers and heat pipes due to its excellent performance and the lack of impact on the environment (zero ODP and GWP < 3) and its physical properties. The thermal behavior of grooved heat pipes (GHPs) and the same grooved heat pipes with a porous layer (GHPPL) experimentally investigated to prove the suggested hypotheses. The set of experiments was performed with Al_2O_3 porous coatings inside aluminum grooved heat pipes. Earlier the copper/propane annular heat pipes with copper sintered porous coating were successfully tested [37]. The evaporator thermal resistance ($R_{e,v}$) of aluminum GHPPL was significantly (1.3–1.4 times) lower (between 0.021 and 0.018 W/K) to compare with GHP (between 0.025 and 0.035 W/K). As a result, substantial intensification of heat transfer was obtained in a new advanced design of the GHPPL. So, the application of NC technology made in this work is encompassed on improving the cooling capability of GHPPLs. Porous coatings of GHP evaporators formed from micro- and NPs enhance heat transfer not only in the thin liquid film, Figure 8.7a, but also in the liquid pool and on flooded surfaces, Figure 8.7b. The GHP and GHPPL were tested simultaneously on the same experimental setup at a temperature range between –30°C and 70°C, typical of the electronic components cooling (Figure 8.8a and b).

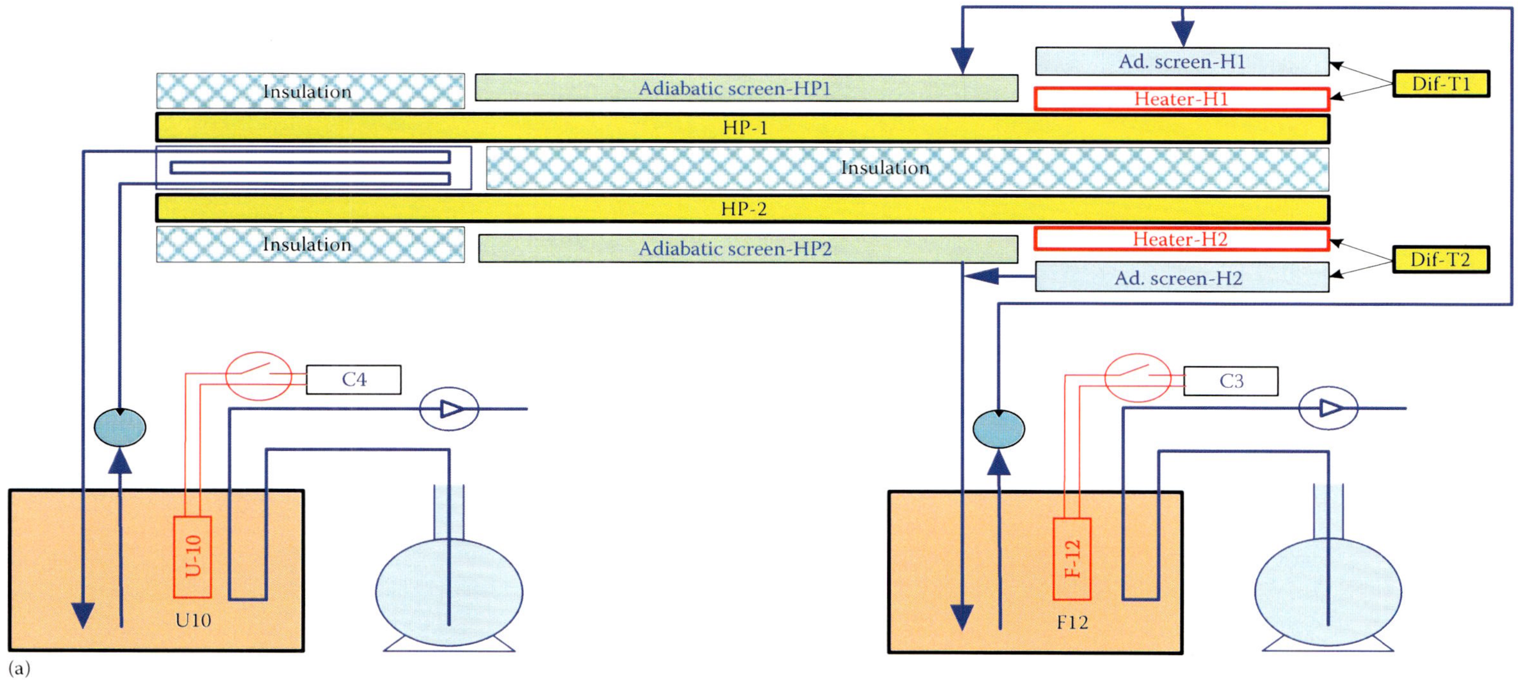

FIGURE 8.8 (a) Experimental setup: heating and cooling parts of two identical heat pipes (first with smooth grooves, second with porous coated grooves); (*Continued*)

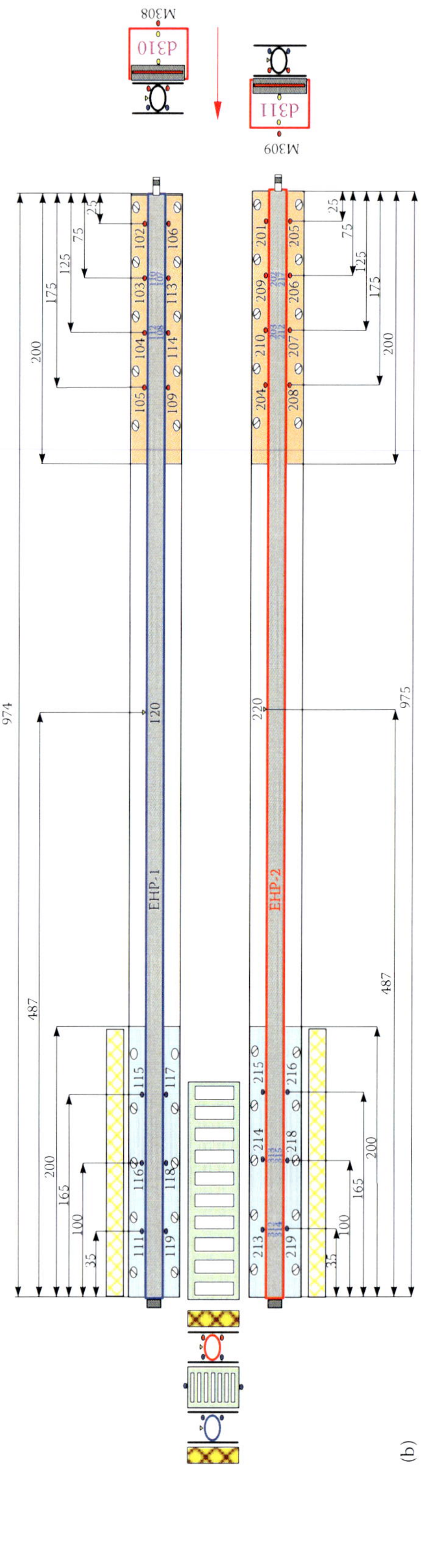

FIGURE 8.8 (*Continued*) (b) Temperature control system (thermocouples) disposed in the evaporator, transport zone, and the condenser of heat pipe.

The NCs play the role of additional centers for stable vapor generation which do not require high superheating of the surface. The NCs (thickness 25–100 μm) on the surface of minigrooves of the aluminum GHPPL allows reducing its thermal resistance and increasing the working fluid capillary pressure and permeability at the same time. Two different GHPs were fabricated and tested: one GHP (TAS-3) evaporator was made with porous coating thickness of 50 μm and the second GHP (TAS-4) evaporator featured the smooth surface of the groove. The experimental plan consists of the determination of next dependences: $\Delta T_{e/v}(Q)$, $R_{e/v}(Q)$, $R_{v/c}(Q)$ at fixed temperatures in adiabatic zone for two GHPs tested in parallel. In all GHP's the vapor temperature in the transport zone was constant for the different heat loads of the evaporator. This procedure was performed by changing the temperature and flow rate of the cooling liquid in the condenser heat exchanger. Unlike the boiling heat transfer on a smooth surface the liquid evaporation/boiling in the porous coating (e.g., like aluminum oxide NCs) is characterized by the constant sources of nucleation (Figures 8.6 and 8.7) due to the limited number of evaporation menisci available inside the porous wick. In porous coatings the liquid/vapor interface consists of menisci, which are situated inside the macropores and numerous menisci existing between macro- and micropores. The menisci of the evaporation disposed on a macropore are typical of low heat flux, when the wick is completely saturated with liquid. For such low heat flux the heat transfer is realized by conduction through the wick. The menisci of the evaporation available in minipores are typical of a high heat flow, when the menisci in macropores recede and are open for the vapor flow. For such a case many nucleation sites become the centers of vapor generation inside the macropore (micromenisci are open on the interface between macro- and micropores). The experimental values of the GHP evaporation heat transfer coefficients h as a function of heat flow and temperature are shown on the Table 8.1.

Following the experimental data, the upper part of the GHPPL groove initiates more intense heat transfer compared to the bottom part of the groove saturated with liquid [36].

The heat flux q through the wall of the GHP can be written as

$$q = \frac{T_w - T_{sat}}{\dfrac{\delta_{wick}}{k_{eff}} + \dfrac{1}{h_e}} \tag{8.1}$$

where $T_w - T_{sat} = \Delta T_t$ and ΔT_t is determined as

$$\Delta T_t = \frac{2\sigma T_{sat}}{h_{lv}\rho_v}\left(\frac{1}{r_v} - \frac{1}{r}\right) \tag{8.2}$$

TABLE 8.1
Experimental Values of the GHP Evaporator Heat Transfer Coefficients via the Heat Flow Value

	Heat flow, W	**40**	**50**	**70**	**80**	**100**	**150**	**170**
$T_{sat} = -10°C$	h_{smooth}, W/(m² K)	6,500	6,500	6,350	6,100	6,000	5,900	5,500
	h_{porous}, W/(m² K)	6,000	6,250	7,000	7,300	7,600	7,000	6,400
	Heat flow, W	**40**	**50**	**70**	**80**	**100**	**150**	—
$T_{sat} = 40°C$	h_{smooth}, W/(m² K)	6,600	6,700	7,100	7,300	7,700	7,100	—
	h_{porous}, W/(m² K)	12,000	12,300	13,000	12,900	12,200	8,200	—
	Heat flow, W	**40**	**50**	**70**	**80**	**100**	**120**	**140**
$T_{sat} = 70°C$	h_{smooth}, W/(m² K)	—	8,500	8,500	8,400	8,200	8,000	7,800
	h_{porous}, W/(m² K)	—	13,000	13,000	13,100	12,900	11,100	9,500

8.3.1 The Effective Thermal Conductivity of the Al_2O_3 Porous System

The effective thermal conductivity of a porous system has been of interest over the last two centuries. As for now, numerous experimental materials have been devised and a large number of formulae have been put forward to calculate the effective thermal conductivity of porous systems [38]. The method of generalized conductivities for the determination of an effective thermal conductivity was suggested for the wick analysis. Here, one assumes the complex of Al_2O_3 particles in the elementary cell on the GHP evaporator to be symmetric and considers only quarters of two particles contacting each other. The schematic of an elementary cell of a capillary porous material and its thermal resistance network is shown in Figure 8.9a and b. Let V be the total volume of an elementary cell, V_1 be the volume of solid phase of the elementary cell, and V_2 be the volume of the vapor phase (macropore). The liquid is situated in a micropore between two solid parts (numbered (1) in Figure 8.9a). In this case, $\Delta/B = X$ and $H/b = X/(0.5 - X)$ [38]. The wick porosity is $\Pi = f(H/b)$.

The effective thermal conductivity k_{eff} of the wick is calculated as

$$\frac{k_{eff}}{k_s} = \frac{1}{\dfrac{1}{(H/B)^2} + A} + v_g(1-B)^2 + \frac{2}{1 + s/b + \dfrac{1}{v_g\, s/B}} \tag{8.3}$$

where

$$A = \frac{1}{\dfrac{k_c}{k_s} + \dfrac{v_g}{4}\left(\dfrac{s}{B}\right)^2 \times 10^3}, \quad B = b + s, \quad s = 2\Delta \tag{8.4}$$

The complex number A characterizes the thermal resistance of the contact between two Al_2O_3 particles; v_g is the ratio of thermal conductivities of the liquid (ammonia) and solid (Al_2O_3); $v_g = k_l/k_s$; k_c is the thermal conductivity of the thermal contact between the particles of Al_2O_3 in vacuum. The thermal conductivity k_s of porous Al_2O_3 particles is equal to 2.1 W/m K. The calculated data of the effective thermal conductivity of the porous deposit saturated with liquid or vapor (ammonia) are listed in Table 8.2.

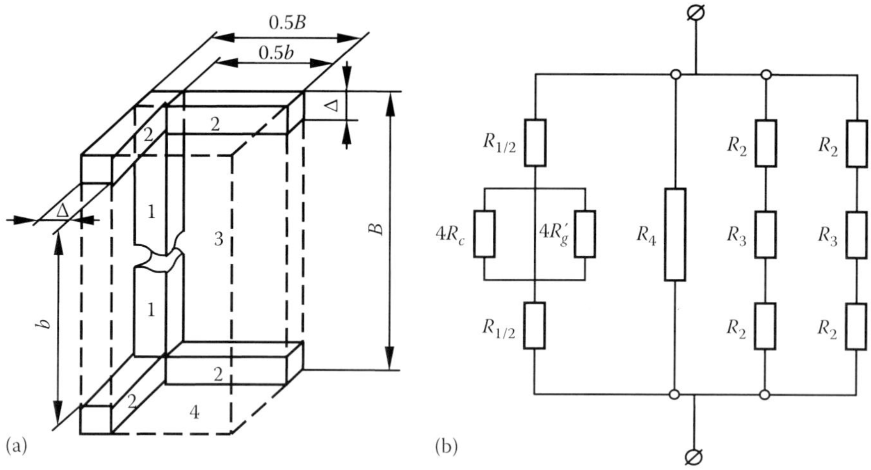

FIGURE 8.9 (a) One quarter of the porous coating elementary cell, (b) diagram of its thermal resistances.

TABLE 8.2
Calculated Values of k_{eff} of the Al_2O_3 Porous Coating (Thickness δ = 50 μm) Completed with Liquid or Vapor

Temperature, °C	−10	30	40
k_{eff}, (liquid), W/(m K)	1.14	1.09	1.07
k_{eff}, (vapor), W/(m K)	0.646	0.656	0.659

The parameters of different aluminum GHPs S-1(S-1–1), S-2(S-2–1), and S-3(S-3–1) as a function of temperature were investigated for a constant heat load [37]. GHPs with smooth grooves are S-1, S-2, and S-3, Figure 8.10.

The characteristics of the GHP with different capillary grooves (smooth GHPs and GHPPL with porous coating) are shown in Table 8.3.

GHPPLs with porous coating are denoted as S-1-1, S-2-1, and S-3-1. The main difference between these heat pipes is its value of fin edge surface coated by the porous layer (Figure 8.7a). The large surface of edge is typical for S-2-1, the medium for S-1-1, and the small surface of edge is typical for S-3-1. This part of the porous coating ensures the most efficient heat transfer intensity by evaporation (no liquid pool over the porous coating).

The difference between heat pipe S-1-1, S-2-1, and S-3-1 heat transfer intensities is 1.3–1.6 times, Figures 8.11 through 8.14. An appreciable increase in the heat transfer intensity was noted for all GHPPLs with porous coating to compare with GHPs with smooth grooves. The evaporator S-2-1 has the highest heat transfer intensity, Figure 8.12. The S-2-1 portion of the evaporating surface with high heat transfer performance is the largest and equal to about 0.6 of the total surface of the porous coating. For S-1-1 it is equal to 0.43 and for S-3-1 to 0.33. These experimental and numerical data are in good agreement with the results published in [39].

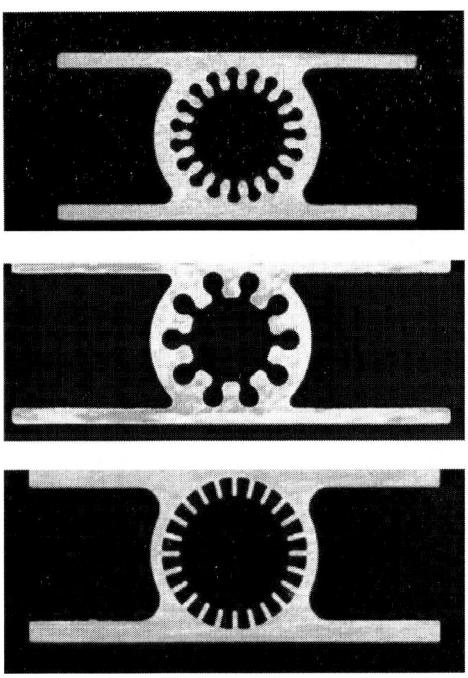

Evaporator cross section S-1

Evaporator cross section S-2

Evaporator cross section S-3

FIGURE 8.10 Cross section of the aluminum heat pipe evaporators S-1, S-2, S-3.

TABLE 8.3
Characteristics of the GHP with Different Capillary Grooves (Smooth GHPs and GHPPL with Porous Coating)

Structure Designation	Groove Shape	Groove Depth, mm	Edge Width at the Crest, mm	Porous Coating Thickness, μm
S-1	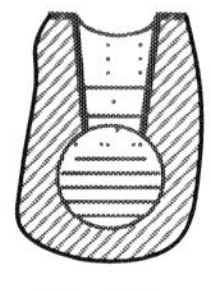	2	0.54	
S-1-1		2	0.54	50
S-2	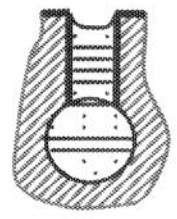	2	1	
S-2-1		2	1	50
S-3	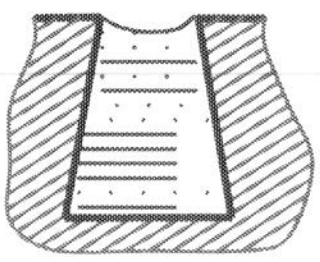	1.3	0.36	
S-3-1		1.3	0.36	50

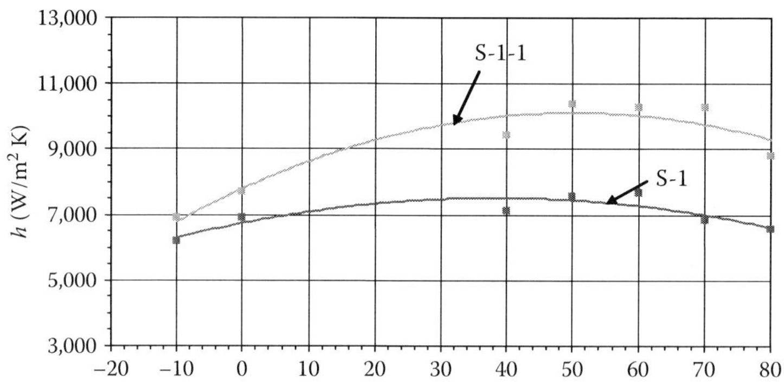

FIGURE 8.11 Heat transfer intensity in evaporators S-1 and S-1-1 (ammonia) as a function of the saturated vapor temperature.

The value of the wetted surface of the GHPPL edge, Figure 8.15, determines the value of the heat flow dissipation (by evaporation) from the porous coating to the vapor channel of the GHPPL. The heat flux removed from the porous layer of certain thickness and length can be determined by the following formula [40]:

$$q = \frac{\dfrac{2\sigma\cos\theta}{R_p} - \rho_l g L \sin\phi}{\left(\dfrac{\mu_l}{\rho_l K A_w} + \dfrac{1}{n_{\max}}\dfrac{8\mu_v}{\rho_v r_v^4}\dfrac{\delta_v}{L}\right)\dfrac{LA_e}{h_{lv}}} \tag{8.5}$$

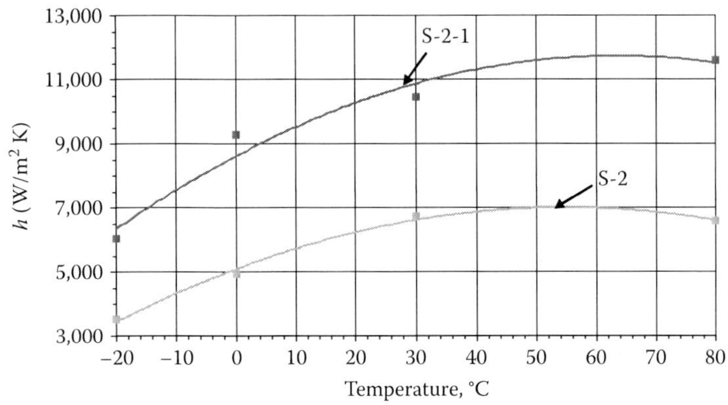

FIGURE 8.12 Heat transfer intensity in evaporators S-2 and S-2-1 (ammonia) as a function of the saturated vapor temperature.

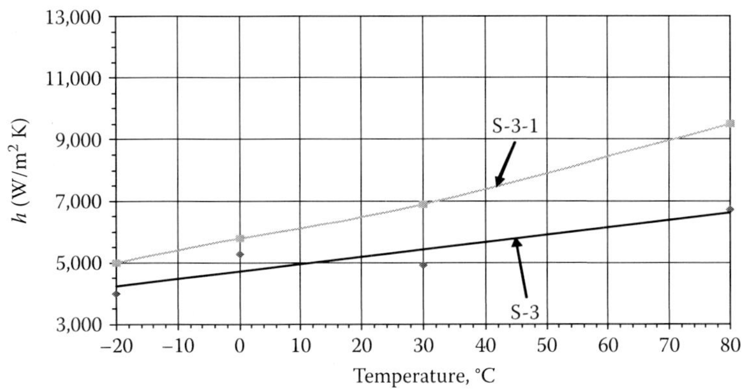

FIGURE 8.13 Heat transfer intensity in evaporators S-3 and S-3-1 (ammonia) as a function of the saturated vapor temperature.

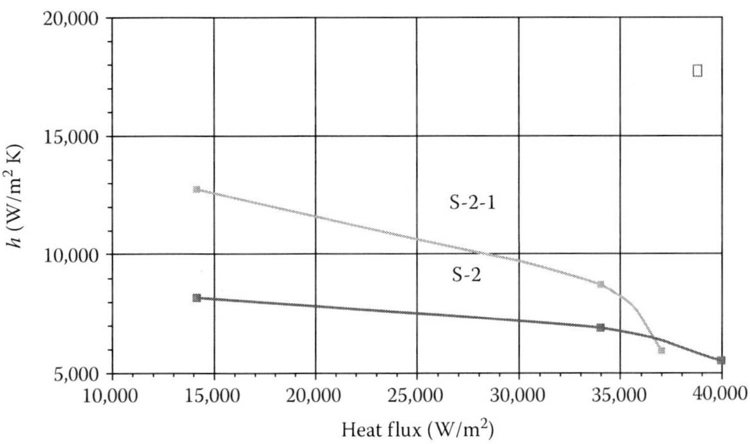

FIGURE 8.14 Heat transfer coefficients as a function of heat load in S-2 and S-2-1. $T_{sat} = 60°C$.

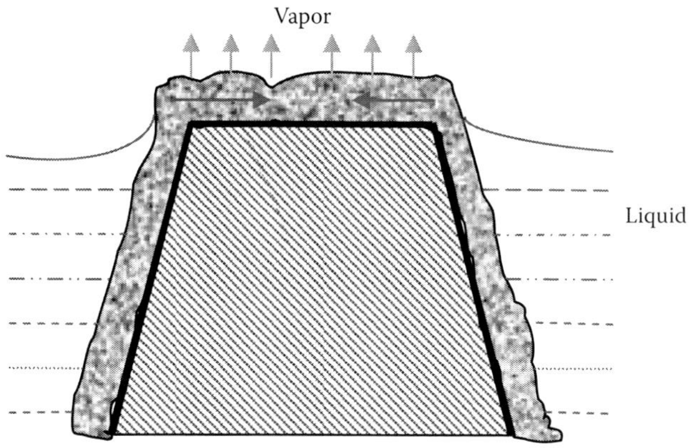

FIGURE 8.15 Schematic of liquid evaporation at the edge of GHP with porous coating.

The total heat transfer coefficient decreases with increasing liquid meniscus radii and increases with increasing of the half width of the fin top surface. With receding of the liquid meniscus in the groove covered by a thin porous layer, the evaporation heat transfer performance is improved significantly. Capillary transport of liquid inside the porous coating (Figure 8.15) becomes an appreciable factor that limits the heat removal value. A maximum liquid curvature of the meniscus depends on the size of micropores and dimensions of NPs. For the fin with a rather wide edge of the fin top surface or in the case when the liquid meniscus is deepening inside the grooves, the capillary potential of the porous coating may turn to be insufficient to moisten all the fins. In this case, a curtain portion of the fin top surface can be dried and the effect of the porous coating is decreasing. The Q_{max} can be dissipated from the top of fin, when the GHP fins are completely filled by liquid. The length of the liquid motion L from the liquid pool through the porous coating is strongly related with the heat flux Q_{max}. Figure 8.16 shows the estimated (calculated) Q_{max} dependence from the temperature.

Figure 8.17 demonstrates the micro/nanocoating (50 μm) of Al_2O_3 particles on the mini fin of aluminum grooved heat pipe GHPPL.

The experimental validation of the heat transfer intensity in GHPPL testifies the heat transfer enhancement (2.5–3 times) compared to the heat transfer occurring on the same GHP with smooth grooves. Porous micro- and nanocoating of mini fins completely modifies the hydrodynamics of a two-phase flow in minigrooves. A microscale effect is essential inside a porous body, and a

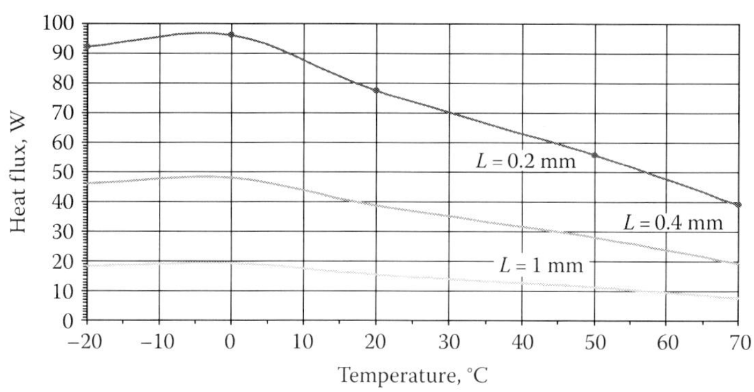

FIGURE 8.16 Heat flow removed from one edge of the capillary groove with porous coating as a function of temperature. L is the width of the edge with porous coating, mm; temperature of ammonia vapor is –20°C.

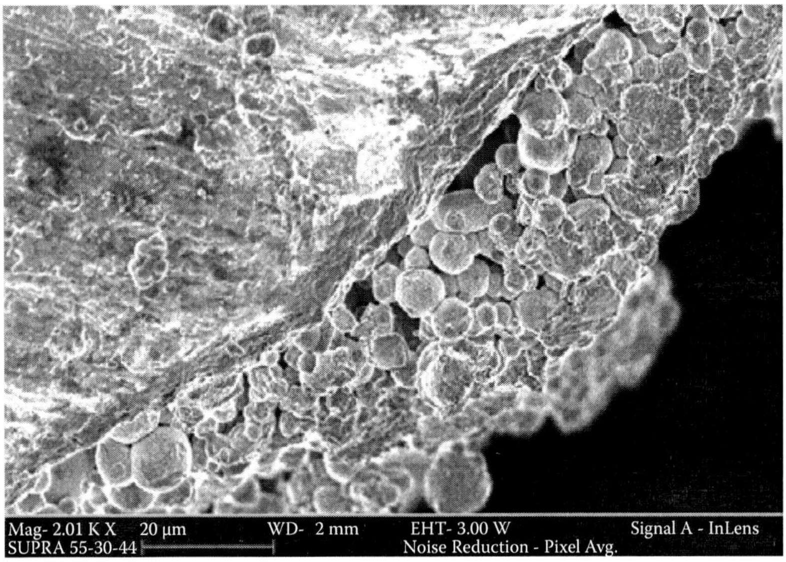

FIGURE 8.17 Aluminum oxide porous coating of GHPPL minigroove.

miniscale effect is dominant in the minigroove. A porous coating with open pores can be considered as a medium in which a large number of "micro heat pipes" with zones of evaporation and condensation are available, Figures 8.6 and 8.7. In the calculation range and experimental validation data, evaporation heat transfer in the groove covered by a thin porous layer is at least two or three times higher than that in the groove without a porous layer. It means that the heat flux increasing may occur with only slight increase in the wall temperature.

8.4 POLYMER COMPOSITES REINFORCED BY NANOWIRES AND NANOPARTICLES AS THE ALTERNATIVE TO METAL HEAT PIPE ENVELOPES

A new technology was developed to design and produce loop polymer thermosyphons capable of long-term operation without permeation by air or working fluids through their walls. Different composite polymer materials with a high effective thermal conductivity are suggested to design a wide range of new heat transfer equipment. Fiber reinforcement increases the mechanical resistance and decreases the thermal expansion, whereas the thermal conductivity tends to increase. Experiments on thermal conductivity $k(T)$ are most useful tool for the investigation of scattering processes and interaction between electrons and phonons in advanced polymers heat exchangers for the magnets cooling.

Three different methods (transient, quasi-stationary, and steady-state) to determine the effective thermal conductivity $k(T)$, thermal diffusivity, $a(T)$, and heat capacity $c_p(T)$ have been suggested and used to analyze thermal properties of polymer complex compounds at low temperature range 10–400 K.

8.4.1 Transient Method

Transient method (adiabatic calorimetry) is based on the analytical solution of the thermal conduction equation for the unlimited cylinder (Figure 8.18a and b), or plate heated by the constant source of energy in adiabatic conditions.

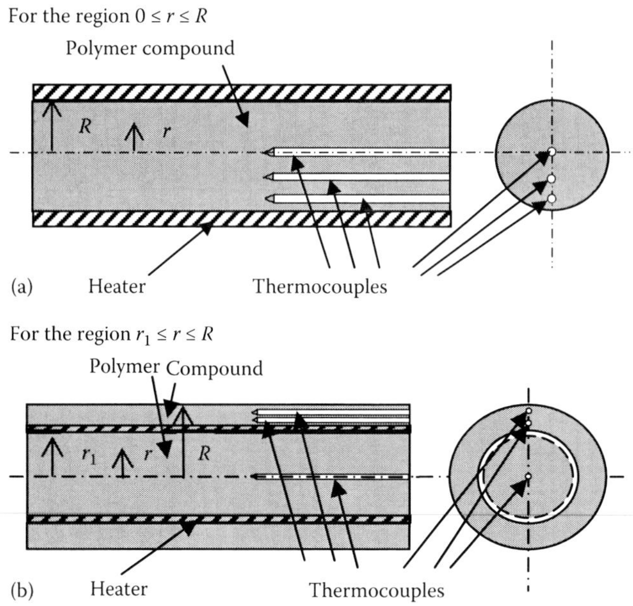

FIGURE 8.18 Two samples of the polymer complex compound cylinders (a and b) used to determine the effective thermal properties (thermal conductivity, thermal diffusivity, and heat capacity) with thermocouples inside.

For the region $0 \le r \le r_1$

$$\frac{\partial T_1}{\partial \tau} = a\left(\frac{\partial^2 T_1}{\partial r^2} + \frac{1}{r}\frac{\partial T_1}{\partial r}\right); \quad \tau > 0 \tag{8.6}$$

$$T_1(r,\tau)_{\tau=0} = 0; \quad \left(\frac{\partial T_1}{\partial r}\right)_{r=0} = 0.$$

For the region $r_1 \le r \le R$

$$\frac{\partial T_2}{\partial \tau} = a\left(\frac{\partial^2 T_2}{\partial r^2} + \frac{1}{r}\frac{\partial T_2}{\partial r}\right); \quad \tau > 0 \tag{8.7}$$

$$T_2(r,\tau)_{\tau=0} = 0; \quad \left(\frac{\partial T_2}{\partial r}\right)_{R} = 0.$$

Boundary conditions on the heater

$$T_{1r=r_1} = T_{2r=r_1},$$

$$\left(\frac{\partial T_1}{\partial r}\right)_{r=r_1} - \left(\frac{\partial T_2}{\partial r}\right)_{r=r_1} = \frac{q}{k}. \tag{8.8}$$

For small *Fo* numbers (0.03–0.05) the temperature distribution at the region

$$0 \le r \le r_1; \quad r = r_1; \quad (r = r_1, q = const)$$

$$T_1 = \frac{qr_1}{k\left[\dfrac{2a\tau}{R^2} + \dfrac{r^2 + r_1^2}{2R^2} + \ln\left(\dfrac{R}{r_1}\right) - \dfrac{3}{4} - 2\sum_{n=1} \dfrac{J_0(r_1\mu / R)}{\mu_n^2 J_0^2(\mu)} \exp\left(\dfrac{-\mu_n^2 a\tau}{R^2}\right)\right]} \tag{8.9}$$

$$T_1 \text{ can be written } T_1 = \frac{qr_1 f(Fo)}{k}, \quad \text{where } \frac{qr_1}{k} = const.$$

For the temperature ratio evolution between the time intervals $n\tau$ and $2n\tau$

$$\frac{T_{2n\tau}}{T_{n\tau}} = \frac{f(2nFo)}{f(nFo)} \tag{8.10}$$

We can calculate $f(nFo)$ and $f(2nFo)$ for two different values of Fo, and define the thermal properties of the testing solid materials as

$$a = \frac{Fo \cdot R^2}{\tau}; \quad k = \frac{qr_1 f(Fo)}{T} = \frac{Qf(Fo)}{2\pi LT} \tag{8.11}$$

8.4.2 Quasi-Stationary Method

For the Fo numbers greater than 0.2 we have the so-called quasi-stationary temperature distribution inside of the polymer cylinder as a function of time:

$$T = \frac{qr_1}{k}\left(\frac{2a\tau}{R^2} + \frac{r^2 + r_1^2}{2R^2} + \ln\frac{R}{r_1} - \frac{3}{4}\right) \tag{8.12}$$

$$0 \le r \le r_1$$

$$b = \frac{dT}{d\tau} = \frac{2aqr_1}{kR^2} = ac\vartheta R^2 \tag{8.13}$$

$$k = \frac{qr_1}{\Delta T}\left(\frac{r_A^2 - r_B^2}{2R^2}\right) = 4V\Delta T \tag{8.14}$$

8.4.3 Steady State Method

For the steady state method the accuracy of the thermal conductivity measurements is good (1%–2%). In this case the heat flow is supplied to one end of the sample and the temperature is recorded at two points. The thermal conductivity of the flat plate of polymer is determined as

$$k = \frac{Ql}{(T_1 - T_2)F} \tag{8.15}$$

where

Q is the heat flow through the sample
F is the cross-section of the sample
l is the length of the sample
$(T_1 - T_2)$ is the temperature drop over the measured sample length

The thermal conductivity, thermal diffusivity, and heat capacity of polymers have been measured using transient method for the short period of time (some minutes). The test samples of the complex polymer composite were investigated for the heat flow direction along the carbon fibers and across the fibers. For the thermal conductivity measurements using steady state method the cylindrical and plate samples were used. Each cylindrical sample was 10 mm in diameter and 15-mm long. The measured section length was 10 mm. The total inaccuracy in the thermal conductivity measurements did not exceeded 3%.

Thermal properties of polymer complex compounds have been measured using the transient, quasi-stationary, and steady-state mode of the sample heating. Test samples were made as cylinders and square plates. The polymer thermal properties were measured in two directions, that is, a heat flow direction was oriented across or along the fiber. For example, the influence of the heat flow Q direction on the effective thermal conductivity, thermal diffusivity, and heat capacity of the carbon fiber reinforced polymer 2763°C in the temperature range 10–400 K is shown on Figure 8.19a and b. The sample dimensions were chosen in such a way that the time taken for the equilibrium to be established and the dilatation effects could be decreased to a minimum. The thermal properties of polymers depend on carbon fiber orientation. The direction of the heat flow and orientation of the carbon fiber are responsible for the conduction and thermal diffusivity value of the polymer composite.

Specific heat does not practically depend on the heat flow direction, since it characterizes the scalar value, that is, energy accumulation. The curves of thermal diffusivity of the complex compound along and across the fibers present some distortions, with a maximum values between 50 and 200 K.

The results of investigation of the polyamide composite polymer with carbon fibers and NPs inside testified its high effective thermal conductivity up to 11 W/m°C at the room temperature.

8.5 POLYMER LOOP THERMOSYPHON WITH FLAT EVAPORATOR AND CONDENSER AND FLEXIBLE TRANSPORT LINES

The second part of the experiments was devoted to the design and test of the polymer loop thermosyphon with flat interface (for heat-generating elements attachment) and flexible transport lines between the flat evaporator and condenser. This thermosyphon passively transfers heat from a heat source (evaporator) to a heat sink (condenser) where the heat is dissipated.

The thermosyphon has rectangular capillary minigrooves as the capillary structure inside the evaporator and condenser, Figures 8.20a and b. The frame of thermosyphon, Figure 8.20b, is made from the polyamide composite with nanocarbon filaments and NPs to increase its effective thermal conductivity [27]. A jacket of polymer composite is formed around a core of carbon fibers to form a heat transfer device with high thermal conductivity. The width and length of the evaporator and condenser is 50 and 250 mm, respectively. The width of the grooved surface inside the evaporator and condenser is 30 mm. The width and depth of the rectangular minigrooves inside the evaporator and condenser are 2.5 mm. The thickness of the evaporator and condenser is 10 mm. There are flexible vapor and liquid transport pipes (200 mm length, 5 mm diameter, respectively) made from pure polyamide to connect the evaporator and condenser which are disposed in parallel one below the other. The experimental setup, Figure 8.21, with the temperature sensors points on the evaporator, condenser, vapor and liquid pipes of thermosyphon was used during experiments.

The width and thickness of rectangular side parts of frames are 10 mm respectively. The working fluid of thermosyphon is isobutene (R 600). The experimental technique was used to perform some necessary operations including temperature measurements (Agilent Data Acquisition Agilent Data Loger HP-34970A with a set of thermocouples connected to the computer), the measurement

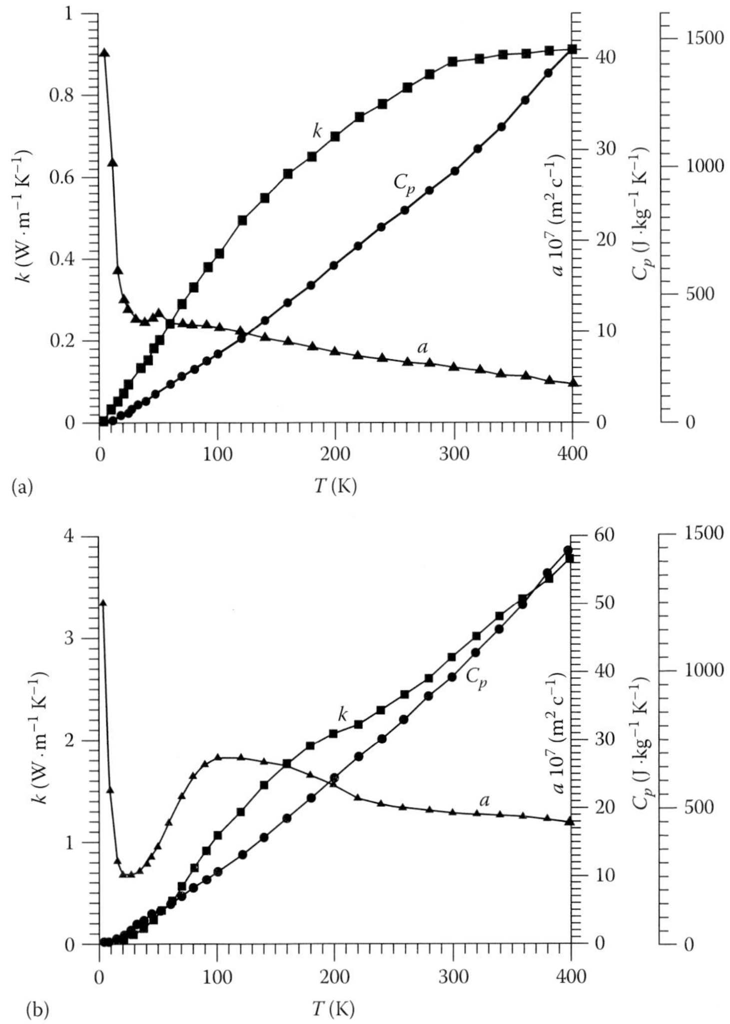

FIGURE 8.19 (a) Thermal properties (k, a, c_p) of the polymer composite 2763°C; direction of the heat flow Q—*across the fibers.* (b) Thermal properties (k, a, c_p) of the polymer composite 2763°C; direction of the heat flow Q—*along the fibers.*

of the maximal heat flow and thermal conductivity of the thermosyphon envelope. The heat flow enters the bottom side of the evaporator. The heat sink is on the top side of the condenser. The temperature difference $T_w - T_{sat}$ between the external walls of the evaporator, T_w, and the saturated temperature of the adiabatic (vapor transport) zone, T_{sat}, was measured directly by four thermocouples, one junction of which was on the evaporator wall, while the other was placed in a thermally controlled liquid bath.

The constant vapor saturation inside the thermosyphon was maintained by regulation of temperature and fluid flow on the condenser heat sink (liquid heat exchanger).

The total thermosyphon thermal resistance, R_{ts}, was calculated as

$$R_{ts} = \frac{t_e - t_c}{Q}, \tag{8.16}$$

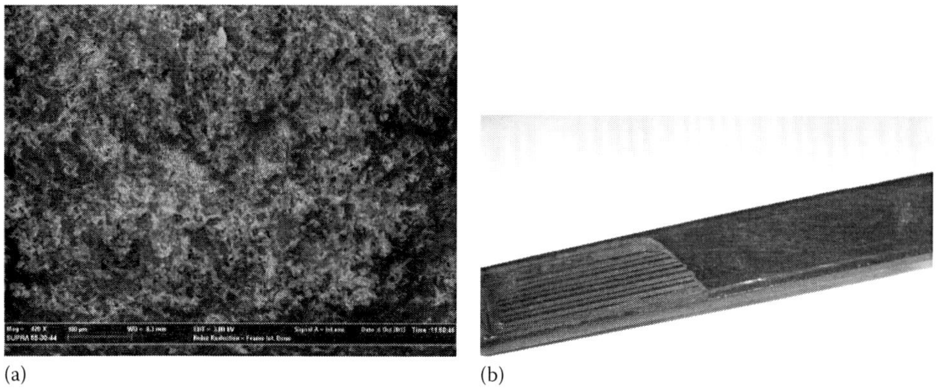

FIGURE 8.20 (a) Flat loop thermosyphon made from polymer composite—polyamide reinforced with nanocarbon filaments and NPs. (b) Flat evaporator (cross section view) with rectangular mini-channels.

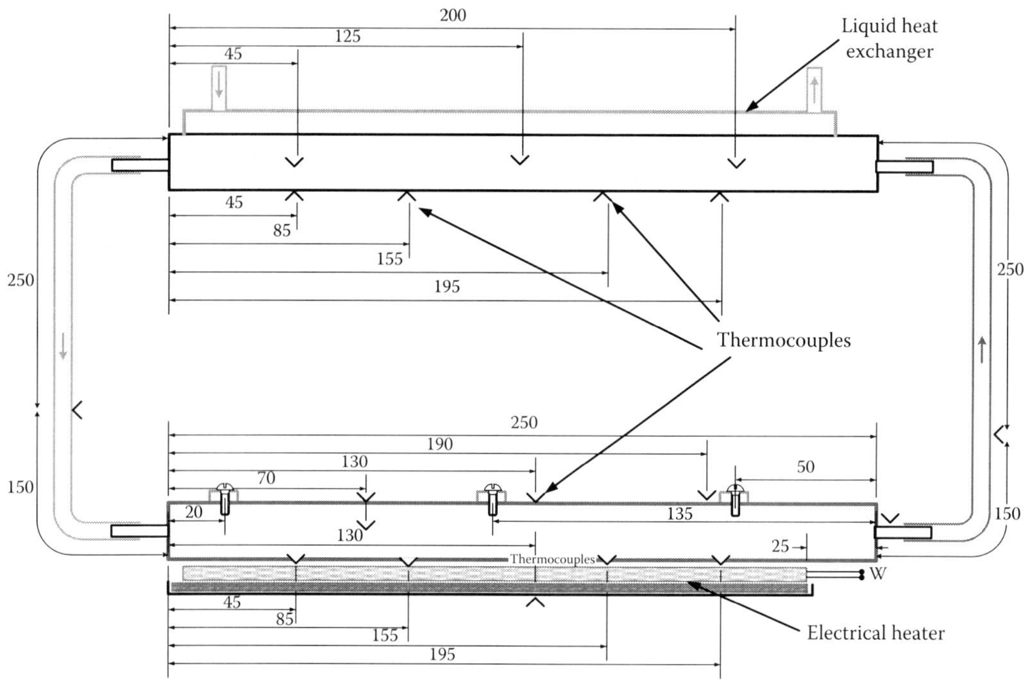

FIGURE 8.21 Experimental setup with thermocouples attachment in the evaporator, condenser, and transport zone (vapor and liquid).

where

T_e (°C) is the mean evaporator temperature
T_c (°C) is the mean condenser temperature
Q (Watts) is the input power

To cool the thermosyphon condenser (liquid loop) a Joulabo F12 recirculation thermal bath with temperature-regulated accuracy of ±0.5°C was used. Heat flow was supplied by the electric cartridge heater, disposed on the bottom surface of the evaporator. The heat sink (liquid heat exchanger)

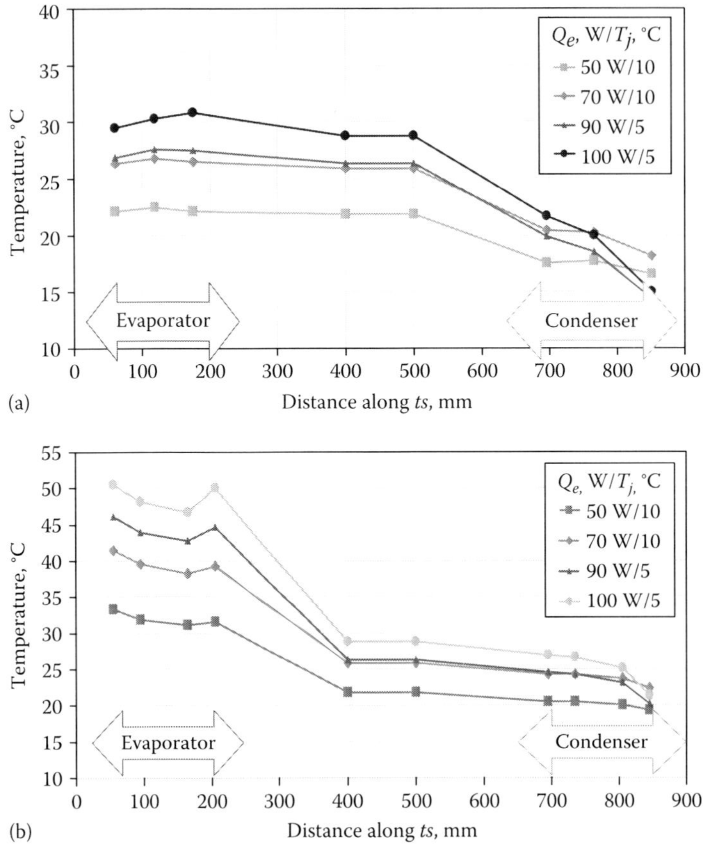

FIGURE 8.22 (a) Mean temperature distribution along the evaporator and condenser *top surface* of thermosyphon as the function of heat input Q. (b) Mean temperature distribution along the evaporator and condenser *bottom surface* of thermosyphon as the function of heat input, Q.

was attached to the top of the condenser. All measurements were performed in a steady-state regime for the heat flux range of 50–100 W. The temperature distribution along the evaporator, transport zone, and condenser as a function of the heat load is shown on Figure 8.22a and b.

The mean temperature profiles on the top surface of the evaporator and condenser are shown on Figure 8.22a, while the temperature profiles on the bottom surface of the evaporator and condenser are shown on Figure 8.22b. The temperature of saturated vapor inside the thermosyphon was varied from 20°C to 40°C with the help of thermostat thermal control. Thermal resistances of evaporator and condenser as a function of heat input are shown in Figure 8.23a and b.

Mean temperature distribution along the *bottom surface* of thermosyphon was used for the thermal resistance R calculations, Figure 8.23b. For this case R_{ts} is nearly two times more to compare with top surface measurements, Figure 8.23a.

The mass of R 600 (13 g) was chosen just to fill the volume of capillary channels in the evaporator, condenser, and liquid transport line at the room temperature for the heat flow 40 W. The analysis of the mean temperature distribution, Figure 8.24, along the thermosyphon for different mass of the working fluid denotes the change of total thermal resistance as a function of fluid mass. For the mass of R 600 equal 13 g the total thermal resistance is optimal while the decreasing of mass down to 11.7 g significantly increases the thermal resistance of the evaporator and is not sufficient to ensure the wettability of total heat loaded surface. Increasing the mass of R 600 up to 14 g does not significantly changes the temperature distribution along the thermosyphon.

1. Recent advances in nanotechnology have allowed development of new NFs and NCs, intensifying heat transfer processes and reducing energy losses. NFs and NCs have been used as the means to increase the heat transfer intensity in original designs of heat pipes and thermosyphons evaporators.
2. Comparative studies of the heat transfer coefficients of the evaporators with different types of capillary grooves were conducted. The regimes of both evaporation and boiling of the working fluid (ammonia) were provided in the set of experiments with the evaporators having smooth capillary grooves and capillary grooves with a porous NC of walls with a thickness of 20–100 μm. Within the entire studied range of temperatures and heat loads, the heat transfer coefficients of all types of evaporators with the porous NC are 1.3–1.8 times higher than of similar evaporators with a smooth surface of capillary grooves (0.015–0.02 K/W for GHP with a porous nanolayer and 0.025–0.035 K/W for classical one).
3. A new type of the horizontal polymer flat loop thermosyphon with nanotechnology application was suggested, designed, and tested. It was found that the evaporator thermal resistance *Rev* of polymer thermosyphon is similar to that of a classical aluminum heat pipe.
4. The combination of the working fluid R 600, polymer (polyamide), and composite material (polyamide reinforced with carbon nanofilaments and NPs) as the thermosyphon envelope is completely compatible.
5. For the future detailed analysis of such heat transfer devices application it is necessary to consider the problem with aging, thermal cycling under pressure (–50°C to 100°C), light gas penetration through the walls of the HP envelope, and phase transition of polymer.

NOMENCLATURE

A	Surface area, m^2
h	Heat transfer coefficient, W/m^2 K
h_{fg}	Latent heat, J/kg
K	Permeability, m^2
k	Thermal conductivity coefficient, W/m K
Q	Heat flux, W
q	Heat flux density, W/m^2
R	Thermal resistance, K/W
r	Radius, m
T	Temperature, °C
L	Length, m

GREEK SYMBOLS

δ	Thickness, mm
μ	Dynamic viscosity, Pa s
φ	Angle
θ	Wetting angle
ρ	Density, kg/m^3
σ	Surface tension, N/m

INDICES

e	Evaporator
eff	Effective

l	Liquid
sat	Saturation
v	Vapor
w	Wall

ACKNOWLEDGMENT

The authors would like to thank D.O. Lapotko, M.I. Rabetsky, L.P. Grakovich, A.S. Zhuravlyov, A.V. Shapovalov, S.P. Bogdanovich, and S.S. Pesetsky for their experimental contribution.

REFERENCES

1. V. Trisak, S. Wongwises, Critical review of heat transfer characteristics of nanofluids, *Renewable and Sustainable Energy Reviews*, 11, 512–523 (2007).
2. W. Yu, D.M. France, J.L. Routbort, S.U.S. Choi, Review and comparison of nanofluid thermal conductivity and heat transfer enhancements, *Heat Transfer Engineering*, 29, 432–460 (2008).
3. F.P. Cabral, G. Ribatski, Theoretical modeling of heat transfer flow boiling of nanofluids inside horizontal micro-scale channels, *Proceedings of (ENCIT 2010) 13th Brazilian Congress of Thermal Sciences and Engineering*, Uberlandia, Brazil, December 5–10, 2010.
4. G. Paul, M. Chopkar, I. Manna, P.K. Das, Technique for measuring the thermal conductivity of nanofluids: A review, *Renewable and Sustainable Energy Reviews*, 14, 1913–1924 (2010).
5. H.A. Mohammad, G. Bhaskaran, N.H. Shuaib, R. Saidur, Heat transfer and fluid flow characteristics in microchannels heat exchanger using nanofluids: A review, *Renewable and Sustainable Energy Reviews*, 15, 1502–1512 (2011).
6. G. Huminic, A. Huminic, Application of nanofluids in heat exchangers: A review, *Renewable and Sustainable Energy Reviews*, 16, 5625–56638 (2012).
7. S. Kakaç Numerical analysis of convective heat transfer with temperature dependant thermal conductivity of nanofluids and thermal dispersion, Ed. M. Rebay, *Proceedings of CONV-14, International Symposium on Convective Heat and Mass Transfer*, Kusadasi, Turkey, June 8–13, 2014.
8. S. Khandekar, Y.M. Joshi, B. Mehta, Thermal performance of closed two-phase thermosyphon using nanofluids, *International Journal of Thermal Sciences*, 47, 659–667 (2008).
9. Y. Lin, S.-W. Kang, H.-L. Chen, Effect of silver nano-fluid on pulsating heat pipe thermal performance, *Applied Thermal Engineering*, 28, 1312–1317 (2008).
10. T. Tsai, H.-T. Chien, P.-H. Chen, Improvement on thermal performance of a disk-shape miniature heat pipe with nanofluid, *Nanoscale Research Letters*, 6(1), 590 (2011).
11. J. Barber, D. Brutun, L. Tadrist, A review on boiling heat transfer enhancement with nanofluids, *Nanoscale Research Letters*, 6, 1–16:280 (2011).
12. A. Mirzamoghadam, I. Catton, A physics model of the evaporating meniscus, *ASME Journal of Heat Transfer*, 110, 201–207 (1988).
13. B. Suman, De Sirshendu, S. DasGupta, Transient modeling of micro-grooved heat pipe, *International Journal of Heat and Mass Transfer*, 48, 1633–1646 (2005).
14. P.C. Stephan, C.A. Busse, Analysis of the heat transfer coefficient of grooved heat pipe evaporator wall, *International Journal of Heat and Mass Transfer*, 35(2), 383–391 (1992).
15. H.B. Ma, G.P. Peterson, Experimental investigation of the maximum heat transport in triangular grooves, *ASME Journal of Heat Transfer*, 118, 740–746 (1996).
16. L. Vasiliev, L. Grakovich, D. Khrustalev, Low temperature axially grooved heat pipe, *Proceedings of the Fourth International Heat Pipe Conference*, London, U.K., 1981, pp. 337–348.
17. J. Wang, I. Catton, Enhanced evaporation heat transfer in triangular grooves covered with a thin fine porous layer, *Applied Thermal Engineering*, 21(17), 1721–1737 (2001).
18. G. Wu, S.-L. Chen, W.-P. Shih, Lamination and characterization of a polyethylenterephthalate flexible micro heat pipe, *Frontiers in Heat Pipes (FHP)*, 3, 023003 (2012). doi: 10.5098/fhp.v3.2.3003, Available at www.ThermalFluidsCentral.org.
19. M. Mochizuki, A. Akbarzadeh, T. Nguyen, A review of heat pipe practical applications and innovative opportunities application for global warming, Eds. L.L. Vasiliev, S. Kakaç, *Heat Pipes and Solid Sorption Transformation: Fundamentals and Practical Applications, Mechanical, Aerospace and Nuclear Engineering*, Taylor & Francis/CRC Press, Boca Raton, FL, 2013, pp. 145–212.

20. L.L. Vasiliev, L.L. Vasiliev, Jr., Heat pipes and thermosyphons for solid sorption machines and fuel cells thermal management, Eds. L.L. Vasiliev, S. Kakaç, *Heat Pipes and Solid Sorption Transformation: Fundamentals and Practical Applications, Mechanical, Aerospace and Nuclear Engineering*, Taylor & Francis/CRC Press, Boca Raton, FL, 2013, pp. 259–282.
21. B. Carlberg, L.L. Ye, J. Liu, Polymer-metal nanofibrous composite for thermal management of microsystems, *Materials Letters*, 75, 229–232 (2012).
22. K.J. Kim, M. Shahinpoor, Ionic polymer–metal composites: II. Manufacturing techniques, *Smart Materials and Structures*, 12, 65–79 (2003).
23. P. Slepička, T. Fidler, A. Vasina, V. Švorčik, Ripple-like structure on PLLA induced by gold deposition and thermal treatment, *Materials Letters*, 79, 4–6 (2012).
24. A.K. Bledzki, J. Gassan, Composites reinforced with cellulose based fibers, *Progress in Polymer Science*, 24, 221 (1999).
25. S. Stankovich, D.A. Dikin, G.H.B. Dommett, K.M. Kohlhaas, E.J. Zimney, E.A. Stach, R.D. Piner, S.T. Nguyen, R.S. Ruoff, Graphene-based composite, *Nature*, 457(7230), 706–710 (2006).
26. Y.S. Lipatov, A.E. Nesterov, T.D. Ignatova, D.A. Nesterov, Effect of polymer–filler surface interactions on the phase separation in polymer blends, *Polymer*, 43, 875 (2002).
27. S.P. Bogdanovich, L.P. Grakovich, L.L. Vasiliev, Heat-conductive polymerous material on basis of thermoelastolayers, *International Conference on Polymeric Composite Materials and Tribology (Polycomtrib-2011)*, Gomel, Belarus, 2011.
28. G.-W. Wu, S.-L. Chen, W.-P. Shih, Lamination and characterization of a polyethylenterephthalate flexible micro heat pipe, *Frontiers in Heat Pipes (FHP)*, 3, 023003 (2012). doi: 10.5098/fhp.v3.2.3003, Available at www.ThermalFluidsCentral.org.
29. A.V. Luikov, L.L. Vasiliev, S.A. Tanaeva, L.S. Domorod,, Experimental investigation of thermal properties of glass-fiber-resin materials from 10 to 400 K, *Letters in Heat and Mass Transfer*, 1, 7–12 (1974).
30. L. Vasiliev, E. Hleb, A. Shnip, D. Lapotko, Bubble generation in micro-volumes of "nanofluids", *International Journal of Heat and Mass Transfer*, 52, 1534–1539 (2009).
31. V.G. Pastukhov, Y.F. Maydanik, Development and experimental investigation of a heat-transfer system on the basis of a loop and pulsating heat pipe, *Proceedings of the 16th International Heat Pipe Conference*, Lyon, France, May 20–24, 2012, pp. 255–260.
32. L. Vasiliev, Jr., Thermally driven heat transfer device. Patent US 20110067843A1, 2011.
33. L. Vasiliev, D. Lapotko, E. Lukianova, A. Zhurablyov, A. Shapovalov, L. Vasiliev, Jr., Two-phase heat transfer enhancement in micro channels and heat pipe evaporators with nano porous structures, *Proceedings of the 14th International Heat Pipe Conference (14th IHPC)*, Florianopolis, Brazil, April 22–27, 2007.
34. D. Lapotko, E. Lukianova, Laser-induced micro-bubbles in cells, *International Journal of Heat and Mass Transfer*, 48(1), 227–234 (2005).
35. D. Lapotko, G. Kuchinsky, Optothermal microscope, Eds. F. Scudiery, M. Bertolotti, *Photoacoustic and Optothermal Phenomena*, AIP, Rome, Italy, 1998, pp. 184–186.
36. L.L. Vasiliev, L.P. Grakovich, M.I. Rabetsky, L.L. Vasiliev, Jr., Grooved heat pipes evaporators with porous coating, *International Heat Pipe Conference (16th IHPC)*, Lyon, France, May 20–24, 2012.
37. L. Vasiliev, A. Zhuravlyov, A. Zhapovalov, Heat transfer in minichannels with micro/nanoparticles deposited on a heat loaded wall, *Journal of Enhanced Heat Transfer* 19(1), 13–24 (2012).
38. A.V. Luikov, A.G. Shashkov, L.L. Vasiliev, Yu.E. Fraiman, Thermal conductivity of porous systems, *International Journal of Heat and Mass Transfer*, 11, 117–140 (1968).
39. D.K. Edwards, I. Catton et al., Evaporation and condensation in circumferential grooves on horizontal tubes, *Presented at the 14th NHTC*, Atlanta, GA, 1973.
40. V. Maziuk, A. Kulakov, M. Rabetsky, L. Vasiliev, M. Vukovich, Miniature heat pipe thermal performance prediction tool-software development, *Applied Thermal Engineering*, 21, 559–571 (2001).

9 Nanofluid Flow Simulation as the Flow through the Porous Media

Amarin Tongkratoke, Anchasa Pramuanjaroenkij, and Sadık Kakaç

CONTENTS

9.1 INTRODUCTION

Nanofluid is an innovation of working fluids to enhance heat transfer in thermal systems, which can be created by combining solid nanoparticles (diameters less than 100 nm) and traditional base fluids. The nanoparticles can be suspended in the base fluids as Brownian motion. In addition, the nanofluids with nanoparticles, which are commonly made from metallic, nonmetallic, or polymer materials, have shown their enhancements in their effective thermal conductivities and their convective heat transfer coefficients [1,2]. Nanofluid is a name conceived by Choi [3] who found the thermal conductivity of nanofluid was higher than the base fluids, resulting in significant enhancements in the heat transfer coefficients. Nanofluid selection is based on heat exchanger materials and the nanofluid properties. Additionally, sizes, shapes, volume fractions, and thermophysical properties of nanofluids, such as thermal conductivity, thermal diffusivity, and viscosity, play important roles on the heat transfer performances.

Nield and Bejan [4] introduced a definition of porous media as a material consisting of a solid with an interconnected void; typical natural porous media are as the human lung tissue, sandstone, and honeycomb. Common properties of the porous media are area to volume ratios and thermal conductivity; when these properties are high, convective heat transfer coefficients are increased. Since structures of the porous media are heterogeneous, the effective thermal property and velocity distribution of the porous media vary. Researches on the porous media behavior and properties have to be extended and continued by integrating various sciences such as heat transfer and fluid dynamics together. More porous media information can advocate many modern sciences and engineering.

There are several approaches that can be used to simulate fluid flow with solid particles such as single-phase flow, two-phase flow, and fluid flow through porous media. In the single-phase approach, the combination is considered as one homogeneous fluid with no slip condition. Although several numerical and experimental studies concerning the heat convection in heated enclosures containing nanofluids were published, most of them concentrated on the nanofluids in cavities and only a few of them considered the nanofluid as a porous medium. The convection in the porous medium has been studied by many groups of researchers including Horton and Rogers [5], Lapwood [6], Nield [7], Rudraiah and Malashetty [8], Murray and Chen [9], Malashetty [10], Pop and Ingham [11], and Vadasz [12]. The dispersion of nano-sized particles in the traditional fluid increased the thermal conductivity of the fluid, and the presence of porous media enhanced the effective thermal conductivity of the base fluid. Thus, the use of nanofluids in porous media would be very much helpful in the heat transfer enhancement. There was a study presenting the steady-state free convective heat transport in a triangular enclosure in which a heater was placed on the vertical wall, saturated with nanofluids and filled with a porous medium [13]. However, the triangular geometry was not directly comparable with the rectangular duct [14]. There were limited numbers of the extended Darcy–Brinkman models dealing with the natural convection in the square porous enclosure that was saturated with a nanofluid and was heated through the bottom wall [15]. Rana et al. [15] focused on the difficulties in these types of problems with the fact that the boundary layers were formed near the heated plates of these configurations and in more complex geometries encountered in modern industrial processes resulting in steep gradients of the field variables. They necessitated appreciable increase of the local resolution in the finite differences method (FDM), the finite element method (FEM), and other classical solution techniques. Their problems stimulated the interest in developing faster and more adaptable solvers of the corresponding thermal and fluid dynamics equations [15]. They found the numerical solution for steady-mixed convection boundary layer flow of a nanofluid along an inclined plate embedded in a porous medium. Even so, they did not account any effects of the parameters influencing the thermal conductivity and dynamic viscosity, such as particle concentrations, particle sizes, temperatures, nature of the base fluids, and the nature of the nanoparticles, which satisfied the experimental data [15]. Nield and Kuznetsov [16] studied the Cheng–Minkowycz problem for natural convection boundary layer flow in a porous medium saturated by a nanofluid. In the modeling of the problem, they simulated nanofluids by incorporating the effects of Brownian motion and thermophoresis. For the porous medium, the Darcy model was taken. Aziz et al. [17] found the numerical solution for the free convection boundary layer flow past a horizontal flat plate embedded in porous medium filled by nanofluid containing gyrotactic microorganisms.

Ghazvini and Shokouhmand [18] analytically and numerically studied forced convection flow of CuO-water nanofluids with 0%–4% volume fraction of nanoparticles as coolant through a microchannel heat sink with constant heat flux (the heat sink at the bottom and the insulated top). Two common analytical approaches were used; the fin and the porous media approaches. The effects of nanoparticle volume fractions, Reynolds number, channel aspect ratios and porosities on temperature distribution, as well as overall heat transfer coefficient were investigated. The results appeared that the fin approach exhibited a bigger value for both dimensionless temperatures for nanofluid and dimensionless temperatures for solid than the porous media approach. From both approaches, an increase in bulk temperature and channel aspect ratios led to particle speed and Brownian motion to increase and due to that, a better heat transport would occur. Additionally, an increase in porosity led to an increase in dimensionless temperatures in both approaches.

Özerinç et al. [19] compared their results that were obtained from the single-phase model by the finite different method with different Al_2O_3 nanofluid volume fractions at 1%, 1.5%, 2%, and 2.5% with the experimental data of the heat transfer coefficients in the Al_2O_3 fully developed laminar flow. Their numerical results, at 2% volume fraction, were in very good agreement with the experimental data at the same nanofluid volume fraction. They presented that taking variable thermal conductivity and variable thermal dispersion into account in nanofluid analysis significantly improved the numerical accuracy. Since the single-phase model is simple, easy to understand, and less complicated, one can develop the single-phase model easily and widely when the nanofluid is assumed to flow similar to other fluid, therefore, they should flow as the single-phase flow. They proved that the simple assumptions and flow model could be used to estimate the heat transfer coefficients of the new nanofluids efficiently.

Tongkratoke et al. [20] applied the single-phase model as the numerical method to analyze the heat transfer performance of the Al_2O_3-water nanofluid, based on the theoretical models. The nanofluid was assumed to flow under laminar fully developed flow condition through a rectangular pipe as in the circuit application. This work also studied the effects of the Maxwell and Yu and Choi models as the thermal conductivity models. The models were used in the simulation domain alternately, named the mixing models, which were different from the works done before which used only one effective model in each simulation. From their calculated results, applying only the Yu and Choi model [21] gave similar results to the mixing thermal conductivity models (Yu and Choi model [21] coupled with Maxwell model [22] at the wall boundary) in Peclet number (Pe) between 4000 and 6000. But, in Pe between 2500 and 4000, the mixing thermal conductivity models provided closer results to the experimental results than those of the single model. They noticed that the thermal conductivity models at the wall boundary affected the accuracy of the calculation in the low Pe range. The lower errors of these mixing models in the low Pe range (from 2500 to 4000) may be caused by heat accumulation, which were calculated, and the errors affected the heat transfer coefficients directly.

Tongkratoke et al. [23] developed the single-phase model as the numerical method to analyze the heat transfer performance of the Al_2O_3-water nanofluid flow as the base flowing through a porous media. In this study, the effects of the porous media model, the permeability and the effective mixing thermal conductivity model were conducted for the Al_2O_3-water nanofluid flow. Firstly, the digit numbers of the porosity values did not affect the heat transfer coefficients. The different permeability occurred in materials, which were manufactured differently and in different shapes, influenced the heat transfer coefficients differently. The more permeability the closer the heat transfer coefficients to the experimental ones. Finally, the mixing thermal conductivity model, which could improve the simulation performance in the previous work, was examined and coupled with the porous model. The final results showed that the mixing thermal conductivity model could improve the numerical results to be closer to the experimental results trivially by the forth digit. From this study, the developed mathematical model was found to demonstrate excellent potential in the nanofluid simulation as the porous media.

In the nanofluid flow, the nanoparticles flowed together with the base fluid and the particles could be assumed to distribute themselves uniformly throughout the base fluid, which is similar to the fluid staying with the solid part of the porous media. The effective thermal conductivity of the nanofluid can be influenced by the thermophysical properties of both nanoparticles and base fluid. So the nanofluid flow could also be assumed as the fluid flow through the uniform porous media with nanofluid properties.

In this chapter, one can apply the single-phase model as the numerical method to analyze the heat transfer performance of the nanofluid. The nanofluid can be assumed to flow under laminar fully developed flow condition through a rectangular pipe as in a circuit application. One can also study the effects of several thermal conductivity models. The models can be used in the simulation domain alternately, named the mixing model, which is different from the works done before that used only one effective model in the simulation. The governing equations written in terms of the primitive variables can be solved through an in-house program using the finite volume method and the SIMPLE algorithm. This chapter emphasized the development of the single-phase models, which can provide the better simulated results, the numerical heat transfer coefficient from the mixing models approach to the coefficients obtained from the experiment. Since the single-phase model is simple, easy to understand, and less complicated, it can be developed easily and widely.

TABLE 9.1
Thermophysical Properties of the Base Fluid (Water), the Cu, and Al_2O_3 Nanoparticles at T = 293.15 K

Thermophysical Properties	C_P (J/kg·K)	ρ (kg/m³)	k (W/m·K)
Water	4181.90	998.20	0.598
Copper	383.1	8930.0	386.000
Aluminum oxide	880.00	3920.00	42.340

Sources: Qiang, L. and Ymin, X., *Sci. China*, 45(4), 408, 2002; Zeinali Heris, S. et al., *Int. J. Heat Fluid Flow*, 28, 203, 2007.

TABLE 9.2
Thermophysical Properties of the Cu-Water and Al_2O_3 Nanofluids at Different Volume Fractions; 0.01, 0.02, and Base Fluid (Water)

	Cu		Al_2O_3
Thermophysical Properties	ϕ = 0.01	ϕ = 0.02	ϕ = 0.02
ρ_{nf}; Pak and Cho [27] (kg/m³)	1061.0	1140.7	1056.6
$C_{p,nf}$; Pak and Cho [27] (J/kg·K)	4150.9	4112.8	4115.862
μ_{nf}; Einstein [28] (kg/m·s)	0.000612	0.000627	0.000953
k_{nf}; Yu and Choi [21] (W/m·K)	0.662	0.682	0.637
k_{nf}; Maxwell [22] (W/m·K)	0.616	0.634	0.633

9.2.2.5 The Effective Mixing Thermal Conductivity Models

Tongkratoke et al. [20] introduced the mixing effective thermal conductivity models, mixing models, or combinations of the two effective thermal conductivity models in only one considered domain. The combinations consisted of the following: (1) the Maxwell [22] model (at the channel nodes) coupled with the Brownian motion model [32] at the wall boundary; (2) the Maxwell [22] model (at the channel nodes) coupled with the Yu and Choi [21] model at the wall boundary; (3) the Brownian motion model [32] (at the channel nodes) coupled with the Maxwell [22] model at the wall boundary; (4) the Brownian motion model [32] (at the channel nodes) coupled with the Yu and Choi [21] model at the wall boundary; (5) the Yu and Choi [21] model (at the channel nodes) coupled with the Maxwell [22] model at the wall boundary; (6) the Yu and Choi [21] model (at the channel nodes) coupled with the Brownian motion model [32] at the wall boundary. From their results, the combinations that consisted of the Yu and Choi [21] model (at the channel nodes) coupled with the Maxwell [22] model at the wall boundary as shown in Figure 9.1 provided the best calculated results when this couple was compared with other (coupled) mixing effective thermal conductivity model. The results were caused by the Maxwell [22] model, which was a static model and suitable for the flow part with low velocity as the channel wall nodes, and the Yu and Choi [21] model, which was a dynamic model considering Brownian motion and suitable for the flow part with high velocity as the inside nodes of the channel.

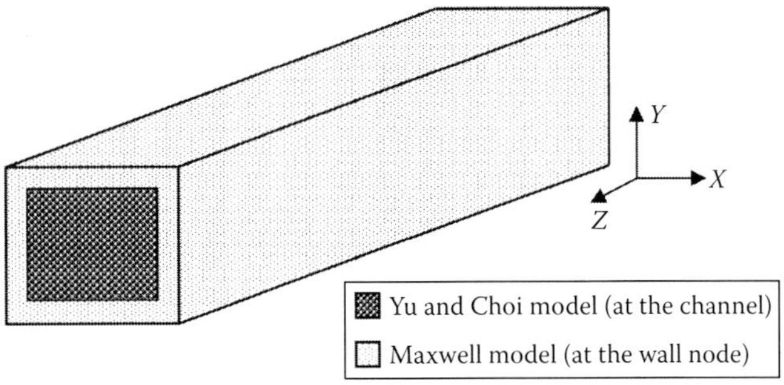

FIGURE 9.1 Geometry of the rectangular pipe for the effective mixing thermal conductivity models (pictures not to scale). (From Tongkratoke, A. et al., *Comput. Therm. Sci.*, 6(1), 1, 2014.)

9.3 RESULTS AND DISCUSSIONS

The governing equations in Section 9.2.1, the Cartesian coordinate, are appropriate for domains with quadrilateral configurations such as rectangular and square channels in electrical applications (Figure 9.2). One can, firstly, assume that the geometry of the considered domain as the channel domain with a width and height of 12 mm and a length of 1 m or a width and height of 6 mm and a length of 1 m. Then the thermophysical properties of nanofluids can be calculated from equations in Section 9.2.2. Finally, one can obtain relationships between heat transfer coefficients (h) and Reynolds number (Re) and validate them with available experimental results. The effect of the grid independence, the effect of different node numbers, which are next to the wall in the mixing thermal conductivity model, and the effect of nanoparticles permeability can also be analyzed.

9.3.1 Validation of Simulation Model

One can find an analysis of a set of experimental results from the works [30,31]. For example, Qiang and Yimin [30] have experimentally studied the fully developed laminar flow of the Cu-water nanofluid in a rectangular pipe (12 mm width and height and 1 m length as shown in Figure 9.2), for the Reynolds number (Re) ranging from 850 to 1050 and the nanofluid volume fraction at 0.01 and 0.02. Their calculated thermophysical properties of the nanofluids were introduced in Section 9.2.2. Figure 9.3 showed calculated heat transfer coefficients, which were obtained from the governing equations in Section 9.2.1, and compared with the experimental heat transfer coefficients from Qiang and Yimin [30]. The simulated results at both

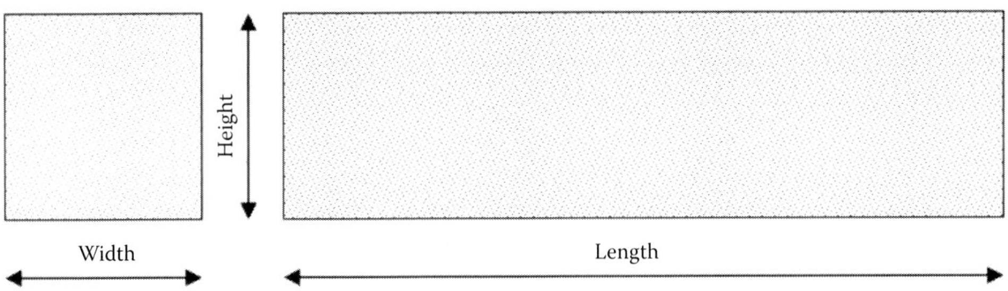

FIGURE 9.2 Geometry of the rectangular pipe (pictures not to scale). (From Tongkratoke, A. et al., Nanofluids flow simulation as the flow through the porous media, *International Symposium on Convective Heat and Mass Transfer*, Kusadasi, Turkey, 2014.)

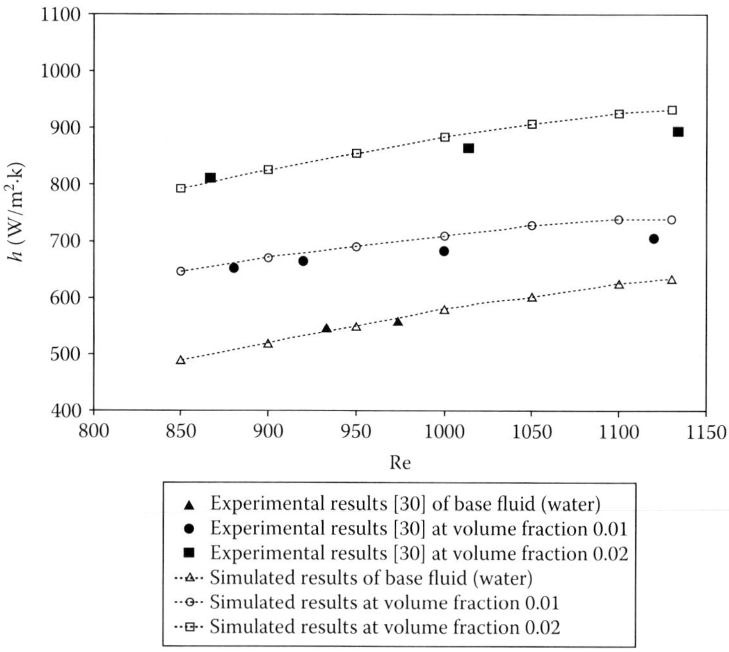

FIGURE 9.3 Comparison between the simulated results and the experimental data in prediction of the heat transfer coefficients of the Cu-water nanofluid. (From Qiang, L. and Ymin, X., *Sci. China*, 45(4), 408, 2002; Tongkratoke, A. et al., The permeability effects of copper-nanofluid flow with using the porous media model, *International Symposium on Advances in Computational Heat Transfer*, Piscataway, NJ, 2015.)

nanofluid volume fractions in the high Reynolds number range (from 1000 to 1050) were close to the experimental ones. Average error percentages between the calculated coefficients and experimental coefficients of the Cu-water nanofluid volume fractions at 0.01 and 0.02 were 3.56 and 2.56, respectively.

Another nanofluid experimental sample, Zeinali Heris et al. [31] who performed an experimental study on the fully developed laminar flow of Al_2O_3-water nanofluid. The thermophysical properties of Al_2O_3-water nanofluid can be obtained with the same procedures as written in Section 9.2.2. Figure 9.4 showed the calculated heat transfer coefficients that were obtained from the governing equations in Section 9.2.1 and compared with the experimental heat transfer coefficients from Zeinali Heris et al. [31] in Reynolds number ranged 400 to 900 and the Al_2O_3-water nanofluid volume fraction at 0.02. From Figure 9.4, in the high Reynolds number range from 600 to 900, the simulation results were close to the experiment results, and the calculated coefficients were different from the experimental coefficients in an average error of 1.55%. But, in the low Reynolds number range from 400 to 600, the calculated values situated away from the experimental values noticeably, and this error may be caused by the effect of the boundary layer thickness for nanofluid flow.

9.3.2 The Effect of the Grid Independence

Since the processing time consumption of the computer plays an important role in simulation works, the grid-independence is one of the parameters that can be used to check grid resolutions and corrections. One can apply three grid sets as 6 × 6 × 100, 30 × 30 × 500, and 60 × 60 × 1000, obtain the results as in Section 9.3.1, and compare the results from three grid sets with the experimental results from Qiang and Yimin [30] as in Figure 9.5. For the nanofluid at the volume fraction

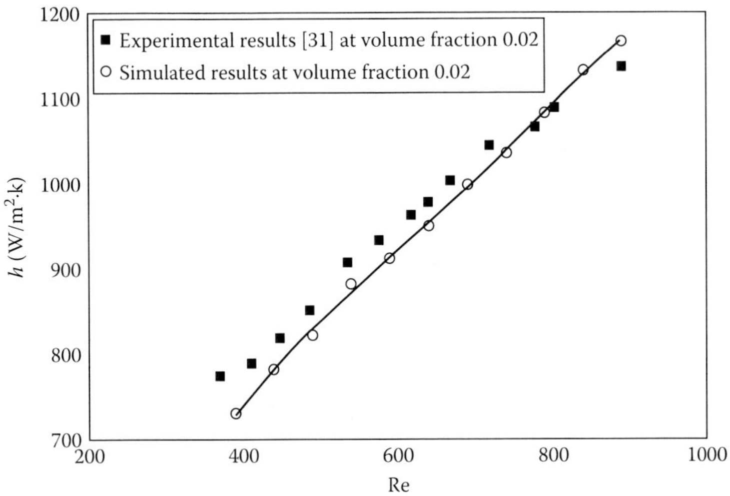

FIGURE 9.4 Comparison between the simulated results and the experimental data in the prediction of the heat transfer coefficients of the Al_2O_3-water nanofluid. (From Zeinali Heris, S. et al., *Int. J. Heat Fluid Flow*, 28, 203, 2007; Tongkratoke, A. et al., The development of mathematical modeling for nanofluid as a porous media in heat transfer technology, *International Conference on Heat Pipes, Heat Pumps, Refrigerators, Power Sources*, Minsk, Belarus, 2015.)

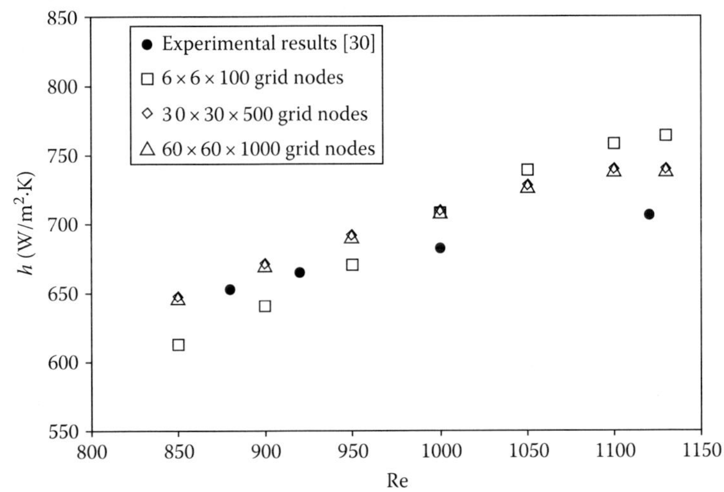

FIGURE 9.5 Comparison between the simulated results with different grid nodes and the experimental data in prediction of the heat transfer coefficients of the Cu-water nanofluid. (From Qiang, L. and Ymin, X., *Sci. China*, 45(4), 408, 2002; Tongkratoke, A. et al., The permeability effects of copper-nanofluid flow with using the porous media model, *International Symposium on Advances in Computational Heat Transfer*, Piscataway, NJ, 2015.)

of 0.02 in Figure 9.5, the 6 × 6 × 100 grid set provided the most diverged results. The average error percentages of the 30 × 30 × 500 and 60 × 600 × 1000 sets were 2.56 and 2.41, respectively, while the time consumptions of these grid sets were 5 and 8 h, respectively. The 30 × 30 × 500 set provided reasonable results in the short time range.

For the Al_2O_3-water nanofluid, one may choose the experimental results from Zeinali Heris et al. [31]. Figure 9.6 showed comparison between the simulated results with three different grid

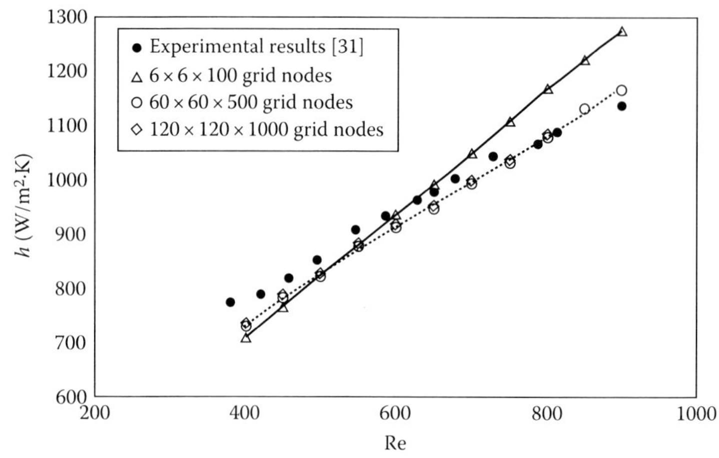

FIGURE 9.6 Comparison between the simulated results with different grid nodes and the experimental data in prediction of the heat transfer coefficients of the Al_2O_3-water nanofluid. (From Zeinali Heris, S. et al., *Int. J. Heat Fluid Flow*, 28, 203, 2007; Tongkratoke, A. et al., The development of mathematical modeling for nanofluid as a porous media in heat transfer technology, *International Conference on Heat Pipes, Heat Pumps, Refrigerators, Power Sources*, Minsk, Belarus, 2015.)

nodes (6 × 6 × 100, 60 × 60 × 500, and 120 × 120 × 1000) and experimental data [31] in the prediction of the heat transfer coefficients for the Al_2O_3-water nanofluid volume fractions at 0.02. The results from the 60 × 60 × 500 and 120 × 120 × 1000 grid sets provided similar trends of the results. They were closer to the experimental results, but they used different time consumptions. The (60 × 60 × 500) set consumed less time. Since the (120 × 120 × 1000) was the grid-independent and the time consumption was considered, so the (60 × 60 × 500) was chosen to be carried on in the next examinations.

9.3.3 The Effect of a Different Number of Nodes Next to the Wall in the Mixing Thermal Conductivity Model

Since in the fully developed laminar flow, the fluid next to the channel wall flows very slowly or stagnates, which is called the boundary layer caused by viscous effects, and the fluid velocity increases continuously to the center of the channel. From Section 9.2.2.5, the mixing thermal conductivity model brought Maxwell model [22] to represent the thermal property of the wall nodes and Yu and Choi model [21] to represent the thermal property of the channel nodes. These two models were different from each other. Because the viscous effect plays an emphasized role in the boundary layer thickness on the channel wall in the fluid flow, Maxwell model [22] as the static model is suitable for the boundary layer character. One can take the experimental results from Qiang and Yimin [30] who performed the experimental study with the Cu-water nanofluid to compare with the simulated results and find the effect of the boundary layer thickness. The simulation could be preformed by altering numbers of the nodes next to the wall, which were occupied by Maxwell model [22] in the mixing thermal conductivity model as 1, 5, 8 and 10. Figure 9.7a and b exhibited the relationships between the heat transfer coefficients and the Reynolds number ranging from 850 to 1050 at 0.01 and 0.02 nanofluid volume fractions. The results obtained from 8 and 10 (Maxwell model) wall nodes or 1.6 and 2 mm (0.2 mm height for each node) did affect the nanofluid heat transfer coefficients in the

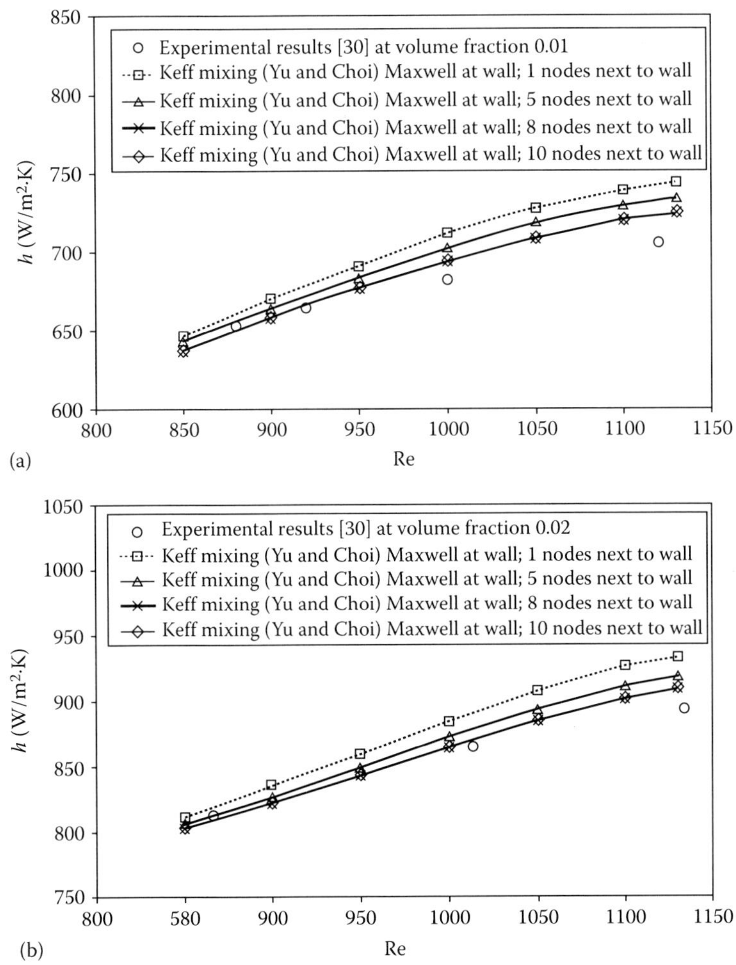

FIGURE 9.7 Comparison between the simulated results with the effective mixing thermal conductivity model containing different nodes next to wall and the experimental data in prediction of the heat transfer coefficients of the Cu-water nanofluid. (a) Volume fraction at 0.01. (b) Volume fraction at 0.02. (From Tongkratoke, A. et al., *Comput. Therm. Sci.*, 6(1), 1, 2014; Tongkratoke, A. et al., The permeability effects of copper-nanofluid flow with using the porous media model, *International Symposium on Advances in Computational Heat Transfer*, Piscataway, NJ, 2015.)

Reynolds number ranging from 850 to 1050; both sets of the simulated results were the same as the experimental results. The boundary layer thickness in this flow field can be estimated traditionally as 1.47–1.67 mm.

For the Al_2O_3-water nanofluid simulation, the experimental results of Zeinali Heris et al. [31] could be brought to compare with the calculated results. The relationships between the heat transfer coefficients and Reynolds number ranging from 400 to 900 at the volume fraction 0.02 could be expressed as in Figure 9.8 for different number of nodes next to the wall. Since the thermal conductivity model [22] in the mixing thermal conductivity model was varied as 1, 10, 15, and 20, one can notice that the simulated results obtained from 15 (Maxwell model) wall nodes or 3 mm (0.2 mm height for each node) were close the experiment result [31]. The boundary layer thickness in this nanofluid flow field can be estimated traditionally as 2.86 mm.

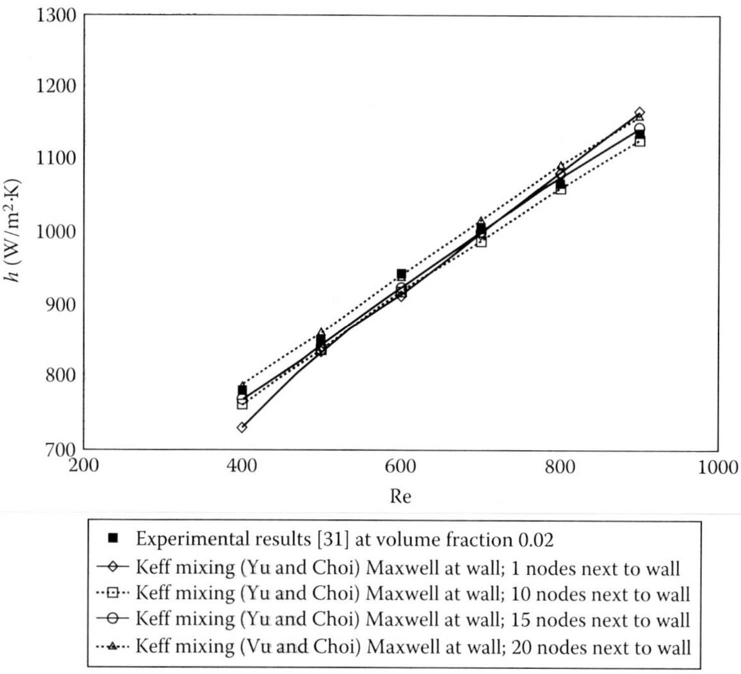

FIGURE 9.8 Comparison between the simulated results with the effective mixing thermal conductivity model containing different nodes next to wall and the experimental data in prediction of the heat transfer coefficients of the Al_2O_3-water nanofluid. (From Tongkratoke, A. et al., *Comput. Therm. Sci.*, 6(1), 1, 2014; Tongkratoke, A. et al., The development of mathematical modeling for nanofluid as a porous media in heat transfer technology, *International Conference on Heat Pipes, Heat Pumps, Refrigerators, Power Sources*, Minsk, Belarus, 2015.)

9.3.4 The Effect of Nanoparticles Permeability

The permeability is one significant parameter in the governing equations in Section 9.2.1 for steady, laminar, forced convection flow in the porous media and represents the fluid flow region among the solid particles in the nanofluid (solid–liquid) flow field, called the nanoparticles permeability. One can obtain the permeability (κ) of the Cu nanoparticles and apply the earlier sections to find the relationships between the heat transfer coefficients and Reynolds number for different volume fractions such as 0.01% and 0.02% nanoparticles. Consequently, the computed results can be compared with the experimental results of Qiang and Yimin [30] by varying the permeability values of the Cu nanoparticles to check the effects of the permeability values. If one carried on the best node set from Section 9.3.3 and applied the mixing thermal conductivity model, Yu and Choi model [21] specified at 44 channel nodes coupled with Maxwell model [22] specified at the 8 wall nodes boundary, one could get the results as in Figure 9.9. One can notice that the permeability for these simulations were varied from 1.17×10^{-11} to 2 m^2, the grid number set was $60 \times 60 \times 500$, and the simulated results obtained from the effective mixing thermal conductivity model with the permeability values from 1.17×10^{-11} to 0.001 m^2 were approaching to the experimental results. The closest simulated results were from the Cu permeability at 1.17×10^{-11} m^2, which was the Cu permeability found in the literature of Mwaba et al. [35].

Additionally, Al_2O_3 nanoparticles could be investigated by comparing the computed results in this section with the experimental results of Zeinali Heris et al. [31]. The relationships between

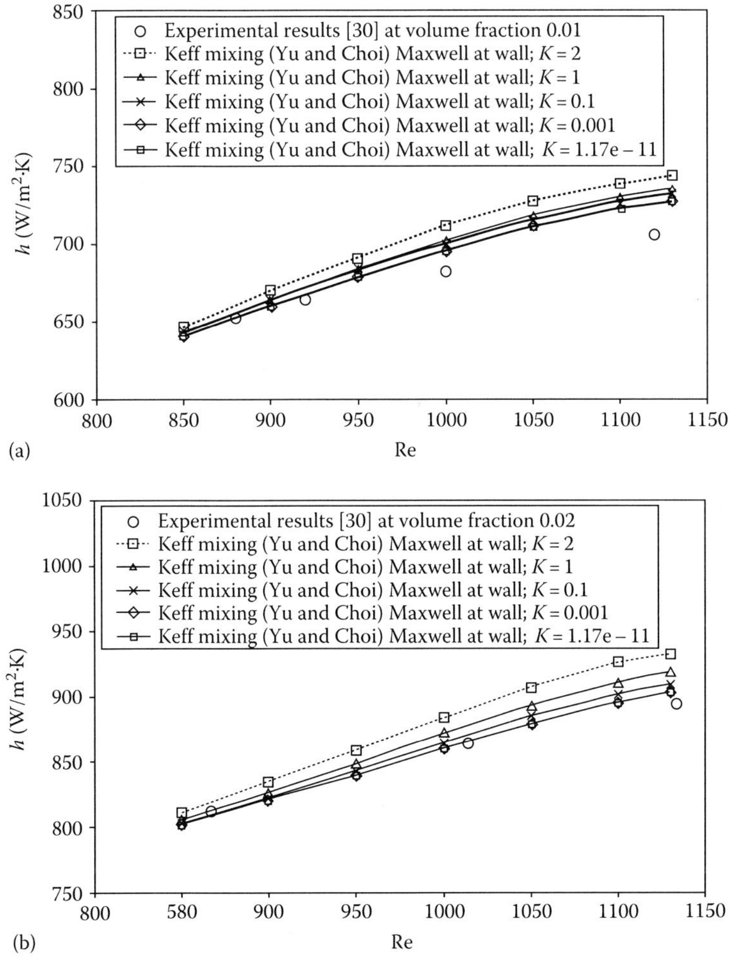

FIGURE 9.9 Comparison between the simulated results with the different permeability values and the effective mixing thermal conductivity model and the experimental data in prediction of the heat transfer coefficients of the Cu-water nanofluid. (a) Volume fraction at 0.01. (b) Volume fraction at 0.02. (From Tongkratoke, A. et al., *Comput. Therm. Sci.*, 6(1), 1, 2014; Tongkratoke, A. et al., The permeability effects of copper-nanofluid flow with using the porous media model, *International Symposium on Advances in Computational Heat Transfer*, Piscataway, NJ, 2015.)

the heat transfer coefficients and Reynolds number for volume fractions at 0.02% nanoparticles and for different Al_2O_3 permeability values with the best grid node (60 × 60 × 500) from Section 9.3.3, as well as applying the mixing thermal conductivity model, Yu and Choi model [21] specified at 30 channel nodes coupled with Maxwell model [22] specified at the 15 wall nodes boundary. The calculated results were shown in Figure 9.10. The permeability for these simulations were varied from 1×10^{-14} to 1 m^2 but only the simulated results obtained from the effective mixing thermal conductivity model with the permeability values from 1×10^{-14} to 0.0001 m^2 were approaching to the experimental results. The closest simulated results were from the Al_2O_3 permeability at 1×10^{-14} m^2, which was the Al_2O_3 permeability found in the literature of Narva and Vin [36].

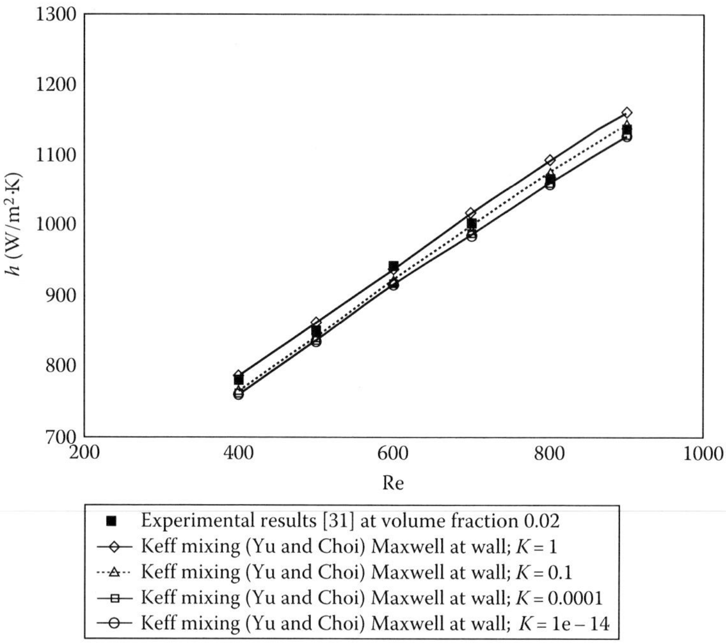

FIGURE 9.10 Comparison between the simulated results with the different permeability values and the effective mixing thermal conductivity model and the experimental data in prediction of the heat transfer coefficients of the Al_2O_3-water nanofluid. (From Tongkratoke, A. et al., *Comput. Therm. Sci.*, 6(1), 1, 2014; Tongkratoke, A. et al., The development of mathematical modeling for nanofluid as a porous media in heat transfer technology, *International Conference on Heat Pipes, Heat Pumps, Refrigerators, Power Sources*, Minsk, Belarus, 2015.)

9.4 CONCLUSION

The in-house program can be generated to model the flow through the porous media by applying the suitable governing equations. The simulation accuracy can be improved by adjusting the parameters of the equations and comparing the calculated results with the experimental ones. The nanofluid flow inside the rectangular pipe as in electronic applications is the example for the fully developed laminar flow of the nanofluids in the simple case. The thermophysical properties of nanofluids should be selected to be placed in the governing equations, so the relationships between the heat transfer coefficients and Reynolds number can be found and analyzed. The validation of the simulation model, the effect of the grid independence, the effect of the different numbers of nodes next to the wall in the mixing thermal conductivity model, and the effect of nanoparticles permeability can be performed. From two examples, Cu and Al_2O_3 nanofluids, which were mentioned before, the simulated results obtained from the effective mixing thermal conductivity model, the Yu and Choi model (at the channel nodes) coupled with the Maxwell model (at the wall nodes) could improve the simulation performance and could be applied with the porous model. The developed model using the mixing thermal conductivity model with the porous media and the suitable permeability assumption could improve the model performance and support its excellent potential in the nanofluid-flow simulation as the flow through the porous media.

REFERENCES

1. S. Kakaç, A. Pramuanjaroenkij, Review of convective heat transfer enhancement with nanofluids, *International Journal of Heat and Mass Transfer*, 52, 3187–3196 (2009).
2. S. Özerinç, S. Kakaç, A.G. Yazıcıoğlu, Enhanced thermal conductivity of nanofluids: A state-of-the-art review, *Microfluidics and Nanofluidics*, 8(2), 145–170 (2010).

3. S.U.S. Choi, Enhancing thermal conductivity of fluids with nanoparticles, *The American Society of Mechanical Engineers*, 66, 99–105 (1995).
4. D.A. Nield, A. Bejan, *Convection in Porous Media*, Springer Science and Business Media, New York, 2006.
5. C.W. Horton, F.T. Rogers, Convection currents in a porous medium, *Journal of Applied Physics*, 16, 367–370 (1945).
6. E.R. Lapwood, Convection of a fluid in a porous medium, *Mathematical Proceedings of the Cambridge Philosophical Society*, 44(4), 508–521 (1984).
7. D.A. Nield, Onset of thermohaline convection in a porous medium, *Water Resources Research*, 4(3), 553–560 (1968).
8. N. Rudraiah, M.S. Malashetty, The influence of coupled molecular diffusion on double-diffusive convection in a porous medium, *Journal of Heat Transfer*, 108(4), 872–876 (1986).
9. B.T. Murray, C.F. Chen, Double-diffusive convection in a porous medium, *Journal of Fluid Mechanics*, 201, 147–166 (1989).
10. M.S. Malashetty, Anisotropic thermoconvective effects on the onset of double diffusive convection in a porous medium, *International Journal of Heat and Mass Transfer*, 36(9), 2397–2401 (1993).
11. I. Pop, D.B. Ingham, *Mathematical and Computational Modelling of Viscous Fluids and Porous Media*, Pergamon, Oxford, United Kingdom, 2001.
12. P. Vadasz, *Emerging Topics in Heat and Mass Transfer in Porous Media: Bioengineering and Microelectronics to Nanotechnology*, Springer, Dordrecht, the Netherlands, 2008.
13. Q. Sun, I. Pop, Free convection in a triangle cavity filled with a porous medium saturated with nanofluids with flush mounted heater on the wall, *International Journal of Thermal Sciences*, 50(11), 2141–2153 (2011).
14. S. Basak, A. Raman, S.V. Garimella, Hydrodynamic loading of microcantilevers vibrating in viscous fluids, *Journal of Applied Physics*, 99(11), 6–10 (2006).
15. P. Rana, R. Bhargava, O.A. Beg, Numerical solution for mixed convection boundary layer flow of a nanofluid along an inclined plate embedded in a porous medium, *Computers and Mathematics with Applications*, 64(9), 2816–2832 (2012).
16. D.A. Nield, A.V. Kuznetsov, The Cheng–Minkowycz problem for natural convective boundary-layer flow in a porous medium saturated by a nanofluid, *International Journal of Heat and Mass Transfer*, 52(25–26), 5792–5795 (2009).
17. A. Aziz, W.A. Khan, I. Pop, Free convection boundary layer flow past a horizontal flat plate embedded in porous medium filled by nanofluid containing gyrotactic microorganisms, *International Journal of Thermal Sciences*, 56, 48–57 (2012).
18. M. Ghazvini, H. Shokouhmand, Investigation of a nanofluid cooled microchannel heat sink using fin and porous media approaches, *Energy Conversion and Management*, 50(9), 2373–2380 (2009).
19. S. Özerinç, A.G. Yazıcıoğlu, S. Kakaç, Numerical analysis of laminar forced convection with temperature-dependent thermal conductivity of nanofluids and thermal dispersion, *International Journal of Thermal Sciences*, 62, 138–148 (2012).
20. A. Tongkratoke, A. Pramuanjaroenkij, A. Chaengbamrung, S. Kakaç, Numerical study of nanofluid heat transfer enhancement with mixing thermal conductivity models, *Computational Thermal Sciences*, 6(1), 1–12 (2014).
21. W. Yu, S.U.S. Choi, The role of interfacial layers in the enhanced thermal conductivity of nanofluids: A renovated Maxwell model, *Journal Nanoparticle Research*, 5, 167–171 (2003).
22. J.C. Maxwell, *Treatise on Electricity and Magnetism*, Oxford University Press, London, U.K., 1940.
23. A. Tongkratoke, A. Pramuanjaroenkij, A. Chaengbamrung, S. Kakaç, Nanofluids flow simulation as the flow through the porous media, *International Symposium on Convective Heat and Mass Transfer*, Kusadasi, Turkey, 2014.
24. G. Dagan, The generalization of Darcy's law for nonuniform flows, *Water Resources Research*, 15(1), 1–7 (1979).
25. G. Dagan, *Flow and Transport in Porous Formations*, Springer-Verlag, New York, 1989.
26. S. Kakaç, Y. Yener, *Convective Heat Transfer*, CRC Press LLC, Boca Raton FL, 1995.
27. B.C. Pak, Y.I. Cho, Hydrodynamic and heat transfer study of dispersed fluids with submicron metallic oxide particles, *Experimental Heat Transfer*, 11(2), 151–170 (1998).
28. A. Einstein, A new determination of the molecular dimensions, *Annals of Physics*, 324(2), 289–306 (1906).
29. V. Trisaksri, S. Wongwises, Critical review of heat transfer characteristics of nanofluids, *Renewable and Sustainable Energy Reviews*, 11(3), 512–523 (2005).

30. L. Qiang, X. Ymin, Convective heat transfer and flow characteristics of Cu-water nanofluid, *Science in China*, 45(4), 408–416 (2002).
31. S. Zeinali Heris, M. Nasr Esfahany, S. Gh. Etemad, Experimental investigation of convective heat transfer of Al_2O_3-water nanofluid in circular tube, *International Journal of Heat and Fluid Flow*, 28, 203–210 (2007).
32. J. Koo, C. Kleinstreuer, A new thermal conductivity model for nanofluids, *Journal of Nanoparticle Research*, 6(6), 577–588 (2004).
33. A. Tongkratoke, A. Pramuanjaroenkij, A. Chaengbamrung, S. Kakaç, The permeability effects of copper-nanofluid flow with using the porous media model, *International Symposium on Advances in Computational Heat Transfer*, Piscataway, NJ, 2015.
34. A. Tongkratoke, A. Pramuanjaroenkij, A. Chaengbamrung, S. Kakaç, The development of mathematical modeling for nanofluid as a porous media in heat transfer technology, *International Conference "Heat Pipes, Heat Pumps, Refrigerators, Power Sources"*, Minsk, Belarus, 2015.
35. M.G. Mwaba, X. Huang, J. Gu, Influence of wick characteristics on heat pipe performance, *International Journal of Energy Research*, 30, 489–499 (2006).
36. V.K. Narva, T. Vin, Technology and properties of porous permeable aluminum base materials, *Journal of Non-Ferrous Metals*, 48(1), 72–75 (2007).

10 Convective Heat Transfer with Gaseous Flow in Microducts

Almila Güvenç Yazıcıoğlu and Sadık Kakaç

CONTENTS

10.1 INTRODUCTION

Over the past couple of decades, there has been significant research and technology advancement in the field of microscale flow and convection heat transfer systems. This is mainly due to the increased requirement for the development of efficient and successful cooling techniques for integrated circuits and microchips. In addition, various other microsystems such as micro reactors, micro fuel cells, and micro valves, among others, involve flow and heat transfer phenomena at the microscale. The current trend in miniaturization, when combined with the overheating of small-scale heat dissipating components, has made the accurate prediction and solution of the microscale convection heat transfer problem essential and critical. As a consequence of detailed and consistent results in the literature, it will be possible to develop reliable thermal packaging systems with effective cooling techniques. Therefore, the fluid flow and heat transfer characteristics of microducts should be investigated and crucial parameters in design, including the pressure drop of the working fluid, fluid flow rate, wall and fluid temperatures, and size (hydraulic diameter, duct shape, number of channels) need to be optimized. Related research has been published in collective works such as books [1–3], edited volumes [4–8], and review articles [9–19]. The readers are referred to these excellent sources to gain in-depth understanding of the microscale flow and convection heat transfer phenomena. This chapter focuses on gaseous slip flow, with particular attention to the published literature in the past 5 years. Major ideas and recent works for liquid flows in microchannels will also be summarized briefly.

10.2 FUNDAMENTALS OF MICRODUCT FLOWS

When the dimensions of channels reduce to below 200 μm, the channels are typically referred to as microchannels [20]. As the characteristic dimensions of channels gets smaller, the data and results in the literature start to deviate from those predicted by conventional continuum theory. For example, transition to turbulence tends to occur at a lower Reynolds number ($Re = u_m D_h/\nu$, with u_m the fluid mean velocity, D_h the hydraulic diameter, and ν the kinematic viscosity) values, and viscous effects and friction factors may be higher than what conventional theory has predicted. The deviations are mainly due to the fact that surface effects, that is, the interactions between the fluid molecules and the solid surface, are relatively significant in micro structures compared to larger-scale systems, since the surface-to-volume ratio is generally high. This affects the velocities, pressure drop, Reynolds number, and eventually convective heat transfer.

For gaseous flow in microducts, the surface effects are even more prominent. Especially at lower pressures, the interactions between gas molecules and the duct surface may become as frequent as the collisions between molecules. As a result, the molecular structure becomes more effective on flow and the flow is referred to as rarefied flow. The dimensionless number that represents the effects of rarefaction is the Knudsen number, which is defined as

$$Kn = \frac{\lambda}{D_h} \tag{10.1}$$

In Equation 10.1

λ is the mean free molecular path for the gas,

D_h is the characteristic flow length, in this case the hydraulic diameter, for ducts

The Knudsen number also represents the degree of deviation from continuum regime. Different flow regimes are identified based on the Knudsen number, a widely accepted classification is provided in Table 10.1.

When the flow is in the continuum regime, the velocity, density, pressure, etc., for the fluid are defined at every point and time in space. Conservation equations are solved with no-velocity slip, no-temperature jump boundary conditions, to obtain the fluid flow and heat transfer characteristic parameters at the macroscale. However as the Kn increases, in the slip-flow regime, the continuum approach may no longer be valid. In that case, the traditional conservation equations may still be utilized, but with modified boundary conditions. In other words, the linear relation between stress and the rate of strain, thus the no-slip velocity condition, and the linear relation between heat flux and temperature gradient, thus the no-temperature jump condition at the solid-fluid interface need to be

TABLE 10.1
Flow Regimes Based on the Knudsen Number

Kn Range	Flow Regime
$Kn \leq 0.001$	Continuum
$0.001 < Kn \leq 0.1$	Slip
$0.1 < Kn \leq 10$	Transition
$Kn > 10$	Free molecular

Source: Gad-El-Hak, M., *J. Fluid Eng.*, 121, 5, 1999.

modified [15]. This is done by defining a finite velocity-slip and temperature-jump at the wall for the fluid molecules. These modified boundary conditions depend on the *Kn* value, some thermophysical properties of the fluid, and accommodation factors. As the *Kn* increases further, in the transition and free molecular regimes, the continuum model breaks, and the molecular model should be utilized to solve the flow, such as direct simulation Monte Carlo method using the Boltzmann transport equation.

10.2.1 Governing Equations

10.2.1.1 Cylindrical Coordinates

For steady, two-dimensional, incompressible flow with constant thermophysical properties in a circular duct, whose geometry is given in Figure 10.1, the governing equations are presented here [21]. An unheated section is provided, so that the fully developed velocity profile may be obtained before thermal development. Therefore, the governing equations are
Continuity

$$\frac{\partial u}{\partial x}+\frac{1}{r}\frac{\partial (rv)}{\partial r}=0 \tag{10.2}$$

x-momentum

$$u\frac{\partial u}{\partial x}+v\frac{\partial u}{\partial r}=-\frac{1}{\rho}\frac{\partial p}{\partial x}+\nu\left[\frac{1}{r}\frac{\partial}{\partial r}\left(r\frac{\partial u}{\partial r}\right)+\frac{\partial^2 u}{\partial x^2}\right] \tag{10.3}$$

r-momentum

$$u\frac{\partial v}{\partial x}+v\frac{\partial v}{\partial r}=-\frac{1}{\rho}\frac{\partial p}{\partial r}+\nu\left\{\frac{\partial}{\partial r}\left[\frac{1}{r}\frac{\partial (rv)}{\partial r}\right]+\frac{\partial^2 v}{\partial x^2}\right\} \tag{10.4}$$

Energy:

$$u\frac{\partial T}{\partial x}+v\frac{\partial T}{\partial r}=\frac{\alpha}{r}\frac{\partial}{\partial r}\left(r\frac{\partial T}{\partial r}\right)+\alpha\frac{\partial^2 T}{\partial x^2}+\frac{1}{\rho c}\phi \tag{10.5}$$

Viscous dissipation is given as follows:

$$\phi=2\mu\left\{\left(\frac{\partial v}{\partial r}\right)^2+\left(\frac{v}{r}\right)^2+\left(\frac{\partial u}{\partial x}\right)^2+\frac{1}{2}\left(\frac{\partial u}{\partial r}+\frac{\partial v}{\partial x}\right)^2\right\} \tag{10.6}$$

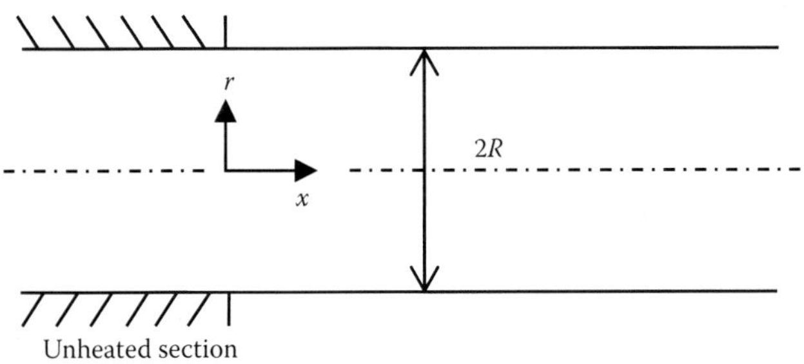

FIGURE 10.1 The problem geometry for two-dimensional flow in a circular duct.

In Equations 10.2 through 10.6

- u and v are the velocity components in x- and r-directions, respectively
- p is the pressure
- ρ is the density
- α is the thermal diffusivity
- ν is the kinematic viscosity
- c is the specific heat
- μ is the dynamic viscosity

Since $u = u(r)$ for steady and hydrodynamically fully developed flow, the velocity component v is zero everywhere. This condition satisfies Equation 10.2 directly. In addition, Equations 10.3 and 10.4 reduce to the following, respectively.

$$-\frac{1}{\rho}\frac{\partial p}{\partial x} + \nu\frac{d^2u}{dr^2} = 0 \tag{10.7}$$

$$-\frac{1}{\rho}\frac{\partial p}{\partial r} = 0 \tag{10.8}$$

Through Equation 10.8, $p = p(x)$ only, therefore Equation 10.7 becomes

$$\frac{dp}{dx} = \mu\frac{d^2u}{dr^2} \tag{10.9}$$

In this equation, both sides may be equal only if they are both equal to the same constant, $-\Delta p/L$, where Δp is the pressure drop in the duct over length L. As a result, Equation 10.9 may be written as follows:

$$\frac{d^2u}{dr^2} = -\frac{\Delta p}{\mu L} \tag{10.10}$$

At the center of the tube, the following boundary condition prevails:

$$r = 0 \quad u = \text{finite} \tag{10.11}$$

At the wall, for continuum flows, the no-slip boundary condition may be applied to Equation 10.10. However when the flow is in the slip-flow regime, the collision frequency is simply not high enough to ensure equilibrium and a certain degree of tangential velocity slip must be allowed. In this case, the following modified boundary condition should be utilized at the wall.

$$r = R \quad u = u_s = -\frac{2 - F_m}{F_m}\lambda\left(\frac{du}{dr}\right)\bigg|_{r=R} \tag{10.12}$$

In Equation 10.12

- u_s is the slip velocity
- F_m is the tangential momentum accommodation factor, which depends on the fluid, the surface material, and the surface finish

For an ideally perfect smooth surface, it has a value of 0.0, while for most surfaces commonly used in engineering applications, it has a value 0.2–1.0, determined experimentally. Taking $F_m = 1$, the slip boundary condition reduces to

$$r = R \quad u = u_s = -\lambda \left(\frac{du}{dr} \right) \bigg|_{r=R} \tag{10.13}$$

By integrating Equation 10.10 twice and applying the boundary conditions, and by using the mean velocity of an incompressible fluid [22], the velocity profile for laminar, hydrodynamically fully developed flow may be written as

$$\frac{u}{u_m} = \frac{2\left[1 - \left(\frac{r}{R}\right)^2\right] + 8Kn}{1 + 8Kn} \tag{10.14}$$

Here, Kn is the Knudsen number defined in Equation 10.1. Note that when $Kn = 0$, which is continuum flow, Equation 10.14 reduces to the well-known velocity profile in a microduct. Figure 10.2 presents the variation of the nondimensional velocity along the radial distance as a function of rarefaction in the flow [23,24]. It can be observed that, for continuum flows ($Kn = 0$), the no-slip velocity is present at the wall. As the degree of rarefaction increases, so does the slip velocity at the wall.

For hydrodynamically fully developed flow, the energy equation (Equation 10.5) reduces to the following:

$$u \frac{\partial T}{\partial x} = \frac{\alpha}{r} \frac{\partial}{\partial r} \left(r \frac{\partial T}{\partial r} \right) + \alpha \frac{\partial^2 T}{\partial x^2} + \frac{\mu}{\rho c} \left(\frac{\partial u}{\partial r} \right)^2 \tag{10.15}$$

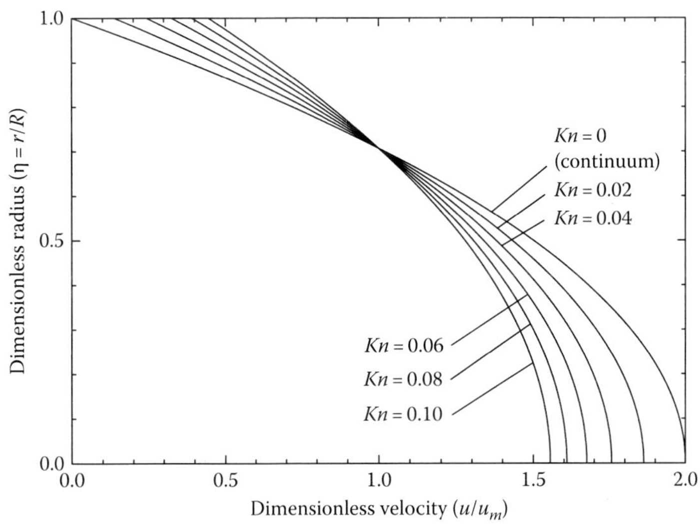

FIGURE 10.2 Dimensionless velocity profile along the radial direction, as a function of the Knudsen number. (With kind permission from Springer Science+Business Media: Yazıcıoğlu, A.G. and Kakaç, S., Convective heat transfer in gaseous slip flow, *Microfluidics Based Microsystems: Fundamentals and Applications*, Kakaç, S., Kosoy, B., Li, D., and Pramuanjaroenkij, A. (eds.), Springer, Dordrecht, the Netherlands, 2010, pp. 15–38.)

To solve the energy equation, the boundary conditions required for uniform wall temperature condition are as follows:

$$r = 0 \quad \frac{\partial T}{\partial r} = 0 \tag{10.16}$$

$$x = 0 \quad T = T_i \tag{10.17}$$

At the wall, for continuum flows, similar to the velocity, the no-jump boundary condition may be applied to Equation 10.15. In contrast, for the slip-flow regime, the following temperature-jump boundary condition should be utilized at the wall:

$$r = R, \quad T = T_s, \quad \text{with } T_s - T_w = -\frac{2 - F_t}{F_t}\frac{2\gamma}{\gamma + 1}\frac{\lambda}{Pr}\left(\frac{\partial T}{\partial r}\right) \tag{10.18}$$

In Equation 10.18
- T_s is the fluid temperature at the surface
- T_w is the duct wall temperature
- γ is the specific heat ratio of the fluid
- Pr is the fluid Prandtl number

F_t is the thermal accommodation factor, which depends on the energy of the incoming stream, the energy carried away by the molecules leaving the surface, and the energy of the molecules leaving the surface at the wall temperature. It may take a value in the range 0.0–1.0, depending on the gas and solid surface, the gas temperature and pressure, the temperature difference between the gas and the surface, and is determined experimentally.

For uniform heat flux, the boundary conditions for the energy equation are provided as follows:

$$r = 0 \quad \frac{\partial T}{\partial r} = 0 \tag{10.19}$$

$$x = 0 \quad T = T_i \tag{10.20}$$

$$r = R \quad \frac{\partial T}{\partial r} = \frac{q''}{k} \tag{10.21}$$

Here
- q'' is the constant and uniform heat flux at the wall,
- k is the thermal conductivity of the fluid

10.2.1.2 Cartesian Coordinates

In this case, the governing equations may be written as follows for steady, two-dimensional, incompressible flow with constant thermophysical properties [21] in the geometry provided in Figure 10.3. Continuity:

$$\frac{\partial u}{\partial x} + \frac{\partial v}{\partial y} = 0 \tag{10.22}$$

x-momentum:

$$u\frac{\partial u}{\partial x} + v\frac{\partial u}{\partial y} = -\frac{1}{\rho}\frac{\partial p}{\partial x} + \nu\left[\frac{\partial^2 u}{\partial x^2} + \frac{\partial^2 u}{\partial y^2}\right] \tag{10.23}$$

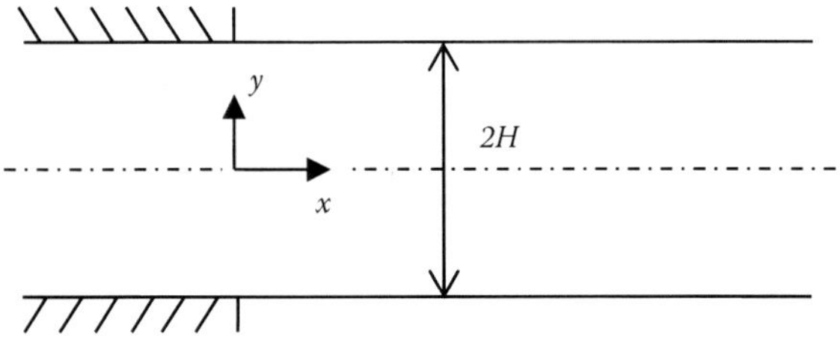

FIGURE 10.3 The problem geometry for two-dimensional flow in a parallel-plate channel.

y-momentum:

$$u\frac{\partial v}{\partial x}+v\frac{\partial v}{\partial y}=-\frac{1}{\rho}\frac{\partial p}{\partial y}+\nu\left\{\frac{\partial^2 v}{\partial x^2}+\frac{\partial^2 v}{\partial y^2}\right\} \tag{10.24}$$

Energy:

$$u\frac{\partial T}{\partial x}+v\frac{\partial T}{\partial y}=\alpha\left(\frac{\partial^2 T}{\partial x^2}+\frac{\partial^2 T}{\partial y^2}\right)+\frac{1}{\rho c}\phi \tag{10.25}$$

And, viscous dissipation is given as follows:

$$\phi=2\mu\left\{\left(\frac{\partial u}{\partial x}\right)^2+\left(\frac{\partial v}{\partial y}\right)^2+\frac{1}{2}\left(\frac{\partial v}{\partial x}+\frac{\partial u}{\partial y}\right)^2-\frac{1}{3}\left(\frac{\partial u}{\partial x}+\frac{\partial v}{\partial y}\right)^2\right\} \tag{10.26}$$

The momentum equations are analyzed in a similar fashion as the cylindrical coordinate system. As a result, the hydrodynamically fully developed velocity profile may be obtained as follows:

$$\frac{u}{u_m}=\frac{3}{2}\frac{\left[1-\left(\frac{y}{H}\right)^2+4Kn\right]}{1+8Kn} \tag{10.27}$$

Here, H is half the distance between parallel plates. With this fully developed velocity profile, the energy equation (Equation 10.25) reduces to

$$u\frac{\partial T}{\partial x}=\alpha\left(\frac{\partial^2 T}{\partial x^2}+\frac{\partial^2 T}{\partial y^2}\right)+\frac{\mu}{\rho c}\left(\frac{\partial u}{\partial y}\right)^2 \tag{10.28}$$

with uniform wall temperature boundary conditions,

$$y=0 \quad \frac{\partial T}{\partial y}=0 \tag{10.29}$$

$$x=0 \quad T=T_i \tag{10.30}$$

$$y=H \quad T=T_s, \quad \text{with } T_s-T_w=-\frac{2-F_t}{F_t}\frac{2\gamma}{\gamma+1}\frac{\lambda}{Pr}\left(\frac{\partial T}{\partial y}\right) \tag{10.31}$$

and uniform wall heat flux boundary conditions,

$$y = 0 \quad \frac{\partial T}{\partial y} = 0 \tag{10.32}$$

$$x = 0 \quad T = T_i \tag{10.33}$$

$$y = H \quad \frac{\partial T}{\partial y} = \frac{q''}{k} \tag{10.34}$$

10.2.2 Dimensionless Numbers

As it is very-well known, convection heat transfer occurring at the surface of the duct is represented by the Nusselt number ($Nu = hD_h/k$, with h the convection heat transfer coefficient). In general, the average Nusselt number is a function of the Reynolds and Prandtl numbers for a given geometry, often represented as a function of the Peclet number ($Pe = RePr$). In the slip-flow regime, the Knudsen number also affects Nusselt number.

When the flow is in the slip-flow regime, viscous heat generation due to the friction between the layers of the fluid may be important. The effect of viscous dissipation is usually neglected in continuum flow, or for low-speed and low-viscosity flows through conventionally-sized channels of short lengths. When the channel lengths are long in comparison to the hydraulic diameter, which is generally the case for microducts, viscous dissipation should be taken into consideration, as has been done in the energy equations presented in the previous sections. These equations need to be nondimensionalized for further analysis to clarify the relative effects of dimensionless numbers. In this section, the cylindrical coordinates are handled, which corresponds to flows in microtubes.

10.2.2.1 Uniform Wall Temperature

For this boundary condition, to nondimensionalize Equation 10.15 and the respective boundary conditions, the following parameters may be defined [24]:

$$\theta = \frac{T - T_w}{T_i - T_w}, \quad \xi = \frac{x}{Pe\, R}, \quad \eta = \frac{r}{R}, \quad u^* = \frac{u}{u_m} \tag{10.35}$$

In Equation 10.35

θ is the nondimensional temperature

ξ and η are the nondimensional axial and radial coordinates

In addition to the aforementioned, the Brinkman number is defined, which is a dimensionless number that represents the effect of viscous heat generation. Its definition varies with the thermal boundary condition at the duct wall. For the uniform wall temperature boundary condition, the Brinkman number is written as follows:

$$Br = \frac{\mu u_m^2}{k(T_i - T_w)} \tag{10.36}$$

With these, Equations 10.15 through 10.18 are rewritten as follows:

$$u^* \frac{\partial \theta}{\partial \xi} = \frac{1}{\eta}\frac{\partial}{\partial \eta}\left(\eta \frac{\partial \theta}{\partial \eta}\right) + \frac{1}{Pe^2}\frac{\partial^2 \theta}{\partial \xi^2} + Br\left(\frac{\partial u^*}{\partial \eta}\right)^2 \tag{10.37}$$

$$\eta = 0 \quad \frac{\partial\theta}{\partial\eta} = 0 \tag{10.38}$$

$$\xi = 0 \quad \theta = 1 \tag{10.39}$$

$$\eta = 1 \quad \theta = -2\kappa Kn\left(\frac{\partial\theta}{\partial\eta}\right) \tag{10.40}$$

In Equation 10.40, κ has been used as a dimensionless parameter that represents the temperature jump at the wall (see Equation 10.18):

$$\kappa = \frac{2-F_t}{F_t}\frac{2\gamma}{\gamma+1}\frac{1}{Pr} \tag{10.41}$$

Solutions have been obtained for the energy equation both numerically [22] and analytically, using general Eigen functions expansion [25]. The reader is referred to these resources for the details. Using the temperature distribution obtained, the local Nusselt number is

$$Nu_x = \frac{h_x D_h}{k} = -\frac{2\left.\dfrac{\partial\theta}{\partial\eta}\right|_{\eta=1}}{\theta_m - \dfrac{4\gamma}{\gamma+1}\dfrac{Kn}{Pr}\left.\dfrac{\partial\theta}{\partial\eta}\right|_{\eta=1}} \tag{10.42}$$

Here, θ_m is the nondimensional mean temperature.

$$\theta_m(\xi) = 2\int_0^1 u^*\theta(\eta,\xi)\eta d\eta \tag{10.43}$$

10.2.2.2 Uniform Wall Heat Flux

In this case, similar boundary conditions are used, with the following exceptions.

$$\theta = \frac{k(T-T_i)}{q''R}, \quad Br = \frac{\mu u_m^2}{q''D_h}, \quad Gz = \frac{2R}{L}Re\,Pr \tag{10.44}$$

Here

Gz is the Graetz number

L is the length

Now, Equation 10.15 may be rewritten as follows:

$$\frac{Gz(1-\eta^2+4Kn)}{2(1+8Kn)}\frac{\partial\theta}{\partial\xi} = \frac{1}{\eta}\frac{\partial}{\partial\eta}\left(\eta\frac{\partial\theta}{\partial\eta}\right) + \frac{1}{Pe^2}\frac{\partial^2\theta}{\partial\xi^2} + \frac{32Br}{(1+8Kn)^2}\eta^2 \tag{10.45}$$

$$\eta = 0 \quad \frac{\partial\theta}{\partial\eta} = 0\theta_m(\xi) = 2\int_0^1 u^*\theta(\eta,\xi)\eta d\eta \tag{10.46}$$

$$\xi = 0 \quad \theta = 1 \tag{10.47}$$

$$\eta = 1 \quad \frac{\partial \theta}{\partial \eta} = 1 \tag{10.48}$$

From a numerical solution of this equation [22], an expression is obtained for the local Nusselt number:

$$Nu_x = \frac{h_x D_h}{k} = -\frac{2}{\theta_s + \frac{4\gamma}{\gamma+1}\frac{Kn}{Pr} - \theta_m} \tag{10.49}$$

where θ_s is the nondimensional temperature of the fluid at the surface.

10.2.2.3 Results

To illustrate the effects of rarefaction and viscous dissipation on fully developed laminar flow for the two boundary conditions listed before, Table 10.2 is provided for $Pr = 0.7$. The Nu values for continuum (without rarefaction and viscous dissipation effects) can be observed as 3.66 (uniform wall temperature) and 4.36 (uniform wall heat flux) in Table 10.2.

As the effect of rarefaction in the flow increases (as Kn increases), Nusselt number decreases due to the temperature jump at the wall. For the uniform wall temperature boundary condition, the presence of viscous dissipation increases heat transfer. Since the definition of the Brinkman number is different for the uniform wall heat flux boundary condition, a positive Br represents a larger fluid temperature than the wall temperature, while a negative Br represents the opposite. Therefore, Nusselt number takes higher values for $Br < 0$ and lower values for $Br > 0$.

The effect of axial conduction in the fluid may be observed by varying the Peclet number, as illustrated in Figure 10.4 for the uniform wall temperature boundary condition. As can be observed, the solution of the classical Graetz problem, $Nu = 3.66$, is obtained for $Pe = 50$, which is negligible axial conduction. On the other hand, when $Pe = 1$, the fully developed value of Nusselt number becomes $Nu = 4.03$. Since the temperature gradient at the wall decreases at low Pe values, the local and fully developed Nu values increase with decreasing Pe.

Local Nusselt number variation along the microtube for the constant wall temperature boundary condition with both viscous dissipation and axial conduction effects are shown in Figure 10.5. As it was explained earlier, this case is for positive Brinkman number, which represents fluid cooling. Local Nu value first decreases due to temperature jump at the wall, then increases to its fully developed value due to the viscous dissipation effect. Because of the definition of Peclet number, as the

TABLE 10.2
Fully Developed Nusselt Numbers in a Microtube, as a Function of Rarefaction and Viscous Dissipation Effects, for the Two Boundary Conditions ($Pr = 0.7$)

	Uniform T_w		Uniform q''		
Kn	$Br = 0.00$	$Br = 0.01$	$Br = 0.00$	$Br = 0.01$	$Br = -0.01$
0.00	3.66	9.60	4.36	4.18	4.56
0.04	3.17	6.03	3.80	3.74	3.87
0.08	2.72	4.36	3.22	3.19	3.24

Source: With kind permission from Springer Science+Business Media: *International Journal of Transport Phenomena*, Gaseous flow in microconduits with viscous dissipation, 8, 2007, 297–315, Çetin, B., Yüncü, H., and Kakaç, S.

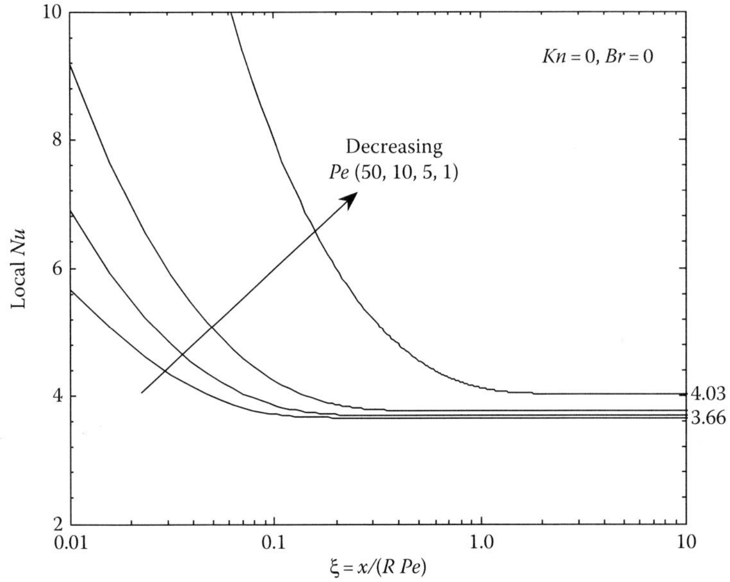

FIGURE 10.4 Local Nusselt number variation with nondimensional axial distance, as a function of Peclet number, with negligible rarefaction and viscous dissipation. (With kind permission from Springer Science+Business Media: Yazıcıoğlu, A.G. and Kakaç, S., Convective heat transfer in gaseous slip flow, *Microfluidics Based Microsystems: Fundamentals and Applications*, Kakaç, S., Kosoy, B., Li, D., and Pramuanjaroenkij, A. (eds.), Springer, Dordrecht, the Netherlands, 2010, pp. 15–38.)

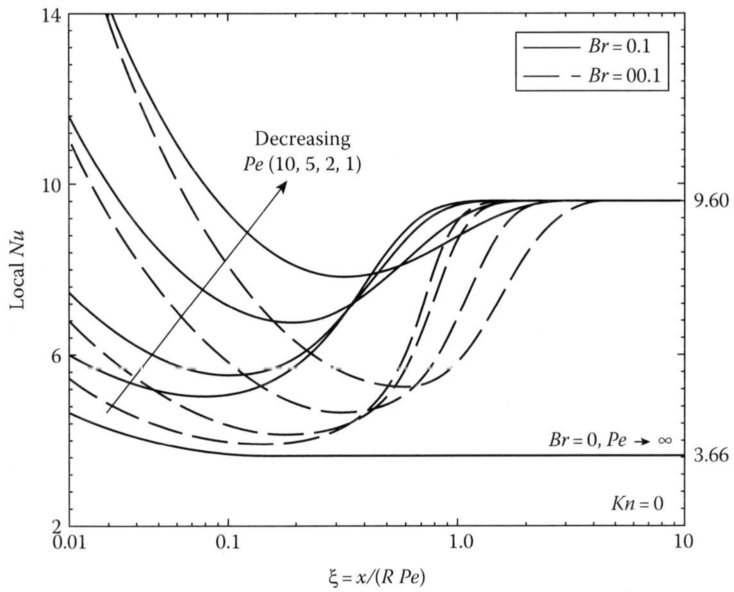

FIGURE 10.5 Local Nusselt number variation with nondimensional axial distance, as a function of Peclet number and with the effect of viscous dissipation (fluid cooling). (With kind permission from Springer Science+Business Media: Yazıcıoğlu, A.G. and Kakaç, S., Convective heat transfer in gaseous slip flow, *Microfluidics Based Microsystems: Fundamentals and Applications*, Kakaç, S., Kosoy, B., Li, D., and Pramuanjaroenkij, A. (eds.), Springer, Dordrecht, the Netherlands, 2010, pp. 15–38.)

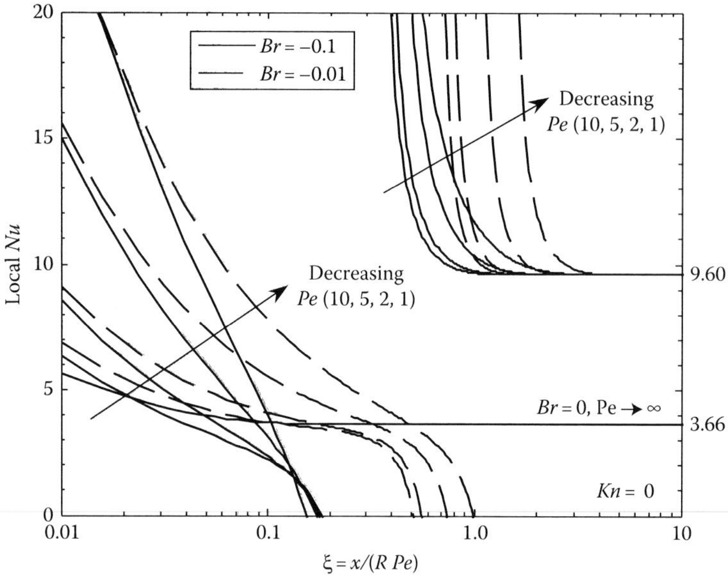

FIGURE 10.6 Local Nusselt number variation with nondimensional axial distance, as a function of Peclet number and with the effect of viscous dissipation (fluid heating). (With kind permission from Springer Science+Business Media: Yazıcıoğlu, A.G. and Kakaç, S., Convective heat transfer in gaseous slip flow, *Microfluidics Based Microsystems: Fundamentals and Applications*, Kakaç, S., Kosoy, B., Li, D., and Pramuanjaroenkij, A. (eds.), Springer, Dordrecht, the Netherlands, 2010, pp. 15–38.)

minima are approached, local *Nu* curves deviate from those for $Br = 0$. It can also be concluded that average *Nu* increases as the effect of axial conduction is more prominent. Finally, the amount of viscous dissipation does not affect the fully developed *Nu* value.

A similar graph is provided in Figure 10.6 for the fluid heating case, that is, negative Brinkman number. This time, the local Nusselt number reaches an asymptotic value when the wall and fluid temperatures are equal. As thermal development continues after this location, the fully developed *Nu* is reached. Similar to fluid cooling cases ($Br > 0$), the amount of viscous dissipation affects the location where the sudden change in local *Nu* occurs, but the fully developed *Nu* is the same for all non-zero *Br* values.

10.3 THE EFFECT OF SURFACE ROUGHNESS

Surface roughness plays an important role in convection heat transfer in microducts. The manufacturing processes determine the shape and size of roughness elements, which usually don't have a well-ordered layout. However, most researchers neglect this since the random distribution and small size of the roughness peaks along a surface are quite difficult to define. In one of the first experimental studies in this area water flow through rough fused silica and stainless steel microtubes was investigated, and deviations from theoretical predictions such as higher friction factor, and early transition from laminar to turbulent flow, were found [27]. Later, heat transfer characteristics were also investigated and smaller *Nu* values were determined for rough surfaces [28]. However, in another work, the Nusselt number and the apparent friction constant both increase with the increase in surface roughness [29].

Turgay and Yazıcıoğlu [30] investigated steady-state, laminar, developing air flow in a parallel-plate microchannel with uniform temperature walls, the bottom wall being rough. The channel schematic is provided in Figure 10.7. Here, the roughness is modeled as two-dimensional equilateral triangular elements placed on the bottom wall surface. The relative surface roughness of the wall may be determined by $e = d/D_h$. In most of the studies in literature, it is stated that silicon

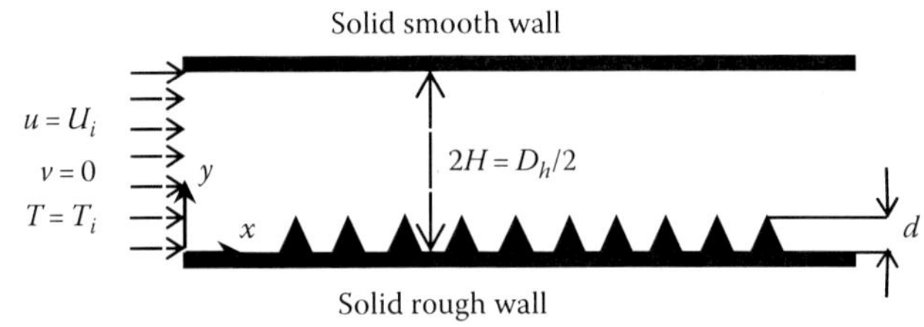

FIGURE 10.7 Rough microchannel schematic. (From Turgay, M.B. and Yazıcıoğlu, A.G., *Numer. Heat Transfer*, 56, 497, 2009.)

microchannels generally have a relative roughness value in the range 0%–4%. Thus, in this work, e = 1.325% and e = 2.0% are considered [30].

The governing energy equation and the boundary conditions for this problem are those that have been provided earlier in Equations 10.28 through 10.31, with the difference of roughness elements on the wall. The solution has been obtained numerically using the finite element method for the slip flow regime. As shown in Figure 10.8 for the two relative roughness values mentioned earlier, for the continuum case (Kn = 0) without viscous dissipation (Br = 0), local Nusselt number has a wavy pattern. Velocity and temperature gradients are higher at the peaks of the elements, thus local Nu is larger there, while at the bottom corners the low gradients result in minimum local Nu.

Figure 10.9 presents the channel-averaged Nusselt number (including smooth inlet/outlet sections) as a function of the degree of rarefaction in the flow, in the slip flow regime. In this analysis, the

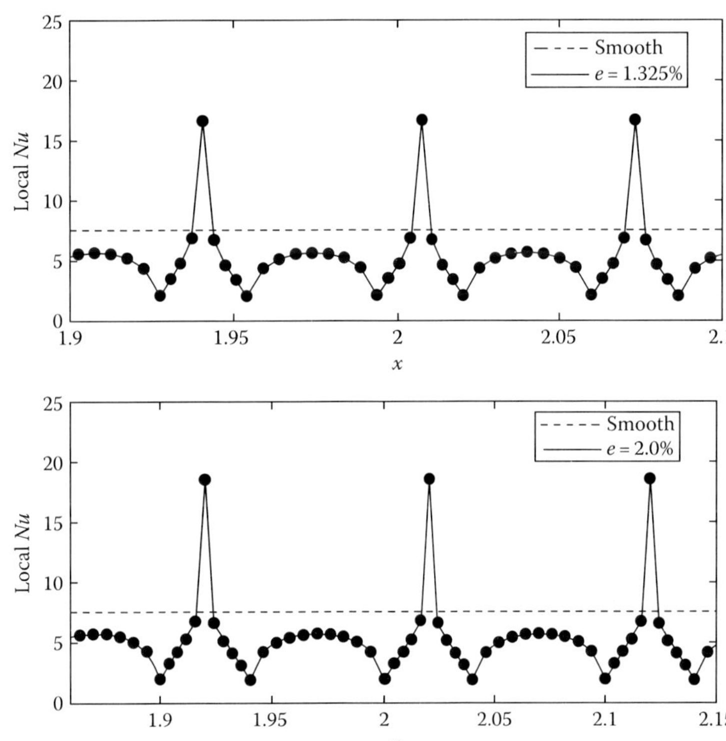

FIGURE 10.8 Local Nu variation over the roughness elements for continuum flow, no viscous dissipation. Re = 100. (With kind permission from Springer Science+Business Media: Yazıcıoğlu, A.G. and Kakaç, S., Convective heat transfer in gaseous slip flow, *Microfluidics Based Microsystems: Fundamentals and Applications*, Kakaç, S., Kosoy, B., Li, D., and Pramuanjaroenkij, A. (eds.), Springer, Dordrecht, the Netherlands, 2010, pp. 15–38.)

Analytical solutions were obtained by Satapathy [73], who investigated steady, laminar, two-dimensional, rarefied slug flow of air in an infinite microtube with mixed boundary conditions. Slip-velocity, temperature-jump, and axial conduction were also taken into account. It was concluded that for the hot fluid and cold wall situation, the local bulk-mean temperature increases as Peclet number or Knudsen number increases in the thermal entrance region. Opposite results were obtained when the flow situation was reversed. Moreover, local Nusselt number increased with an increase in Peclet number, but decreased with an increase in Knudsen number.

The effects of variable physical properties was investigated by Zade et al. [74] numerically, for simultaneously developing slip-flow in rectangular microchannels with constant wall temperature, at $Re = 0.1$. Different channel aspect ratios were studied at various Knudsen numbers. Several results were obtained: (1) As density varied, lower friction coefficients occurred, except at the channel inlet; (2) Nusselt number was only affected by the change of Knudsen number and was higher in variable-density slip-flow cases compared to constant-density simulations; (3) Depending on the relative importance of variable-properties, gas density and viscosity had opposing effects in the prediction of friction factor at the channel inlet; (4) With variable-properties, Nusselt number increased, compared to constant-property solutions.

Further investigations on variable properties were conducted by Sun and Jaluria [75] analytically and numerically. Convective heat transfer in two-dimensional, pressure-driven Nitrogen flow in long microchannels (length to diameter ratio = 500) was considered. In addition, rarefaction effects (slip flow, temperature jump, and thermal creep), pressure work, and viscous dissipation were all taken into account with uniform wall temperature and uniform wall heat flux boundary conditions. Their results indicated that for thermally developing flow at uniform wall temperature, Nusselt number decreased as thermal resistance increased due to temperature jump. However, in the fully developed region, rarefaction leaded to larger Nusselt number. For the uniform wall heat flux condition, Nusselt number decreased as the combined effects of pressure work and viscous dissipation increased. The authors recommended examining pressure work and viscous dissipation together in order to get better results for internal gas flows.

Yang et al. [76] investigated single phase air flow inside three microtubes whose dimensions are 86, 308, and 920 μm, respectively. Their results showed that the frictional coefficient of gas flow in a microtube is the same as that in conventional larger tubes, if the effect of gaseous flow compressibility is taken into consideration.

The effects of thermal creep on the development of gaseous slip-flow and heat transfer were investigated by Amiri-Jaghargh et al. [77]. The Navier–Stokes and energy equations coupled with velocity-slip and temperature-jump conditions applied at the constant temperature rectangular microchannel walls were solved using a control-volume method. It was observed that due to the temperature-jump, tangential temperature gradients form in the gas layer adjacent to the wall, despite the constant wall temperature boundary condition. The effects of variable properties, velocity-slip, temperature-jump, and thermal creep on friction and heat transfer coefficients were studied for different channel aspect ratios and Knudsen and Reynolds numbers.

Conjugate heat transfer in parallel-plate microchannels was investigated by Cole et al. [78] who included fluid and wall axial conduction in their analysis. The flow was in the slip-flow regime with constant-property air. The temperature distribution and heat transfer coefficient were obtained for various Knudsen and Peclet number values, and wall materials, through an analytical procedure in the form of integrals using Green's functions. The addition of slip flow reduced the average heat transfer coefficient on the heater. At low gas flows, the wall temperature was larger than the fluid temperature downstream of the heater, and the range of Peclet values defining this behavior depended on the interplay of axial wall conduction and fluid convection.

The analytical investigation of velocity, mass flow rate, and pressure distribution for steady, laminar, incompressible, fully developed gaseous slip flow in nonsymmetrical and symmetric parabolic cross-section microchannels was obtained by Tahmouresi and Das [79]. The channels had various aspect ratios and the method of separation of variables was utilized. A simple model was proposed to predict the Poiseuille number for slip flow in parabolic microchannels. The equation, given here, presents the Poiseuille number, normalized with respect to the no-slip value.

$$\frac{Po}{Pr_{ns}} = \frac{1}{1+\left(2.538+5.507\varepsilon+0.011\varepsilon^2\right)\beta} \tag{10.62}$$

In Equation 10.62

ε is the aspect ratio of the parabolic microchannel

β is a slip parameter

For symmetric parabolic microchannels ($\varepsilon = 1$), the results showed good agreement with other works for rectangular microchannels. The authors concluded that the developed model could be used to predict mass flow rate and pressure distribution of slip flow in parabolic microchannels.

Very recently, Knupp et al. [80] utilized the GITT to solve the transient behavior of conjugated heat transfer in laminar microchannel flow. Axial conduction effect and preheating or precooling of the region upstream of the heat exchange section were included in the analyses. The authors illustrated the technique for microchannels with polymeric walls of different thicknesses, compared their results, proposed with user-prescribed accuracy, to purely numerical solutions obtained through the commercial simulation tool *COMSOL Multiphysics*, and achieved excellent agreement.

10.6 CONCLUSIONS

In light of the fundamentals and relevant research in the literature for convective heat transfer in microducts, with particular emphasis on slip-flow, effects of rarefaction, axial conduction, and viscous dissipation, surface condition, and variation of thermophysical properties with temperature, the major conclusions may be summarized listed as follows:

- Depending on the values of Knudsen, Brinkman, Peclet, and Prandtl numbers, convective heat transfer in microducts may be enhanced and quite different from predictions of macroscale correlations. Flow geometry (channel shape and size, aspect ratio, length to diameter ratio) also affects the heat transfer characteristics, which can be significantly different from conventionally sized channels. As a result, correlations for Nusselt number may include Knudsen and Brinkman numbers, as well as specific parameters related to flow geometry, such as aspect ratio.
- Although the experimental analyses on liquid convective heat transfer in microducts have been in the continuum regime, and thus the conventional Navier–Stokes equations are applicable, the data indicate significant departure from the classical correlations for the conventionally sized tubes.
- In general, for constant wall temperature and constant wall heat flux conditions, velocity slip and temperature jump affect the heat transfer in opposite ways: A large slip on the wall will increase the convection along the surface. On the other hand, a large temperature jump will decrease the heat transfer by reducing the temperature gradient at the wall. Therefore, neglecting temperature jump in analyses will result in the overestimation of the heat transfer coefficient.
- As the flow becomes increasingly rarefied, or as Knudsen number takes higher values, a reduction in Nusselt number is observed, since the flow deviates from the continuum behavior.
- For high values of rarefaction and temperature jump, the effect of axial conduction is generally negligible. However for lower rarefaction and temperature jump values, as Peclet number decreases (axial conduction effect increases), the fully developed Nusselt number increases more significantly. Also, the difference between simultaneously and thermally

developing Nusselt number becomes more significant with the increase in Peclet number. It may be concluded from these observations that the effect of axial conduction should not be neglected for low-rarefied flows and with low values of temperature jump.

- The Prandtl number also affects heat transfer characteristics. The effect of temperature jump reduces as Prandtl number increases, resulting in an increase in Nusselt number.
- With turbulent flow in microducts, the significant increase in the pumping power overcomes the improvement in the thermal performance. Therefore, analysis and designs gravitate toward laminar flow.
- The behavior of convective heat transfer coefficient in the laminar and transition regimes is very different when compared to the behavior at macroscale. For example, in the laminar regime, Nusselt number decreases with increasing Reynolds number. This unusual behavior may alter the status of thermal development and hence the conventional thermal entry length, since the variation of the heat transfer coefficient along the flow is a variation of the boundary condition.
- The flow transition point and range are affected by the fluid temperature, velocity, and geometric parameters of the microchannel. The transitions are also a direct result of the large fluid temperature rise in the microchannels, which causes significant fluid thermophysical property variations and, hence, significant increases in the relevant flow parameters, such as the Reynolds number. The flow regime boundaries from laminar-to-transition and from transition-to-turbulent are affected by the Brinkman number as well, though to a lesser degree, in addition to Reynolds number.
- Variations in the thermophysical properties alter the flow significantly, specifically the Reynolds number, which in turn affects the convection heat transfer coefficient. Property variation effects should be included in numerical solutions in microchannels when the entry length is comparable to the channel length. However, constant property analyses will yield satisfactory results for long channels, in which entry length is negligible compared to the overall channel length.
- Viscous dissipation is an important physical effect in microducts, since they have high length-to-diameter ratios. This is valid even for low values of Brinkman number. It is especially important for laminar flow, where considerable temperature gradients exist. The increase in fully developed Nusselt number with the added effect of viscous dissipation suggests that this effect should not be neglected for long channels.
- In general, roughness can reduce or increase Nusselt number, depending on the distribution, spacing, and geometry of the obstructions. However one common conclusion is that roughness is more effective at low *Kn* values. Thus, the fully developed Nusselt number increases mostly with respect to the smooth channel value for rarefied flows, but not for continuum. The increase in Peclet number increases Nusselt number more for low values of rarefaction.

REFERENCES

1. G. Karniadakis, A. Beskok, N. Aluru, *Microflows and Nanoflows: Fundamentals and Simulation*, Springer, New York, 2005.
2. C.B. Sobhan, G.P. Peterson, *Microscale and Nanoscale Heat Transfer: Fundamentals and Engineering Applications*, CRC Press, Boca Raton, FL, 2008.
3. L.P. Yarin, A. Mosyak, G. Hetsroni, *Fluid Flow, Heat Transfer and Boiling in Micro-Channels*, Springer, New York, 2008.
4. S. Kakaç, H. Yüncü, K. Hijkata (eds.), *Cooling of Electronic Systems*, NATO ASI Series E, 258, Kluwer Academic Publishers, Dordrecht, the Netherlands, 1992.
5. Y. Bayazıtoğlu, S. Kakaç, Flow regimes in microchannel single-phase gaseous flow, *Microscale Heat Transfer: Fundamentals and Applications in Biological Systems and MEMS*, S. Kakaç, L. Vasiliev, Y. Bayazıtoğlu, Y. Yener (eds.), Kluwer Academic Publishers, Dordrecht, the Netherlands, 2005.

6. Y. Yener, S. Kakaç, M. Avelino, T. Okutucu, Single phase forced convection in microchannels: State-of art-review, *Microscale Heat Transfer: Fundamentals and Applications in Biological Systems and MEMS*, S. Kakaç, L. Vasiliev, Y. Bayazıtoğlu, Y. Yener (eds.), Kluwer Academic Publishers, Dordrecht, the Netherlands, 2005.
7. R.M. Cotta, S. Kakaç, M.D. Mikhailov, F.V. Castellos, C.R. Cardoso, Transient flow and thermal analysis in microfluidics, *Microscale Heat Transfer: Fundamentals and Applications in Biological Systems and MEMS*, S. Kakaç, L. Vasiliev, Y. Bayazıtoğlu, Y. Yener (eds.), Kluwer Academic Publishers, Dordrecht, the Netherlands, 2005.
8. Y. Bayazıtoğlu, G. Tunç, K. Wilson, I. Tjahjono, Convective heat transfer for single phase gases in microchannel slip flow: Analytical solutions, *Microscale Heat Transfer: Fundamentals and Applications in Biological Systems and MEMS*, S. Kakaç, L. Vasiliev, Y. Bayazıtoğlu, Y. Yener (eds.), Kluwer Academic Publishers, Dordrecht, the Netherlands, 2005.
9. A. Bar-Cohen, State of the art and trends in the thermal packaging of the electronic equipment, *ASME Journal of Electronic Packaging* 114, 257–270 (1992).
10. M. Gad-El-Hak, The fluid mechanics of microdevices, *Journal of Fluids Engineering* 121, 5–33 (1999).
11. G.L. Morini, Single-phase convective heat transfer in microchannels: A review of experimental results, *International Journal of Thermal Sciences* 43, 631–651 (2004).
12. G. Hetsroni, A. Mosyak, E. Pogrebnyak, L.P. Yarin, Heat transfer in micro-channels: Comparison of experiments with theory and numerical results, *International Journal of Heat and Mass Transfer* 25–26, 5580–5601 (2005).
13. P. Rosa, T.G. Karayiannis, M.W. Collins, Single-phase heat transfer in microchannels: The importance of scaling, *Applied Thermal Engineering* 29, 3447–3468 (2009).
14. N.T. Obot, Toward a better understanding of friction and heat/mass transfer in microchannels: A literature review, *Microscale Thermophysical Engineering* 6, 155–173 (2002).
15. A. Beşkök, G.E. Karniadakis, W. Trimmer, Rarefaction, compressibility effects in gas microflows, *Journal of Fluids Engineering* 118, 448–456 (1996).
16. S.T. Kadam, R. Kumar, Twenty-first century cooling solution: Microchannel heat sinks, *International Journal of Thermal Sciences* 85, 73–92 (2014).
17. R. Dey, T. Das, S. Chakraborty, Frictional and heat transfer characteristics of single-phase microchannel liquid flows, *Heat Transfer Engineering* 33(4–5), 425–446 (2012).
18. G.L. Morini, Y. Yang, H. Chalabi, M. Lorenzini, A critical review of the measurement techniques for the analysis of gas microflows through microchannels, *Experimental Thermal and Fluid Science* 35(6), 849–865 (2011).
19. G.L. Morini, Y. Yang, Guidelines for the determination of single-phase forced convection coefficients in microchannels, *Journal of Heat Transfer* 135(10), 101004 (2013).
20. S.G. Kandlikar, Heat transfer mechanisms during flow boiling in microchannels, *Transactions of the ASME* 126, 8–16 (2004).
21. S. Kakaç, Y. Yener, A. Pramuanjaroenkij, *Convective Heat Transfer*, 3rd edn., CRC Press, Boca Raton, FL, 2014.
22. W. Sun, S. Kakaç, A.G. Yazıcıoğlu, A numerical study of single-phase convective heat transfer in microtubes for slip flow, *International Journal of Thermal Sciences* 46, 1084–94 (2007).
23. B. Çetin, Analysis of single phase convective heat transfer in microtubes and microchannels, MSc Thesis, Middle East Technical University, Ankara, Turkey, 2005.
24. A.G. Yazıcıoğlu, S. Kakaç, Convective heat transfer in gaseous slip flow, *Microfluidics Based Microsystems: Fundamentals and Applications*, S. Kakaç, B. Kosoy, D. Li, A. Pramuanjaroenkij (eds.), Springer, Dordrecht, the Netherlands, 2010.
25. M. Barışık, Analytical solution for single phase microtube heat transfer including axial conduction and viscous dissipation, MSc Thesis, Middle East Technical University, Ankara, Turkey, 2008.
26. B. Çetin, H. Yüncü, S. Kakaç, Gaseous flow in microconduits with viscous dissipation, *International Journal of Transport Phenomena* 8, 297–315 (2007).
27. Gh.M. Mala, D. Li, Flow characteristics of water in microtubes, *International Journal of Heat and Fluid Flow* 20, 142–148 (1999).
28. W. Qu, Gh.M. Mala, D. Li, Heat transfer for water flow in trapezoidal silicon microchannels, *International Journal of Heat and Mass Transfer* 43, 3925–3936 (2000).
29. H.Y. Wu, P. Cheng, An experimental study of convective heat transfer in silicon microchannels with different surface conditions, *International Journal of Heat and Mass Transfer* 46, 2547–2556 (2003).
30. M.B. Turgay, A.G. Yazıcıoğlu, Effect of surface roughness in parallel-plate microchannels on heat transfer, *Numerical Heat Transfer* 56, 497–514 (2009).

31. B. Dai, M. Li, Y. Ma, Effect of surface roughness on liquid friction and transition characteristics in micro- and mini-channels, *Applied Thermal Engineering* 67, 283–293 (2014).
32. Z. Li, X. Huai, Y. Tao, H. Chen, Effects of thermal property variations on the liquid flow and heat transfer in microchannel heat sinks, *Applied Thermal Engineering* 27, 2803–2814 (2007).
33. S.P. Guidice, C. Nonino, S. Savino, Effects of viscous dissipation and temperature dependent viscosity in thermally and simultaneously developing laminar flows in microchannels, *International Journal of Heat and Fluid Flow* 28, 15–27 (2007).
34. J.T. Liu, X.F. Peng, B.X. Wang, Variable-property effect on liquid flow and heat transfer in microchannels, *Chemical Engineering Journal* 141, 346–353 (2008).
35. M.S. El-Genk, I. Yang, Numerical analysis of laminar flow in micro-tubes with a slip boundary, *Energy Conversion and Management* 50, 1481–1490 (2009).
36. S. Kakaç, A.G. Yazıcıoğlu, A.C. Gözükara, Effect of variable thermal conductivity and viscosity on single phase convective heat transfer in slip flow, *Heat and Mass Transfer* 47, 879–891 (2011).
37. A.C. Gözükara, Heat transfer analysis of single phase forced convection in microchannels and microtubes with variable property effect, MSc Thesis, Middle East Technical University, Ankara, Turkey, 2010.
38. D.B. Tuckerman, R.F.W. Pease, Optimized convective cooling using micromachined structure, *Journal of the Electrochemical Society* 129(3), 98C (1982).
39. X.F. Peng, G.P. Peterson, Convective heat transfer and flow friction for water flow in microchannel structures, *International Journal of Heat and Mass Transfer* 39, 2599–2608 (1996).
40. H.S. Park, J. Punch, Friction factor and heat transfer in multiple microchannels with uniform flow distribution, *International Journal of Heat and Mass Transfer* 51, 4435–4443 (2008).
41. P.Y. Wu, W.A. Little, Measurement of heat transfer characteristics of gas flow in fine channels heat exchangers used for microminiature refrigerators, *Cryogenics* 24(5), 415–420 (1984).
42. S.B. Choi, R.F. Barron, R.O. Warrington, Fluid flow and heat transfer in microtubes, *Micromechanical Sensors, Actuators, and Systems* ASME DSC-32, 123–134 (1991).
43. D. Yu, R.O. Warrington, R.F. Barron, T.A. Ameel, An experimental and theoretical investigation of heat transfer in microtubes, *Proceedings of ASME/JSME Thermal Engineering Conference* 1, 523–530 (1995).
44. T.A. Ameel, X. Wang, R.F. Barron, R.O. Warrington, Laminar forced convection in a circular tube with constant heat flux and slip flow, *Microscale Thermophysical Engineering* 1(4), 303–320 (1997).
45. C.P. Tso, S.P. Mahulikar, The use of the Brinkman number for single phase forced convective heat transfer in microchannels, *International Journal of Heat and Mass Transfer*, 41(12), 1759–1769 (1998).
46. C.P. Tso, S.P. Mahulikar, The role of the Brinkman number in analyzing flow transitions in microchannels, *International Journal of Heat and Mass Transfer* 42, 1813–1833 (1999).
47. G. Tunç, Y. Bayazıtoğlu, Heat transfer for gaseous flow in microtubes with viscous heating, *Proceedings of the ASME Heat Transfer Division* HTD 366-2, 299–306 (2000).
48. G. Tunç, Y. Bayazıtoğlu, Heat transfer in microtubes with viscous dissipation, *International Journal of Heat and Mass Transfer* 44, 2395–2403 (2001).
49. G. Tunç, Y. Bayazıtoğlu, Heat transfer in rectangular microchannels, *International Journal of Heat and Mass Transfer* 45, 765–773 (2002).
50. S. Yu, T.A. Ameel, Slip-flow heat transfer in rectangular microchannels, *International Journal of Heat and Mass Transfer* 44, 4225–4234 (2001).
51. M.D. Mikhailov, R. Cotta, S. Kakaç, Steady-state and periodic heat transfer in microconduits, *Microscale Heat Transfer: Fundamentals and Applications in Biological Systems and MEMS*, S. Kakaç, L. Vasiliev, Y. Bayazıtoğlu, Y. Yener (eds.), Kluwer Academic Publishers, Dordrecht, the Netherlands, 2005.
52. M. Renksizbulut, H. Niazmand, G. Tercan, Slip-flow and heat transfer in rectangular microchannels with constant wall temperature, *International Journal of Thermal Sciences* 45, 870–881 (2006).
53. H. Herwig, S.P. Mahulikar, Variable property effects in single-phase incompressible flows through microchannels, *International Journal of Heat and Mass Transfer* 45, 977–981 (2006).
54. M. Avcı, O. Aydın, Laminar forced convection slip-flow in a micro-annulus between two concentric cylinders, *International Journal of Heat and Mass Transfer* 51, 3460–3467, 2008.
55. H.D. Madhawa Hettiarachchi, M. Golubovic, W.M. Worek, W.J. Minkowycz, Three-dimensional laminar slip-flow and heat transfer in a rectangular microchannel with constant wall temperature, *International Journal of Heat and Mass Transfer* 51, 5088–5098, 2008.
56. M. Wang, X. Lan, Z. Li, Analyses of gas flows in micro- and nanochannels, *International Journal of Heat and Mass Transfer* 51, 3630–3641 (2008).

57. B. Çetin, A.G. Yazıcıoğlu, S. Kakaç, Fluid flow in microtubes with axial conduction including rarefaction and viscous dissipation, *International Communications in Heat and Mass Transfer*, 35, 535–544 (2008).
58. B. Çetin, A.G. Yazıcıoğlu, S. Kakaç, Slip-flow heat transfer in microtubes with axial conduction and viscous dissipation—An extended Graetz problem, *International Journal of Thermal Sciences* 48(9), 1673–1678 (2009).
59. N. Tsuzuki, M. Utamura, T.L. Ngo, Nusselt number correlations for a microchannel heat exchanger water supplier with s-shaped fins, *Applied Thermal Engineering* 29, 3299–3308 (2009).
60. G. Wang, L. Hao, P. Cheng, An experimental and numerical study of forced convection in a microchannel with a negligible axial heat conduction, *International Journal of Heat and Mass Transfer* 52, 1070–1074 (2009).
61. J.P. Mchale, S.V. Garimella, Heat transfer in trapezoidal microchannels of various aspect ratios, *International Journal of Heat and Mass Transfer* 53, 365–375 (2010).
62. P. Gunnasegaran, H.A. Mohammed, N.H. Shuaib, R. Saidur, The effect of geometrical parameters on heat transfer characteristics of microchannels heat sink with different shapes, *International Communications in Heat and Mass Transfer* 37, 1078–1086 (2010).
63. S. Barlak, S. Yapıcı, O.N. Sara, Experimental investigation of pressure drop and friction factor for water flow in microtubes, *International Journal of Heat and Mass Transfer* 50, 361–368 (2011).
64. J. Ma, L. Li, Y. Huang, X. Liu, Experimental studies on single-phase flow and heat transfer in a narrow rectangular channel, *Nuclear Engineering and Design* 241, 2865–2873 (2011).
65. C. Wang, P. Gao, S. Tan, Z. Wang, C. Xu, Experimental study of friction and heat transfer characteristics in narrow rectangular channel, *Nuclear Engineering and Design* 250, 646–655 (2012).
66. C.Y. Huang, C.M. Wu, Y.N. Chen, T.M. Liou, The experimental investigation of axial heat conduction effect on the heat transfer analysis in microchannel flow, *International Journal of Heat and Mass Transfer* 70, 169–173 (2014).
67. B. Dai, M. Li, C. Dang, Y. Ma, Q. Chen, Investigation on convective heat transfer characteristics of single phase liquid flow in multi-port micro-channel tubes, *International Journal of Heat and Mass Transfer* 70, 114–118 (2014).
68. J. Zhang, Y.H. Diao, Y.N. Zhao, Y.N Zhang, An experimental study of the characteristics of fluid flow and heat transfer in the multiport microchannel flat tube, *Applied Thermal Sciences* 65, 209–218 (2014).
69. L. Kuddusi, E. Çetegen, Thermal and hydrodynamic analysis of gaseous flow in trapezoidal silicon microchannels, *International Journal of Heat and Mass Transfer* 48, 353–362 (2009).
70. J. van Rij, T. Ameel, T. Harman, The effect of viscous dissipation and rarefaction on rectangular microchannel convective heat transfer, *International Journal of Heat and Mass Transfer*, 48, 271–281 (2009).
71. J. van Rij, T. Ameel, T. Harman, An evaluation of secondary effects on microchannel frictional and convective heat transfer characteristics, *International Journal of Heat and Mass Transfer* 52, 2792–2801 (2009).
72. A. Demsis, B. Verma, S.V. Prabhu, A. Agrawal, Heat transfer coefficient of gas flowing in a circular tube under rarefied condition, *International Journal of Thermal Sciences* 49, 1994–1999 (2010).
73. A.K. Satapathy, Slip flow heat transfer in an infinite microtube with axial conduction, *International Journal of Thermal Sciences* 49, 153–160 (2010).
74. A.Q. Zade, M. Renksizbulut, J. Friedman, Heat transfer characteristics of developing gaseous slip-flow in rectangular microchannels with variable physical properties, *International Journal of Heat and Fluid Flow* 32, 117–127 (2011).
75. Z. Sun, Y. Jaluria, Convective heat transfer in pressure-driven nitrogen slip flows in long microchannels: The effects of pressure work and viscous dissipation, *International Journal of Heat and Mass Transfer* 55, 3488–3497 (2012).
76. C. Yang, C. Chen, T. Lin, S.G. Kandlikar, Heat transfer and friction characteristics of air flow in microtubes, *Experimental Thermal and Fluid Sciences* 37, 12–18 (2012).
77. A. Amiri-Jaghargh, H. Niazmand, M. Renksizbulut, Entrance effects of thermal creep on fluid heating in rectangular microchannels, *International Journal of Modern Physics C* 24(8), 1350054 (2013).
78. K.D. Cole, B. Çetin, L. Brettmann, Microchannel heat transfer with slip flow and wall effects, *Journal of Thermophysics and Heat Transfer* 28(3), 455–462 (2014).
79. F. Tahmouresi, S.K. Das, Analytical modeling of gaseous slip flow in parabolic microchannels, *Journal of Fluids Engineering: Transactions of the ASME* 136(7), 071201 (2014).
80. D.C. Knupp, R.M. Cotta, C.P. Naveira-Cotta, S. Kakaç, Transient conjugated heat transfer in microchannels: Integral transforms with single domain formulation, *International Journal of Thermal Sciences* 88, 248–257 (2015).

11 Numerical Simulation of Combined Microscale Effects on Convective Heat Transfer in Single-Phase Flows

Mourad Rebay, Yassine Kabar, and Sadık Kakaç

CONTENTS

11.1 INTRODUCTION

Convective heat transfer through parallel-plate microchannels heat sink is often encountered in electronics cooling, chemical and bio-engineering industries, military applications, and energy conversion. Heat transfer and fluid flow in microchannels has been studied extensively during the past two decades, mainly as a result of the search for efficient methods to cool electronic devices. Since the pioneering works of Tuckerman and Pease in early 1980s [1], microchannels have revealed their capabilities of removing high heat fluxes in very compact exchangers. Along with being compact in size, microchannel heat sinks have the added benefit of providing high heat transfer coefficients. This performance is based on the premise that the heat transfer coefficient can be inversely proportional to the hydraulic diameter of the channel. This is also related to the fact that the surface/volume ratio is definitely higher in micro systems. In particular, because of confinement, the friction of the fluid induces heating release likely to modify the thermal field [2].

Heat and fluid flow in microscale has been widely studied in the literature where the microscale effects have been considered separately. The reader may refer to Chapter 10 for literature review of analytical, numerical, and experimental studies and the key results obtained. For an in-depth understanding of fundamentals and analysis of convective heat transfer in microscale, different sources are available in the literature. One can quote references [3–18] among a long list of interesting studies. These studies have been conducted for the characterization of heat transfer and fluid flows in microchannel systems where micro-scale effects cannot be neglected as it is often done in macroscale heat transfer problems. Let us quote among these effects: viscous dissipation, heat conduction in the wall of microtubes or microchannels, rarefaction, compressibility, slip velocity and temperature jump at the wall, and roughness of the wall surface. It has been demonstrated experimentally or numerically that viscous dissipation plays a premier role in heat transfer even in laminar flow with low velocity. Among numerous authors, Tso and Mahulikar [19,20], Koo and Kleinstreuer [21], Morini [22], Celata et al. [23], and Guidice et al. [24] have conducted numerical or experimental studies for analyzing the effect of viscous dissipation in microtubes and microchannels. Axial conduction has been taken into account in many studies when the thickness of the wall becomes in the same order of magnitude of the hydraulic diameter of the passage section of the fluid, and in particular for short microchannels. The studies of Maranzana et al. [25], Nonino et al. [26] have considered adiabatic thermal wall boundary conditions in the entrance and in the outlet of a microchannel. They showed the importance of the effects of axial conduction in thick walls for short two parallel-plate microchannels. Tiselj et al. [27] considered nonadiabatic condition in the inlet and outlet of the solid wall. The thermal wall boundary conditions have a dominant role in such a problem. Kabar et al. [28] have studied numerically the case where both fluid and wall have the same temperature at the entry of the microchannel, and the obtained results are different from the previous works. This test case is important as the microchannel can be a part of a heat exchanger with multiple microchannels with common inlet and outlet plena, manifolds, or headers. The fluid collector at the entrance of the microchannels makes the temperature on the side of the wall equal to the fluid temperature.

In gaseous flow, when the dimensions of the microchannels are in the order of magnitude of micron, the flow is rarefied because the characteristic length of the system becomes comparable to the mean free path of gas molecules. Knudsen number ($Kn = \lambda/L_0$) is usually used for quantifying the degree of rarefaction, where λ is the mean free path of the molecules, and L_0 is the characteristic length of the geometry. According to the classification given in [29,30], the fluid is considered continuum for very small Knudsen numbers ($Kn < 0.001$) and free molecular flow for large values of Kn ($Kn > 10$). The gaseous flow is in the transition flow regime when $0.1 < Kn < 10$. A continuum fluid with slip-flow regime can be adopted when the Knudsen number is in the range $0.001 < Kn < 0.1$. As the microchannel dimensions are limited by the micro-fabrication technologies, the hydraulic diameter cannot have an unlimited value. In heat transfer and energy conversion systems, microchannel size is of the order of 10 μm and air is the most available gas encountered in fluid flow applications. In such flows, the Knudsen number is in the range $0.001 < Kn < 0.1$. That is why the research works in gaseous flow and heat transfer in microchannels have been focused on continuum flow with slip-flow regime. In such a flow, the molecular collisions with the walls dominate over intermolecular collisions. This results in a breakdown of the continuum assumption in a thin region adjacent to the walls. For modeling this type of flow, it is permissible to use the conventional continuum momentum and energy equations with modified boundary conditions for slip velocity and temperature jump on the interface gas/solid.

Heat transfer in rarefied flows in microchannels has also been widely studied in the past two decades. For instance, Larrode et al. [31] have proved the effect of slip-flow and temperature jump at the wall on the modification of the heat transfer rate. Barron et al. [32] and Ameel et al. [33] showed that slip flow and temperature jump conditions are essential in the analysis of heat transfer in gaseous flow with moderate rarefaction. Indeed, the slip velocity increases the fluid velocity near the wall and therefore the convective heat transfer between the fluid and the wall is enhanced. In contrast, the temperature jump decreases the temperature gradient at the wall, which leads to decrease in the heat transfer. The effects of rarefaction on heat transfer depend also on gas-wall, usually represented by the parameter $\beta = \beta_T/\beta_v$ (ratio of the global temperature jump coefficient by the velocity slip coefficient). Ghodoossi and Egrican [34] found that, when $\beta < 0.5$ the slip velocities at the wall are dominant and the heat transfer increases. For $\beta \geq 0.5$, the effects of the temperature jump are dominant compared to slip velocity, consequently, the heat transfer decreases. $\beta = 1.667$ is a typical value for engineering applications with air as the working fluid. Mikhailov and Cotta [35] have presented an analytical analysis of a gaseous developed laminar flow in parallel-plate microchannel with slip flow condition. They have used the integral transform method with an exact solution of the corresponding eigen value problem and obtained the value of fully developed Nusselt number Nu = 1.67 for $\beta = 1$. Chen [36] deduced for a similar case, with $\beta = 2$, a Nusselt number value Nu = 3.33. Tunç and Bayazitoglu [37,38], Jeong and Jeong [39], and Çetin et al. [40] have studied the slip-flow and temperature jump combination with viscous dissipation in the hydrodynamically fully developed with thermally developing flow (Graetz problem). They provided the Nusselt distribution for different flow parameters and Knudsen number with and without viscous dissipation. Van Rij et al. [41,42] studied numerically the effects of viscous dissipation and axial conduction on slip flow in rectangular microchannel. They highlighted their significant effects in the simultaneous development of hydrodynamic and thermal boundary layers as well as in the fully developed flow region.

The aim of this chapter is to simulate numerically the laminar flow of a Newtonian fluid in a two-dimensional parallel-plate microchannel with combined microscale effects. The plates are subjected to constant heat flux density q_0 or constant temperature T_0 conditions on their outer surfaces. Combined effects of the axial conduction in the fluid, axial, and transversal conduction in the plates, simultaneous developing of heat and fluid flows, slip velocity, temperature jump, and viscous dissipation are taken into account simultaneously. The finite volume method is used for solving the dimensionless conservation equations in an in-house FORTRAN program including the mentioned effects and is valid for liquid as well as gaseous flow.

11.2 MATHEMATICAL MODELING OF CONVECTIVE MICROSCALE HEAT TRANSFER IN LIQUID AND GASEOUS FLOWS

The schematic of the problem and coordinate system considered in this analysis are shown in Figure 11.1. The height of the channel is 2H and the plate thickness is E. The flow is considered along the x-axis and the channel length L is chosen high enough that hydrodynamically and thermally developed flow is reached at the exit. A constant temperature T_0 or constant heat flux density q_0 boundary condition is assumed at the outer surface of the walls.

Some typical results for water and air flow will be presented in this chapter. However, the built computational program is valid whatever the working fluid, the wall material and thickness are.

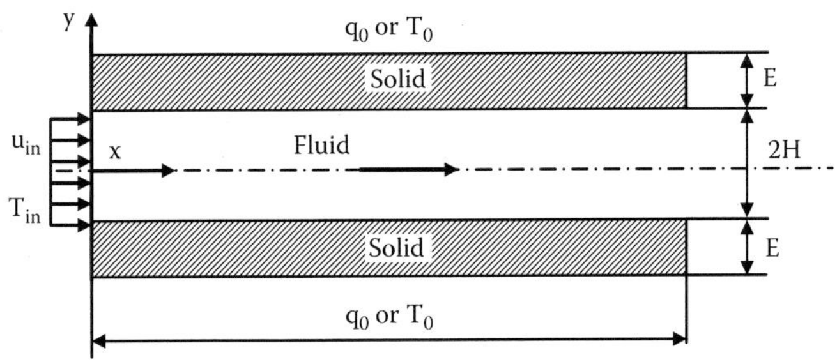

FIGURE 11.1 Geometry of problem.

Results are given for different conductivities solid/fluid ratio, $K = k_s/k_f$, wall thickness to half channel height ratio $\Delta = E/H$, Brinkman number Br, and Knudsen number Kn.

11.2.1 Governing Equations

The numerical model for fluid flow and heat transfer in the microchannel was developed under the following assumptions:

- Steady state fluid flow and heat transfer
- Incompressible and laminar flow
- Constant fluid properties
- Negligible radiation heat transfer
- Rarefaction effects set velocity slip and temperature-jump at the fluid–wall interface when the fluid is gas

Based on the previous assumptions, the governing equations describing the fluid flow and the heat transfer in the microchannel can be written as

$$\frac{\partial u}{\partial x} + \frac{\partial v}{\partial y} = 0 \tag{11.1}$$

$$u\frac{\partial u}{\partial x} + v\frac{\partial u}{\partial y} = -\frac{\partial p}{\partial x} + \frac{\partial}{\partial x}\left[\mu\left(\frac{\partial u}{\partial x} + \frac{\partial v}{\partial y}\right)\right] \tag{11.2}$$

$$u\frac{\partial v}{\partial x} + v\frac{\partial v}{\partial y} = -\frac{\partial p}{\partial y} + \frac{\partial}{\partial y}\left[\mu\left(\frac{\partial u}{\partial x} + \frac{\partial v}{\partial y}\right)\right] \tag{11.3}$$

When viscous dissipation is taken into account, the energy equation in the fluid with no heat generation is expressed as

$$\rho\, c_p \left[u\frac{\partial T}{\partial x} + v\frac{\partial T}{\partial y}\right] = \frac{\partial}{\partial x}\left(k_f \frac{\partial T}{\partial x}\right) + \frac{\partial}{\partial y}\left(k_f \frac{\partial T}{\partial y}\right) + \Phi \tag{11.4}$$

Where, $\Phi = 2\mu\left[\left(\frac{\partial u}{\partial x}\right)^2 + \left(\frac{\partial v}{\partial y}\right)^2 + \frac{1}{2}\left(\frac{\partial u}{\partial y} + \frac{\partial v}{\partial x}\right)^2\right]$ is the viscous dissipation term.

Equation of energy in the solid is

$$\frac{\partial}{\partial x}\left(k_s \frac{\partial T}{\partial x}\right) + \frac{\partial}{\partial y}\left(k_s \frac{\partial T}{\partial y}\right) = 0 \tag{11.5}$$

In order to generalize the solution results, these governing equations are nondimensionalized using the following dimensionless variables:

$$Y = \frac{2y}{D_h};\quad U = \frac{u}{u_{in}};\quad V = \frac{v}{u_{in}};\quad P = \frac{p}{\rho u_{in}^2}; \tag{11.6}$$

For continuity and momentum equations, the following dimensionless coordinate is used:

$$X^+ = \frac{2x}{D_h \, Re} \quad (11.7)$$

Whereas, the following dimensionless coordinate is used for the equation of energy:

$$X^* = \frac{2x}{D_h \, Re \, Pr} = \frac{X^+}{Pr} \quad (11.8)$$

D_h is the hydraulic diameter, $D_h = 2H$ and Re is the Reynolds number associated to the flow $Re = u_{in} D_h / \nu$, where ν is the cinematic viscosity.

For imposed temperature T_0 on the outer surface of the wall, the nondimensional temperature θ is defined by the following relation:

$$\theta = \frac{T - T_0}{T_{in} - T_0} \quad (11.9)$$

And for uniform heat flux density q_0 condition, the expression of the dimensionless temperature θ is

$$\theta = \frac{(T - T_{in}) k_f}{D_h q_0} \quad (11.10)$$

Finally, the equation of continuity, momentum, and energy in the fluid in dimensionless form become

$$\frac{1}{Re} \frac{\partial U}{\partial X^+} + \frac{\partial V}{\partial Y} = 0 \quad (11.11)$$

$$\frac{U}{Re} \frac{\partial U}{\partial X^+} + V \frac{\partial U}{\partial Y} = -\frac{1}{Re} \frac{\partial P}{\partial X^+} + 2 \left(\frac{\partial^2 U}{(Re)^3 \partial X^{+2}} + \frac{\partial^2 U}{Re \, \partial Y^2} \right) \quad (11.12)$$

$$\frac{U}{Re} \frac{\partial V}{\partial X^+} + V \frac{\partial V}{\partial Y} = -\frac{\partial P}{\partial Y} + 2 \left(\frac{\partial^2 V}{(Re)^3 \partial X^{+2}} + \frac{\partial^2 V}{Re \, \partial Y^2} \right) \quad (11.13)$$

$$\frac{1}{Pe} U \frac{\partial \theta}{\partial X^*} + V \frac{\partial \theta}{\partial Y} = \frac{2}{Pe} \left(\frac{\partial^2 \theta}{Pe^2 \partial X^{*2}} + \frac{\partial^2 \theta}{\partial Y^2} \right)$$
$$+ 4 \frac{Br}{Pe} \left[\left(\frac{1}{Pe} \frac{\partial U}{\partial X^*} \right)^2 + \left(\frac{\partial V}{\partial Y} \right)^2 + \frac{1}{2} \left(\frac{\partial U}{\partial Y} + \frac{1}{Pe} \frac{\partial V}{\partial X^*} \right)^2 \right] \quad (11.14)$$

Where Pe is the Péclet number ($Pe = Re \cdot Pr$) and Br is the Brinkman number defined by

$$Br = \frac{\mu u_{in}^2}{k_f (T_{in} - T_0)} \quad (11.15)$$

for uniform surface temperature problem, and by

$$Br = \frac{\mu u_{in}^2}{D_h q_0} \tag{11.16}$$

for uniform heat flux problem.

The presence of viscous dissipation increases the heat transfer rate between the plate and the fluid. Br = 0 corresponds to the case of negligible viscous dissipation. For uniform surface temperature, Br < 0 is associated with the heating of the fluid as $T_0 > T_{in}$, and Br > 0 is associated with the cooling of the fluid ($T_0 < T_{in}$). Since the definition of the Brinkman number is different for uniform heat flux problem (see Equation 11.16), positive Br values represent heating of the fluid as the heat flux q_0 entering in the solid wall is positive, and negative values are associated with a cooling of the fluid.

The energy equation in the solid wall can be written in dimensionless form for both boundary conditions as

$$\left(\frac{\partial^2 \theta_s}{\partial X^{*2}} + \frac{\partial^2 \theta_s}{\partial Y^2}\right) = 0 \tag{11.17}$$

11.2.2 Boundary Conditions

In the entry of the channel, the wall temperature is assumed to be equal to the fluid temperature (T_{in}). As mentioned before, this assumption is justified by the fact that the microchannel is usually a part of a heat exchanger with multiple microchannels and common inlet plena or manifold. At the outlet, fully developed flow and heat transfer is adopted in both fluid and wall. The boundary conditions can be expressed mathematically as

$$x = 0, -H \le y \le H: u = u_{in}, v = 0, T = T_{in} \tag{11.18}$$

$$x = 0, -(H + E) \le y \le -H \quad \text{and} \quad H \le y \le (H + E): u = 0, v = 0, T = T_{in} \tag{11.19}$$

$$x = L, -H \le y \le H: \quad \left(\frac{\partial u}{\partial x}\right) = \left(\frac{\partial v}{\partial x}\right) = \left(\frac{\partial t}{\partial x}\right) = 0 \tag{11.20}$$

The imposed condition on the external surfaces of the wall is

$$x \ge 0, y = \pm(H + E): T = T_0 \text{ for uniform temperature, and} \tag{11.21}$$

$$-k_s\left(\frac{\partial T}{\partial y}\right) = q_0 \quad \text{for uniform heat flux} \tag{11.22}$$

Using the dimensionless variables given by Equations 11.6, 11.9, and 11.10, we obtain the following dimensionless boundary conditions:

At the entrance:

$$U = 1, V = 0 \quad \text{and} \quad \theta = 0 \tag{11.23}$$

At the exit:

$$\left(\frac{\partial U}{\partial X^+}\right) = \left(\frac{\partial V}{\partial X^+}\right) = \left(\frac{\partial \theta}{\partial X^*}\right) = 0 \tag{11.24}$$

At $Y = \pm(0.5 + \Delta)$:

$$\theta = 0 \quad \text{for uniform temperature, and} \tag{11.25}$$

$$-\left(\frac{\partial \theta}{\partial Y}\right) = \frac{1}{2K} \quad \text{for uniform heat flux density} \tag{11.26}$$

For gaseous as well as liquid flow, the continuity of heat flux at the fluid–wall interface is expressed by

$$k_f\left(\frac{\partial T_f}{\partial y}\right)_i = k_s\left(\frac{\partial T_s}{\partial y}\right)_i \tag{11.27}$$

This condition is expressed in the dimensionless coordinates by

$$\left(\frac{\partial \theta_f}{\partial Y}\right)_i = K\left(\frac{\partial \theta_s}{\partial Y}\right)_i \tag{11.28}$$

The subscript i denotes the interface wall–fluid.

11.2.3 Slip Flow and Temperature Jump Conditions

Rarefaction effects are significant only for gaseous flows. In this study, the standard expressions of the continuity, Navier–Stokes and energy equations rely on the assumption that the fluid may be treated as a continuum flow with slip flow and temperature jump when the Knudsen number is in the range $0.001 < Kn < 0.1$. Liquid is always considered in continuum flow in normal conditions. Indeed, the typical mean free path of molecules under ambient conditions, λ, is in the range 0.1–1 nm, so the Kn number is higher than 0.001 only if the hydraulic diameter of the microchannel is less than 1μm, which is not attainable in engineering applications [5]. In heat transfer studies in macroscale channels and tubes, no-slip condition at a fluid–solid interface is enforced in the momentum equation, and an analogous no temperature jump condition is applied in the energy equation. The notion underlying the no-slip/no-jump condition is that within the fluid there cannot be any finite discontinuities of velocity/temperature. The interaction between a fluid particle and a wall is similar to that between neighboring fluid particles, and therefore no discontinuities are allowed at the fluid–solid interface either. In other words, the fluid velocity must be zero relative to the surface, and the fluid temperature must be equal to that of the surface. But strictly speaking these two boundary conditions are valid only if the fluid flow adjacent to the surface is in thermodynamic equilibrium. This requires an infinitely high frequency of collisions between the fluid and the solid surface. In practice, the no-slip/no-jump condition leads to fairly accurate predictions as long as $Kn < 0.001$ (for gases). Beyond that, the collision frequency is simply not high enough to ensure equilibrium and a certain degree of tangential-velocity slip and temperature jump must be allowed [17].

Analysis of mathematical modeling of slip flow and temperature jump has been given in several sources, as for example, the references [7,8,11,17]. For moderate temperature gradient, the creep effect can be neglected at the interface wall–gas, and the first-order slip and jump conditions derived by Maxwell can be adopted. Therefore, slip-velocity and temperature jump are respectively expressed by

$$u_{slip} - u_w = \frac{2-\sigma_v}{\sigma_v}\lambda \left.\frac{\partial u}{\partial y}\right|_i \tag{11.29}$$

$$T_{slip} - T_i = \frac{2-\sigma_T}{\sigma_T} \frac{2\gamma}{(\gamma+1)} \frac{k}{C_p\mu} \lambda \left.\frac{\partial T}{\partial y}\right|_i \quad (11.30)$$

where the subscript slip denotes velocity or temperature in the fluid. σ_v is the tangential momentum accommodation coefficient. It describes the interaction of the fluid molecules with the wall. This coefficient varies from zero to unity from specular to diffuse reflections, respectively. Its value depends on the type of gas and solid and on the surface roughness. σ_T is the thermal accommodation factor, which depends on the energy of the incoming stream and the energy carried away by the molecules leaving the surface at the wall temperature. It takes a value in the range 0.0–1.0, depending on the gas and solid surface, the gas temperature, the pressure, and the temperature difference between the gas and the solid surface. These accommodation coefficients are determined experimentally. For most of the gas–solid couples used in engineering applications, values of both accommodation coefficients in ambient conditions are close to unity. We have considered in this study $\sigma_v = \sigma_T = 1$.

11.3 NUMERICAL SIMULATION OF CONJUGATED PROBLEM WITH MICROSCALE EFFECTS

11.3.1 Solution Method

11.3.1.1 Discretization of Equations

The governing equations (Equation 11.11 through 11.14), with the associated boundary conditions (Equations 11.23 through 11.28), were solved through an in-house program by using the finite volume method. The pressure–velocities coupling is solved by the SIMPLER algorithm [43]. Convection fluxes are evaluated by the power-law scheme, and diffusion fluxes by the central differences scheme. The discretized algebraic equations are solved by the line-by-line tridiagonal matrix algorithm (TDMA). Dimensionless velocity components U and V are computed in each node in the domain (X^+, Y); they are therefore transposed into the domain (X^*, Y) using Equation 11.8.

The algebraic equations involving the unknown values of the dependent variable at the chosen grid points are derived from the differential equation. The solution methodology is based on the control volume method applied to a single domain formulation proposed for modeling the heat transfer phenomena at both the fluid stream and the microchannel wall. By making use of coefficients represented as space dependent functions with abrupt transitions occurring at the fluid–wall interfaces, the mathematical model allows unifying the model into a single domain formulation with variable coefficients. For example, the fluid viscosity is attributed to the fluid region in the domain, and a specific very high value of "solid viscosity" is given to the solid region. The computational domain is divided into a number of nonoverlapping control volumes shown in Figure 11.2, such that one control volume surrounds each grid point. Then the differential equation is integrated over each control volume. Piecewise profiles between the grid points are used to evaluate the integrals. Integrating the energy equation over the control volume and using a fully implicit scheme, we obtain the following algebraic form in an internal control volume of the fluid domain:

$$a_p\theta_p = a_E\theta_E + a_W\theta_W + a_N\theta_N + a_S\theta_S + b \quad (11.31)$$

Where the coefficients a_p, a_E, a_W, a_N, and a_S represent the convection and diffusion influence at the four faces of the boundary volume as shown in Figure 11.2.

The evaluation of thermal conductivities at the interface of two neighboring nodes is carried out using a harmonic function [43].

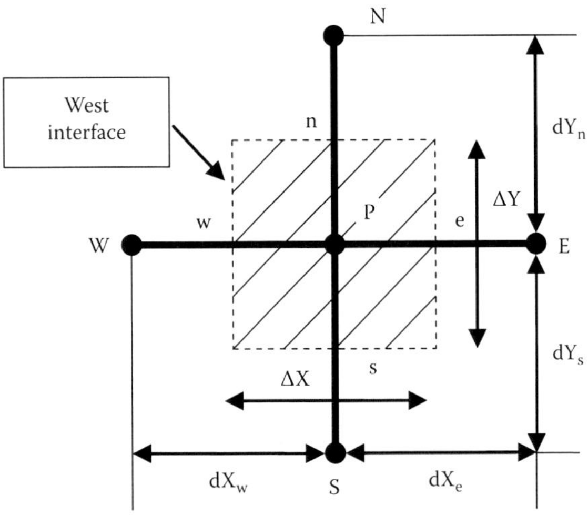

FIGURE 11.2 Volume of control.

The convergence criterion used in the performed numerical procedure is defined as follows:

$$\max\left|\frac{\phi^{n+1}-\phi^{n}}{\phi^{n}}\right|\leq\varepsilon \tag{11.32}$$

The convergence criterion is based on the sum of normalized residuals for each variable ϕ. We assume convergence when all residues are less than 10^{-4}. As the problem is symmetric, one half of microchannel is modeled in order to optimize the computational time.

11.3.1.2 Meshing

A nonuniform grid in X^+-, X^*-, and Y-directions have been used in order to take into account the high velocity and temperature gradients at the entrance region of the microchannel and close to the wall–fluid interface. The gradients have been computed by the means of second order differences. The study domain is divided into N × M nodes. Our numerical experiments indicate that the iterative procedure has converged with a maximum of twenty iteration steps. In the present method, the mesh-point distribution in the spatial coordinate directions is computed using a trigonometric progressing function. For example, in X^*-direction, the following relation is used for the evolution of the grid size:

$$\text{ang}=\frac{\pi}{2(N-2)} \tag{11.33}$$

$$\Delta X_j^* = X_{max}^*\left[\cos\left((j-1)\ \text{ang}\right)-\cos\left(j\ \text{ang}\right)\right] \tag{11.34}$$

N is the number of grid points in X^*-direction, and j is an integer representing the node number. Similar relations are also used in X^+- and Y-directions. Figure 11.3 illustrate the meshing map.

11.3.1.3 Implementation of Slip Flow and Temperature Jump Conditions

Numerical implementation of slip velocity and temperature jump at the interface is detailed in the papers of Madhawa et al. [44] and Kabar et al. [45]. The dimensionless slip velocity and temperature jump on the fluid/solid interface are expressed respectively as

$$U_{slip}-U_i=2Kn\frac{2-\sigma_v}{\sigma_v}\left.\frac{\partial U}{\partial Y}\right|_i \tag{11.35}$$

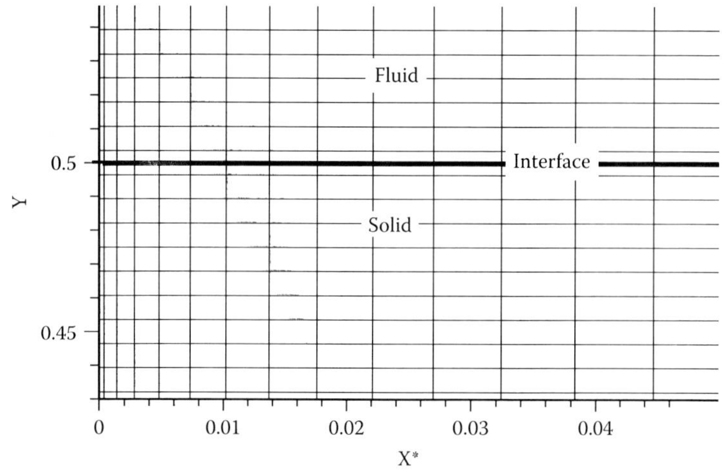

FIGURE 11.3 Meshing of the domain.

$$\theta_{slip} - \theta_i = 2\ \mathrm{Kn}\frac{2-\sigma_T}{\sigma_T}\frac{2\gamma}{(\gamma+1)}\frac{1}{\mathrm{Pr}}\left.\frac{\partial\theta}{\partial Y}\right|_i \quad (11.36)$$

These conditions are incorporated into the numerical resolution procedure as a coupled boundary condition.

A control volume adjacent to the wall–fluid interface is shown in Figure 11.4. The discretized energy equation for a fluid control volume adjacent to the interface can be written in the following algebraic form:

$$a_N^*(\theta_N - \theta_p) + a_S(\theta_S - \theta_p) + a_E(\theta_E - \theta_p) + a_W(\theta_W - \theta_p) + b = 0 \quad (11.37)$$

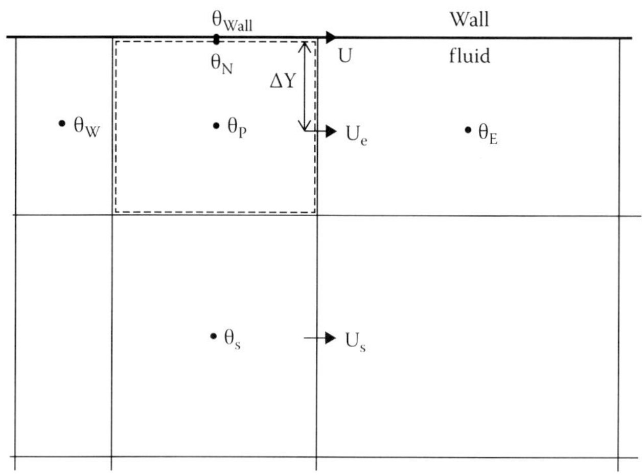

FIGURE 11.4 Control volume adjacent to the wall.

The coefficient a_N in the discretized energy equation in an internal control volume (Equation 11.31) is replaced at the interface by a_N^*:

$$a_N^* = \frac{a_N}{\left(1 + \frac{2\gamma \, Kn}{(\gamma + 1) Pr}\right)} \tag{11.38}$$

Thus, the discretized equations for boundary control volumes can be written similar to internal control volumes (Equation 11.31) with a simple adjustment of the boundary convection–diffusion coefficient, a_N.

11.3.2 Validation

11.3.2.1 Grid Independency

In order to demonstrate the numerical convergence of the proposed numerical scheme, a mesh independence study has been carried out before validating the predicted results with the benchmark solutions. The Nusselt number and maximum velocity in the fully developed flow region are considered the parameters for the comparison purpose. Figure 11.5 depicts a comparison of the Nusselt number evolution along the microchannel for different meshes. Results for three meshes: 200 × 172 (34,400 grid points), 300 × 172 (51,600 nodes), and 300 × 372 (111,600 nodes) among six tested meshes (between 34,400 and 111,600 nodes) are presented on that figure. The accuracy of the numerical predictions of the flow parameters increases in consistent with mesh refinement. The grid-independent examination shows that satisfactory results can be obtained by using a grid number of 300 × 272 whatever is the value of Brinkman number, Knudsen number, wall to channel thicknesses ratio, and solid to fluid conductivities ratio. Note that the length of the channel L has been fixed great enough to obtain fully developed flow and heat transfer at the exit of the channel. With such mesh, the asymptotic value of Nusselt number obtained at the exit is $Nu_\infty = 7.52$ for $Kn = 0$ (no rarefaction effect), $Br = 0.0$ (no viscous dissipation), and $\Delta = 0.0$ (negligible wall thickness). This is conformed to the known value for two parallel-plate microchannel flow ($Nu_\infty = 7.54$). It has been observed that maximum velocity depends further on the mesh size than the Nusselt number. The maximum deviation between the meshes was noted to be less than 1%.

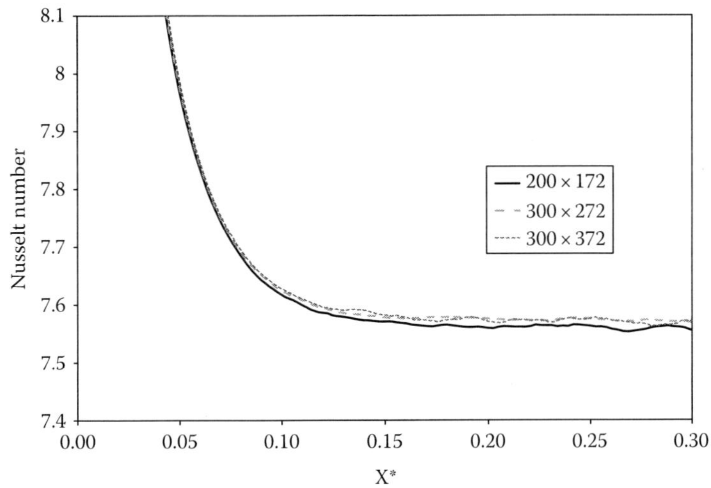

FIGURE 11.5 Comparison of the Nusselt number for different meshes.

11.3.2.2 Comparison with Solutions in the Fully Developed Flow Region

The numerical procedure developed for this study has been validated with the analytical solutions in the fully developed flow region of parallel-plate microchannel. For the hydrodynamic validation, both Poiseuille number and velocity profiles for different Kn numbers are compared with the analytical solutions. Kandlikar et al. [30] suggested the following correlation for Poiseuille number:

$$\mathrm{Po} = \frac{24}{1 + 12\dfrac{2-\sigma_v}{\sigma_v}\mathrm{Kn}} \tag{11.39}$$

Figure 11.6 shows a very good agreement between the numerical results and the analytical solution. The discrepancy did not exceed 0.5%.

Van Rij et al. have performed in [42] an analytical study on slip-flow in the fully developed region and have suggested the following expression for the axial velocity profile:

$$\frac{u(y)}{u_m} = \frac{u_s}{u_m} + 6\left(1 - \frac{u_s}{u_m}\right)\left(\frac{y}{H} - \frac{y^2}{H^2}\right) \tag{11.40}$$

where

$$\frac{u_s}{u_m} = 1 - \frac{1}{\left(1 + 12\sigma_v \mathrm{Kn}^2\right)} \tag{11.41}$$

The velocity profiles obtained from the numerical resolution are compared in Figure 11.7 to those computed from Equation 11.40. An excellent matching can be noted between the profiles.

A comparison of the numerical results obtained for uniform surface temperature with the correlation given by Shah and Bhatti [46] in their analytical study is shown in Figure 11.8. Our numerical

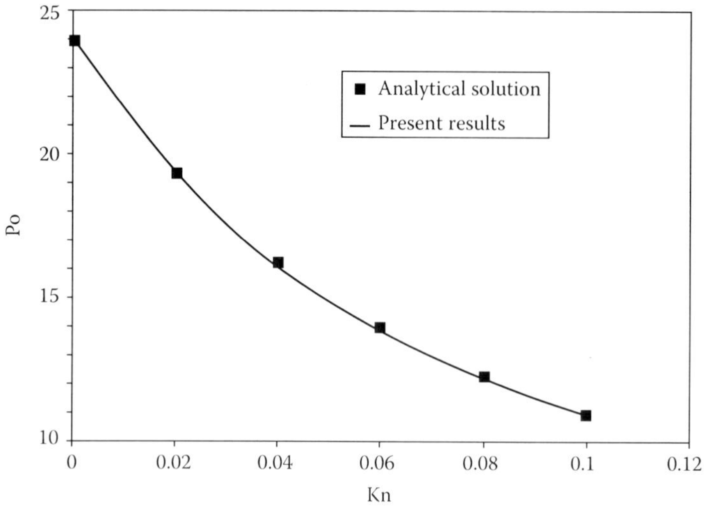

FIGURE 11.6 Comparison of fully developed flow Poiseuille number with analytical solution.

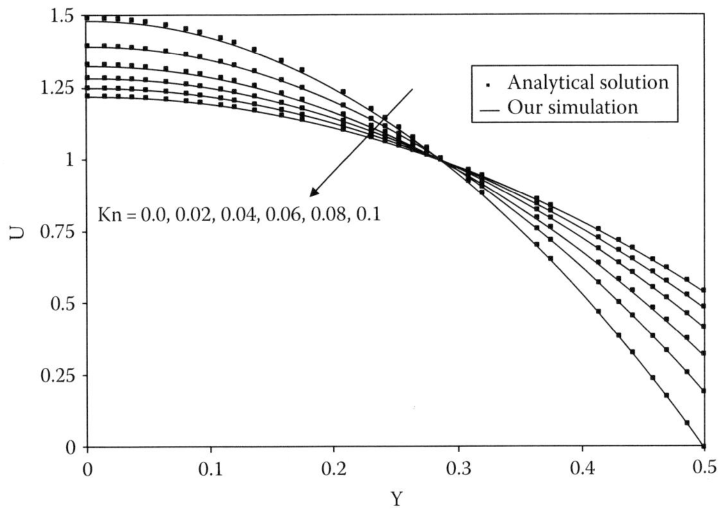

FIGURE 11.7 Comparison of the axial velocity profiles with those obtained analytically. (From Van Rij, J. et al., *Int. J. Therm. Sci.*, 46, 1095, 2007.)

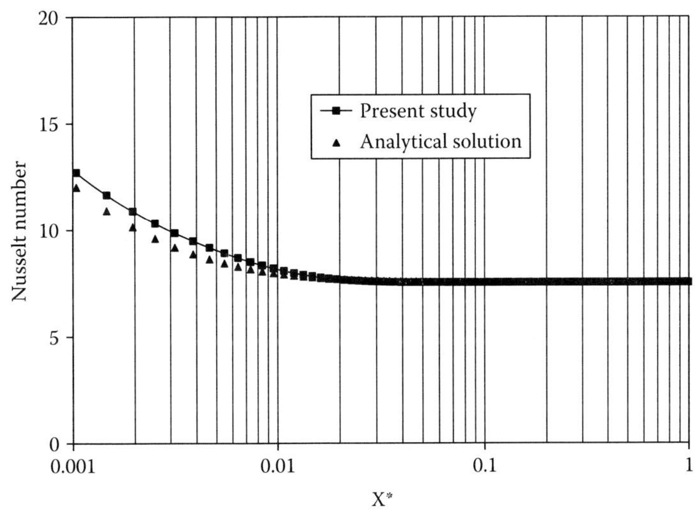

FIGURE 11.8 Comparison of Nusselt number distribution with analytical solution obtained by Shah and Bhatti [46].

solution is performed for Pr = 7, Br = 0, and Δ = 0. Shah and Bhatti recommended the following analytical expression in a wide range of Prandt Number ($0.1 \leq Pr \leq 1000$):

$$Nu_{X^*} = 7.55 + \frac{0.024(X^*)^{-1.14}\left(0.0179\,Pr^{0.17}(X^*)^{-0.64} - 0.14\right)}{\left(1 + 0.0358\,Pr^{0.17}(X^*)^{-0.64}\right)^2} \tag{11.42}$$

Relative deviation between the numerical solution and the analytical one is less than 0.5%, except in the entrance region ($X^* < 0.02$) where a deviation of about 6% can be observed. This can be attributed to the analytical solution deviation with the Prandtl number as mentioned by the authors. In the fully developed flow region, the asymptotic Nusselt number (Nu = 7.54) is obtained.

Numerical simulations are performed in a wide range of parameters variations (Pr, Kn, Br, Δ, and K). In the following sections, some typical results for simultaneous developing of thermally and hydraulically flows are given for water (Pr = 7) and air (Pr = 0.71) as typical liquid and gas with a Reynolds number Re = 100. In microchannel heat sink or heat exchangers, the wall can be as thick as the microchannel, and sometimes thicker than the microchannel. Thus, five wall thickness to half channel height ratios have been considered in the numerical simulations; Δ = E/H = 0, 0.2, 1, 2, and 4. Δ = 0 is associated to the case of non-conjugated problem, where the plates have no influence on the heat transfer and solutions should match the well-known results of constant temperature or constant heat flux problem. Computational results are given for three solid/fluid conductivities ratio; K = 1, 10, and 100. These values cover some couples of solid/fluid usually used in micro-heat exchangers, fuel cell, and electronics cooling. Indeed, the thermal conductivities ratio may be of the order of 1 for the combination of glass/water (k_{water} = 0.6 and k_{glass} = 1.1) and of order 10^4 for the couple Copper/air (k_{copper} = 400 and k_{air} = 0.026).

11.4 COMBINED MICROSCALE EFFECTS IN GASEOUS FLOW

11.4.1 Hydrodynamically Developing Flow Field

The numerical results of the evolution of Poiseuille number for different Knudsen number in the range 0–0.1 is plotted in Figure 11.9. As expected, because of the developing flow region at the entrance of the microchannel, the Poiseuille number is decreasing from a high values and tends toward the fully developed flow values as X^+ increases. Note that in fully developed flow region Po = 24 for Kn = 0 (no-slip condition). Po decreases when Kn number increases because of the slip of gas molecules on the wall. The velocity gradient decreases with increasing scarcity by increasing the Kn number. Thus, depletion reduces the friction factor of the fluid in the microchannel. Poiseuille number in the fully developed region varies from 24 for Kn = 0 to 10 for Kn = 0.1. Its values are indicated in the Figure 11.9.

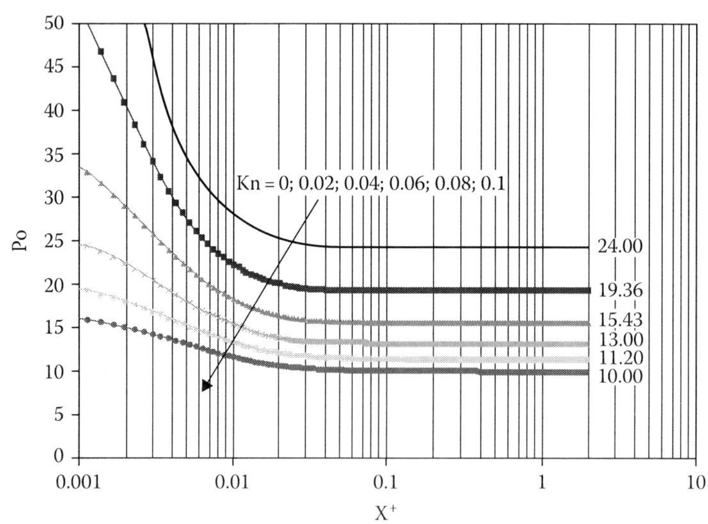

FIGURE 11.9 Poiseuille number evolution for different Knudsen number.

11.4.2 Thermally Developing Flow Field for a Uniform Outer Wall Temperature

The temperature profiles for the hydrodynamically and thermally developing flow are illustrated in Figure 11.10a through c for Br = 0, 0.1, and −0.1, respectively. The computational fixed parameters are $\Delta = 4$, $K = 100$, and Kn = 0.04. One can note on these profiles that, close to the entry of the channel (i.e., Figure 11.10a at $X^* = 0.0005$), the temperature in the fluid are merged at the central region of the microchannel. They are becoming more dissociated as they go up to the interface with the wall. The effect of the viscous dissipation and the temperature jump at the interface is apparent. When the distance from the entry grows and the fluid progresses along the microchannel, the profiles are becoming more and more dissociated (Figure 11.10b). In the fully developed region (i.e., Figure 11.10c, $X^* = 0.8$), the temperature profile for Br = 0 is flat, indicating a uniform temperature in the fluid as well as in the wall (thermodynamic equilibrium between the wall and the fluid). For Br = 0.1 (cooling of the fluid), the temperature in the fluid is higher than at the outer surface of the wall because of the heating of fluid generated by viscous dissipation, which leads to positive value of the dimensionless temperature θ in both fluid and wall. Whereas, in the case of the heating of fluid (Br = −0.1), the fluid temperature can be higher than the surface temperature, this is inducing negative values of the dimensionless temperature θ. Temperature distribution varies gradually from the outer surface of the wall to the centre of the microchannel, except at the fluid/wall interface where a temperature jump occurs.

Temperature profiles for different Knudsen numbers are plotted on Figure 11.11 for Br = 0.1. The effect of rarefaction is well noticeable on these curves and the temperature jump is high as Kn number is large.

The Nusselt number is defined by

$$\mathrm{Nu}_{X^*} = \left. 2\frac{\partial\theta}{\partial Y}\right|_i \Big/ (\theta_b - \theta_i) \tag{11.43}$$

Where θ_b is the bulk temperature and θ_i is the fluid/wall interface temperature. The bulk temperature is computed with the following relation

$$\theta_b = \frac{\int_0^{0.5} U\theta dY}{\int_0^{0.5} UdY} \tag{11.44}$$

In Figure 11.12a through c, the effect of Kn number on the Nusselt number distribution is shown for, respectively, Br = 0, 0.1 and −0.1. As expected, when Kn increases, the Nusselt number decreases whatever is the Br number. The increase in the Kn number shifts the entrance length to the right and leads to an augmentation of the establishment length of the fully developed heat transfer and fluid flows. For Br = −0.1 (Figure 11.12c), increasing Kn number shifts also the jump-point to the right. Indeed, when Br < 0 (heating of the fluid), the evolution of Nu number presents a discontinuity at a certain transition coordinate X^* around 0.1 for Kn = 0 and around 0.4 for Kn = 0.1. The transitional coordinate increases by increasing Kn number. These coordinates correspond to where the bulk temperature θ_b equals the inner surface temperature of the wall θ_i. In other words, the Nusselt number is becoming meaningless and tends to minus infinity at these locations. Then, it abruptly takes plus infinity and subsequently decreases gradually to the typical fully developed values for

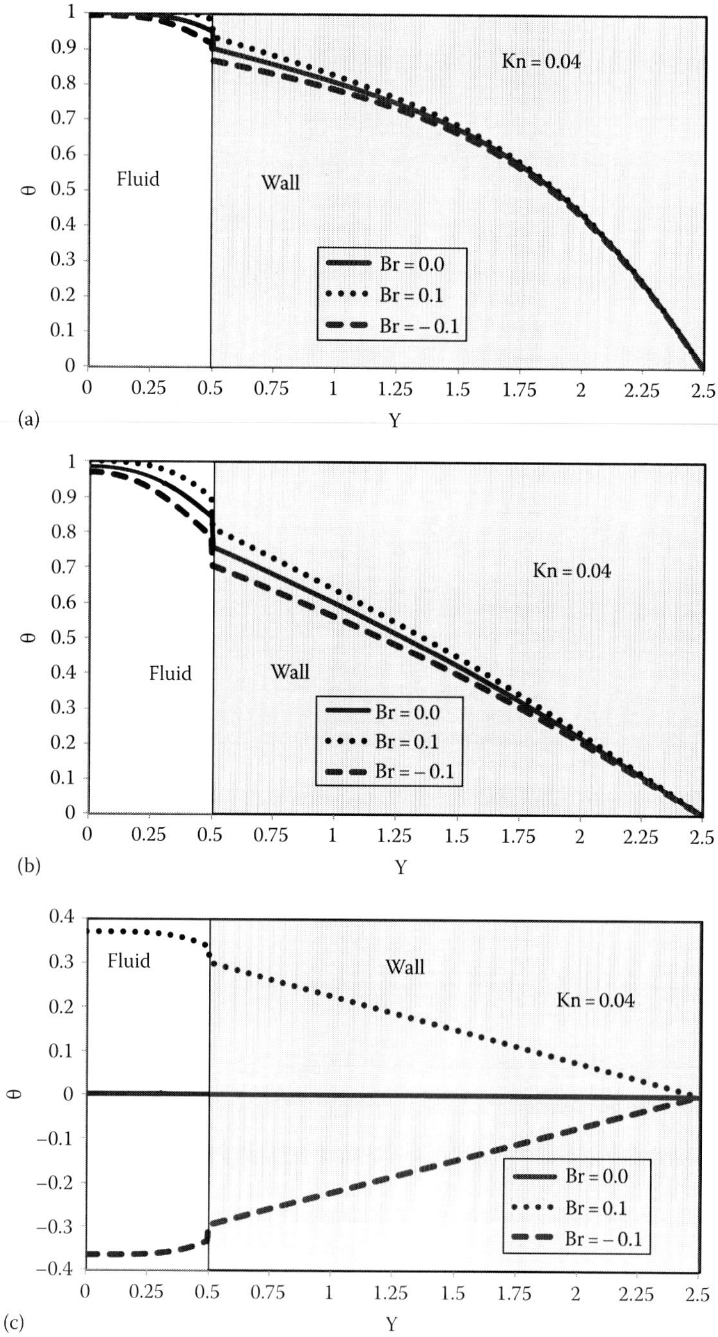

FIGURE 11.10 Temperature profiles at different locations for $\Delta = 4$. (a) $X^* = 0.0005$; (b) $X^* = 0.001$; (c) $X^* = 0.8$.

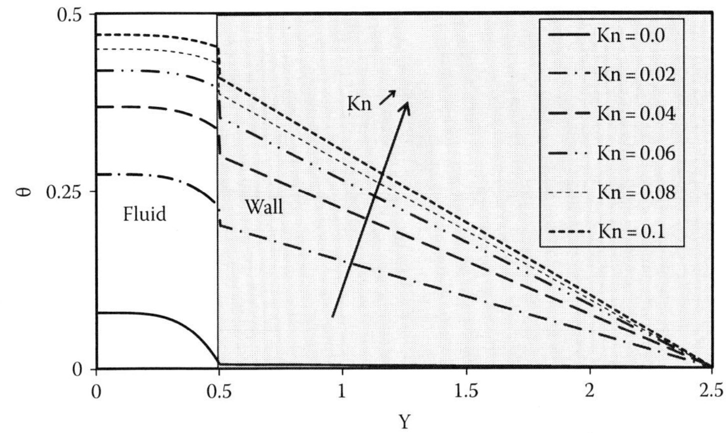

FIGURE 11.11 Temperature profiles at the exit section of the microchannel for different Kn number (Br = 0.1).

each Kn number. This is due to the fact that heat generation by viscous dissipation makes the fluid hotter than the wall. For Br = 0.1 (Figure 11.12b), the Nusselt number decreases from the entry of the microchannel, reaches a minimum value, and then increases gradually to the same value as for negative Br number in the fully developed flow region.

Fully developed flow Nusselt number Nu_∞ obtained with and without viscous dissipation are presented on Tables 11.1 and 11.2 for different Kn number and Δ values for a fixed K value (K = 100). These results are compared with the results of Çetin et al. [18], who have studied hydrodynamically developed thermally developing gaseous laminar flow in parallel-plate microchannel with negligible wall conduction (Greatz problem). The tables depict the results for different wall thickness to channel height ratios: Δ = 0.0 (which is similar to the studied case in [18]), Δ = 0.2 and Δ = 4. They showed the influence of heat diffusion in the wall combined with viscous dissipation and rarefaction. The Nusselt number Nu_∞ decreases when Kn increases. It varies from 7.54 for Kn = 0 (no slip/no jump) to 5.18 for Kn = 0.1 in the case of Br = 0, and from 17.51 to 7.23 when Br $\neq$ 0. Combination of rarefaction with wall conduction and viscous dissipation has significant influence on the heat transfer rate. Indeed, deviation of the numerical results from Çetin's ones for the case Δ = 0 is not exceeding 1.5% for negligible viscous dissipation (Table 11.1), which may be attributed to the numerical error. However, this deviation is about 13% for thick with high conductivity wall (Δ = 4 and K = 100) in all cases of viscous dissipation (negligible or not). Computational results for thick wall with low conductivities lead to higher deviations [28].

Figure 11.13a and b illustrates the Nusselt number distribution obtained with Kn = 0.04, Br = 0 (a) and Br = −0.1 (b) for different thicknesses ratio Δ. It is obvious from these figures that increasing the wall thickness induces an increase in the Nusselt number in both developing and fully developed regions. It is clear that Δ is a key parameter influencing the Nusselt number distribution. Contrary to the case of negligible wall thickness, a high wall thickness affects the thermal resistance of the wall; it makes the inner surface temperature of the channel varying like it is under a condition somewhat deviating from uniform temperature and approaching a uniform heat flux condition. This trend should be highlighted in future research works with different solid/fluid conductivities ratio.

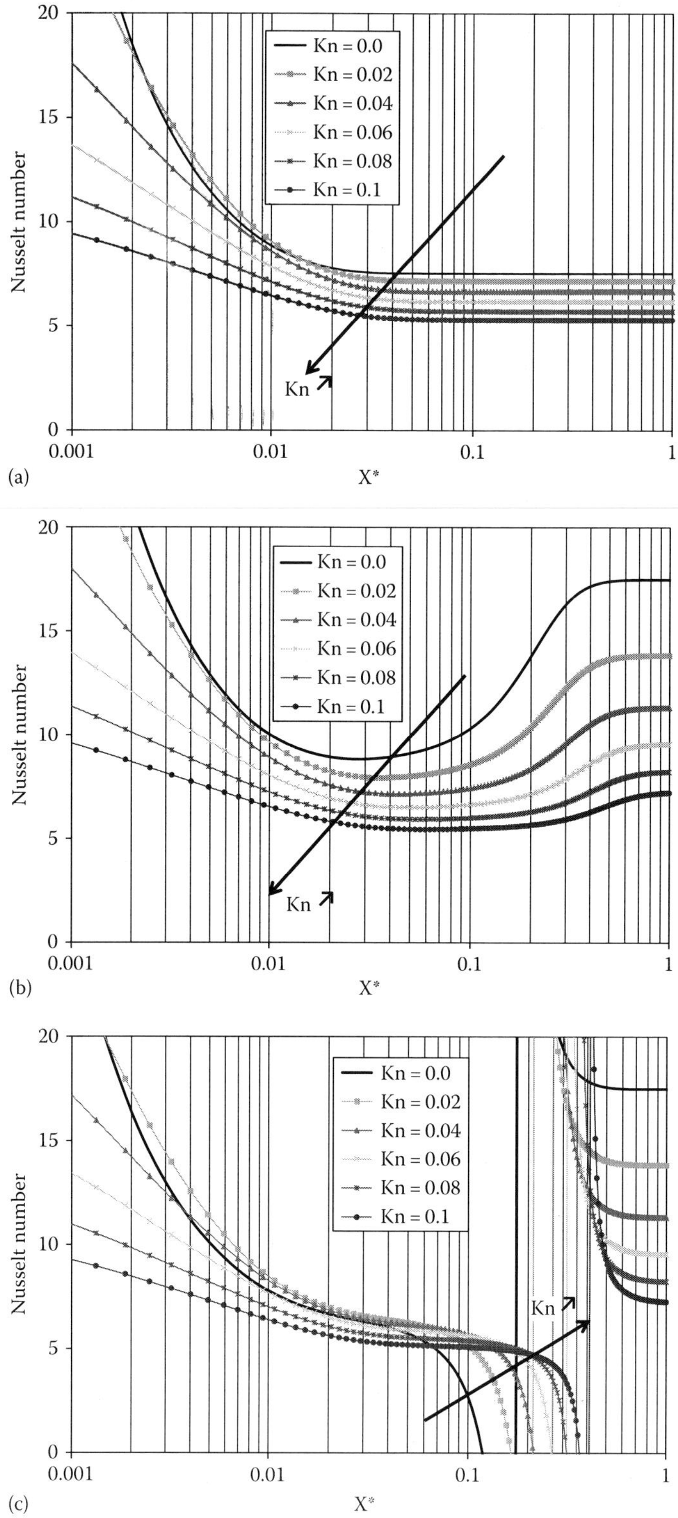

FIGURE 11.12 Nusselt number distributions for different Brinkman numbers. (a) Br = 0; (b) Br = 0.1; and (c) Br = −0.1.

TABLE 11.1
Fully Developed Nusselt Number Nu_∞ without Viscous Dissipation

	Δ = 0.0	Δ = 0.2	Δ = 4.0	Çetin et al. [18]
Kn = 0.00	7.542	7.620	7.621	7.541
Kn = 0.02	7.063	7.236	7.522	6.925
Kn = 0.04	6.572	6.750	7.184	6.374
Kn = 0.06	5.946	6.147	6.492	5.882
Kn = 0.08	5.601	5.771	6.010	5.445
Kn = 0.10	5.188	5.338	5.581	5.058

TABLE 11.2
Fully Developed Nusselt Number Nu_∞ with Viscous Dissipation

	Δ = 0.0	Δ = 0.2	Δ = 4.0	Çetin et al. [18]
Kn = 0.00	17.508	17.627	17.628	17.497
Kn = 0.02	13.827	14.112	14.128	13.688
Kn = 0.04	11.325	11.420	12.155	11.222
Kn = 0.06	9.548	9.689	10.562	9.499
Kn = 0.08	8.236	8.342	9.253	8.229
Kn = 0.10	7.233	7.316	8.212	7.256

11.4.3 Thermally Developing Flow Field for a Uniform Outer Wall Heat Flux

The contours of isotherms for the hydrodynamically and thermally developing flow with uniform heat flux on the outer surface of the plates, obtained with Δ = 2 and Br = 0, are illustrated in Figure 11.14a through c for three Knudsen numbers: (a) Kn = 0.0; (b) Kn = 0.04; and (c) Kn = 0.06. They depict temperature jump at the solid-fluid interface and the change in the temperature profile with the Knudsen number. The slip velocity and temperature jump dominate the heat transfer in the microchannel and by consequence heat diffusion within the wall. Results for additional values of Δ and K (not presented here) confirm the same trend of the combined thermal diffusion in the wall with rarefaction in the gas.

Figure 11.15 represents the variation of the local Nusselt number with X* for Δ = 2, K = 100 and Br = 0. As for uniform outer wall surface temperature condition, the effect of slip velocity and temperature jump decreases the heat transfer coefficient and thereby the Nusselt number. When Kn increases, Nu decreases. This decay is more important than in case of uniform temperature, and is more pronounced at the entrance region. Nu number with uniform heat flux is higher than with uniform temperature. The fully developed flow Nusselt number for each Knudsen number obtained numerically are in good agreement with the analytical solution suggested by Van Rij [41] and expressed as

$$Nu = 420\left[51 + 420\left(\frac{2-\sigma_T}{\sigma_T}\right)\left(\frac{2\gamma}{1+\gamma}\right)\left(\frac{1}{Pr}\right)Kn - 2\frac{u_s}{u_m}\left(9-\frac{u_s}{u_m}\right)\right]^{-1} \quad (11.45)$$

The ratio u_s/u_m is defined in Equation 11.41.

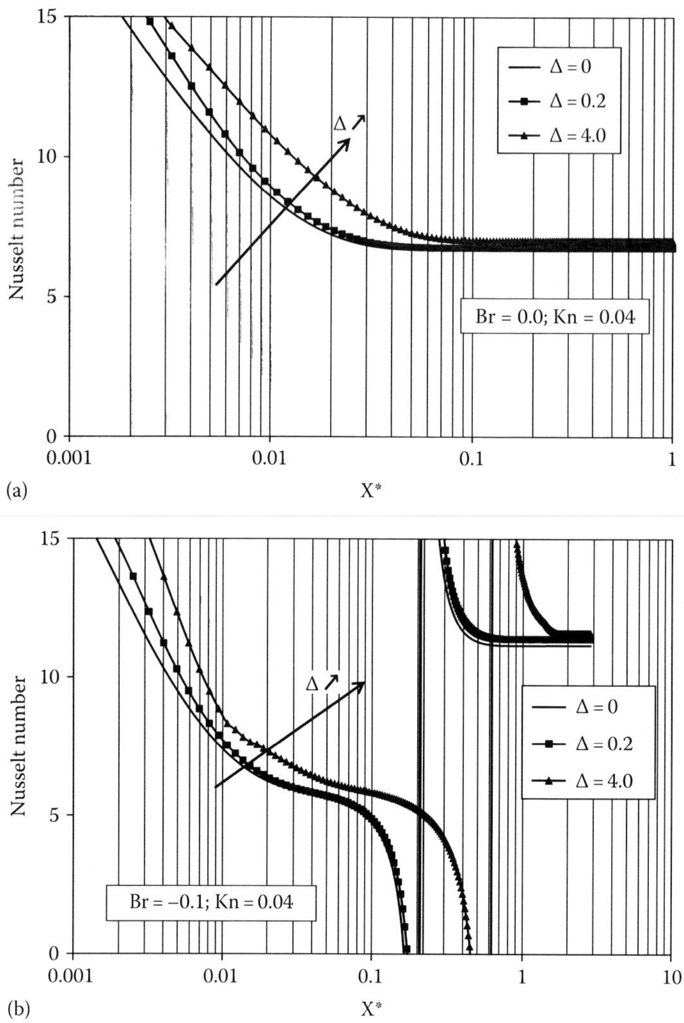

FIGURE 11.13 (a) Nusselt number distributions for different Δ(Br = 0). (b) Nusselt number distributions for different Δ(Br = −0.1).

11.5 COMBINED MICROSCALE EFFECTS IN LIQUID FLOW

Figure 11.16 illustrates the evolution of Nusselt number along the microchannel for different Δ in the case of uniform temperature at the outer surface of wall. The increase of the Nusselt number with the thicknesses ratio Δ can be easily observed. The effect of the wall thickness is particularly located at the entrance region of the microchannel ($X^* < 0.02$). Downstream this location, Nu goes toward the fully developed flow values and the same trend as with uniform temperature is obtained.

The fluid bulk temperature along the microchannel length X*, computed with the Equations 11.7 through 11.10, is shown in Figure 11.17 for different conductivities ratio $K = k_s/k_f$. The bulk temperature, is given for Br = 0; it varies from 1 at the entry cross section to 0 at the outlet section. As K decreases, the bulk temperature is shifted toward the high X* values. This means that a low conductivity of the wall tends to delay the establishment of the fully developed flow, and therefore to increase the thermal development length.

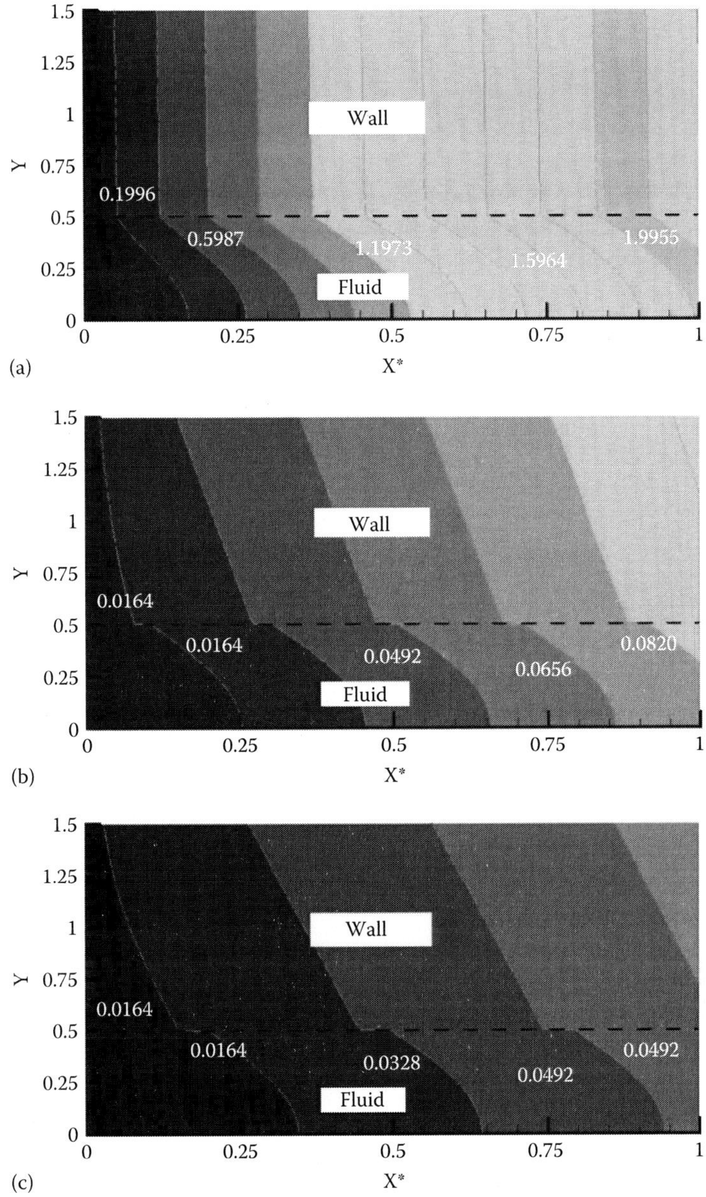

FIGURE 11.14 Isotherms for $\Delta = 2$ and $Br = 0$. (a) $Kn = 0.0$; (b) $Kn = 0.04$; (c) $Kn = 0.06$.

11.6 CONCLUDING REMARKS

A numerical study of conjugated heat transfer in two parallel-plate microchannel has been performed using the control volume method. The built in-house program is based on resolution strategy with single domain including the wall and the fluid and valid for both liquid and gas. We have taken into account the combined effects of axial and transversal conductions, viscous dissipation and rarefaction. Significant distortions in the temperature profiles and the Nusselt number evolutions have been noted due to these effects. Due to the slip flow and temperature jump, the heat transfer rate decreases by increasing Kn number. It was established that the

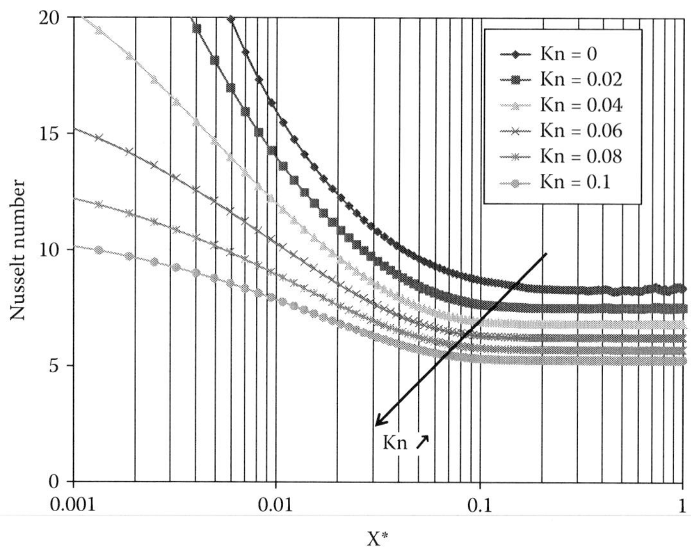

FIGURE 11.15 Variation of local Nusselt number along the microchannel for different Kn numbers (Br = 0 and Δ = 1).

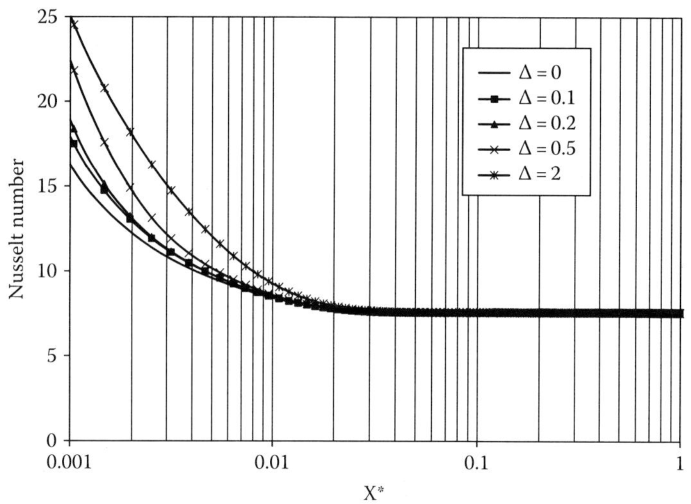

FIGURE 11.16 Distribution of Nusselt number for different Δ(Pr = 7, Br = 0).

conduction effect is more pronounced for the wall with low thermal conductivity/high thickness, and particularly, at the entrance region of the channel. In the fully developed heat and fluid flow region, low conductivity or/and high wall thickness lead to an increase in the Nusselt number. The effects of viscous dissipation induce also an increase of the local Nusselt number, and a longer entrance length. This trend is particularly of great interest in engineering applications using short microchannel with a very thick wall. Average Nusselt number could be significantly different from the obtained results in conventional dimensions, where the effects of wall conduction are negligible.

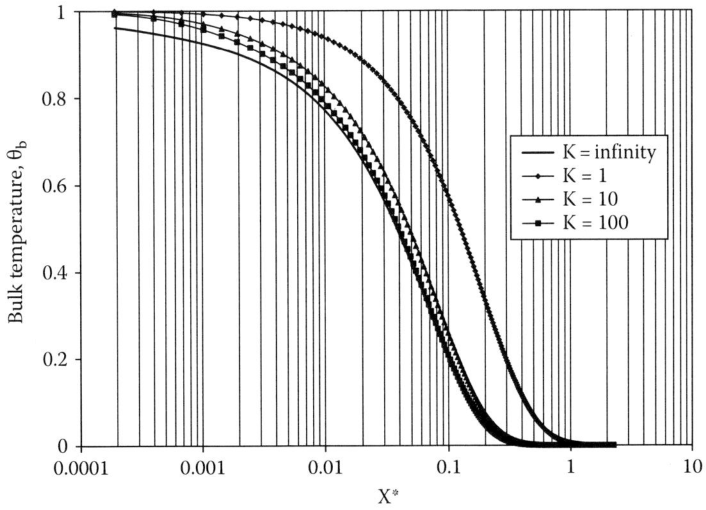

FIGURE 11.17 Distribution of the bulk temperature (with $\Delta = 1$ and $Br = 0$).

REFERENCES

1. D.B. Tuckerman, R.F.W. Pease, Optimized convective cooling using micromachined structure, *Journal of the Electrochemical Society*, 129(3), 98C (1982).
2. X.F. Peng, G.P. Peterson, Convective heat transfer and flow friction for water flow in microchannel structures, *International Journal of Heat and Mass Transfer*, 39, 2599–2608 (1996).
3. C. Yang, C. Chen, T. Lin, S.G. Kandlikar, Heat transfer and friction characteristics of air flow in microtubes, *Experimental Thermal and Fluid Sciences*, 37, 12–18 (2012).
4. Z.Y. Guo, Z.X. Li, Size effect on single-phase channel flow and heat transfer at microscale, *International Journal of Heat and Fluid Flow*, 24, 284–298 (2003).
5. G.L. Morini, Single-phase convective heat transfer in microchannels: A review of experimental results, *International Journal of Thermal Sciences*, 43, 631–651 (2004).
6. C.Y. Huang, C.M. Wu, Y.N. Chen, T.M. Liou, The experimental investigation of axial heat conduction effect on the heat transfer analysis in microchannel flow, *International Journal of Heat and Mass Transfer*, 70, 169–173 (2014).
7. G. Karniadakis, A. Beskok, N. Aluru, *Microflows and Nanoflows: Fundamentals and Simulation*, Springer, New York, 2005.
8. Y. Bayazıtoglu, G. Tunç, K. Wilson, I. Tjahjono, Convective heat transfer for single phase gases in microchannel slip flow: Analytical solutions, *Microscale Heat Transfer: Fundamentals and Applications in Biological Systems and MEMS*, S. Kakaç, L. Vasiliev, Y. Bayazıtoglu, Y. Yener (eds.), Kluwer Academic Publishers, Dordrecht, the Netherlands, 2005.
9. R.M. Cotta, S. Kakaç, M.D. Mikhailov, F.V. Castellos, C.R. Cardoso, Transient flow and thermal analysis in microfluidics, *Microscale Heat Transfer: Fundamentals and Applications in Biological Systems and MEMS*, S. Kakaç, L. Vasiliev, Y. Bayazıtoglu, Y. Yener (eds.), Kluwer Academic Publishers, Dordrecht, the Netherlands, 2005.
10. D.C. Knupp, C. Naveira-Cotta, R.M. Cotta, Theoretical–experimental analysis of conjugated heat transfer in nanocomposite heat spreaders with multiple microchannels, *International Journal of Heat and Mass Transfer*, 74, 306–318 (2014).
11. P. Rosa, T.G. Karayiannis, M.W. Collins, Single-phase heat transfer in microchannels: The importance of scaling, *Applied Thermal Engineering*, 29, 3447–3468 (2009).
12. S.X. Zhang, Y.L. He, G. Lauriat, W.Q. Tao, Numerical studies of simultaneously developing laminar flow and heat transfer in microtubes with thick wall and constant outside wall temperature, *International Journal of Heat and Mass Transfer*, 53, 3977–3989 (2010).

13. P. Lopez, Y. Bayazitoglu, An extended thermal lattice Boltzmann model for transition flow, *International Journal of Heat and Mass Transfer*, 65, 374–380 (2013).
14. D.C. Knupp, R.M. Cotta, C.P. Naveira-Cotta, S. Kakaç, Transient conjugated heat transfer in microchannels: Integral transforms with single domain formulation, *International Journal of Thermal Sciences*, 88, 248–257 (2015).
15. G.L. Morini, Y. Yang, H. Chalabi, M. Lorenzini, A critical review of the measurement techniques for the analysis of gas microflows through microchannels, *Experimental Thermal and Fluid Science*, 35(6), 849–865 (2011).
16. G.L. Morini, Y. Yang, Guidelines for the determination of single-phase forced convection coefficients in microchannels, *Journal of Heat Transfer*, 135(10), 101004 (2013).
17. W. Sun, S. Kakaç, A. Guvenç Yazıcıoglu, A numerical study of single-phase convective heat transfer in microtubes for slip flow, *International Journal of Thermal Sciences*, 46, 1084–1094 (2007).
18. B. Çetin, H. Yuncu, S. Kakaç, Gaseous flow in microconduits with viscous dissipation, *International Journal of Transport Phenomena*, 8, 297–315 (2007).
19. C.P. Tso, S.P. Mahulikar, The use of the Brinkman number for single phase forced convective heat transfer in microchannels, *International Journal of Heat and Mass Transfer*, 41(12), 1759–1769 (1998).
20. C.P. Tso, S.P. Mahulikar, The role of the Brinkman number in analyzing flow transitions in microchannels, *International Journal of Heat and Mass Transfer*, 42, 1813–1833 (1999).
21. J. Koo, C. Kleinstreuer, Viscous dissipation effects in microtubes and microchannels, *International Journal of Heat and Mass Transfer*, 47, 3159–3169 (2004).
22. G.L. Morini, Viscous heating in liquid flows in micro-channels, *International Journal of Heat and Mass Transfer*, 48(17), 3637–3647 (2005).
23. G.P. Celata, G.L. Morini, V. Marconi, S.J. McPhail, G. Zummo, Using viscous heating to determine the friction factor in microchannels—An experimental validation, *Experimental Thermal and Fluid Science*, 30, 725–731 (2006).
24. S.P. Guidice, C. Nonino, S. Savino, Effects of viscous dissipation and temperature dependent viscosity in thermally and simultaneously developing laminar flows in microchannels, *International Journal of Heat and Fluid Flow*, 28, 15–27 (2007).
25. G. Maranzana, I. Perry, D. Maillet, Mini- and micro-channels: Influence of axial conduction in the walls, *International Journal of Heat and Mass Transfer*, 47, 3993–4004 (2004).
26. C. Nonino, S. Savino, S. Del Giudice, L. Mansutti, Conjugate forced convection and heat conduction in circular microchannel, *International Journal of Heat and Fluid Flow*, 30, 823–830 (2009).
27. I. Tiselj, G. Hetsroni, B. Mavko, A. Mosyak, E. Pogrebhyak, Z. Segal, Effect of axial conduction on the heat transfer in micro-channels, *International Journal of Heat and Mass Transfer*, 47, 2551–2565 (2004).
28. Y. Kabar, M. Rebay, M. Kadja, C. Padet, Numerical resolution of conjugate heat transfer problem in a parallel-plates micro-channel, *Heat Transfer Research*, 41(3), 247–263 (2010).
29. M. Gad-el-Hak, *MEMS Introduction and Fundamentals*, Taylor & Francis Group, Boca Raton, FL, 2002.
30. S.G. Kandlikar, S. Garimella, D. Li, S. Colin, M.R. King, *Heat Transfer and Fluid Flow in Minichannels and Microchannels*, Elsevier Ltd, Oxford, UK (2006).
31. F.E. Larrode, C. Housiadas, Y. Drossinos, Slip-flow heat transfers in circular tubes, *International Journal of Heat and Mass Transfer*, 43, 2669–2680 (2000).
32. R.F. Barron, X.M. Wang, T.A. Ameel, R.O. Warrington, The Graetz problem extended to slip-flow, *International Journal of Heat and Mass Transfer*, 40(8), 1817–1823 (1997).
33. T.A. Ameel, X. Wang, R.F. Barron, R.O. Warrington, Laminar forced convection in a circular tube with constant heat flux and slip flow, *Microscale Thermophysical Engineering*, 1(4), 303–320 (1997).
34. L. Ghodoossi, N. Egrican, Prediction of heat transfer characteristics in rectangular microchannels for slip flow regime and H1 boundary condition, *International Journal of Thermal Sciences*, 44, 513–520 (2005).
35. M.D. Mikhailov, R.M. Cotta, Mixed symbolic-numerical computation of convective heat transfer with slip flow in microchannels, *International Communications in Heat and Mass Transfer*, 32, 341–348 (2005).
36. Y. Chen, C. Zhang, M. Shi, J. Wu, Three-dimensional numerical simulation of heat and fluid flow in noncircular microchannel heat sinks, *International Communications in Heat and Mass Transfer*, 36, 917–920 (2009).
37. G. Tunç, Y. Bayazıtoglu, Heat transfer in microtubes with viscous dissipation, *International Journal of Heat and Mass Transfer*, 44, 2395–2403 (2001).

38. G. Tunç, Y. Bayazıtoglu, Heat transfer in rectangular microchannels, *International Journal of Heat and Mass Transfer*, 45, 765–773 (2002).
39. H.E. Jeong, J.T. Jeong, Extended Graetz problem including streamwise conduction and viscous dissipation in microchannel, *International Journal of Heat and Mass Transfer*, 49, 2151–2157 (2006).
40. B. Çetin, A.G. Yazıcıoglu, S. Kakaç, Slip-flow heat transfer in microtubes with axial conduction and viscous dissipation—An extended Graetz problem, *International Journal of Thermal Sciences*, 48(9), 1673–1678 (2009).
41. J. Van Rij, T. Ameel, T. Harman, The effect of viscous dissipation and rarefaction on rectangular microchannel convective heat transfer, *International Journal of Heat and Mass Transfer*, 48, 271–281 (2009).
42. J. Van Rij, T. Harman, T. Ameel, The effect of creep flow on two-dimensional isoflux microchannels, *International Journal of Thermal Sciences*, 46, 1095–1103 (2007).
43. S.V. Patankar, *Numerical Heat Transfer and Fluid Flow*, Hemisphere, Washington, DC, 1980.
44. H.D. Madhawa Hettiarachchi, M. Golubovic, W.M. Worek, W.J. Minkowycz, Three-dimensional laminar slip-flow and heat transfer in a rectangular microchannel with constant wall temperature, *International Journal of Heat and Mass Transfer*, 51, 5088–5098 (2008).
45. Y. Kabar, R. Bessaïh, M. Rebay, Conjugate heat transfer with rarefaction in parallel-plate microchannel, *Superlattices and Microstructures*, 60, 370–388 (2013).
46. R.K. Shah, M.S. Bhatti, Laminar convective heat transfer in ducts, *Handbook of Single-Phase Convective Heat Transfer*, S. Kakac, R.K. Shah, W. Aung (eds.), Wiley, New York, 1987.

12.1 INTRODUCTION

Microchannel fluid flow and heat transfer was first put forward at the beginning of 1980s (1981 and 1982) by Tuckerman and Pease [1,2], who employed the direct water circulation in microchannels (with channel width of 300 μm and depth of 50 μm) fabricated in silicon chips as a new cooling technology. They were able to reach the highest heat flux of 7.9 MW m^{-2} with the maximum temperature difference between substrate and inlet water of 71°C. Soon Wu and Little published two papers in 1983 and 1984, reporting their measurement results of the friction factors and heat transfer characteristics of gas flow in microchannel heat exchangers used for micro/miniature refrigerators [3,4]. Their channels have a width of 130–200 μm and a depth of 30–60 μm. Since then a large number of experimental and numerical studies have been conducted for the flow and heat transfer in microchannels. In the development of measurement techniques, the contributions given by Ho and his coworkers are especially important. They developed micropressure transducers to measure the local pressure in a microchannel and confirmed the nonlinear pressure distribution in uniform microchannel [5–7].

As for the size classification of microchannel, it is still quite arbitrary. For single-phase flow and heat transfer, the following classifications are generally accepted [8]: channels with diameter less than 1 mm are classified as microchannels; channels with dimensions in the range of a few millimeters are regarded as mini channels; and those above this are called conventional channels. For noncircular channels, the hydraulic diameter may be used for its classification.

In recent years, considerable progress has been made to miniaturize all kinds of devices. As indicated before, in 1980s these achievements gave rise to a new field: microelectromechanical systems (MEMS); it has been ever increasing in many fields due to the rapid development of fabrication technology. Devices having dimensions of the order of microns have being developed for many applications including microelectronic cooling systems, bipolar plates of fuel cell, microreactors for a range of processes, advanced propulsion systems, flow sensors, complex systems consisting of pumps, valves, and other components, etc. These devices or systems employ fluid flows in microchannels/tubes under unusual conditions, and this has led to the creation of a new discipline: microfluidics [9,10]. Both MEMS and microfluidics are based on the understanding of the behavior of fluid flow and heat transfer in microscale.

Historically, severe diversities and deviations once existed between the measurement results for friction factor (and Nusselt number) of microchannels by different authors, even for the very basic problem such as the friction constant (and Nusselt number) of laminar flow in circular microchannels. With the progresses in fabrication method, measurement, and data acquisition techniques, the reasons behind such deviations and diversities are gradually revealed. And it is now generally accepted that, except those size effects (rarefaction and compressibility effects of gases, the effect of surface roughness, etc.), conventional solutions and correlations for fluid flow and heat transfer of single phase flow can still be used for microchannels [11–14]. One important outcome of this understanding is that the numerical methods based on the continuum assumption, such as FVM, can still be adopted to predict the fluid flow and heat transfer behavior in microchannels.

In this chapter, experimental and numerical results of single-phase fluid flow and heat transfer in microchannel and nozzles will be presented. For the simplicity of presentation, those diversities will not be introduced and restated again, and what will be presented are those generally accepted results, even though obtained mainly by the authors' group.

12.2 EXPERIMENTAL AND NUMERICAL RESULTS OF GAS FLOW IN MICROCHANNELS

In the flow of a gas through microchannel, great changes in properties of the gas occur because of the large pressure drop caused by the friction effects even when the inlet flow velocity is small. For example, pressure drops of several atmospheres were observed in a 4 mm long channel due to the viscous friction [6]. Friction contributes to large pressure gradient, which in turn generates changes in gas density. The great density change may lead to two important effects: rarefaction and compressibility. These two effects are verified by our experimental results presented here.

12.2.1 Experimental Setup and Test Tubes

A test rig is set up to verify the rarefaction effect, compressibility effect, and roughness effect [15]. The descriptions of the test tubes are listed in Table 12.1, and the working fluids are nitrogen and helium gas. For details of the experimental setup, refer [15].

The tested microtubes shown in Table 12.1 are classified into four types: four circular glass microtubes (fused silicon tube-FST: D = 50–201 μm), three square glass microchannels (fused silicon square tube-FSCT: D_h = 52–100 μm), four stainless steel microtubes (SST: D = 119–300 μm), and three circular glass microtubes (microfused silicon tube-MFST: D = 10–20 μm). The measured surface relative roughness of the circular glass microtubes and square glass microchannels is significantly below 1%. However, for the four stainless steel tubes of D_i = 300, 260, 172, and 119 μm,

TABLE 12.1
Descriptions of Test Tubes

Tube/Channel	Hydraulic Diameter $D \pm 1$ μm	Length $L \pm 0.02$ mm	Re	Outlet Ma	Outlet Kn
FST1	201.44	100.40	28–6200	0.017–0.6	1.6×10^{-4}–9.8×10^{-4}
FST2	102.74	93.02	33–3100	0.04–0.6	2.8×10^{-4}–1.9×10^{-3}
FST3	74.36	49.42	7–2700	0.012–0.6	3.5×10^{-4}–2.6×10^{-3}
FST4	50.55	50.28	6–1200	0.015–0.6	8.1×10^{-4}–3.9×10^{-3}
FSC1	100.25	74.14	60–3600	0.05–0.6	2.6×10^{-4}–1.4×10^{-3}
FSC2	75.34	49.66	7–2700	0.01–0.6	3.5×10^{-4}–2.6×10^{-3}
FSC3	52.03	33.62	24–2000	0.05–0.6	4.3×10^{-4}–3.7×10^{-3}
SST1	300.00	100.00	81–5200	0.22–0.6	2.1×10^{-4}–2.2×10^{-4}
SST2	260.00	100.00	81–5000	0.19–0.6	1.8×10^{-4}–2.6×10^{-4}
SST3	172.00	100.02	14–2700	0.015–0.58	2×10^{-4}–8.8×10^{-4}
SST4	119.09	99.76	14–4900	0.01–0.6	3.5×10^{-4}–1.6×10^{-3}
MFST1	19.78 ± 0.5	27.50	4–230	0.025–0.44	0.0028–0.01
MFST2	14.54 ± 0.3	29.92	9–230	0.075–0.52	0.0035–0.014
MFST3	10.06 ± 0.3	29.72	3–70	0.04–0.3	0.0065–0.02

the measured relative roughness is ε/D_i is 2.3%, 2.7%, 4.1%, and 5.9%, respectively. Some photos by a scanning electron microscope (SEM) are shown in Figure 12.1.

12.2.2 Rarefaction Effect

The Knudsen number, Kn, is defined by Equation 12.1 as

$$\mathrm{Kn} = \lambda/L \tag{12.1}$$

It is a direct indication on the flow regime of gases, where λ is the mean free path of molecules, and L is the characteristic length of the space wherein the gas exists. According to [16–18], there are four regimes of gases flow as defined by Equation 12.2:

$$\begin{aligned} &\mathrm{Kn} \le 0.001, \quad \text{continuum region} \\ &0.001 < \mathrm{Kn} < 0.1, \quad \text{velocity slip, temperature jump region} \\ &0.1 < \mathrm{Kn} < 10, \quad \text{transitional region} \\ &\mathrm{Kn} > 10, \quad \text{free molecule region} \end{aligned} \tag{12.2}$$

For Kn less than 0.001, the continuum assumption is valid for both governing equation and boundary conditions. When this condition is violated, a certain degree of tangential velocity slip in the streamwise direction must be considered [17–19]. The range of $0.001 < \mathrm{Kn} < 0.1$ is the slip flow regime, where the Navier–Stokes is valid, but slip boundary condition for velocity and temperature jump boundary conditions at the wall should be used. For $\mathrm{Kn} > 10$, the system can be considered as a free-molecular flow and direct solution of the Boltzmann equation or gas dynamic model such as DSMC is required. The intermediate region ($0.1 < \mathrm{Kn} < 10$) is the transitional flow regime. For numerical simulation of gases flow, the appropriate numerical methods for the noncontinuum regimes can be summarized as follows: the Navier–Stokes equation with slip boundary condition can be adopted up to $\mathrm{Kn} = 0.1$, the lattice Boltzmann method can be used up to Kn about 1, and in the range of Kn larger than about 0.1 the direct simulation of Monte Carlo (DSMC) can be used [20].

In the velocity slip region, the tangential slip velocity reduces the wall shear stress; hence, the pressured drop decreases and the mass flow rate increases at the same pressure drop. These are called rarefaction effect. To verify the rarefaction effect, experimental study was conducted in our group for microtubes MFST1, MFST2, and MFST3 as shown in Table 12.1 with smooth inner surface. It is well known that for the fully developed laminar flow in tubes, the product of Reynolds number and friction factor equals constant: $f\mathrm{Re} = 64$. Therefore, the ratio of $f\mathrm{Re}/64$ may be taken as an indication to show whether rarefaction effect plays a role: rarefaction reduces the friction factor hence this ratio will be less than one. The test results of helium and nitrogen in the three tubes are shown in Figure 12.2. In the test range, the tube outlet Knudsen numbers vary from 0.028 to 0.3. It can be clearly seen that friction factors are reduced significantly in the three microtubes, showing an appreciable rarefaction effect.

12.2.3 Compressibility Effect

As indicated earlier for the gas flow in microchannels, great changes in density of the gas occur because of the large pressure drop caused by the friction effects even when the flow velocity is small. The gas expands its volume and the local velocity increases. Thus, the gas pressure difference in the microchannel is made up of two parts: viscous friction and acceleration loss. This means that the total friction factor will be larger than the conventional value; hence, the product of Re and f will be larger than the constant 64 of the conventional duct.

FIGURE 12.1 SEM images of three typical tested tubes. (a) Fused silica microtube with D_h = 102 μm, (b) fused silica square microchannel with D_h = 52 μm, and (c) stainless steel microtube with D = 172 μm. (Reprinted from *Int. J. Heat Mass Transfer*, 50, Tang, G.H., Li, Z., He, Y.L., and Tao, W.Q., Experimental study of compressibility, roughness and rarefaction influences on microchannel flow, 2282–2295, Copyright 2007, with permission from Elsevier.)

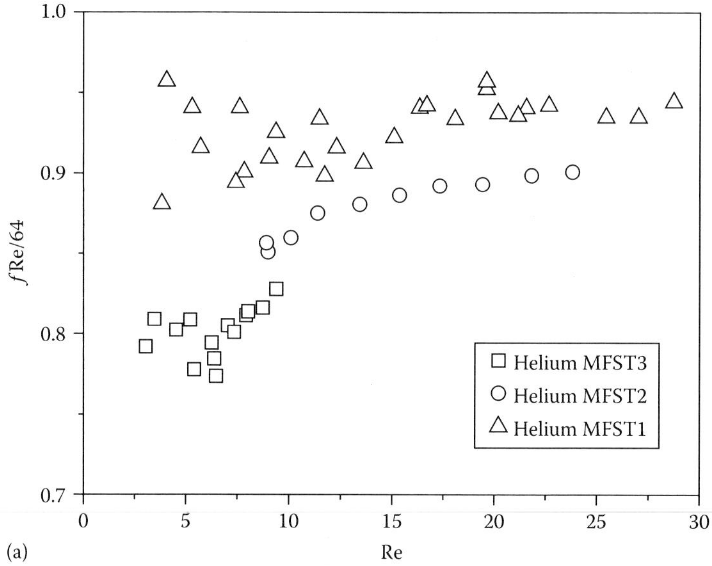

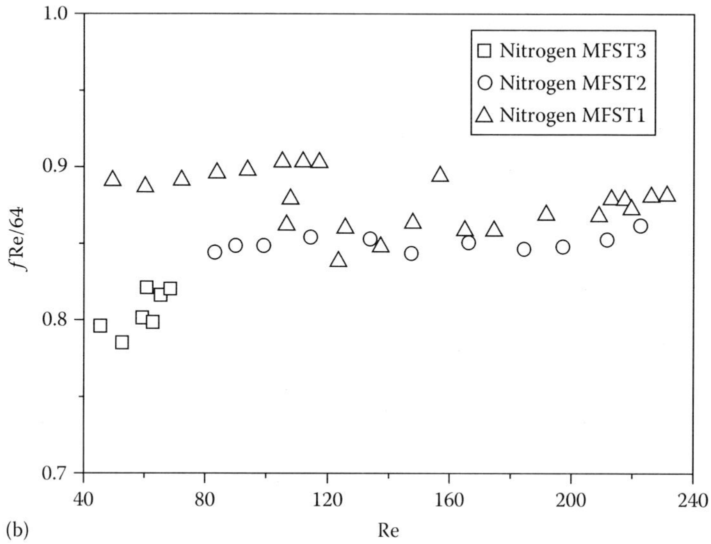

FIGURE 12.2 Test results showing rarefaction effect for (a) helium and (b) nitrogen. (Reprinted from *Int. J. Heat Mass Transfer*, 50, Tang, G.H., Li, Z., He, Y.L., and Tao, W.Q., Experimental study of compressibility, roughness and rarefaction influences on microchannel flow, 2282–2295, Copyright 2007, with permission from Elsevier.)

Figure 12.3 shows the measured compressibility effect on friction factor. The experimental value of *f*Re is normalized by the theoretical prediction and is plotted against mean Mach number of the channel inlet and outlet. Presented in the figures are all the experimental data in circular glass tubes and square glass channels for Reynolds number less than 2200. As the Mach number increases, the normalized value has a slight increase in the small Mach number region. The increase in the ratio accelerates when the mean Mach number is higher than 0.25. The test data in Figure 12.3 has been curved fitted by an equation, indicated by the line with closed circles, and compared with the solid line depicting the analytical solutions provided in [21]. It can be seen that the agreement between the analytical solution and the experimentally curved fitting is very good.

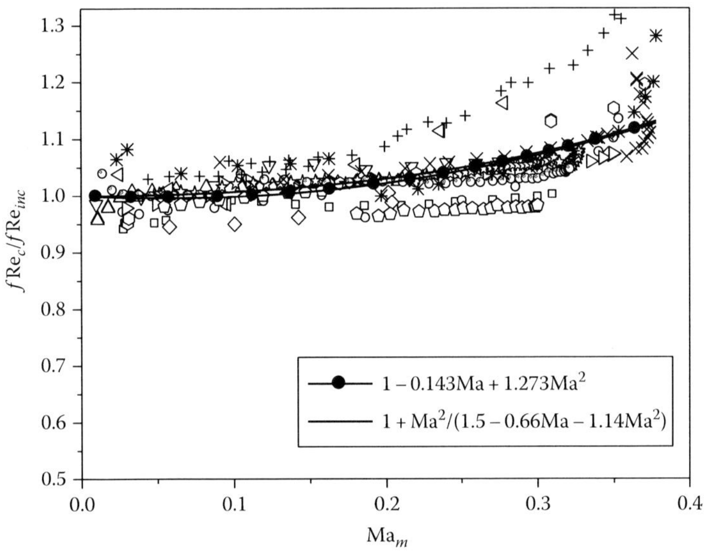

FIGURE 12.3 Test data showing compressibility effect. (Reprinted from *Int. J. Heat Mass Transfer*, 50, Tang, G.H., Li, Z., He, Y.L., and Tao, W.Q., Experimental study of compressibility, roughness, and rarefaction influences on microchannel flow, 2282–2295, Copyright 2007, with permission from Elsevier.)

12.2.4 Roughness Effect

The gas flow process in a microchannel is assumed to be isothermal. In fact, due to viscous dissipation effect, some gas temperature rise from channel inlet to outlet was observed and the temperature increase was less than 3°C in the whole test. The mean value of the measured inlet and outlet was used to determine fluid properties. For a one-dimensional, isothermal and compressible flow, the average friction factor f of a duct with length L and inlet and out let Mach numbers Ma_1 and Ma_2, respectively, is determined by the following equation [22]

$$f_c = \frac{D_h}{L}\left[\frac{\text{Ma}_2^2 - \text{Ma}_1^2}{\gamma \text{Ma}_1^2 \text{Ma}_2^2} - \ln\left(\frac{\text{Ma}_2}{\text{Ma}_1}\right)^2\right] \quad (12.3)$$

where γ is the gas-specific heat ratio.

It should be noted that the first term in the right part of Equation 12.3 represents the viscous effect, while the second one is the acceleration effect. The measured friction factors are partially presented in Figures 12.4 and 12.5. All the tested data were obtained with Kn less than 0.001. For incompressible fully developed laminar flow in a conventional-size circular tube, $f = 64/\text{Re}$. For turbulent flow Filonenko equation [23] of $f = (1.82 \log \text{Re} - 1.64)^{-2}$ is used for comparison. It can be clearly observed that for all the smooth tubes, the friction test data agree with the conventional correlation quite well, including the transition Reynolds number from laminar flow to turbulent flow; while for the stainless steel tubes, the test data are significantly higher than the conventional one. As indicated before for FST tubes, the relative surface roughness is less than 1% but for the SST it is larger than 2%. It is well known that for conventional tube or duct the relative roughness has effect on the friction factor only for the turbulent flow, but not for laminar flow [24]. Thus, this roughness effect on the laminar flow friction factor should be regarded as one feature of the microchannel fluid flow.

From the previous experimental results, some conclusions can be made for gaseous flow in microchannels as follows: (1) when Knudsen number is less than 0.001 and the relative roughness

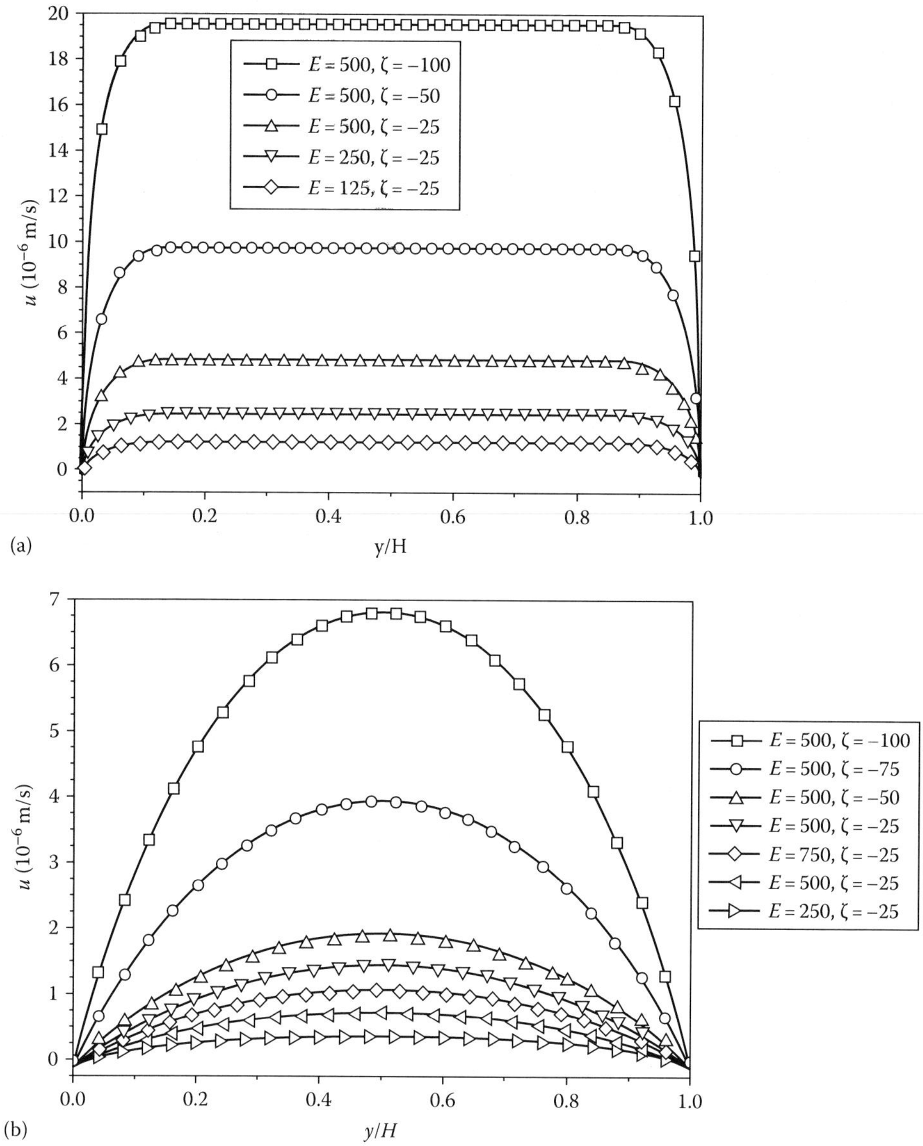

FIGURE 12.15 Streamwise velocity profiles of electro-osmotic flow (a) $c_\infty = 10^{-4}$mol/L and (b) $c_\infty = 10^{-7}$mol/L. (From Tang, G.H. et al., *J. Appl. Phys.*, 100, 094908, 2006. With permission.)

TABLE 12.2
Parameters of Two Stainless Steel Tubes

Tube	D_i/D_o (μm)	L (mm)	L_h (mm)	$\frac{D_o - D_1}{2D_1}$	ε_{ave} (μm)	ε_{ave}/D_h (%)	k W/(m K)	Q (W)
No. 4	1570/1810	300	270	7.6	15	0.95	16.3	12
No. 5	624.4/1000	300	270	30	9	1.4		9
No. 6	373/670	300	270	40	9	2.4		9

electron microscope (SEM) are shown in Figure 12.16. The mean temperature of the deionized water was used as the reference temperature to determine its thermophysical properties. For the details of the test system refer [34].

The heat transfer rate, Q, and heat flux, q, of the tube wall can be obtained from Equations 12.4 and 12.5:

$$Q = mc_p(T_{out} - T_{in}) \tag{12.4}$$

$$q = \frac{Q}{A_w} = \frac{mc_p(T_{out} - T_{in})}{\pi D_i L} \tag{12.5}$$

In addition, the measure heat transfer rate was evaluated by the electrical power input (Equation 12.6):

$$\Phi = U \cdot I \tag{12.6}$$

where
- m is the water mass flow rate
- T_{in}, T_{out} are the inlet and outlet water temperatures
- D_i and L are the tube inner diameter and length, respectively
- U and I are the input electrical voltage and current, respectively

It should be noted that for any data an appropriate heat balance should be made and the relative difference between Q, Φ should be less than 5%.

Then the local heat transfer coefficient and Nusselt number can be determined by Equations 12.7 through 12.10:

$$h_x = \frac{qL_x/L}{\Delta T_x} \tag{12.7}$$

$$\text{Nu}_x = \frac{mc_p(T_{out} - T_{in})D_i L_x/L}{A_w \Delta T_x k_l} \tag{12.8}$$

$$\Delta T_x = T_{w,x} - T_{b,x} \tag{12.9}$$

$$T_{b,x} = T_{in} + \frac{QL_x/L}{mc_p} \tag{12.10}$$

12.4.2 Numerical Simulation

The heat conduction in the tube wall and the fluid convective heat transfer are governed by the equations in axisymmetric cylindrical coordinates; hence, the problem is 2-D in nature. Figure 12.17 presents the computational domain. The following assumptions are adopted: (1) the fluid is incompressible with constant properties and the fluid flow is in steady state; (2) the flow is laminar; (3) the radiation heat transfer is neglected; (4) the body force is neglected; and (5) the electrokinetic phenomena are neglected. The Navier–Stokes and energy equations in the axisymmetrical coordinates are well documented in literature [39]; hence, they are neglected here for simplicity. Only the boundary conditions are described as follows: at the tube inlet, both

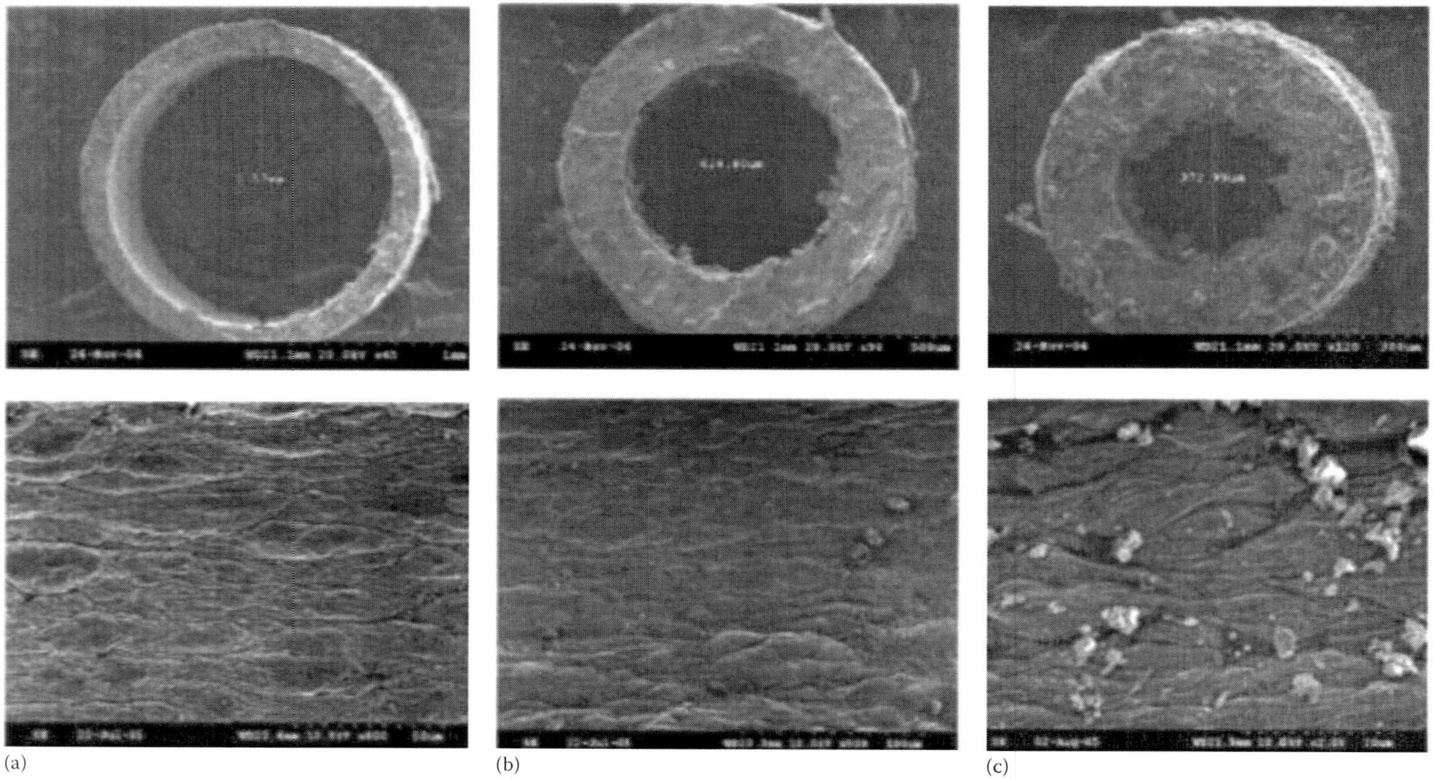

FIGURE 12.16 SEM images of the three tested tubes: (a) $D_i = 1570$ μm, (b) $D_i = 624.4$ μm, and (c) $D_i = 373$ μm. (From Li, Z. et al., *Int. J. Heat Mass Transfer*, 50, 3447, 2007. With permission.)

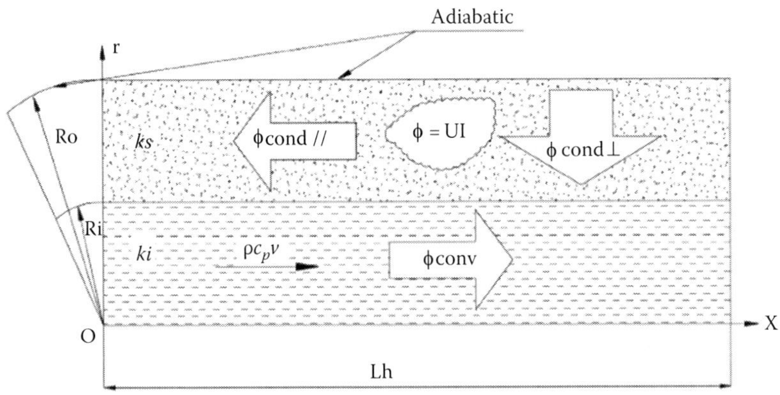

FIGURE 12.17 The domain for conjugated computation. (From Li, Z. et al., *Int. J. Heat Mass Transfer*, 50, 3447, 2007. With permission.)

temperature and velocity are given; at the tube outlet, fully developed flow and heat transfer are assumed; at the tube center, symmetric condition is adopted; at the outer surface of the tube, adiabatic condition is adopted; and in the tube wall, an inner heat source term is specified, which equals the total power input divided by the tube wall volume.

The finite volume method (FVM) [39,40] was used to discretize the governing equations and the SIMPLER algorithm was adopted to deal with the linkage between velocity and pressure. The grid system was 152 (x direction) and 32 (r direction). The relative mass unbalance in the entire domain less than 10^{-6} was taken as the convergence criterion. The problem is of conjugated type and the entire region (including the solid wall) was treated as the computational domain [39]. A special code was developed for the problem in the authors group.

12.4.3 Experimental and Numerical Results

12.4.3.1 Experimental Friction Factor

The friction factors in the fully developed region were measured, and the results show that the friction factor of tube No. 4 with surface roughness less than 1% agrees with the conventional prediction quite well while those of tubes No. 5 and 6 with surface roughness more than 1.5% are higher than that of conventional prediction. These results are consistent with our previous ones.

12.4.3.2 Experimental and Computational Heat Transfer Results

In the reduction of data for the local heat transfer, the local fluid bulk temperature in Equation 12.9 has been calculated by three methods. The variation of the local Nusselt number obtained by the three bulk temperatures are shown in Figure 12.18 for tubes 4 and 6.

The numerical actual local bulk temperature denoted by $T_{b,ave,num}$ as

$$T_{b,ave,num} = \frac{\sum_j \rho_l \cdot u(i,j) \cdot c_p \cdot T(i,j) \cdot R(j) \cdot \Delta r}{\sum_j \rho_l \cdot u(i,j) \cdot cp \cdot R(j) \cdot \Delta r} \tag{12.11}$$

The corresponding results are shown by triangles in the figure:

1. $T_{b,\text{int},num}$, numerical result by linear interpolation between inlet and outlet temperature (shown by circles).
2. $T_{b,\text{int},\text{exp}}$, experimental result by linear interpolation between inlet and outlet temperature (shown by solid squares).

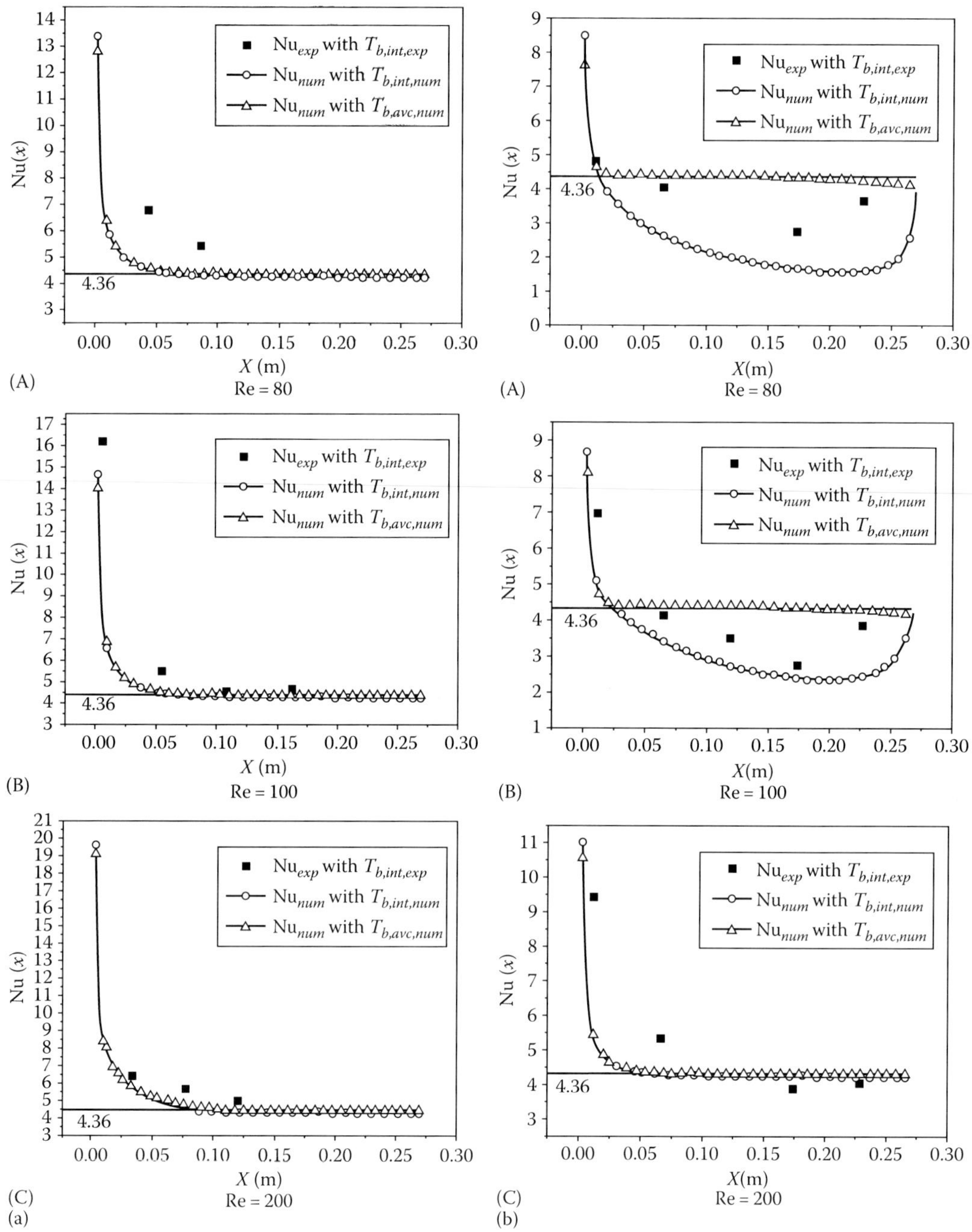

FIGURE 12.18 Variation of local Nusselt number of the two tubes: (a) D_i/D_o = 1570/1810(No. 4) and (b) D_i/D_o = 373/670(No. 6). (From Li, Z. et al., *Int. J. Heat Mass Transfer*, 50, 3447, 2007. With permission.)

It is well known that for the constant heat flux condition, the fully developed Nusselt number of laminar flow is 4.36 [41]. In the figure, this value is set as a comparison horizon. From the test results following features may be noted. (1) Difference in the determination of T_b leads to different local Nusselt number distribution. (2) For test tube No. 4, whose ratio of wall thickness to hydrodynamic diameter of tube is 0.076, the local Nusselt number variation curves of the three bulk temperature do not differ too much. With the increase in Reynolds number, the differences between the three curves become smaller and when Re = 200, three curves coincide with each other quite well. This may imply that the axial heat conduction of tube No. 4 does not make significant impact on the convective heat transfer characteristics. (3) For test tube No. 6, whose ratio of wall thickness to hydrodynamic diameter of tube is 0.4, the local Nusselt number variation curves of the three bulk temperature differ more significantly. However, the variation of these curves also exhibits such a common tendency: the higher the Reynolds number, the smaller the difference, and the local Nusselt number gradually approaches the constant value of 4.36 in the downstream part of the test tube. (4) For the three tubes and three Reynolds number, the local Nusselt variation trends coincide with the conventional results well when the $T_{b,ave,num}$ is taken for data reduction, while those by linear interpolation, $T_{b,\text{int},num}$ deviate from the conventional results significantly for tube No. 6. This implies that when the ratio of tube wall thickness over the diameter is high, the linear interpolation does not reflect actual situation. Hence, it is not appropriate for data reduction. This is possibly one of the reasons that some microchannel heat transfer data deviate from conventional results. (5) The larger the tube wall thickness over the tube inner diameter, the more significant the tube wall heat conduction; the larger the fluid Reynolds number, the weaker the effect of tube wall axial heat conduction; and the longer the distance from the tube inlet, the smaller the difference of the local Nusselt number from the conventional result (Equation 4.36). The results of tube No. 5 are very similar to those of No. 6.

The numerical results show that the local Nusselt number deviates from the conventional ones comes from two factors (1) the actual local heat flux is not equal to the nominal heat flux, shown by Equation 12.6 and (2) the local fluid bulk temperature is not equal to the linearly interpolated fluid temperature. Both of the two factors are related to the heat conduction in the tube wall. Thus, for the heat transfer in microchannels tube wall axial heat conduction plays an important role, and the effect of the axial heat conduction in the tube wall depends on the relative importance of heat conduction through the tube wall and the convective heat transfer at the tube surface.

To further reveal the effect of axial heat conduction for the cases studied, streamwise bulk temperature distributions are shown in Figure 12.19 for tube No. 6 with the highest ratio of wall thickness over diameter. In the figure, the three dashed lines represent the profiles obtained by numerically

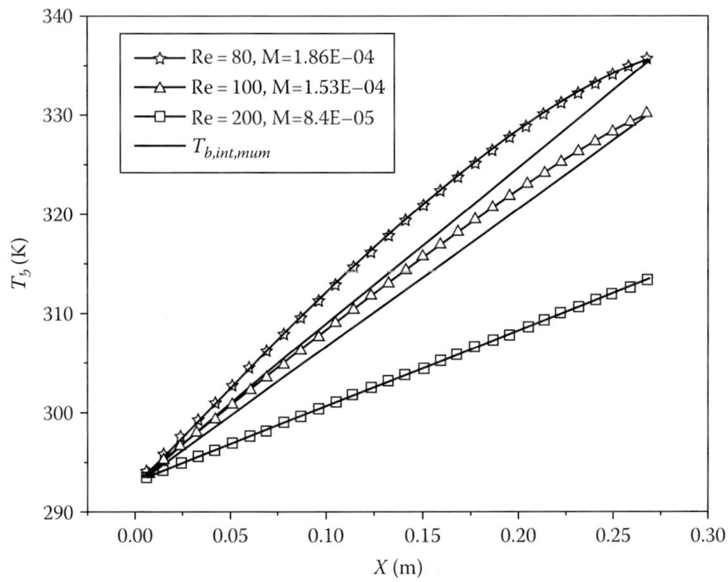

FIGURE 12.19 Fluid bulk temperature distributions. (From Li, Z. et al., *Int. J. Heat Mass Transfer*, 50, 3447, 2007. With permission.)

linear interpolation between the outlet and the inlet temperature, while the three shaped points are the numerically actual local bulk temperature calculated by Equation 12.11. For Re = 200, the actual bulk temperature coincides with the lineal interpolation, meaning that the inner surface being at uniform heat flux condition. With the decrease in Re the deviation between the distribution increases, and the actual bulk temperature becomes higher than that of the linear interpolation, except at the two ends. This implies that when the convective heat transfer is not strong enough, the heat conducts in the tube wall from tube exit to tube inlet, making the local inner surface heat flux nonuniform. Thus, it becomes clear that for the cases studied, where the tube wall has uniform inner heat source caused by electric heating, the role of the tube wall axial conduction is to make inner surface heat flux nonuniform. Only when the inner surface convective heat transfer is strong enough such effect becomes negligible.

12.5 NUMERICAL STUDIES OF SEVERAL SPECIAL CASES

In this section, examples of fluid flow and heat transfer in four microstructures are presented. The four structures are 3-D wavy channels, microporous medium, micronozzle, and microtubes with thick wall. The first two examples are simulated by FVM, and the last two examples are studied by LBM and DSMC method, respectively.

12.5.1 Liquid Flow and Heat Transfer in 3-D Wavy Microchannels

12.5.1.1 Physical Model and Numerical Methods

Wavy channel is an effective way for enhancing convective heat transfer and is easy to manufacture. Two different types of wavy channels were studied [42], and they are schematically shown in Figure 12.20: one with crests and troughs facing each other alternately by a phase of $\Phi = 180°$ (Figure 12.20a), and the other with crests and troughs facing each other, that is, $\Phi = 0°$ (Figure 12.20b). There are four parameters: wave amplitude (A), wavelength (λ), channel aspect ratio (α), and Re. The channel length was 20 mm with 2 mm straight sections at the inlet and the outlet. The wavy section spanned the middle 16 mm length of the channel with a hydraulic diameter of 500 μm. The Reynolds numbers considered are 50, 100, and 150. The 3-D forced convection flow and heat transfer were modeled using FLUENT. The assumptions made are the same as those in

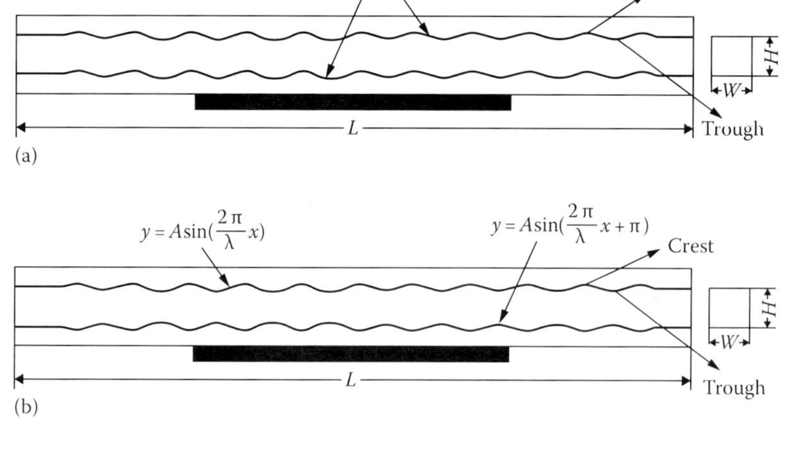

FIGURE 12.20 Two types of wavy channels (a) $\phi = 180°$ and (b) $\phi = 0°$. (From Gong, L. et al., *ASME J. Heat Transfer*, 133, 051702-1, 2011.)

Section 12.4.2. The boundary conditions are as follows: a uniform fluid inlet velocity was specified, along with the outlet pressure; a uniform fluid inlet temperature was specified and all the solid outer walls were assumed as insulated; and a constant wall heat flux condition was assumed at the channel center on the one side straight surface (the heavy black lines shown in Figure 12.20).

12.5.1.2 Major Numerical Results

For heat transfer enhancement study, performance evaluation is an important part to classify under what condition the enhanced structure can enhance heat transfer [43]. In this study, heat transfer enhancement under identical pumping power is adopted as the criterion, and the performance factor can be expressed by the following equation [43]:

$$PF = Nu_{enh}/Nu_{ref})/(f_{enh}/f_{ref})^{1/3} \quad (12.12)$$

where the subscripts *enh* and *ref* represent the enhanced results and the referenced results, respectively. For the cases studied, the straight duct is the referenced structure. Comparison of the numerical results shows that the wavy channel with $\Phi = 180°$ performs better than that of $\Phi = 0°$. Thus, in the following only the results of the former structure are presented.

12.5.1.2.1 Velocity Field and Local Nusselt Number Variation

Streamwise velocity vector and local Nusselt number distributions are provided in Figures 12.21 and 12.22, respectively. From Figure 12.21, it can be clearly seen that as the amplitude is increased, the velocity boundary is thinner due to the geometry governed biased shift of the maximum velocity from the centerline toward the wave peaks. This results in significantly enhanced thermal performance, which can be observed in Figure 12.22, where the local Nusselt number can be found to be larger for $A = 200$ μm near the wave peaks compared with $A = 50$ μm at the same axial locations. The highest ratio of surface areas for the wavy and straight channels for all the cases studied being about 1.05, far less than the Nusselt number ratio with the highest being about 1.52. This implies that increased surface area provides only a minor contribution to heat transfer improvement, but it is the boundary layer thinning which is more important.

12.5.1.2.2 Effects of Parameters on Performance

The effects of the three parameters on the heat transfer performance were studied, and Figure 12.23 shows the effect of the amplitude on PF as an example. It can be observed that as large as about 50% enhancement can be obtained under the identical pumping power constraint. Hence, microwavy channel with an appropriate design is an efficient tool for enhancing heat transfer.

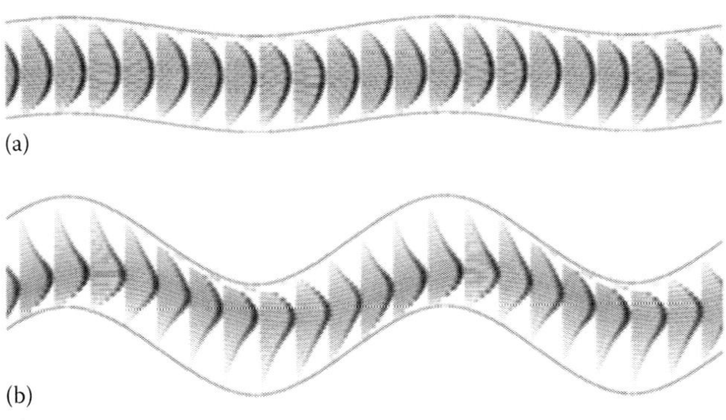

FIGURE 12.21 Velocity vector distributions (Re = 150): (a) $A = 50$ μm and (b) $A = 200$ μm. (From Gong, L. et al., *ASME J. Heat Transfer*, 133, 051702-1, 2011.)

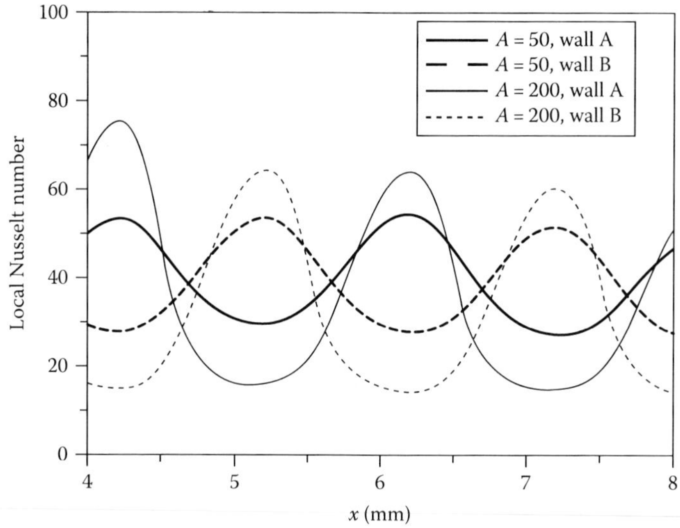

FIGURE 12.22 Local Nusselt number distribution (Re = 150). (From Gong, L. et al., *ASME J. Heat Transfer*, 133, 051702-1, 2011.)

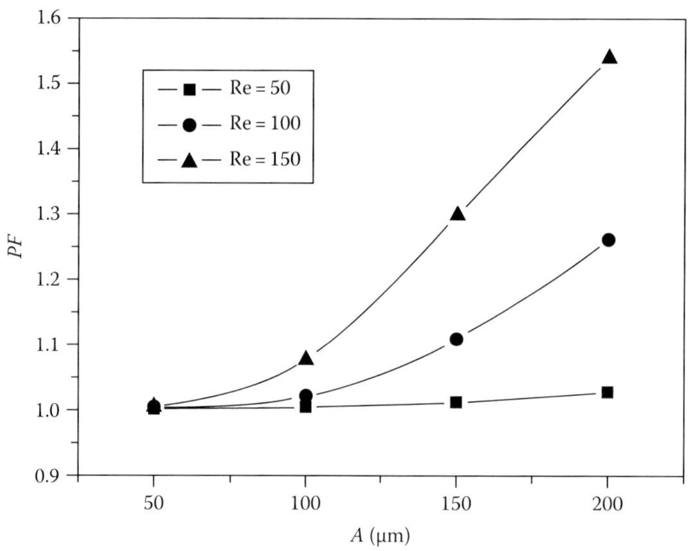

FIGURE 12.23 Effect of amplitude. (From Gong, L. et al., *ASME J. Heat Transfer*, 133, 051702-1, 2011.)

12.5.2 Simultaneously Developing Flow and Heat Transfer in Microtubes with Thick Wall and Constant Outside Surface Temperature

In Section 12.4.3, we have seen that for a microtube heated by electric current through its wall, the role of the axial heat conduction is to rearrange the tube inner surface heat flux from otherwise uniform to nonuniform. We may ask that if we change the boundary condition, say given constant outside wall temperature, what will be the role of the axial heat conduction for a microtube with a thick wall? This problem had been numerically studied in [44], and its major results are presented here.

12.5.2.1 Physical Model and Numerical Methods

The cross section of a thick wall tube is shown in Figure 12.24. Partial boundary conditions are also shown in the figure. The central line is symmetric and the flow outlet is assumed to be fully developed. The tube wall thermal conductivity is k_s, and that of the fluid is k_f. The basic assumptions for the problem are the same as those presented in Section 12.4.3; therefore, they are neglected here. The whole region (including the solid wall) is treated as the computational domain. The conduction in the wall and convection in the fluid are solved simultaneously; hence, computation is of conjugated type. The total length of x direction is taken as five times of the entrance length theoretically determined for the tube of conventional size. The grid-independent examination shows that satisfactory results can be obtained by using a grid spacing number of 16–62 in the r direction with 16 in the fluid region and 0–46 in the solid according to the wall thickness. A total number of grid spacing in the x direction is determined according to the ratio of length to width being taken, in which the grid spacing is ranged from one to four times of the grid spacing in the radial direction.

The problem has four effective parameters, Re, Pr, ratio of solid/fluid thermal conductivity, denoted by k_{sf}, and the ratio of wall thickness over tube inner radius, denoted by δ/r_i. The variation range of the four parameters is shown in Table 12.3. A self-developed code was used to do simulation.

12.5.2.2 Major Numerical Results

The major findings of this study are as follows:

1. Under the conditions of RePr $\geq$ 30, $k_{sf} \geq 25$, the effects of the ratio of wall thickness over tube radius on the local Nusselt number, the thermal entrance length and asymptotic Nusselt number can be neglected.
   The results of the predicted thermal entrance length and the asymptotic Nusselt number are in good agreement with the results of convective heat transfer in a conventional tube with uniform inner wall temperature.

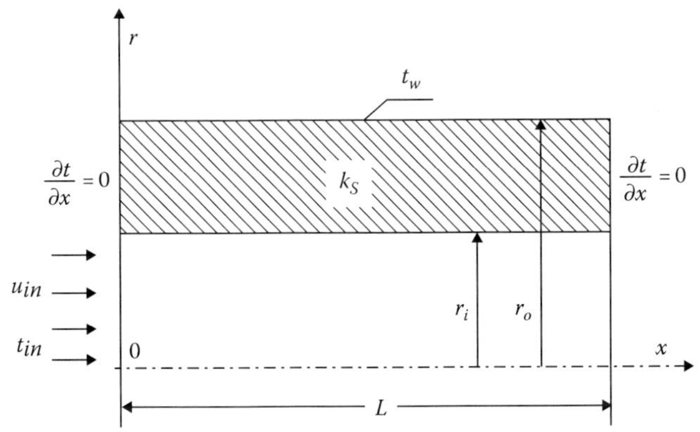

FIGURE 12.24 Physical model. (From Zhang, S.X. et al., *Int. J. Heat Mass Transfer*, 53, 3977, 2010. With permission.)

TABLE 12.3
Parameters of Numerical Simulations

Re	Pr	k_{sf}	δ/R_i
10, 30, 50, 100, 200, 2000, 2280	1, 3, 5, 7	1, 10, 15, 25, 50, 100, 300, 500, 664, 10,000	0, 0.08, 0.42, 0.84, 1.69, 2.53, 3.37, 3.79

2. If RePr $\geq$ 30 but k_{sf} < 25, the effects of wall thickness over tube radius on the local Nusselt number, the thermal entrance length and the asymptotic Nusselt number should be considered. As δ/R_i increases, that is, when the axial conduction becomes more important, the thermal entrance length and asymptotic Nusselt number increase and are larger than those of convective heat transfer in a conventional tube with uniform inner wall temperature. The maximum asymptotic Nusselt numbers numerically predicted is about 4.3. For the same three cases but with k_{sf} = 1, the streamwise variations of the local Nusselt number are presented in Figure 12.25, where it can be seen that for the most thick situation (δ/R_i = 3.79), the asymptotic value of Nusselt number is about 4.3; it is very likely that as the value of δ/R_i further increases and the value of k_{sf} further decreases, the heat transfer in a thick-walled tube with constant outside-wall temperature undergoes the transformation of the inner surface thermal boundary condition from uniform temperature to uniform heat flux.

Finally from the two examples of axial heat conduction effect presented in Section 12.4.3 and the present section, one may conclude that the effect of tube axial heat conduction is problem dependent. It may be described as follows: the role of the tube axial heat conduction is to change the inner surface thermal boundary condition which the tube otherwise would have. For example, for the situation shown in Section 12.4.3, if the axial heat conduction can be neglected, the tube inner surface would have uniform heal flux condition, while the axial heat conduction makes the inner surface heat flux nonuniform; For the situation studied in this section, if the axial heat conduction can be neglected, the inner tube wall would have uniform wall temperature condition, while the tube axial heat conduction makes it approaching uniform heat flux condition.

12.5.3 Gaseous Flow in Microscale Porous Media

In this section, the LBM is adopted to simulate 3-D fluid flow in a simplified porous medium model, and some useful conclusions can be obtained [45].

12.5.3.1 Physical Model and Numerical Method

A simplified microscale 3-D porous medium is shown in Figure 12.26, which consists of 30 identical cubes with a dimension of 1.2 μm and is included in the domain of $H = W = L$ = 8.4 μm. The whole domain is divided into 112 × 112 × 112 cubic lattices in the x, y, and z directions. The simulations were conducted for four gases: hydrogen (H_2), nitrogen (N_2), helium (He), and carbon dioxide (CO_2). Their Knudsen numbers are equal to 0.1, 0.055, 0.16, and 0.0236, respectively.

12.5.3.2 Major Numerical Results

The predicted results of the velocity vectors at Z = 0.5 plane for the carbon dioxide and nitrogen are presented in Figures 12.27 and 12.28, respectively. In the two figures, the length of the vectors stands for the magnitude of the velocity, and the length is normalized with an equal scale for both figures. It can be clearly observed from the figure that the fluid detours the cubes and goes through open space. By carefully expecting the velocity distributions between two neighboring cubes in the two figures, it can be found that in Figure 12.28 the velocity profile of nitrogen with Kn = 0.55 is flatter than that of carbon dioxide (Figure 12.27, Kn = 0.0236). This is because a larger Knudsen number leads to a larger rarefaction influence.

One important characteristic parameter of porous medium is its permeability [46–48]. It describes the capability of a porous medium to transport a fluid through it at a given pressure drop. For a given kind of gas but at different pressure, the go-through capability should be different. Obviously, the gas at the velocity slip regime will go through more easily than that at the continuum

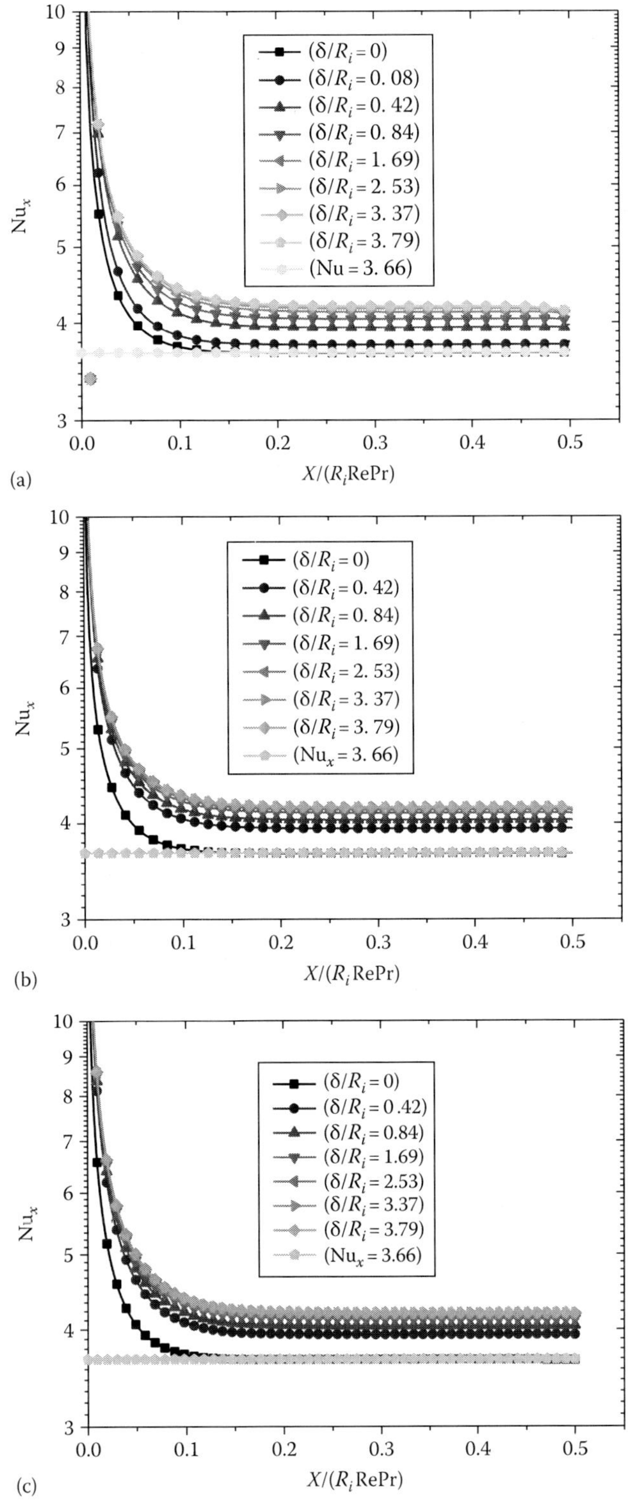

FIGURE 12.25 Streamwise variation of local Nusselt number for $k_{sf} = 1$. (a) Re = 50, Pr = 1.0, $k_{sf} = 1$, (b) Re = 50, Pr = 7.0, $k_{sf} = 1$, and (c) Re = 2280, Pr = 1.0, $k_{sf} = 1$. (From Zhang, S.X. et al., *Int. J. Heat Mass Transfer*, 53, 3977, 2010. With permission.)

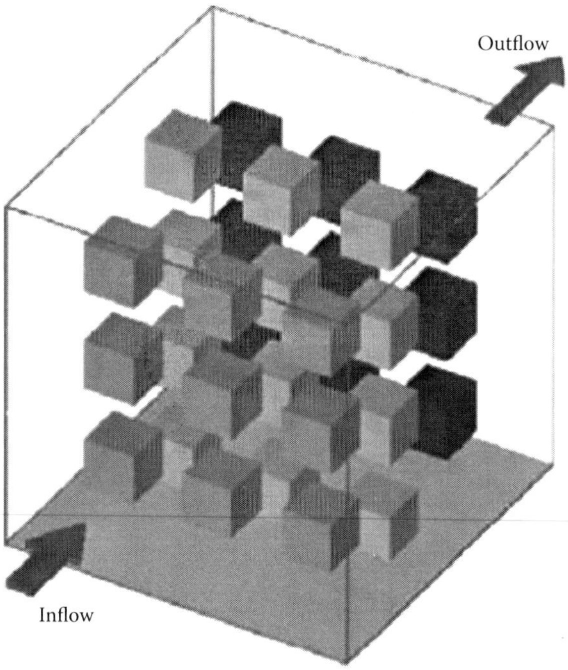

FIGURE 12.26 A simplified microscale porous medium model. (From Tang, G.H. et al., *J. Appl. Phys.*, 97, 104918, 2005. With permission.)

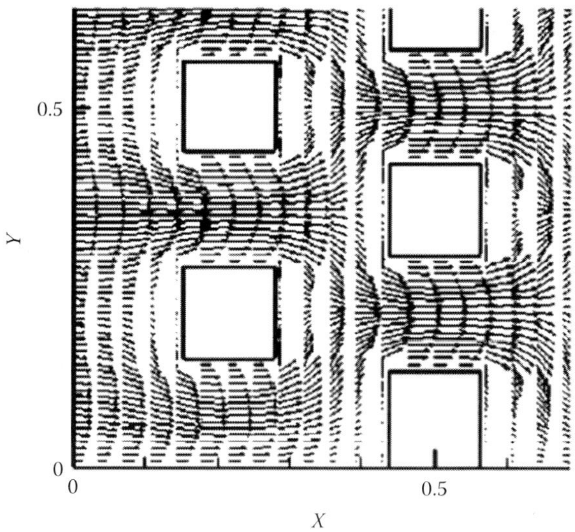

FIGURE 12.27 Velocity vectors on $Z = 0.5$ plane for carbon dioxide. (From Tang, G.H. et al., *J. Appl. Phys.*, 97, 104918, 2005. With permission.)

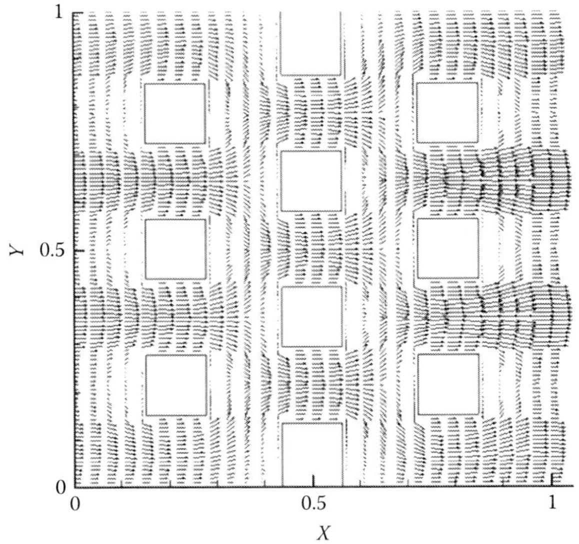

FIGURE 12.28 Velocity vectors on $Z = 0.5$ plane for nitrogen. (From Tang, G.H. et al., *J. Appl. Phys.*, 97, 104918, 2005. With permission.)

regime. Considering the gas slip effect, Klinkenberg [46] proposed the following gas permeability equation for flow through a porous medium:

$$k_g = k_\infty \left(1 + \frac{b}{\bar{p}}\right) \quad (12.13)$$

where
$\bar{p}$ is the average pressure of the inlet and outlet
k_∞ is the permeability unaffected by the velocity slip

Coefficient b should be larger than zero to reflect the positive effect of velocity slip on the gas permeability. Equation 12.13 has been verified by our LBM results. For the details refer [45].

12.5.4 Gas Flow and Heat Transfer in Micronozzle

12.5.4.1 Physical and Numerical Models

The micronozzle is used to accurately control satellites altitude and trajectory. Recently studies on small satellites or microsatellites have been developed very fast for the advantage of small size, low cost, and timesaving in manufacturing. Free molecular microelectrothermal resist jet (FMMR) is one kind of the micronozzle. The configuration of a typical FMMR is shown in Figure 12.29 [49]. The working principles of FMMR are explained here. The inlet propellant of low temperature flows over the heating surface, and the propellant molecules knock with heating surface. After the collision, macroscopically the temperature of propellant raises meaning that the microscopic velocity of molecules increases. The propellant then rushes out of the nozzle with high speed and thrust is

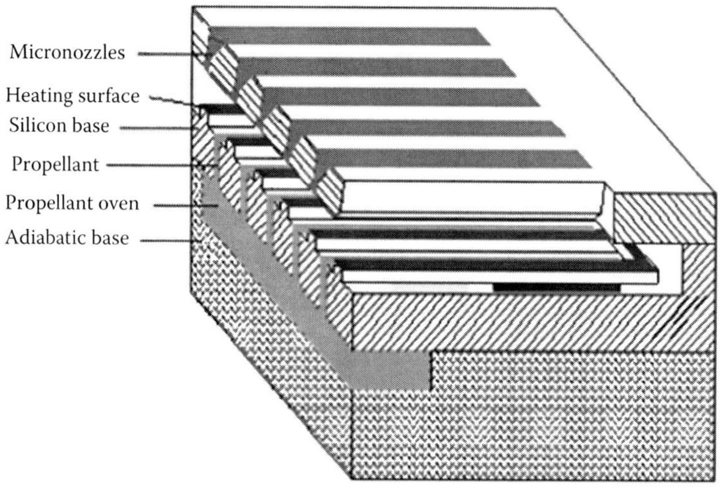

FIGURE 12.29 Configuration of FMMR. (From Sun, Z.X. et al., *Microfluid Nanofluid*, 7, 621, 2009. With permission.)

generated. For the purpose of enlarging collision rate between molecules and the heating surface while reducing collision rate between molecules, the dimension of nozzle should be in the order of the mean free path of the propellant molecules so that the rarefaction effect in the nozzle flow field becomes significant.

Usually the Knudsen number at the nozzle is greater than 0.1 and the flow at the nozzle falls within the scope of transition flow or free molecule flow. Navier–Stokes equation becomes invalid in the transition and free molecule regime. DSMC method developed by Bird [50] is the most useful method to simulate the flow field in these regimes, and it is adopted in [49].

A DSMC–FVM-coupled simulation is conducted in [49]. The temperature of solid area is calculated with the finite volume method based on energy equation and the field of flow area is simulated by DSMC method. The temperatures of solid area and flow area are coupled together at the solid–fluid interface. As shown in Figure 12.29, several micronozzles are connected in parallel to increase the thrust and each nozzle can be regarded identical. Thus, only one nozzle is adopted for simulation. The cross-sectional dimensions of the nozzle are much less than the axial length; hence, a 2-D model is adopted as shown in Figure 12.30. Because of the symmetry of the structure, only half of the cross section is used for computation. The structure of the nozzle cross section is quite complex so unstructured grid system is utilized in this simulation to perfectly simulate the tilted solid wall of the nozzle.

In order to obtain accurate results by DSMC, the cell dimensions should be less than the mean free path and the time step must be less than mean free time between two collisions. In addition, at least seven particles should be placed in a cell [51]. All these points were kept in the simulation. The gas–wall interface boundary condition was treated by a coupled method: the Dirichlet boundary conditions, that is, the surface temperature has to be preset before DSMC process for the fluid region and the Neumann boundary condition, that is, the flux at the surface can be obtained from the DSMC simulation. This heat flux is then used as the input at the interface for the solid wall heat conduction simulation. Such Dirichlet–Neumann transferring method is used to couple the temperature and heat flux at the fluid–solid interface boundary. The propellant used is argon and the material of solid area is silicon with the thermal conductivity of 149 W/(m·K). The variable hard sphere (VHS) model [50] is used. The working conditions used in simulation is listed in Table 12.4. Details of DSMC simulation can be found in [49–51] and are not described here.

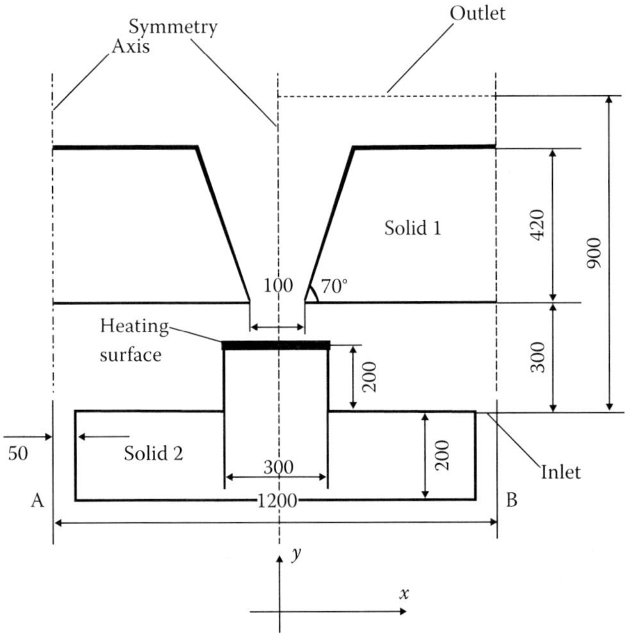

FIGURE 12.30 Cross section details of FMMR (unit: μm). (From Sun, Z.X. et al., *Microfluid Nanofluid*, 7, 621, 2009. With permission.)

TABLE 12.4
Working Conditions Used in This Simulation

Case	1	2	3	4	5
Inlet pressure (Pa)	500	1,000	2,000	5,000	10,000
Outlet pressure (Pa)	0	0	0	0	0
Inlet temperature (K)	300	300	300	300	300

12.5.4.2 Major Numerical Results

The streamlines, temperature contours of the solid regions and fluid part for Case 3 are taken as example, and presented in Figures 12.31 through 12.33, respectively. From the three figures, following features may be noted. First, as expected, propellant comes into the field from inlet and then flows through the throat. Such general flow picture is the same for all the five cases studied. Second, the temperature difference within the solid regions 1 and 2 is very small due to low density of the propellant and high thermal conductivity of silicon.

So a "general temperature" may be assumed for each solid area. Third, along the flow path the fluid temperature increases first due to the heat transfer between solid area and propellant, then decreases sharply in the region of the nozzle throat, and at the nozzle exit the fluid temperature reduces to the lowest value.

The predicted results show that the simulated solid wall pressure varies with the inlet temperature significantly. With the same other given conditions but different prespecified inlet pressure, the mass flow rate and the specific impulse change appreciably. Previous studies often prespecified the wall temperature, without coupling between fluid flow in nozzle and wall heat conduction. From our results, it can be concluded that the prespecified wall temperature method is not feasible and the

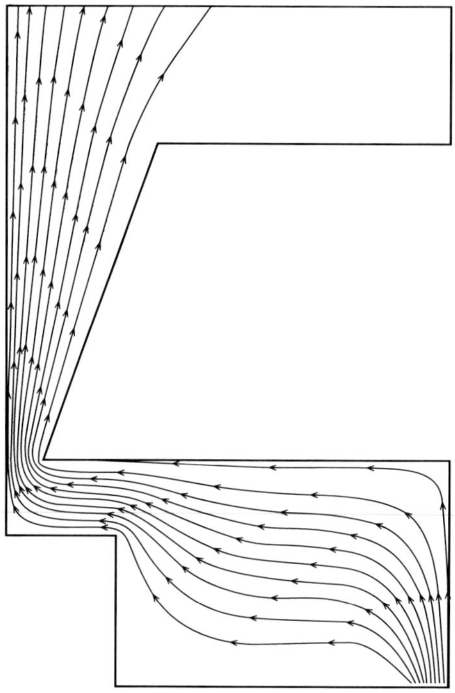

FIGURE 12.31 Stream lines of flow field. (From Sun, Z.X. et al., *Microfluid Nanofluid*, 7, 621, 2009. With permission.)

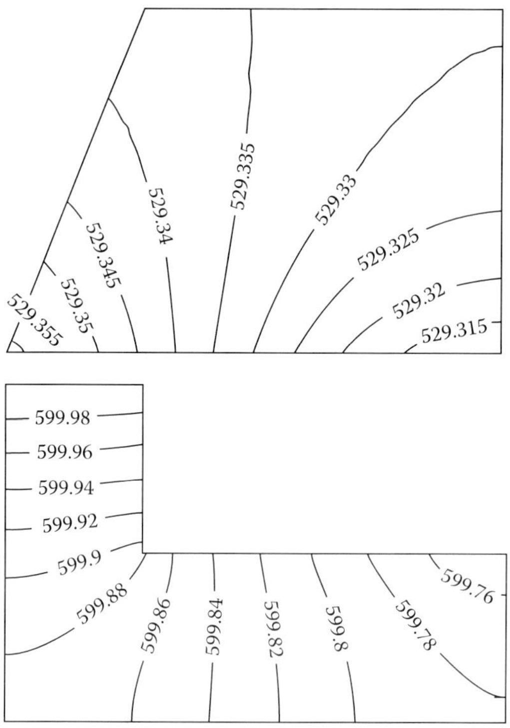

FIGURE 12.32 Solid temperature distribution (unit: K). (From Sun, Z.X. et al., *Microfluid Nanofluid*, 7, 621, 2009. With permission.)

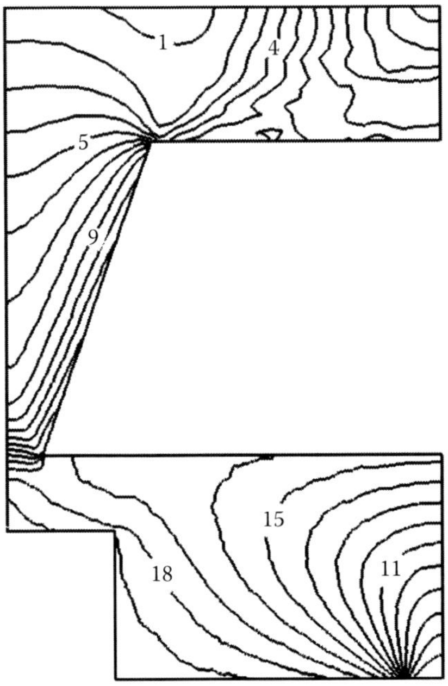

Level	T (K)
19	580
18	560
17	540
16	520
15	500
14	480
13	460
12	440
11	420
10	400
9	380
8	360
7	340
6	320
5	300
4	280
3	260
2	240
1	220

FIGURE 12.33 Fluid temperature distribution (unit: K). (From Sun, Z.X. et al., *Microfluid Nanofluid*, 7, 621, 2009. With permission.)

DSMC–FVM-coupled simulation is necessary for performance prediction of the low-temperature micronozzle.

12.6 APPLICATION EXAMPLE: DESIGN OF A WATER-COOLED MICROCHANNEL HEAT SINK WITH HIGHLY UNIFORM SURFACE TEMPERATURE

With a continuously increasing integration of electronic circuit, thermal management rises to be a great challenge in the design of electronic devices. Micro/minichannel cooler has been considered as an effective heat removal tool [52,53]. Uniform temperature distribution is highly required to obtain high performance and reliable stability of many electronic devices. Usually, uniform temperature distribution can be obtained by appropriately designing the flow field. The typical flow headers are U-type or Z-type. In [53] a micro/minichannel heat sink has been numerically designed in conjunction with thermal resistance analysis for which the nonuniformity of temperature distribution could be controlled about 2 K for a high heat flux of 2 MW m^{-2}. The major results are presented in this section.

12.6.1 Physical and Mathematical Model

A pictorial view of the micro/minichannel heat sink and its one unit is outlined in Figure 12.34. The heat generated in the heating surface conducts through the substrate, and one part dissipates to the liquid through the bottom of the channel, and the rest transfers into the fin and is brought away by the cooling fluid. One unit can be regarded as the representative and was studied in [53]. In order to simulate the fluid flow and heat transfer characteristics, the following assumptions were made: (1) natural convection within the channels is neglected; (2) the flow and heat transfer are steady and in turbulent regime; and (3) all the lateral walls are adiabatic. The governing equations are 3-D continuity, momentum, and energy equations together with the standard k–ε model in Cartesian

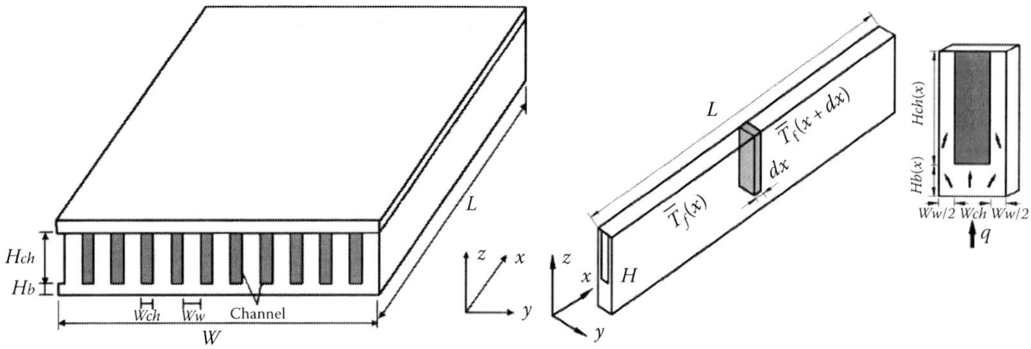

FIGURE 12.34 Micro/minichannel heat sink. (a) Schematic view of the micro/mini-channel heat sink and (b) one unit model. (From Mu, Y.T. et al., *Int. J. Heat Mass Transfer*, 85, 147, 2015. With permission.)

coordinates, which are well documented in literature [39] and omitted here for simplicity. The related boundary conditions are: symmetry condition for the two lateral walls, no-slip condition for all velocities on the solid walls, $u = U_{in}$ for the entrance of the distributor, and local one-way condition for the exit [39].

Fluid temperature $T_{in} = 293$ K for the entrance, $q = 2$ MW·m^{-2} at the substrate bottom surface. The material of the substrate is pure copper, and the cooling medium is water. Since the heat flux is very large, the streamwise temperature variation of the fluid is obvious; thus, the water properties varying with temperature are taken into account.

12.6.2 Numerical and Thermal Resistance Analysis Methods

Finite volume method is used to discretize the governing equations, and FLUENT software is used for simulation. Grid-independence tests have been performed, and the final control volumes are about 4.0 million.

In order to get an appropriate streamwise channel shape, a detailed analysis of the thermal resistance from the heating element to the heat sink surface was conducted in [53]. The major conclusion is that increasing the cooling channel height streamwisely is helpful to increase the temperature uniformity of the heat sink. Thus, the variable channel height technique was proposed in [53] to improve heat sink temperature uniformity (Figure 12.35).

12.6.3 Results and Analysis

The major results of [53] are presented here.

12.6.3.1 Flow Head Comparison

The temperature nonuniformity and thermal resistance of the heating surface are highly influenced by the uniformity of flow distribution. Among the four heads (distributors) studied, the

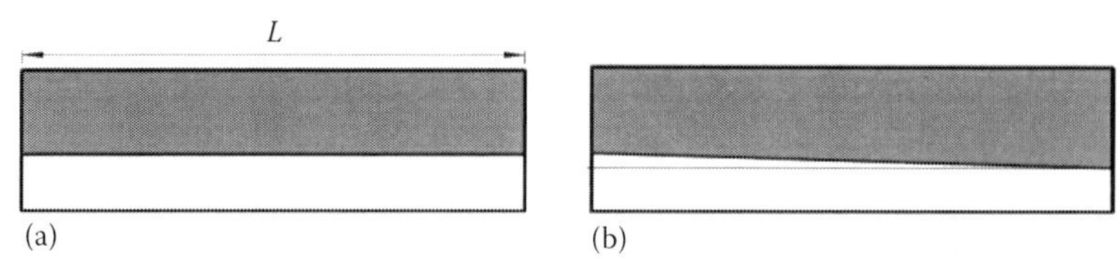

FIGURE 12.35 Channels with constant and variable channel height. (a) Constant height and (b) variable height. (From Mu, Y.T. et al., *Int. J. Heat Mass Transfer*, 85, 147, 2015. With permission.)

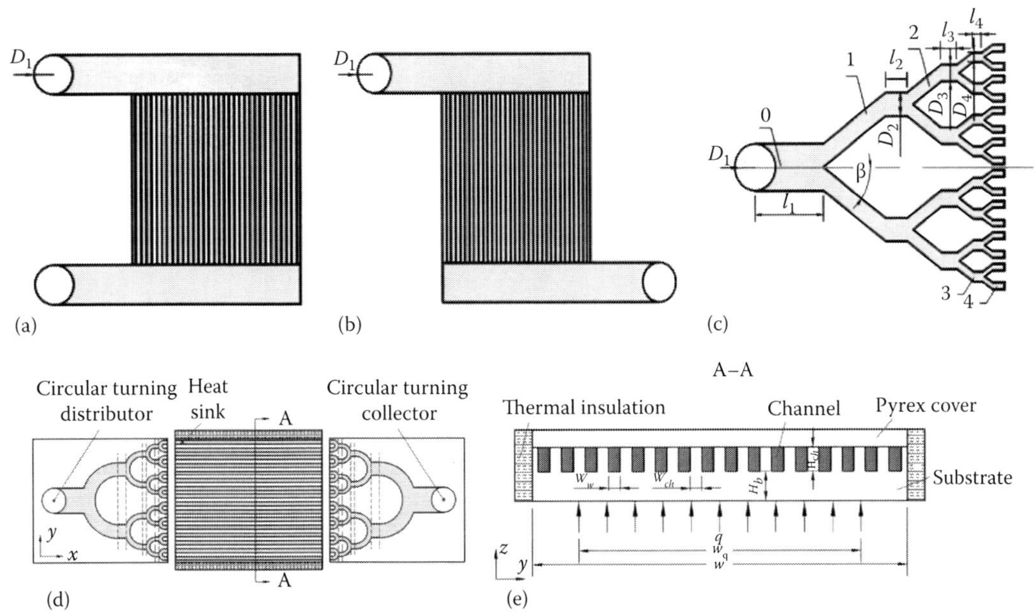

FIGURE 12.36 Heat sink with different channel head configurations. (a) U-type, (b) Z-type, (c) tree type, (d) circular turning type (top view), and (e) A–A view. (From Mu, Y.T. et al., *Int. J. Heat Mass Transfer*, 85, 147, 2015. With permission.)

circular turning one (Figure 12.36) performs the best. Following results are only for the heat sink with circular turning type distributor.

12.6.3.2 Effect of Aspect Ratio of Channel on the Temperature Uniformity

The temperature nonuniformity and thermal resistance of the heating surface decrease with the aspect ratio. The temperature nonuniformity drops from 7.0 to 3.8 K with the aspect ratio ranging from 2 to 7 when the inlet velocity equals 4 $m \cdot s^{-1}$.

12.6.3.3 Effect of Heating Area

The dimensions of the heat sink should be the same as the heating surface, for which the temperature difference $\Delta T = T_{max} - T_{min}$ and thermal resistance θ is the smallest for the same other conditions.

12.6.3.4 Effect of Channel Variable Height

The effect of streamwise variation of channel height is depicted in Figure 12.37. In the figure, a control coefficient α is introduced. $\alpha = 0$ corresponds to the conventional channels with uniform height. The larger the coefficient, the bigger the streamwise variation of channel height. When α ranges from 1 to 1.5, the temperature nonuniformity reaches the lowest value under different velocities. For example, when the inlet velocity is 4 $m \cdot s^{-1}$ and the channel height decreases linearly, the temperature nonuniformity drops from 4.75 to 1.29 K for the control coefficient α ranging from 0 to 1.5. As shown in the figure, the temperature nonuniformity of heat sink with conventional channel is 4.75 K when inlet velocity equals 4 $m \cdot s^{-1}$, while those of the novel channels are quite small with the value of 0.92–1.44 K for the case of $\alpha = 1$. The related pumping power and pressure drop are 4.28 W and 32 kPa, respectively, which are acceptable.

Therefore, the three novel heat sinks with variable height channel in conjunction with the circular turning type distributor can be applicable to such special cases for which the high heat flux (2 $MW \cdot m^{-2}$) and small temperature nonuniformity (less than 1 K) are demanded.

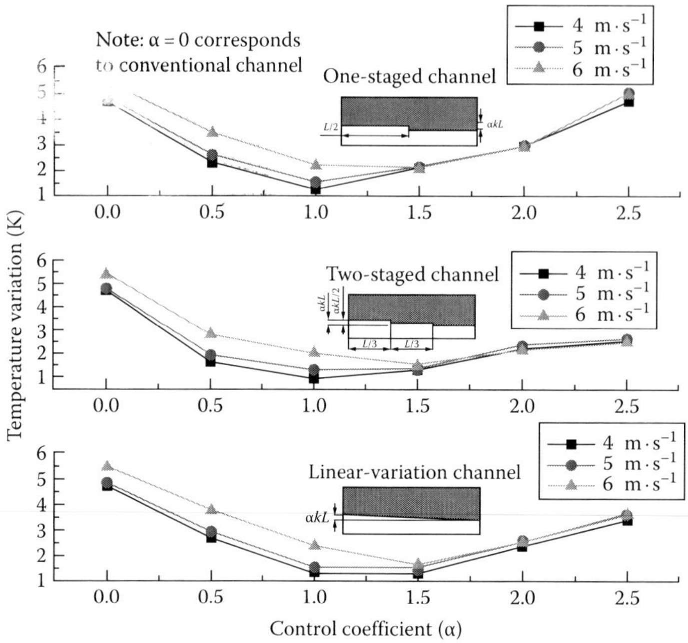

FIGURE 12.37 Effect of channel streamwise variation of height. (From Mu, Y.T. et al., *Int. J. Heat Mass Transfer*, 85, 147, 2015. With permission.)

12.7 BRIEF SUMMARY

In this chapter, we presented the results of our experimental and numerical investigations for gas/liquid flow and heat transfer in microchannels/nozzle with different shapes and different dimensions. From the broad existing literature and our results, some important conclusions are made. This section consists of two parts. First, some general conclusions are provided about microscale single-phase fluid flow and heat transfer, and the size effects on microscale fluid flow and heat transfer are summarized. Second, measurement uncertainties in microscale fluid flow and heat transfer are analyzed, which may partially account for the test data diversity and discrepancy in the early stage of development of microscale fluid flow and heat transfer.

12.7.1 General Conclusions on Single-Phase Microscale Heat Transfer

For microscale channel with dimensions below 1 mm to about several micrometers, except the size effects described later, the conventional macroscale correlations for fluid friction factor and heat transfer and the transition Reynolds number from laminar to turbulent can still be applied.

The so-called size effects include two aspects: the first is the ratio of surface area over volume. This ratio becomes larger and larger with the decrease in channel dimension; hence, those factors related to surface area will become more important or be affected. The friction factor, the friction-induced gas compressibility, and the role of EDL of liquid belong to this aspect. The second is the ratio of length over length. The effect of the axial heat conduction becomes significant when the ratio of wall thickness over tube diameter (radius) becomes large. In addition, if the ratio of the mean free path of gas molecule over the characteristic length of channel, that is, if the Knudsen number is larger than 0.001, the gas rarefaction should be taken into account. Thus, following five size effects should be noted.

1. *Friction factor*: For the fully developed laminar flow in microchannels of gas and liquid if the relative surface roughness is larger than 1%–1.5%, the friction factor will be larger than macroscale correlation 64/Re, while for the macroscale laminar flow surface roughness does not play a role.
2. *Gas compressibility*: For microchannel gas flow because of great pressure drop induced by the large surface friction, the compressibility often should be taken into account when the tube average Ma number is larger than 0.25. This effect tends to increase gas pressure drop.
3. *Gas rarefaction*: When gas is rarefied, velocity slip and temperature jump occur at wall, which tend to decrease pressure drop. But heat transfer may be enhanced or deteriorated compared with the continuum region, depending on the accommodation coefficients, which were not mentioned in this chapter. The accommodation coefficients describe the gas–surface interactions. Its value ranges from near zero (corresponding to specular reflection) to 1 (corresponding to diffuse scattering). Interested readers can refer [54,55].
4. *Axial heat conduction*: We presented two examples of the axial heat conduction effect. It seems that the role of axial heat conduction tends to change the channel inner surface thermal boundary condition, which the inner surface otherwise would have. For example, for given wall outside constant temperature condition, the role of tube axial conduction tends to change the inner surface being at uniform heat flux.
5. *Role of EDL*: If some electrolyte, say water with ions, is flowing in a microtube made by dielectric materials, such as silicon, the electric double layer will play a role if tube dimension is less than 1 μm. But if tube dimension is in the order of several microns, the electrokinetic effects, such as electroviscous effect, may be neglected.

12.7.2 Inadequacies in Measurement and Data Reduction for Single-Phase Microscale Heat Transfer

Finally discussion is made for the inadequacies in the measurement and data reduction for single-phase microscale flow and heat transfer, which may, at least partially, account for the diversity and discrepancy of results in the early development stage. These include the following:

1. Measurement errors of the inner size of the microchannel. An accurate measurement of microchannel size is of significant importance, because $\mathrm{Re} \sim D_i^1$, $u \sim D_i^2$, $f \sim D_i^5$.
2. Developing entrance effect. If the flow is not fully developed when the test microchannel is not long enough, data reduction by fully developed equation will leads to a wrong result.
3. Inlet and exit losses. These losses should be taken off when only the friction factor in the duct is studied.
4. Nonuniform flow rate distribution. When multimicrochannel system is applied in the experiment this nonuniformity will cause some uncertainty when the average flow rate is taken as individual tube flow rate.
5. Roughness effect (as discussed earlier).
6. Rarefaction and compressibility effect (as discussed earlier).
7. Axial heat conduction (as discussed earlier).
8. The possible dissolved gas in liquid. Usually, a high gas pressure is applied to push the liquid flowing through the microchannels. The dissolved gas will affect test steadiness and friction factor.
9. Viscous dissipation effect. The dissipation heat increases the liquid temperature; hence, its viscosity varies in the whole microchannel.
10. Electrokinetic effects (as discussed earlier).

REFERENCES

1. D.B. Tuckerman, R.E. Pease, High-performance heat sinking for VLSI, *Electron Device Letters, IEEE*, 2(5), 126–129 (1981).
2. D.B. Tuckerman, R.E. Pease, Optimized convective cooling using micromachined structure, *Journal of Electrochemical Society*, 129(3), C98 (1982).
3. P.Y. Wu, W.A. Little, Measurement of the friction factors for the flow of gases in very fine channels used for microminiature Joule-Thomson refrigerators, *Cryogenics*, 23, 273–277 (1983).
4. P.Y. Wu, W.A. Little, Measurement of the heat-transfer characteristics of gas-flow in fine channel heat-exchangers used for micro/miniature refrigerators, *Cryogenics*, 24, 415–420 (1984).
5. J.C. Shih, C.M. Ho, J. Liu, Y.C. Tai, Monatomic and polyatomic gas flow through uniform microchannels, in: *Proceedings of 1996 ASME International Mechanical Engineering Congress and Exposition*, New York.
6. K.C. Pong, C.M. Ho, J. Liu, Y.C. Tai, Nonlinear pressure distribution in uniform microchannels, *ASME FED*, 197, 51–56 (1994).
7. C.M. Ho, Y.C. Tai, Micro-electro-mechanical systems (MEMS) and fluid flows, *Annual Review of Fluid Mechanics*, 30, 579–612 (1998).
8. C.B. Sobhan, G.P. Peterson, *Microscale and Nanoscale Heat Transfer*, CRC Press, Boca Raton, FL (2008).
9. P. Tabeling, *Introduction to Microfluidics*, Oxford University Press, Oxford, U.K. (2006).
10. G.H. Ji, H.M. Ji, *Microfluidic Theory and Elements*, Higher Education Press, Beijing, China (2009) (in Chinese).
11. N.T. Obot, Towards a better understanding of friction and heat/mass transfer in microchannels—A literature review, in: *Proceedings of International Conference on Heat Transfer and Transport Phenomena in Microscale*, G.P. Celetaet et al., eds., Banff, Alberta, Canada, October, 15–20, 2000. Begell-House, New York.
12. S.V. Garimella, C.B. Sobhan, Transport in microchannels—A critical review, *Annual Review of Heat Transfer*, 13, 1–50 (2003).
13. W.Q. Tao, Y.L. He, G.H. Tang, Z. Li, No new physics in single phase fluid flow and heat transfer in mini-and-micro channels. Is it a conclusion?, in: *Proceedings of MNHT-2008, Micro/Nanoscale Heat Transfer International Conference*, Tainan, Taiwan, January 6–9, 2008.
14. V. Kumar, M. Paraschivoiu, K.D.P. Nigam, Single-phase fluid flow and mixing in microchannels, *Chemical Engineering Science*, 66, 139–1372 (2011).
15. G.H. Tang, Z. Li, Y.L. He, W.Q. Tao, Experimental study of compressibility, roughness and rarefaction influences on microchannel flow, *International Journal of Heat and Mass Transfer*, 50, 2282–2295 (2007).
16. H. Tsien, Super aerodynamics, the mechanism of rarefied gas, *Journal of Aeronautical Science*, 13, 653–664 (1946).
17. F.M. Devienne, Low density heat transfer, in: Hartnett J.P., Irvine T.F., Jr., eds., *Advances in Heat Transfer*, Vol. 2, Academic Press, New York (1965).
18. G.S. Springer, Heat transfer in rarefied gases, in: Hartnett J.P., Irvine T.F., Jr., eds., *Advances in Heat Transfer*, Vol. 7, Academic Press, New York (1971).
19. M. Gad-el-Hak, The fluid mechanics of microdevices—The freeman scholar lecture, *ASME Journal of Fluids Engineering*, 121, 5–33 (1999).
20. W.Q. Tao, Y.L. He, Multiscale simulations of heat transfer and fluid flow problems, *ASME Journal of Heat Transfer*, 134, 031018-1–031018-13 (2012).
21. Z.Y. Guo, Z.X. Li, Size effect on microscale single-phase flow and heat transfer, *International Journal of Heat and Mass Transfer*, 46, 149–159 (2003).
22. M.A. Saad, *Compressible Fluid Flow*, Englewood Cliffs, Prentice-Hall, NJ (1985).
23. V. Gnielinski, New equations for heat and mass transfer in turbulent pipe and channel flows, *International Chemical Engineering*, 16, 359–368 (1976).
24. F.M. White, *Fluid Mechanic*, 5th edn., McGraw-Hill, Boston, MA (2003).
25. C.S. Chen, S.M. Lee, J.D. Sheu, Numerical analysis of gas flow in microchannels, *Numerical Heat Transfer, Part A*, 33, 749–762 (1998).
26. G.X. Li, W.Q. Tao, Z.Y. Li, B. Yu, Direct numerical simulation of rarefied turbulent microchannel flow, *Microfluid Nanofluid*, 2, 106–116 (2006).
27. G.H. Tang, W.Q. Tao, Y.L. He, Simulation of fluid flow and heat transfer in a plate channel using the lattice Boltzmann method. *International Journal of Modern Physics B*, 17, 183–187 (2003).

28. G.H. Tang, W.Q. Tao, Y.L. He, Lattice Boltzmann method for simulating gas flow in microchannels, *International Journal of Modern Physics C*, 2, 335–347 (2004).
29. S. Succi, *Lattice Boltzmann Equation for Fluid Dynamics and Beyond*, Clarendon Press, Oxford, U.K. (2001).
30. Y.L. He, Y. Wang, Q. Li, *Lattice Boltzmann Method: Theory and Applications*, Science Press, Beijing, China (2009).
31. A.A. Mohamad, *Lattice Boltzmann Method, Fundamentals and Engineering Applications with Computer Codes*, Springer-Verlag London Limited, London, U.K. (2011).
32. E.B. Arkilic, M.A. Schmidt, K.S. Breuer, Gaseous slip flow in long microchannels, *Journal of Microelectromechanical System*, 6, 167–175 (1997).
33. J.C. Shih, C.M. Ho, J.O. Liu, Y.C. Tai, Monatomic and polyatomic gas flow through uniform microchannels, *ASME MEMS*, DSC-59, 197–203 (1996).
34. Z. Li, Y.L. He, G.H. Tang, W.Q. Tao, Experimental and numerical studies of liquid flow and heat transfer in microtubes. *International Journal of Heat and Mass Transfer*, 50, 3447–3460 (2007).
35. D.O. Li, Electro-viscous effects on pressure-driven liquid flow in microchannels, *Colloids and Surface A*, 195, 35–57 (2001).
36. G.H. Tang, Z. Li, C.Y. Zhao, W.Q. Tao, Experimental observations and lattice Boltzmann method study of the electroviscous effect for liquid flow in microchannels, *Journal of Micromechanical Microengineering*, 17, 539–550 (2007).
37. G.H. Tang, Z. Li, J.K. Wang, Y.L. He, W.Q. Tao, Electroosmotic flow and mixing in microchannels with the lattice Boltzmann method, *Journal of Applied Physics*, 100, 094908 (2006).
38. L.O. Ren, W.L. Qu, D.Q. Li, Interfacial electrokinetic effects on liquid flow in microchannels *International Journal of Heat and Mass Transfer*, 44, 3125–3134 (2001).
39. W.Q. Tao, *Numerical Heat Transfer*, 2nd edn., Xi'an Jiaotong University Press, Xi'an, China (2001).
40. S.V. Patankar, *Numerical Heat Transfer and Fluid Flow*, McGraw-Hill, New York (1980).
41. S.M. Yang, W.Q. Tao, *Heat Transfer*, 4th edn., Higher Education Press, Beijing, China (2006).
42. L. Gong, K. Kosta, W.Q. Tao, Y. Joshi, Parametric numerical study of flow and heat transfer in microchannels with wavy walls, *ASME Journal of Heat Transfer*, 133, 051702-1–051702-10 (2011).
43. J.F. Fan, W.K. Ding, Y.L. He, W.Q. Tao, A performance evaluation plot of enhanced heat transfer techniques oriented for energy-saving, *International Journal of Heat and Mass Transfer*, 52, 33–44 (2009).
44. S.X. Zhang, Y.L. He, G. Lauriat, W.Q. Tao, Numerical studies of simultaneously developing laminar flow and heat transfer in microtubes with thick wall and constant outside wall temperature, *International Journal of Heat and Mass Transfer*, 53, 3977–3989 (2010).
45. G.H. Tang, W.Q. Tao, Y.L. He, Three-dimensional lattice Boltzmann model for gaseous flow in rectangular microtubes and microscale porous media, *Journal of Applied Physics*, 97, 104918 (2005).
46. L.J. Klinkenberg, The permeability of porous media to liquids and gases, in: *Drilling and Production Practices*, American, Petroleum Institute, New York (1941).
47. A.E. Scheidrgger, *The Physics of Flow through Porous Media*, 3rd edn., University of Toronto Press, Toronto, Ontario, Canada (1972).
48. J.L. Ge, *Porous Flow Mechanics in Gas and Oil Reservoir*, Petroleum Industry Press, Beijing, China (1982) (in Chinese).
49. Z.X. Sun, Z.Y. Li, Y.L. He, W.Q. Tao, Coupled solid (FVM)-fluid (DSMC) simulation of micro-nozzle with unstructured-grid, *Microfluid Nanofluid*, 7, 621–631 (2009).
50. G. Bird, *Molecular Gas Dynamics and the Direct Simulation of Gas Flows*, Clarendon Press, Oxford, U.K. (1994).
51. G. Bird, Sophisticated DSMC, Notes prepared for a short course at the *DSMC07 Meeting*, Santa Fe, NM, September 30, 2007.
52. X.L. Xie, Z.J. Liu, Y.L. He, W.Q. Tao, Numerical study of laminar heat transfer and pressure drop characteristics in a water-cooled minichannel heat sink, *Applied Thermal Engineering*, 29(1), 64–74 (2009).
53. Y.T. Mu, L. Chen, Y.L. He, W.Q. Tao, Numerical study of temperature uniformity in a novel minichannel heat sink with different configuration, *International of Journal of Heat and Mass Transfer*, 85, 147–157 (2015).
54. G. Tunc, Y. Bayazitoglu, Heat transfer in rectangular microchannels, *International Journal of Heat and Mass Transfer*, 45, 765–773 (2002).
55. C.H. Chen, Slip-flow heat transfer in a microchannel with viscous dissipation, *Heat Mass Transfer*, 43, 853–860 (2006).

13 Analytical–Numerical Solutions for Conjugated Heat Transfer in Multistream Microsystems

J.L.Z. Zotin, Diego C. Knupp, and Renato M. Cotta

CONTENTS

13.1 INTRODUCTION

Since the past few decades, several studies were carried out to the miniaturization of process equipment, such as microheat exchangers, microreactors, and micromixers [1,2], aimed at improved efficiency and cost-effectiveness in different applications. In this context, researchers were concerned with observed discrepancies between experimental heat and mass transfer results at the microscale and classical macroscale models and correlations [3,4].

Among other issues that should be taken into consideration toward the reduction of such discrepancies, it is highlighted the formulation of conjugated conduction–convection heat transfer problems to accurately describe the effects of the substrates on which the microchannels are etched [5–8], in opposition to what generally occurs in macroscale applications, that is, when conjugation effects are usually disregarded, allowing for the direct treatment of convective heat transfer with simplified boundary conditions of either prescribed temperature or heat flux at the walls. Another common important scaling effect that results from the miniaturization of process equipment is the increasing relative importance of the axial conduction term in the fluid energy equation. Whereas this term is almost always neglected in macroscale internal convection problems, with the exception of thermal devices with liquid metal as the working fluid, its consideration can be of crucial importance for accurately modeling heat and fluid flow in microsystems, as a result of the generally low Péclet numbers involved [8,9].

Recently, Knupp et al. [10] proposed a hybrid numerical–analytical solution for conjugated heat transfer problems through a combination of a single-domain formulation and integral transforms. In this methodology, the conjugated problem is reformulated employing a single governing energy equation for the whole spatial domain (fluid stream and solid walls), which is obtained by making use of space-variable coefficients with abrupt transitions at the solid–fluid interfaces, in order to take into account the actual solid and fluid regions' physical properties and velocity profiles. Then, the generalized integral transform technique (GITT) [11–14] is employed for the solution of the resulting partial differential problem with space-variable coefficients. This reformulation and solution methodology was originally presented and verified in [10] and subsequently advanced to tackle several relevant issues in microscale heat transfer, such as axial conduction and preheating effects [15–17], three-dimensional conjugated problems and irregular geometries [18], and a few applications involving microheat exchangers [19,20].

Multistream microfluidic devices are quite common and desired in order to enhance productivity and thermal efficiency, such as with micromixers, microreactors, and microheat exchangers [19–24], and eventually, the analysis of multiphase flows may also be an important issue [25–27]. There is no need to recall that the combined effect of multiple substrates conjugation and fluid streams axial diffusion can be essential in properly modeling these and related microdevices. Here, the above-referenced hybrid solution methodology is employed in the analysis of microsystems with multiple streams, including stratified multiphase flows, illustrating alternative solution paths and the flexibility and robustness of the proposed approach.

13.2 PROBLEM FORMULATION AND SOLUTION METHODOLOGY

Consider a laminar incompressible internal multiphase flow of stratified immiscible Newtonian fluids between parallel plates, at steady state, and undergoing convective heat transfer. The channel walls are considered to participate on the heat transfer problem through both transversal and axial heat conduction, and the external surfaces exchange heat with the surrounding environment, which is at temperature T_∞, with known external heat transfer coefficients at the two bounding walls of the whole microsystem, h_0 and h_L. Figure 13.1 shows a schematic representation of this problem

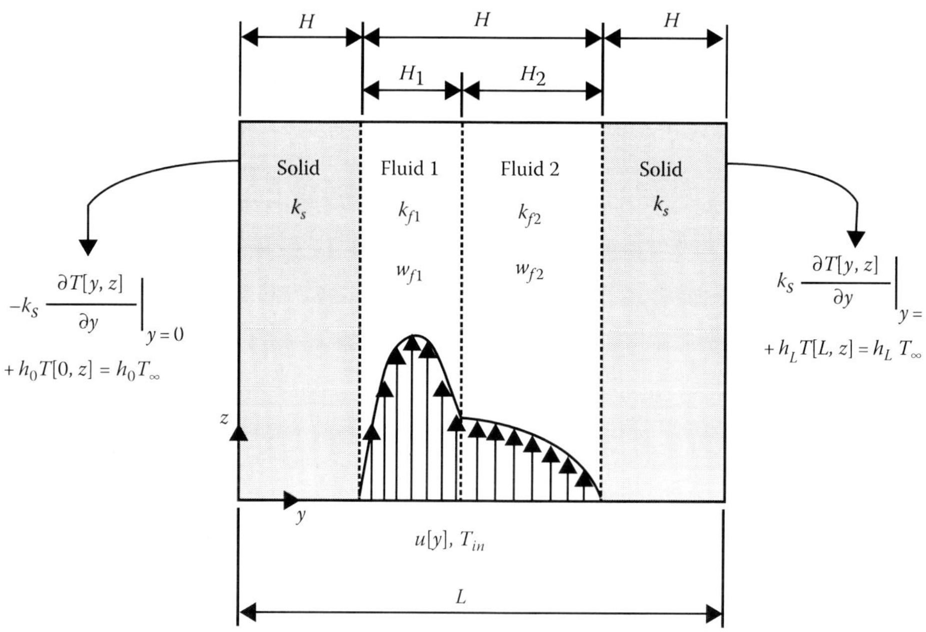

FIGURE 13.1 Schematic representation of a single channel with two-phase stratified flow.

considering a single channel with a two-phase stratified flow. The single-domain formulation, for the solid and fluid regions, is written as follows [15]:

$$u(y)w(y)\frac{\partial T(y,z)}{\partial z} = k(y)\frac{\partial^2 T(y,z)}{\partial z^2} + \frac{\partial}{\partial y}\left(k(y)\frac{\partial T(y,z)}{\partial y}\right), \quad 0 < y < L,\ z > 0 \tag{13.1}$$

$$T(y,0) = T_{in}, \quad 0 < y < L \tag{13.2}$$

$$\left.\frac{\partial T}{\partial z}\right|_{z\to\infty} = 0, \quad 0 < y < L \tag{13.3}$$

$$-k(0)\left.\frac{\partial T}{\partial y}\right|_{y=0} + h_0 T(0,z) = h_0 T_\infty, \quad z > 0 \tag{13.4}$$

$$k(L)\left.\frac{\partial T}{\partial y}\right|_{y=L} + h_L T(L,z) = h_L T_\infty, \quad z > 0 \tag{13.5}$$

Where $u(y)$, $w(y)$, and $k(y)$ are, respectively, the axial velocity, the thermal capacity, and the thermal conductivity. One may observe that all these space-variable coefficients already taken into account, through abrupt transitions at the interfaces, the different properties of each subregion of the original domain, that is, the different fluids and solid walls, including null velocity in the solid regions, where convection does not exist.

The following dimensionless formulation is then achieved, already filtered in order to yield homogeneous boundary conditions in Y direction:

$$U(Y)W(Y)\frac{\partial \theta^*(Y,Z)}{\partial Z} = \frac{K(Y)}{\mathrm{Pe}^2}\frac{\partial^2 \theta^*}{\partial Z^2} + \frac{\partial}{\partial Y}\left(K(Y)\frac{\partial \theta^*}{\partial Y}\right), \quad 0 < Y < 1,\ Z > 0 \tag{13.6}$$

$$\theta^*(Y,0) = -1, \quad \left.\frac{\partial \theta^*}{\partial Z}\right|_{Z\to\infty} = 0, \quad 0 < Y < 1 \tag{13.7}$$

$$-\left.\frac{\partial \theta^*}{\partial Y}\right|_{Y=0} + \mathrm{Bi}_0\theta^*(0,Z) = 0, \quad \left.\frac{\partial \theta^*}{\partial Y}\right|_{Y=1} + \mathrm{Bi}_L\theta^*(1,Z) = 0, \quad Z > 0 \tag{13.8}$$

with the filtering chosen as

$$\theta(Y,Z) = \theta^*(Y,Z) + 1 \tag{13.9}$$

where the following dimensionless groups have been employed:

$$Z = \frac{z}{L\text{Pe}}; \quad Y = \frac{y}{L}; \quad U = \frac{u(y)}{4u_{av}}; \quad \theta = \frac{T - T_{in}}{T_\infty - T_{in}}; \quad K = \frac{k(y)}{k_{f1}};$$
$$W = \frac{w(y)}{w_{f1}}; \quad \text{Bi}_0 = \frac{h_0 L}{k_s}; \quad \text{Bi}_L = \frac{h_L L}{k_s}; \quad \text{Pe} = \frac{4u_{av} L w_{f1}}{k_{f1}} \tag{13.10}$$

where

u_{av} is the average velocity inside the channel

k_{f1} and w_{f1} are, respectively, the thermal conductivity and heat capacity of "fluid 1," which are arbitrarily chosen for the reference values

The space-variable piecewise functions $U(Y)$, $W(Y)$, and $K(Y)$ are, respectively, the dimensionless axial velocity, which is zero in the solid wall regions, the dimensionless heat capacity, and the dimensionless thermal conductivity, all of them having different values at the solids and fluids regions, with abrupt transitions at the interfaces. The velocity profile of a two-phase stratified flow depends on the flow rate of each phase and the viscosity ratio and is derived in Bird et al. [28].

It should be mentioned that even though this problem has been formulated upon the schematic representation shown in Figure 13.1, with a single channel and two fluids, the problem given by Equations 13.6 through 13.8 can be readily employed to describe any number of channels and fluids in stratified flow by just adequately specifying the space-variable coefficients $U(Y)$, $W(Y)$, and $K(Y)$.

In the next section, two alternative paths are described for the solution of the problem given by Equations 13.6 through 13.8. The first one, presented in Section 13.2.1, employs a nonclassical eigenvalue problem [29–31] that allows for the exact transformation the above-mentioned problem, yielding a fully analytical solution of the resulting system. The second solution path, presented in Section 13.2.2, employs a partial transformation scheme, after the introduction of a pseudotransient term, with the proposition of a space-variable classical Sturm–Liouville eigenvalue problem. The resulting system of differential equations is then numerically solved. In both cases, the generalized integral transform technique (GITT) is employed in the solution of the space-variable eigenvalue problems, which do not allow for explicit exact analytical solutions.

13.2.1 Solution via Nonclassical Eigenvalue Problem

The following nonclassical eigenvalue problem can be obtained by application of the separation of variables method to Equations 13.6 through 13.8 [31]:

$$\frac{d}{dY}\left(K(Y)\frac{d\psi_i(Y)}{dY}\right) + \left(\frac{K(Y)}{\text{Pe}^2}\mu_i^4 + U(Y)W(Y)\mu_i^2\right)\psi_i(Y) = 0 \tag{13.11}$$

$$-\left.\frac{d\psi_i(Y)}{dY}\right|_{Y=0} + \text{Bi}_0\psi_i(0) = 0 \tag{13.12}$$

$$\left.\frac{d\psi_i(Y)}{dY}\right|_{Y=1} + \text{Bi}_L\psi_i(1) = 0 \tag{13.13}$$

where

ψ_i are the eigenfunctions

μ_i are the corresponding eigenvalues

Then, the generalized integral transform technique itself is employed in the solution of this eigenvalue problem, considering the following transform–inverse pair:
Transform:

$$\bar{\psi}_{i,n} = \int_0^1 \tilde{\Omega}_n(Y)\psi_i(Y)dY \tag{13.14}$$

Inverse:

$$\psi_i(Y) = \sum_{n=1}^{\infty} \tilde{\Omega}_n(Y)\bar{\psi}_{i,n} \tag{13.15}$$

where $\Omega_n(Y)$ are the eigenfunctions from a simpler eigenvalue problem, which allows for explicit exact analytical solution:

$$\frac{d^2\Omega_n(Y)}{dY^2} + \lambda_n^2\Omega_n(Y) = 0 \tag{13.16}$$

$$-\left.\frac{d\Omega_n(Y)}{dY}\right|_{Y=0} + \mathrm{Bi}_0\Omega_n(0) = 0 \tag{13.17}$$

$$\left.\frac{d\Omega_n(Y)}{dY}\right|_{Y=1} + \mathrm{Bi}_L\Omega_n(1) = 0 \tag{13.18}$$

where λ_n^2 are the corresponding eigenvalues and the normalized eigenfunctions are given by the following equation:

$$\tilde{\Omega}_n(Y) = \frac{\Omega_n(Y)}{\sqrt{M_n}}, \quad M_n = \int_0^1 \Omega_n^2(Y)\,dY \tag{13.19}$$

Operating on Equation 13.11 with $\int_0^1 \tilde{\Omega}_n(Y)(\cdot)dY$, one obtains the following algebraic system, given in matrix form:

$$(\mathbf{A} + \mathbf{D})\bar{\psi} = \mu^4\mathbf{E} + \mu^2\mathbf{B} \tag{13.20}$$

where the corresponding coefficient matrices are given by the following equations:

$$A_{mn} = \int_0^1 \frac{d^2\tilde{\Omega}_n(Y)}{dY^2}\tilde{\Omega}_m(Y)dY + \int_0^1 K(Y)\frac{d\tilde{\Omega}_n(Y)}{dY}\frac{d\tilde{\Omega}_m(Y)}{dY}dY$$
$$- K(1)\tilde{\Omega}_m(1)\left.\frac{d\tilde{\Omega}_n(Y)}{dY}\right|_{Y=1} + K(0)\tilde{\Omega}_m(0)\left.\frac{d\tilde{\Omega}_n(Y)}{dY}\right|_{Y=0} \tag{13.21}$$

$$D_{mn} = \lambda_m^2\delta_{mn} \tag{13.22}$$

$$E_{mn} = \int_0^1 \frac{K(Y)}{\mathrm{Pe}^2} \tilde{\Omega}_n(Y)\tilde{\Omega}_m(Y)\, dY \tag{13.23}$$

$$B_{mn} = \int_0^1 U(Y)W(Y)\tilde{\Omega}_n(Y)\tilde{\Omega}_m(Y)\, dY \tag{13.24}$$

Where δ_{mn} is the Kronecker delta. To solve the problem given by Equations 13.20 through 13.24, the following transformation was employed [31]:

$$\nu^2 \mathbf{G}\phi = \mathbf{H}\phi \tag{13.25}$$

where the matrices **G** and **H** are given by the following equations:

$$\mathbf{G} = \begin{bmatrix} \mathbf{0} & \mathbf{E} \\ \mathbf{E} & \mathbf{B} \end{bmatrix}, \quad \mathbf{H} = \begin{bmatrix} \mathbf{E} & \mathbf{0} \\ \mathbf{0} & \mathbf{F} \end{bmatrix}, \quad \text{with } \mathbf{F} = \mathbf{A} + \mathbf{D} \tag{13.26}$$

Equations 13.25 and 13.26 can be readily solved with the built-in function *Eigensystem* of the *Mathematica* system [32], which gives the eigenvector:

$$\phi = \begin{bmatrix} \phi_1 \\ \phi_2 \end{bmatrix} \tag{13.27}$$

where the eigenvectors ϕ_2 and their associated eigenvalues ν correspond, respectively, to the expansion coefficients $\bar{\psi}$ in Equation 13.15, and the associated eigenvalues μ of the original eigenvalue problem (Equations 13.11 through 13.13). It must be highlighted that the algebraic eigenvalue problem given by Equations 13.20 through 13.24 forms an infinite system, which must be truncated at the N_{aux}-th column and row, retaining a sufficiently high number of terms to ensure the user-prescribed accuracy.

Once the original eigenfunctions $\psi_i(Y)$ and the corresponding eigenvalues μ_i are available, the solution of Equations 13.6 through 13.8 can be written as follows:

$$\theta^*(Y,Z) = \sum_{i=1}^{N} C_i e^{-\mu_i^2 Z} \psi_i(Y) \tag{13.28}$$

where the coefficients C_i must be determined from the application of the initial condition to Equation 13.28, and making use of the eigenfunctions orthogonality property, which for the eigenvalue problem considered (Equations 13.11 through 13.13) can be stated as [15] follows:

$$\int_0^1 \left[U(Y)W(Y) + \left(\mu_i^2 + \mu_j^2\right) \frac{K(Y)}{\mathrm{Pe}^2} \right] \psi_i(Y)\psi_j(Y) dY = N_i \delta_{ij} \tag{13.29}$$

$$N_i = \int_0^1 \left[U(Y)W(Y) + 2\mu_i^2 \frac{K(Y)}{\mathrm{Pe}^2} \right] \psi_i^2(Y)\, dY \tag{13.30}$$

Then, the following algebraic system is obtained.

$$C_j N_j - \int_0^1 \left(U(Y)W(Y) + \mu_j^2 \frac{K(Y)}{\mathrm{Pe}^2} \right) \psi_j(Y)\theta^*(Y,0)dY$$
$$- \sum_{i=1}^{N} C_i \int_0^1 \mu_i^2 \frac{K(Y)}{\mathrm{Pe}^2} \psi_i(Y)\psi_j(Y)dY = 0, \quad j = 1,2,\ldots,N \tag{13.31}$$

which can be analytically solved through symbolic computation for the coefficients C_i, $i = 1,2,\ldots,N$.

Once $\theta^*(Y,Z)$ is determined, Equation 13.9 is recalled to provide the desired dimensionless temperature field, $\theta(Y,Z)$.

13.2.2 Solution via Partial Transformation

In this alternative path, a pseudotransient term is included in the formulation for the solution through a partial transformation scheme with a classical eigenvalue problem [33]. The pseudotransient formulation, in dimensionless form and already filtered with Equation 13.10, is given by the following:

$$\frac{\partial\theta^*(Y,Z,t)}{\partial t} + U(Y)W(Y)\frac{\partial\theta^*(Y,Z,t)}{\partial Z} = \frac{K(Y)}{\mathrm{Pe}^2}\frac{\partial^2\theta^*}{\partial Z^2} + \frac{\partial}{\partial Y}\left(K(Y)\frac{\partial\theta^*}{\partial Y} \right), \quad 0 < Y < 1, Z > 0, t > 0 \tag{13.32}$$

$$\theta^*(Y,Z,0) = f(Y,Z), \quad 0 < Y < 1, Z > 0 \tag{13.33}$$

$$\theta^*(Y,0,t) = -1, \quad 0 < Y < 1, t > 0 \tag{13.34}$$

$$\left.\frac{\partial\theta^*}{\partial Z}\right|_{Z\to\infty} = 0, \quad 0 < Y < 1, t > 0 \tag{13.35}$$

$$-\left.\frac{\partial\theta^*}{\partial Y}\right|_{Y=0} + \mathrm{Bi}_0\theta^*(0,Z,t) = 0, \quad Z > 0, t > 0 \tag{13.36}$$

$$\left.\frac{\partial\theta^*}{\partial Y}\right|_{Y=1} + \mathrm{Bi}_L\theta^*(1,Z,t) = 0, \quad Z > 0, t > 0 \tag{13.37}$$

For the solution of the problem given by Equations 13.32 through 13.37, the following transform-inverse pair is employed.

Transform:

$$\bar{\theta}_i^*(Z,\mathrm{t}) = \int_0^1 \tilde{\xi}_i(Y)\theta^*(Y,Z,t)dY \tag{13.38}$$

Inverse:

$$\theta^*(Y,Z,t) = \sum_{i=1}^{\infty} \tilde{\xi}_i(Y)\bar{\theta}_i^*(Z,t) \tag{13.39}$$

where the eigenfunctions $\xi_i(Y)$ come from the solution of the following eigenvalue problem:

$$\frac{d}{dY}\left(K(Y)\frac{d\xi_i(Y)}{dY}\right) + \beta_i^2\xi_i(Y) = 0 \tag{13.40}$$

$$-\left.\frac{d\xi_i(Y)}{dY}\right|_{Y=0} + \mathrm{Bi}_0\xi_i(0) = 0 \tag{13.41}$$

$$\left.\frac{d\xi_i(Y)}{dY}\right|_{Y=1} + \mathrm{Bi}_L\xi_i(1) = 0 \tag{13.42}$$

Where β_i are the eigenvalues corresponding to $\xi_i(Y)$. The normalized eigenfunctions, $\tilde{\xi}_i(Y)$, are readily calculated as follows:

$$\tilde{\xi}_i(Y) = \frac{\xi_i(Y)}{\sqrt{P_i}}, \quad P_i = \int_0^1 \xi_i^2(Y)dY \tag{13.43}$$

Again the GITT itself is employed for the solution of the eigenvalue problem given by Equations 13.40 through 13.43, through an expansion in terms of a chosen basis, since it does not allow for explicit exact analytical solution. After integral transforming Equation 13.40, again making use of the auxiliary eigenvalue problem, given by Equations 13.16 through 13.18, the following algebraic system is obtained [15,17]:

$$(\mathbf{K} - \alpha^2\mathbf{L})\bar{\zeta} = 0 \tag{13.44}$$

where

$$\alpha_i = \beta_i, \quad L_{mn} = -\int_0^1 \tilde{\Omega}_m(Y)\tilde{\Omega}_n(Y)dY \tag{13.45}$$

$$K_{mn} = -\int_0^1 K(Y)\frac{d\tilde{\Omega}_m(Y)}{dY}\frac{d\tilde{\Omega}_n(Y)}{dY}dY + K(1)\tilde{\Omega}_m(1)\left.\frac{d\tilde{\Omega}_n(Y)}{dY}\right|_{Y=1} - K(0)\tilde{\Omega}_m(0)\left.\frac{d\tilde{\Omega}_n(Y)}{dY}\right|_{Y=0} \tag{13.46}$$

This algebraic eigenvalue problem can be readily handled by numerical routines, such as the *Eigensystem* built-in routine of the *Mathematica* system [32]. Again, it is important to note that the rows and columns of system (Equations 13.44 through 13.46) must be truncated with a finite number of terms N_{aux}, which must be high enough to achieve the prescribed accuracy.

Once the solution of the considered eigenvalue problem (Equations 13.40 through 13.42) is available, the transformed problem is obtained by operating on Equation 13.32 with $\int_0^1 \tilde{\xi}_i(Y)(\cdot)dY$, yielding:

$$\frac{\partial\bar{\theta}_i^*(Z,t)}{\partial t} + \beta_i^2\bar{\theta}_i^*(Z,t) = g_i(Z,t,\bar{\theta}^*) \tag{13.47}$$

where

$$\bar{\theta}^* = \left\{\bar{\theta}_1^*, \bar{\theta}_2^*, \ldots, \bar{\theta}_N^*\right\} \tag{13.48}$$

$$g_i(Z,t,\bar{\theta}^*) = -\sum_{j=1}^{N}\frac{\partial\bar{\theta}_j^*}{\partial Z}\int_0^1 U(Y)W(Y)\tilde{\xi}_i(Y)\tilde{\xi}_j(Y)dY + \sum_{j=1}^{N}\frac{\partial^2\bar{\theta}_j^*}{\partial Z^2}\int_0^1 \frac{K(Y)}{Pe^2}\tilde{\xi}_i(Y)\tilde{\xi}_j(Y)dY \tag{13.49}$$

with the following transformed initial and boundary conditions:

$$\bar{\theta}_i^*(Z,0) = \int_0^1 \tilde{\xi}_i(Y)f(Y,Z)dY \tag{13.50}$$

$$\bar{\theta}_i^*(0,t) = \int_0^1 \tilde{\xi}_i(Y)\theta^*(Y,0,t)dY, \quad \left.\frac{\partial\bar{\theta}_i^*(Z,t)}{\partial Z}\right|_{Z\to\infty} = 0 \tag{13.51}$$

The system of coupled one-dimensional partial differential equations given by Equations 13.47 through 13.51 can be readily solved numerically for the transformed potentials $\bar{\theta}_i^*$, with absolute and relative error control, through the built-in function *NDSolve* of the *Mathematica* system [32]. Once the transformed potentials are available, the inverse formula (Equation 13.39) is employed to obtain $\theta^*(Z,Y,t)$ and then Equation 13.10 is recalled to provide the original dimensionless temperature field, $\theta(Y,Z)$. In this study, only the steady-state solution is of interest, and therefore, any feasible function can be employed as the initial condition, $f(Y,Z)$, preferably an estimate of the steady-state solution.

13.3 RESULTS AND DISCUSSION

The results obtained with the methodologies described in Sections 13.2.1 and 13.2.2 are now presented considering Pe = 0.025 and a prescribed temperature boundary condition, which can be readily obtained by letting Bi_0, $Bi_L \to \infty$ in the general formulation presented. Two cases are considered, the first one, in Section 13.3.1, considers a substrate with a single microchannel, whereas the second case, in Section 13.3.2, considers a substrate containing four parallel microchannels. In all cases, it has been considered a two-phase stratified flow inside the microchannels.

13.3.1 Single Channel

The schematic representation of this case is shown in Figure 13.1. The channel height is $H = L/3$ and the layer thickness ratio between the two fluids is $b = H_1/H_2$. The velocity profile of this stratified two-phase flow is determined by the flow ratio and the viscosity ratio between the two fluids [28].

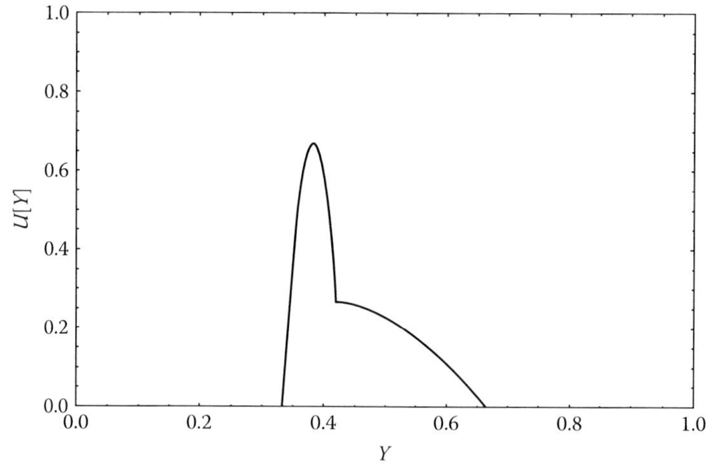

FIGURE 13.2 Dimensionless velocity profile for the two-phase stratified flow in a single microchannel.

In order to report results, two hypothetical fluids are considered with a flow ratio of $Q_1/Q_2 = 1$ and a viscosity ratio of $r_\mu = 1/84$, which is equivalent to the viscosity ratio between water and olive oil, yielding $b = 0.3483$. The dimensionless velocity profile obtained in this case is depicted in Figure 13.2.

The space-variable coefficients $U(Y)$, $K(Y)$, and $W(Y)$ are given, in this test case, by the following:

$$U(Y) = \begin{cases} 0, & 0 \le Y < \dfrac{1}{3} \\ \dfrac{u(y)}{4u_{mean}}, & \dfrac{1}{3} \le Y < \dfrac{2}{3}, \\ 0, & \dfrac{2}{3} \le Y \le 1 \end{cases} \quad K(Y) = \begin{cases} 0.25, & 0 \le Y < \dfrac{1}{3} \\ 1, & \dfrac{1}{3} \le Y < \dfrac{1}{3} + \dfrac{1}{3}b \\ 0.6, & \dfrac{1}{3} + \dfrac{1}{3}b \le Y < \dfrac{2}{3} \\ 0.25, & \dfrac{2}{3} \le Y \le 1 \end{cases}$$

$$W(Y) = \begin{cases} 0, & 0 \le Y < \dfrac{1}{3} \\ 1, & \dfrac{1}{3} \le Y < \dfrac{1}{3} + \dfrac{1}{3}b \\ 3, & \dfrac{1}{3} + \dfrac{1}{3}b \le Y < \dfrac{2}{3} \\ 0, & \dfrac{2}{3} \le Y \le 1 \end{cases} \tag{13.52}$$

In order to give an overview of the convergence behavior of the solution, it is first illustrated in Table 13.1, the first five eigenvalues of the nonclassical eigenvalue problem (Equations 13.11 through 13.13) calculated with different truncation orders (N_{aux}) of the transformed algebraic eigensystem given by Equations 13.20 through 13.24. It can be observed a consistent convergence of at least three significant digits for $N_{aux} < 50$.

The convergence of the computed eigenfunctions is also illustrated, by means of Figure 13.3, which shows the 15th eigenfunction, $\psi_{15}(Y)$, calculated with different number of terms in the expansion given by Equation 13.15. One can observe from the curves for the truncation orders $N_{aux} = 40$ and $N_{aux} = 50$, that the considered eigenfunction is fully converged to the graph scale.

TABLE 13.1
Convergence Behavior of the First Five Eigenvalues of the Nonclassical Eigenvalue Problem, for Different Truncation Orders

Eigenvalues μ_i	$N_{aux} = 10$	$N_{aux} = 20$	$N_{aux} = 30$	$N_{aux} = 40$	$N_{aux} = 50$
μ_1	0.05592	0.05568	0.05560	0.05557	0.05555
μ_2	0.17444	0.17396	0.17378	0.17365	0.17360
μ_3	0.24429	0.24085	0.23987	0.23948	0.23925
μ_4	0.28477	0.28315	0.28278	0.28259	0.28239
μ_5	0.42467	0.42119	0.42052	0.42020	0.42007

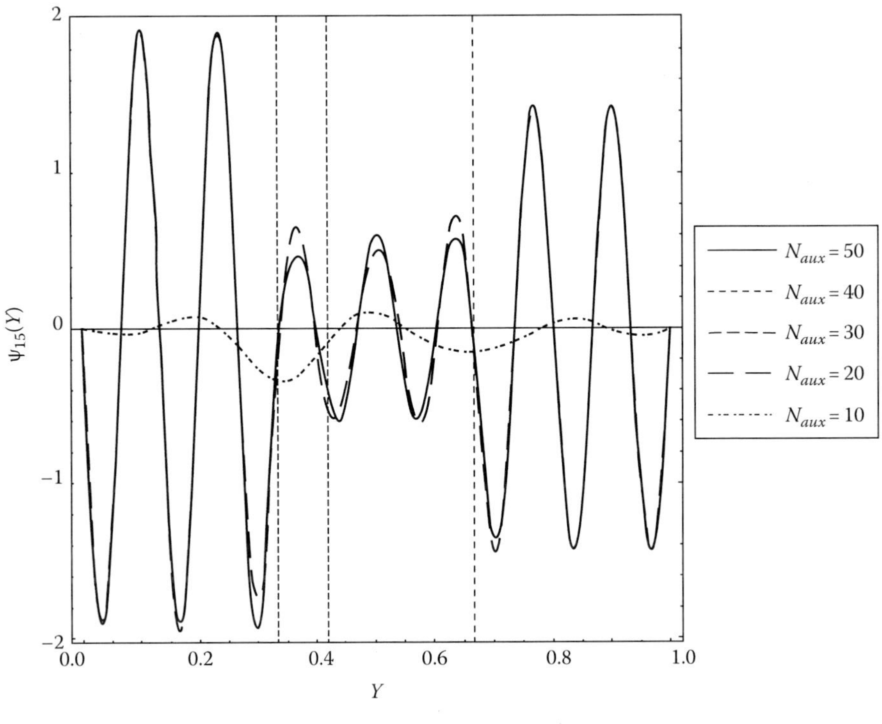

FIGURE 13.3 Convergence behavior of the 15th eigenfunction of the nonclassical eigenvalue problem, $\psi_{15}(Y)$.

Keeping $N_{aux} = 50$ terms in the solution of the nonclassical eigenvalue problem, it is now illustrated the convergence behavior of the temperature field expansion given by Equation 13.28 employing different number of terms N. The results are presented in Table 13.2 for two different longitudinal positions, ZPe = 0.1 and ZPe = 0.2, where one may observe a consistent convergence of at least three significant digits for all selected positions at ZPe = 0.1 and at least five significant digits at ZPe = 0.2, confirming the expected better convergence for larger values of the longitudinal coordinate.

Similar results are now presented for the solution of the same problem but carried out through the partial transformation scheme, described in Section 13.2.2. Table 13.3 presents the convergence behavior of the first five eigenvalues β_i of the space-variable coefficients eigenvalue problem (Equations 13.40 through 13.42) employing different truncation orders, $N_{aux} < 60$, in the transformed algebraic eigensystem (Equations 13.44 through 13.46), illustrating a convergence of two significant digits in the worst case, for the fifth eigenvalue, β_5.

Table 13.4 illustrates the convergence behavior of the temperature field calculated with the expansion given by Equation 13.39 for different number of terms, N, keeping fixed the number of terms

TABLE 13.2
Convergence Behavior of the Dimensionless Temperatures (Solution via the Nonclassical Eigenvalue Problem—Section 13.2.1) for Different Transversal Positions

	Order	$Y = 0.1$	$Y = 0.25$	$Y = 0.5$	$Y = 0.75$	$Y = 0.9$
ZPe = 0.1	$N_P = 5$	0.52253	0.21617	0.14437	0.23535	0.51231
	$N_P = 10$	0.49445	0.23362	0.15653	0.23986	0.49652
	$N_P = 15$	0.49781	0.23373	0.15779	0.23996	0.49974
	$N_P = 20$	0.49801	0.23335	0.15770	0.23961	0.50004
	$N_P = 25$	0.49793	0.23344	0.15768	0.23968	0.49996
ZPe = 0.2	$N_P = 5$	0.70557	0.41465	0.30288	0.4275	0.70714
	$N_P = 10$	0.70201	0.41763	0.30461	0.42831	0.70562
	$N_P = 15$	0.70208	0.41761	0.30466	0.42829	0.70568
	$N_P = 20$	0.70208	0.41761	0.30465	0.42829	0.70568
	$N_P = 25$	0.70208	0.41761	0.30465	0.42829	0.70568

TABLE 13.3
Convergence Behavior of the First Five Eigenvalues β_i of the Space-Variable Coefficients Eigenvalue Problem, Equations 13.40 through 13.42

Eigenvalues β_i	$N_{aux} = 20$	$N_{aux} = 30$	$N_{aux} = 40$	$N_{aux} = 50$	$N_{aux} = 60$
β_1	1.6035	1.6023	1.6018	1.6015	1.6013
β_2	3.8916	3.8848	3.8798	3.8775	3.8756
β_3	5.2814	5.2631	5.2559	5.2517	5.2473
β_4	7.2012	7.2004	7.1992	7.1989	7.1985
β_5	9.3117	9.2762	9.2616	9.2539	9.2448

TABLE 13.4
Convergence Behavior of the Temperature Field for the Solution Through the Partial Transformation Scheme, Section 13.2.2

	Order	$Y = 0.1$	$Y = 0.25$	$Y = 0.5$	$Y = 0.75$	$Y = 0.9$
ZPe = 0.1	$N_P = 10$	0.49651	0.23740	0.15742	0.24157	0.49800
	$N_P = 20$	0.49830	0.23499	0.15903	0.24058	0.50019
	$N_P = 30$	0.49808	0.23411	0.15810	0.24007	0.50003
	$N_P = 40$	0.49800	0.23368	0.15775	0.23985	0.49999
	$N_P = 50$	0.49796	0.23349	0.15762	0.23975	0.49997
ZPe = 0.2	$N_P = 10$	0.70345	0.42246	0.30834	0.43146	0.70655
	$N_P = 20$	0.70260	0.41892	0.30604	0.42905	0.70601
	$N_P = 30$	0.70230	0.41822	0.30511	0.42863	0.70580
	$N_P = 40$	0.70220	0.41783	0.30474	0.42841	0.70575
	$N_P = 50$	0.70214	0.41768	0.30459	0.42832	0.70571

employed in the eigenvalue problem solution, $N_{aux} = 60$. One may observe a consistent convergence behavior of at least three significant digits throughout.

Figure 13.4 presents the transversal temperature profiles from $Y = 0$ to $Y = 1$ at different longitudinal positions, obtained through both methodologies described in Sections 13.2.1 and 13.2.2, confirming the overall good adherence between the two alternative solutions throughout.

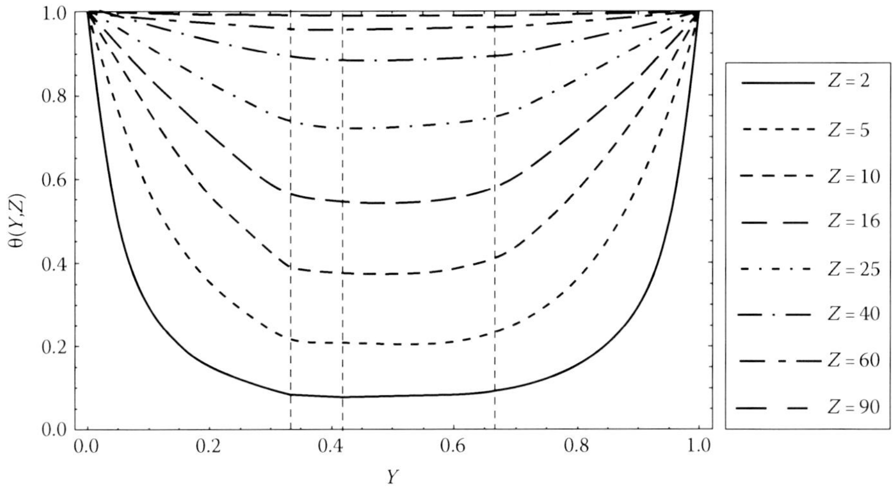

FIGURE 13.4 Transversal temperature profiles at different longitudinal positions, calculated with both methodologies of Sections 13.2.1 and 13.2.2.

TABLE 13.5
Direct Comparison between the Solution Paths Described in Sections 13.2.1 and 13.2.2 for the Single Microchannel Case

	*Z*Pe = 0.1		*Z*Pe = 0.2	
	Section 13.2.1	Section 13.2.2	Section 13.2.1	Section 13.2.2
$Y = 0.1$	0.49793	0.49796	0.70208	0.70214
$Y = 0.25$	0.23344	0.23349	0.41761	0.41768
$Y = 0.5$	0.15768	0.15762	0.30465	0.30459
$Y = 0.75$	0.23968	0.23975	0.42829	0.42832
$Y = 0.9$	0.49996	0.49997	0.70568	0.70571

Table 13.5 consolidates the two solutions, allowing for a closer comparison, confirming an agreement between them to plus or minus one in the fourth significant digit. Finally, Figure 13.5 presents some isotherms, where it is possible to clearly distinguish the solid walls from the fluid streams region, as well as to observe the effects of the two-phase stratified flow, yielding nonsymmetrical isotherms with respect to the channel centerline.

13.3.2 Multiple Channels

Some results are now presented for a multichannels case, considering multiphase stratified flows inside four channels, aimed at further demonstrating the flexibility and robustness of the methodology employed. The schematic representation of the investigated problem is shown in Figure 13.6. Figure 13.7 depicts the dimensionless velocity profiles employed in the single-domain model, already accounting for zero velocity at the solid walls and different velocity profiles at each fluid. The other space-variable coefficients, here omitted for brevity, take into account the different thermophysical properties at each region of the original model (walls and different fluids).

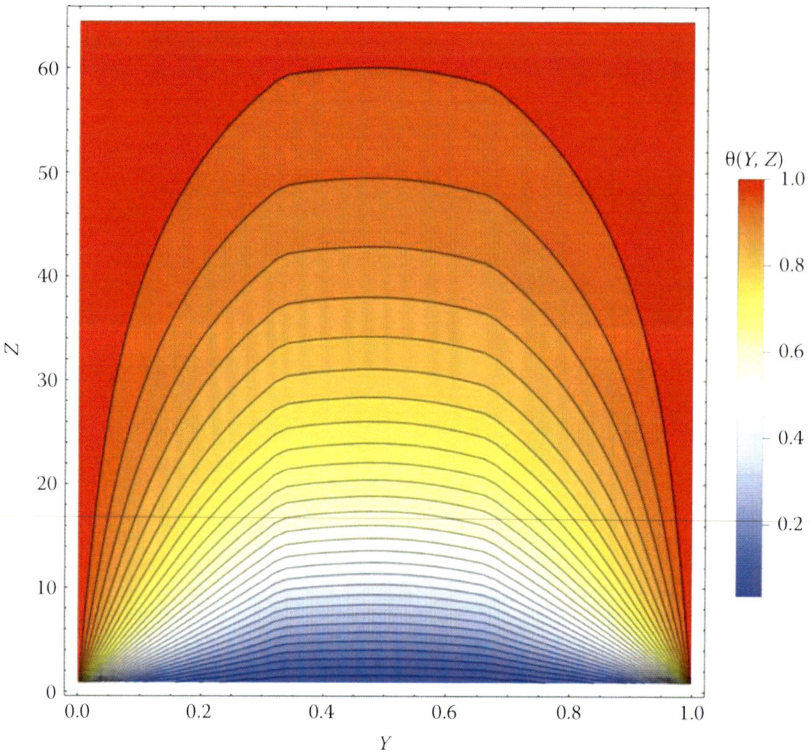

FIGURE 13.5 Calculated isotherms for the case of two-phase stratified flow in a single microchannel.

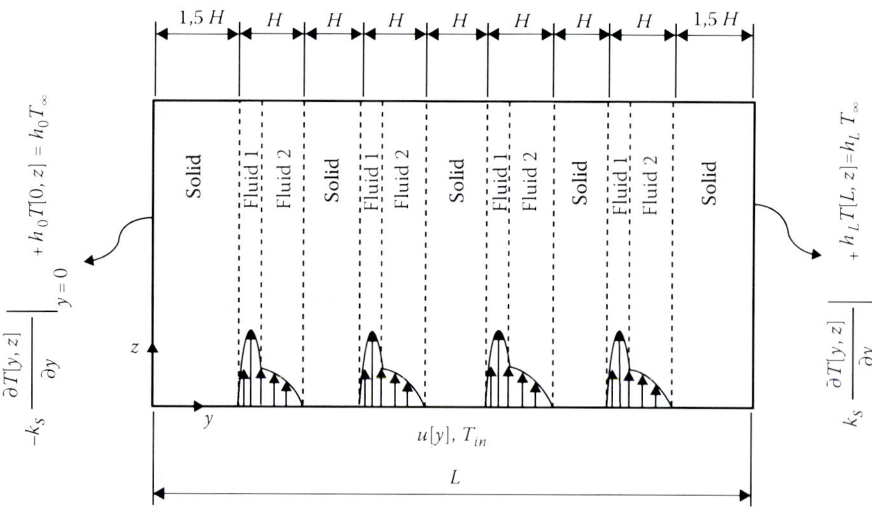

FIGURE 13.6 Schematic representation of the two-phase stratified flow in a four-channel microdevice.

Table 13.6 presents the convergence behavior of the eigenvalue problem solution through GITT, by presenting the first five eigenvalues computed for different truncation orders, N_{aux}, through the solution paths via the nonclassical eigenvalue problem (Section 13.2.1) and via the partial transformation scheme (Section 13.2.2). One may observe that a slightly higher truncation order is needed in this case to achieve the same convergence level, in comparison with the single-channel

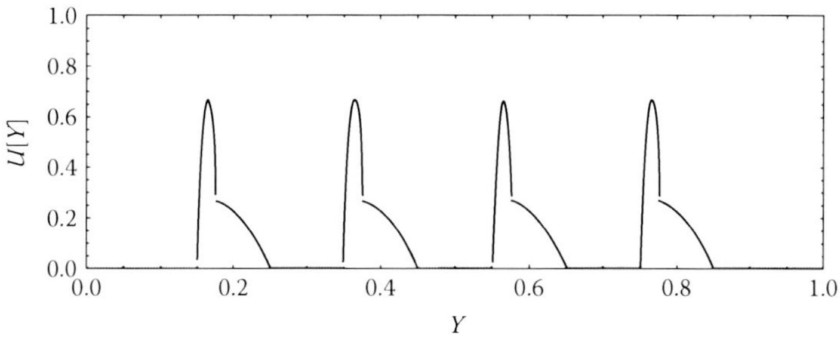

FIGURE 13.7 Schematic representation of the two-phase stratified flow in a four-channel device.

TABLE 13.6
Convergence Behavior of the First Five Eigenvalues in the Multichannel Case

	Eigenvalues	$N_{aux} = 30$	$N_{aux} = 40$	$N_{aux} = 50$	$N_{aux} = 60$	$N_{aux} = 70$
Nonclassical space-variable coefficients eigenvalue problem (Section 13.2.1)	μ_1	0.25214	0.25171	0.25152	0.25139	0.25134
	μ_2	0.35801	0.35720	0.35694	0.35668	0.35662
	μ_3	0.43855	0.43706	0.43681	0.43630	0.43625
	μ_4	0.49850	0.49646	0.49627	0.49553	0.49550
	μ_5	0.68450	0.68405	0.68327	0.68297	0.68280
Classical space-variable coefficients eigenvalue problem (Section 13.2.2)	β_1	1.7587	1.7532	1.7503	1.7488	1.7479
	β_2	3.5420	3.5284	3.5225	3.5186	3.5170
	β_3	5.3655	5.3341	5.3258	5.3168	5.3146
	β_4	7.1831	7.1352	7.1248	7.110	7.1080
	β_5	10.261	10.227	10.187	10.177	10.165

case. This is naturally expected once the space-variable coefficients have a more complex functional behavior for the multichannel problem. Nevertheless, for $N_{aux} < 70$, a consistent convergence behavior to at least three significant digits is observed.

Table 13.7 presents the convergence behavior of the dimensionless temperatures at different transversal positions and at the longitudinal position $ZPe = 0.1$, keeping the truncation order of the

TABLE 13.7
Convergence Behavior of the Temperature Field for the Multichannel Case

		$ZPe = 0.1$				
	Order	$Y = 0.1$	$Y = 0.25$	$Y = 0.5$	$Y = 0.75$	$Y = 0.9$
Solution via nonclassical eigenvalue problem (Section 13.2.1)	$N_p = 5$	0.49675	0.22273	0.15825	0.24224	0.45792
	$N_p = 10$	0.44766	0.23707	0.16522	0.23641	0.45488
	$N_p = 15$	0.4513	0.23671	0.16708	0.23659	0.45825
	$N_p = 20$	0.45143	0.23655	0.16707	0.23642	0.45843
	$N_p = 25$	0.45140	0.23659	0.16704	0.23646	0.45834
Solution via partial transformation (Section 13.2.2)	$N_p = 20$	0.45448	0.23911	0.16777	0.23597	0.45958
	$N_p = 30$	0.45167	0.23612	0.16511	0.23501	0.45822
	$N_p = 40$	0.45126	0.23633	0.16631	0.23715	0.45866
	$N_p = 50$	0.45135	0.23677	0.16717	0.23735	0.45869
	$N_p = 60$	0.45116	0.23626	0.16712	0.23612	0.45811

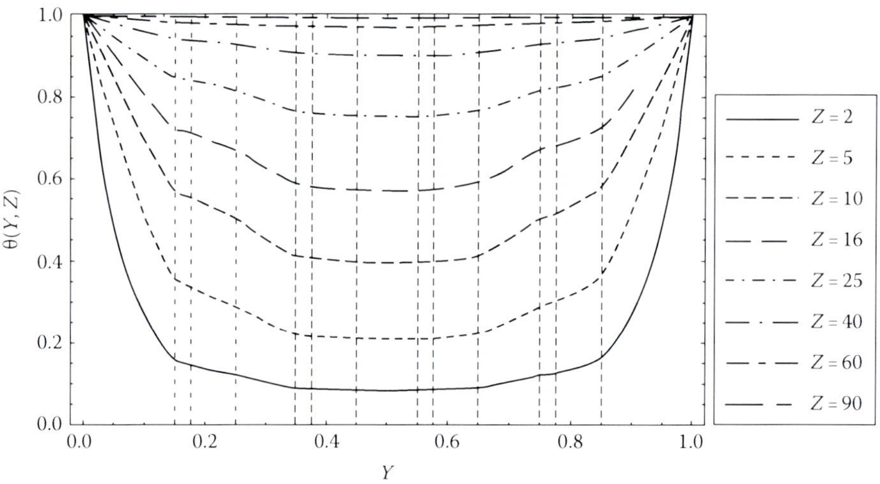

FIGURE 13.8 Transversal temperature profiles at different longitudinal positions, calculated through the two methodologies described in Sections 13.2.1 and 13.2.2 for the multichannel geometry.

eigenvalue problem solution fixed with $N_{aux} = 70$, and varying the number of terms in the expansion of the temperature field, Equation 13.15, via the nonclassical eigenvalue problem—and Equation 13.39 via the partial transformation scheme. The first solution path presents a faster convergence behavior, yielding practically four converged significant digits for $N<25$. The partial transformation scheme yields at least three converged significant digits in practically all selected positions for $N<60$.

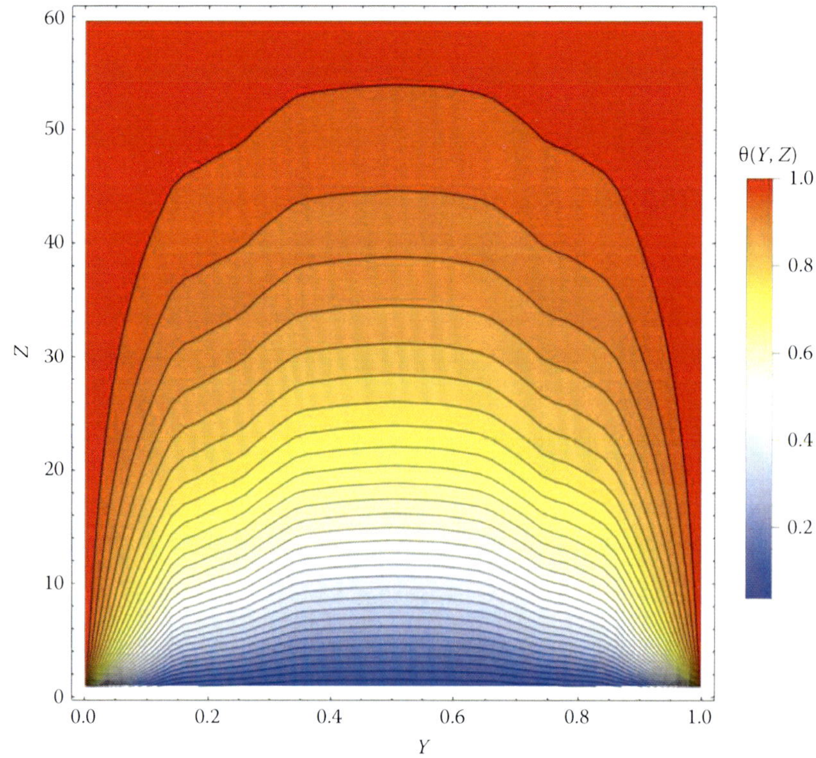

FIGURE 13.9 Calculated isotherms for the two-phase stratified flow in a four-channels device.

Figure 13.8 presents a few transversal temperature profiles at different longitudinal positions, illustrating the good adherence between the two alternative solutions throughout. Finally, Figure 13.9 depicts some isotherms for this multichannels case, which clearly illustrate the different subregions of the whole original domain, composed of solid walls and four microchannels with different fluids in stratified flow in each of them.

13.4 CONCLUSIONS

The generalized integral transform technique (GITT) is employed in combination with a single-domain reformulation strategy, to provide hybrid numerical–analytical solutions to conjugated heat transfer in microfluidic systems composed of multiple channels, fluid streams and walls, such as in stacks of direct contact microheat exchangers and stratified flow microreactors with immiscible fluids. The governing steady-state energy equations for both the walls and fluid streams, within the thermally developing region and accounting for axial heat diffusion, are then combined into a single-domain formulation that is written in terms of space variable coefficients, representing the transitions of the thermophysical properties and velocity profiles.

After this reformulation, a nonclassical eigenvalue problem with variable coefficients is first adopted, which allows for one single integral transformation over the entire spatial domain, including the axial diffusion term, offering an analytical solution to the posed problem and thus markedly simplifying the computational task. Second, a pseudotransient partial transformation scheme is also implemented for comparison purposes, which employs a classical Sturm–Liouville problem with variable coefficients, that allow for obtaining hybrid numerical–analytical results for the temperature field, along the longitudinal coordinate, from just one transformed system.

After illustrating the consistent convergence characteristics of the proposed eigenfunction expansions in each alternative solution path, a couple of physical situations are explored more closely, first for a single-channel case and second for a multiple channels case, both including direct contact heat transfer between immiscible fluids. Within the range of parameters and expansion truncation orders considered, the two alternative solutions reach an agreement to plus or minus one in the fourth significant digit of the computed dimensionless temperatures throughout the domain.

The proposed approach can be readily extended to three-dimensional and irregular geometries, as recently demonstrated for conjugated problems in both single and multiple channels configurations with just one fluid stream per channel [18–20], as well as in gas flows within the slip flow regime, also recently shown to be solvable by the same problem reformulation strategy through the introduction of a fictitious layer concept [34].

REFERENCES

1. C.B. Sobhan, G.P. Peterson, *Microscale and Nanoscale Heat Transfer: Fundamentals and Engineering Applications*, CRC Press, Boca Raton, FL, 2008.
2. N. Kockmann, *Transport Phenomena in Micro Process Engineering*, Springer-Verlag, Berlin, Germany, 2008.
3. G.L. Morini, Single-phase convective heat transfer in microchannels: A review of experimental results, *Int. J. Thermal Sci.*, 43, 631–651 (2004).
4. Y. Yener, S. Kakaç, M. Avelino, T. Okutucu, Single-phase forced convection in microchannels—A state-of-the-art review, in: S. Kakaç, L.L. Vasiliev, Y. Bayazitoglu, Y. Yener (eds.), *Microscale Heat Transfer—Fundamental Applications*, pp. 1–24, Springer, The Netherlands, 2005.
5. A. Koşar, Effect of substrate thickness and material on heat transfer in microchannel heat sinks, *Int. J. Thermal Sci.*, 49(4), 635–642 (2010).
6. G. Maranzana, I. Perry, D. Maillet, Mini and microchannels: Influence of axial conduction in the walls, *Int. J. Heat Mass Transfer*, 47, 3993–4004 (2004).

7. J.S. Nunes, R.M. Cotta, M. Avelino, S. Kakaç, Conjugated heat transfer in microchannels, in: *Microfluidics Based Microsystems: Fundamentals and Applications*, NATO Science for Peace and Security Series A: Chemistry and Biology, Springer, Dordrecht, the Netherlands, pp. 61–82, Springer, The Netherlands, 2010.
8. H. Herwig, O. Hausner, Critical view on "new results in micro-fluid mechanics": An example, *Int. J. Heat Mass Transfer*, 46(5), 935–937 (2003).
9. B. Çetin, A. Güvenç Yazicioğlu, S. Kakaç, Slip-flow heat transfer in microtubes with axial conduction and viscous dissipation—An extended Graetz problem, *Int. J. Thermal Sci.*, 48(9), 1673–1678 (2009).
10. D.C. Knupp, C.P. Naveira-Cotta, R.M. Cotta, Theoretical analysis of conjugated heat transfer with a single domain formulation and integral transforms, *Int. Commun. Heat Mass Transfer*, 39(3), 355–362 (2012).
11. R.M. Cotta, Hybrid numerical-analytical approach to nonlinear diffusion problems, *Num. Heat Transfer Part B*, 127, 217–226 (1990).
12. R.M. Cotta, Benchmark results in computational heat and fluid flow: The integral transform method, *Int. J. Heat Mass Transfer*, 37, 381–393 (1994).
13. R.M. Cotta, *Integral Transforms in Computational Heat and Fluid Flow*, CRC Press, Boca Raton, FL, 1993.
14. R.M. Cotta, M.D. Mikhailov, *Heat Conduction: Lumped Analysis, Integral Transforms, Symbolic Computation*, John Wiley & Sons, Chichester, U.K., 1997.
15. D.C. Knupp, C.P. Naveira-Cotta, R.M. Cotta, Conjugated convection-conduction analysis in microchannels with axial diffusion effects and a single domain formulation, *J. Heat Transfer*, 135(9), 091008 (2013).
16. D.C. Knupp, R.M. Cotta, C.P. Naveira-Cotta, Heat transfer in microchannels with upstream–downstream regions coupling and wall conjugation effects, *Numer. Heat Transf. Part B*, 64(5), 365–387 (2013).
17. D.C. Knupp, R.M. Cotta, C.P. Naveira-Cotta, S. Kakaç, Transient conjugated heat transfer in microchannels: Integral transforms with single domain formulation, *Int. J. Thermal Sci.*, 88, 248–257 (2015).
18. D.C. Knupp, R.M. Cotta, C.P. Naveira-Cotta, Fluid flow and conjugated heat transfer in arbitrarily shaped channels via single domain formulation and integral transforms, *Int. J. Heat Mass Transfer*, 82, 479–489 (2015).
19. D.C. Knupp, C.P. Naveira-Cotta, A. Render, M.K. Tiwari, R.M. Cotta, D. Poulikakos, Analysis of conjugated heat transfer in micro-heat exchangers via integral tranforms and non-intrusive optical techniques, *Int. J. Numer. Meth. Heat Fluid Fl.*, 25(6), 1444–1462 (2015).
20. D.C. Knupp, C.P. Naveira-Cotta, R.M. Cotta, Theoretical–experimental analysis of conjugated heat transfer in nanocomposite heat spreaders with multiple microchannels, *Int. J. Heat Mass Transfer*, 74, 306–318 (2014).
21. S. Zimmermann, I. Meijer, M.K. Tiwari, S. Paredes, B. Michel, D. Poulikakos, Aquasar: A hot water cooled data center with direct energy reuse, *Energy*, 43, 237–245 (2012).
22. C.S. Sharma, S. Zimmermann, M.K. Tiwari, B. Michel, D. Poulikakos, Optimal thermal operation of liquid-cooled electronic chips, *Int. J. Heat Mass Transfer*, 55, 1957–1969 (2012).
23. W. Escher, T. Brunschwiler, B. Michel, D. Poulikakos, Experimental investigation of an ultrathin manifold microchannel heat sink for liquid-cooled chips, *J. Heat Transfer*, 132, 081402 (2010).
24. T.-W. Lee, H.-S. Kim, T.-H. Tran, J. Jang, D.S. Yoon, J.-H. Kim, E.-K. Kim, Y.-M. Koo, S.W. Lee, W.-J. Chang, Diffusion-based multi-stream bioluminescent reaction in a microfluidic device, *Chem. Eng. J.*, 185–186, 321–327 (2012).
25. C.L. Ong, J.R. Thome, Macro-to-microchannel transition in two-phase flow: Part 1—Two-phase flow patterns and film thickness measurements, *Exp. Therm. Fluid Sci.*, 35(1), 37–47 (2011).
26. P. Valiorgue, S.N. Ritchey, J.A. Weibel, S.V. Garimella, Design of a non-intrusive electrical impedance-based void fraction sensor for microchannel two-phase flows, *Meas. Sci. Technol.*, 25(9), 095301 (2014).
27. A. Asthana, I. Zinovik, C. Weinmueller, D. Poulikakos, Significant Nusselt number increase in microchannels with a segmented flow of two immiscible liquids: An experimental study, *Int. J. Heat Mass Transfer*, 54(7–8), 1456–1464 (2011).
28. R.B. Bird, W.E. Stewart, E.N. Lightfoot, *Transport Phenomena*, 2nd edn. John Wiley & Sons, New York, 2002.
29. B. Vick, M.N. Özişik, Y. Bayazitoğlu, A method of analysis of low Peclet number thermal entry region problems with axial conduction, *Lett. Heat Mass Transfer*, 7(4), 235–248 (1980).
30. B. Vick, M.N. Özişik, An exact analysis of low Peclet number heat transfer in laminar flow with axial conduction, *Lett. Heat Mass Transfer*, 8(1), 1–10 (1981).

31. M.C. Oliveira, R. Ramos, R.M. Cotta, On the eigenvalues basic to the analytical solution of convective heat transfer with axial diffusion effects, *Commun. Numer. Methods Eng.*, 11(4), 287–196 (1995).
32. S. Wolfram, *The Mathematica Book*, Wolfram Media, Cambridge, U.K., 2005.
33. R.M. Cotta, D.C. Knupp, C.P. Naveira-Cotta, L.A. Sphaier, J.N.N. Quaresma, The Unified Integral Transforms (UNIT) algorithm with total and partial transformation, *Comput. Therm. Sci.*, 6(6), 507–524 (2014).
34. D.C. Knupp, C.P. Naveira-Cotta, R.M. Cotta, Conjugated heat transfer in microchannels with slip flow regime via single domain formulation and integral transforms, *Proceedings of the Seventh International Symposium on Advances in Computational Heat Transfer*, CHT-15, ICHMT, Piscataway, NJ, May 24–29, 2015.

14 Direct-Inverse Problem Analysis in the Thermal Characterization of Microsystems

Carolina P. Naveira-Cotta

CONTENTS

14.1 INTRODUCTION

The proper simulation of microfluidic systems, in different applications, requires the accurate knowledge of physical properties of both working fluids and substrates, transport and other constitutive parameters inherent to the specific modeling employed, geometries and dimensions, besides the acting domain and boundaries source functions. Every and each of these model input information may affect the simulation accuracy and the corresponding adherence to experimental predictions at the microscale. While specific experimental techniques and correlations are available in many situations to estimate to a certain extent these quantities, it is also quite desirable to identify at least those more uncertain parameters and functions, ideally, under actual working conditions.

Nonintrusive temperature measurementcan be a powerful tool in the thermal characterization of microsystems. For such an experimental characterization, it is of crucial importance to employ reliable temperature measurement techniques capable of capturing the essential thermal phenomena that take place throughout the domain, providing substantial amount of information with high spatial resolution and high frequency, such as achievable through infrared camera thermography, with or without microscopic lenses [1–6].

Such a technique can be especially powerful in combination with an accurate direct problem solution approach and a robust inverse problem analysis methodology, allowing for parametric and functional identifications in different working conditions and classes of microsystems. This aspect has been fairly recently demonstrated in the thermal characterization of heterogeneous media, such as nanocomposites and functionally graded materials, toward the identification of space-variable thermophysical properties and source functions in different heat conduction applications [7–12].

The common aspects on these previous developments are the employment of the Generalized Integral Transform Technique (GITT) [13–18] in the direct problem solution, a hybrid numerical-analytical technique for convection-diffusion problems with automatic global accuracy control, and the use in the inverse problem solution of a robust Bayesian inference approach based on the Markov Chain Monte Carlo (MCMC) method [19–22]. When combined, the resulting direct-inverse problem analysis permits the identification of parameters in the transformed domain, thus providing a marked collapse of experimental or simulated data, besides allowing for the representation of the unknown functions to be estimated in terms of eigenfunctions themselves. These two novel aspects of this advanced direct-inverse problem analysis are responsible for both cost effectiveness and robustness in the reconstruction of the space-variable properties and sources.

Also quite recently, this same approach has been proposed to the thermal characterization of microsystems [23,24], with promising results. Thus, the present chapter illustrates the combination of integral transforms for the direct problem solution and Bayesian inference for the inverse problem analysis, as applied to the thermal characterization of distinct microsystems, namely, a microchannel with slip flow conditions and a micromodel of a porous medium with liquid flow. In both situations, the fluid is considered to enter the microsystem at known temperature and mass flow rate, exchanging heat by convection with the external environment, and accounting for conjugation effects with the microsystem walls. From the external wall temperature measurements along the channel length at steady state, governing parameters of the heat and fluid flow processes in the microsystem can be identified, as verified by an *a priori* sensitivity analysis. The Generalized Integral Transform Technique (GITT) is employed in the solution of the conjugated conduction-convection problem in the microchannels, providing an accurate solution to the formulated direct problem in each case, and yielding simulated experimental results with prescribed noise. Then, the Markov Chain Monte Carlo method (MCMC) is utilized, with the Metropolis–Hastings sampling algorithm, to achieve the identification of the governing parameters in each situation. Numerical results illustrate the capabilities of the proposed characterization approach, providing estimated parameters and associated confidence intervals for a fairly wide range of conditions.

14.2 DIRECT PROBLEM SOLUTION

In order to illustrate the direct-inverse problem analysis, we consider the 2D situation of a parallel-plates microchannel, with steady thermally developing laminar flow. In the applications that follow, both a gas flow between parallel plates in the slip flow regime and a liquid flow in a porous media filled microchannel are analyzed. The fluid is assumed to enter the channel with a fully developed velocity profile and a uniform temperature, exchanging heat by convection with the surroundings with an external heat transfer coefficient that is not known *a priori* in the inverse problem analysis. Thermophysical properties are assumed to be constant, while axial conduction and viscous dissipation are neglected. Although more involved formulations can be handled by the proposed approach, the direct problem solution here illustrated suffices to demonstrate the applicability of the proposed direct-inverse problem analysis. In dimensionless form, the energy equation for the fluid within the microchannel, accounting for the symmetry of the heat transfer process, either for the slip flow model or the micromodel of a porous media, can be written as

$$W(Y)\frac{\partial\theta(Y,Z)}{\partial Z}=\frac{\partial^2\theta(Y,Z)}{\partial Y^2},\quad 0<Y<1,\ Z>0 \tag{14.1}$$

$$\theta(Y,0)=1,\quad 0\le Y\le 1 \tag{14.2}$$

$$\left.\frac{\partial\theta(Y,Z)}{\partial Y}\right|_{Y=0}=0 \tag{14.3}$$

$$\frac{\partial\theta(Y,Z)}{\partial Y}\Big|_{Y=1} = -Bi_{eff}\theta(1,z), \quad Z > 0 \tag{14.4}$$

where the corresponding dimensionless groups shall be later on defined for each application here considered. The function $W(Y)$ is related to the fully developed velocity profile across the microchannel, while the parameter Bi_{eff} will include the temperature jump condition, and in the case of a slip flow regime, the external convective heat transfer and the conjugation effect represented by the heat conduction across the wall.

The solution of the dimensionless problem (Equations 14.1 through 14.4) is then a special case from the general solution given in [25] and can be written as

$$\theta(Y,Z) = \sum_{i=1}^{\infty} \bar{f}_i \tilde{\psi}_i(Y) e^{-\mu_i^2 Z} \tag{14.5}$$

$$\bar{f}_i = -\frac{\tilde{\psi}_i'(1)}{\mu_i^2} \tag{14.6}$$

where $\psi_i(Y)$ are eigenfunctions of the following Sturm–Liouville problem

$$\frac{d^2\psi_i(Y)}{dY^2} + \mu_i^2 W(Y)\psi_i(Y) = 0, \quad 0 < Y < 1 \tag{14.7}$$

$$\frac{d\psi_i(Y)}{dY}\Big|_{Y=0} = 0 \tag{14.8}$$

$$\frac{d\psi_i(Y)}{dY}\Big|_{Y=1} = -Bi_{eff}\psi_i(1) \tag{14.9}$$

with the corresponding normalization integral and normalized form of the eigenfunction:

$$N_i = \int_0^1 W(Y)\psi_i^2(Y)dY \tag{14.10}$$

$$\tilde{\psi}_i(Y) = \frac{\psi_i(Y)}{N_i^{1/2}} \tag{14.11}$$

The solution of problem (Equations 14.7 through 14.9) was obtained in [26–27] in terms of the confluent hypergeometric function, also known as Kummer function, which was readily available in the *Mathematica system* [28]. Here, a more convenient path for the corresponding inverse problem analysis was chosen, in light of the intensive computational task required by the parameters identification algorithm, so as to provide a solution in terms of simpler functions. Thus, following the ideas in the Generalized Integral Transform Technique, GITT [13–18], the solution of problem (14.7 through 14.9) is provided as an eigenfunction expansion from a simpler eigenvalue problem [25] that retains the same boundary conditions of the original problem, but avoids the variable coefficient corresponding to the dimensionless velocity field, in the form

$$\frac{d^2\Omega_n(Y)}{dY^2} + \lambda_n^2\Omega_n(Y) = 0, \quad 0 < Y < 1 \tag{14.12}$$

$$\frac{d\Omega_n(Y)}{dY}\Big|_{Y=0} = 0 \quad (14.13)$$

$$\frac{d\Omega_n(Y)}{dY}\Big|_{Y=1} = -Bi_{eff}\Omega_n(1) \quad (14.14)$$

which is readily solved as

$$\Omega_n(Y) = \cos(\lambda_n Y) \quad (14.15)$$

$$\lambda_n \tan(\lambda_n) = Bi_{eff}, \quad n = 1,2,\ldots \quad (14.16)$$

Once the auxiliary eigenfunctions and eigenvalues have been obtained, we may express the desired eigenfunction of the original problem as an expansion of these simpler functions:

$$\psi_i(Y) = \sum_{n=1}^{\infty} \tilde{\Omega}_n(Y)\bar{\psi}_{i,n}, \quad \textit{inverse} \quad (14.17)$$

$$\bar{\psi}_{i,n} = \int_0^1 \tilde{\Omega}_n(Y)\psi_i(Y)dY, \quad \textit{transform} \quad (14.18)$$

The integral transformation of the original eigenvalue problem is then performed by employing the operator $\int_0^1 \tilde{\Omega}_n(Y) - dY$, over Equation 14.7, which results in the following algebraic eigenvalue problem for the original problem eigenvalues and corresponding eigenvectors:

$$\left(\mathbf{A} - \mu^2\mathbf{B}\right)\bar{\psi} = 0 \quad (14.19)$$

$$\bar{\psi} = \left\{\bar{\psi}_{n,m}\right\} \quad (14.20)$$

$$\mathbf{B} = \{B_{n,m}\}, \quad B_{n,m} = \int_0^1 W(Y)\tilde{\Omega}_n(Y)\tilde{\Omega}_m(Y)dY \quad (14.21)$$

$$A_{n,m} = \lambda_n^2\delta_{n,m}, \quad \text{where } \delta_{n,m} = 1, \text{ for } n = m, \quad \text{or} \quad \delta_{n,m} = 0, \text{ for } n \neq m \quad (14.22)$$

The algebraic problem (Equations 14.19 through 14.22) can be numerically solved to provide results for the eigenvalues μ^2 and eigenvectors $\bar{\psi}$ from this matrix eigenvalue problem analysis [25], which will be combined within the inverse formula (14.17) to provide the desired eigenfunctions of the original eigenvalue problem.

The fluid temperature at the interface with the internal wall is then directly computed from the inverse formula as

$$\theta(1,Z) = \sum_{i=1}^{\infty} \bar{f}_i\tilde{\psi}_i(1)e^{-\mu_i^2 Z} \quad (14.23)$$

and the corresponding internal and external wall temperatures, accounting or not for the temperature jump, for the heat conduction across the wall and heat convection with the external environment, can then be computed for each specific application, in what follows.

The fluid average temperature and the local Nusselt number based on the external wall temperature, along the channel length, are then determined from

$$\theta_{av}(Z) = \sum_{i=1}^{\infty} \frac{(\tilde{\psi}_i'(1))^2}{\mu_i^4} e^{-\mu_i^2 Z} \tag{14.24}$$

$$Nu(Z) = \frac{\sum_{i=1}^{\infty} \frac{(\tilde{\psi}_i'(1))^2}{\mu_i^2} e^{-\mu_i^2 Z}}{\theta_{av}(Z) - \theta_w(Z)} \tag{14.25}$$

14.3 INVERSE PROBLEM SOLUTION

In this work, we employ Bayesian inference techniques for the estimation of parameters in both the microchannel with slip flow and the porous medium filled microchannel forced convection problem, via the MCMC method [19–22], as applied to the simultaneous identification of the hydraulic and thermal parameters in the original problem formulation. The Metropolis–Hastings algorithm is employed for the sampling procedure, implemented in the *Mathematica* platform [7–9].

Consider the vector of parameters appearing in the present physical model, $\mathbf{P}^T$, where N_p is the number of parameters. For the solution of the inverse problem of estimating **P**, we assume the available measured temperature data given by

$$(\mathbf{Y}-\mathbf{T})^T = \left(\vec{Y}_1 - \vec{T}_1, \vec{Y}_2 - \vec{T}_2, \ldots, \vec{Y}_{N_x} - \vec{T}_{N_x}\right) \tag{14.26}$$

where $\vec{Y}_i$ contains the measured temperatures for each of the N_x sensors at time t_i, $i = 1, \ldots, N_t$, so that we have $N_{meas} = N_x \cdot N_t$ measurements in total. In the present steady-state estimation procedure, the sensors are assumed to be distributed along the channel wall length and the measurements are taken as an average in a time interval within the steady period ($N_t = 1$), so $N_{meas} = N_x$.

Bayes' theorem can then be stated as [19,20]

$$p_{posterior}(\mathbf{P}) = p(\mathbf{P}|\mathbf{Y}) = \frac{p(\mathbf{P})p(\mathbf{Y}|\mathbf{P})}{p(\mathbf{Y})} \tag{14.27}$$

where

- $p_{posterior}(\mathbf{P})$ is the posterior probability density, that is, the conditional probability of the parameters **P** given the measurements **Y**
- $p(\mathbf{P})$ is the prior density, that is, a statistical model for the information about the unknown parameters prior to the measurements
- $p(\mathbf{Y}|\mathbf{P})$ is the likelihood function, which gives the relative probability density of different measurement outcomes **Y** with a fixed **P**;
- $p(\mathbf{Y})$ is the marginal probability density of the measurements, which plays the role of a normalizing constant

In this work, we assume that the measurement errors are Gaussian random variables, with known (modeled) means and covariances, and that the measurement errors are additive and independent of the unknowns. With these hypotheses, the likelihood function can be expressed as [19,20]

$$p(\mathbf{Y}|\mathbf{P}) = (2\pi)^{-M/2}|\mathbf{W}|^{-1/2}\exp\left\{-\frac{1}{2}[\mathbf{Y}-\mathbf{T}(\mathbf{P})]^T\mathbf{W}^{-1}[\mathbf{Y}-\mathbf{T}(\mathbf{P})]\right\} \tag{14.28}$$

where **W** is the covariance matrix of the measurement errors.

Unfortunately, for most problems of practical relevance, it is complicated to generate the posteriori $p(\mathbf{P}|\mathbf{Y})$. Therefore, more sophisticated methods are required to obtain a sample of $p(\mathbf{P}|\mathbf{Y})$, for example, the simulation technique based on Markov chains [20]. The numerical method most used to explore the space of states of the posteriori is the Monte Carlo approach. The Monte Carlo simulation is based on a large number of samples of the probability density function (in this case, the function of the posterior probability density $p(\mathbf{P}|\mathbf{Y})$). Several sampling strategies are proposed in the literature, including the Monte Carlo method with Markov Chain (MCMC), adopted in this work, where the basic idea is to simulate a "random walk" in the space of $p(\mathbf{P}|\mathbf{Y})$ that converges to a stationary distribution. The Metropolis–Hastings algorithm, which is here employed, is the most commonly used Monte Carlo method with Markov Chain algorithm [19,20]. In this algorithm, a value is generated from an auxiliary distribution and accepted with a given probability. This correction mechanism ensures the convergence of the chain to the equilibrium distribution. That is, the algorithm now includes an additional step, where the transition mechanism depends on a proposal for a transition and a stage of assessing the equilibrium density, but this is represented by the global transition via the probability of acceptance.

The implementation of the Metropolis–Hastings algorithm starts with the selection of the transition core $q(\mathbf{P}^*|\mathbf{P})$, which is used to draw the new candidate, state $\mathbf{P}^*$, given the current state **P** of the Markov chain. The new value $\mathbf{P}^*$ is accepted with a certain probability given by the ratio of Hastings $RH(\mathbf{P}, \mathbf{P}^*)$. In practical terms, the transition core $q(\mathbf{P}^*|\mathbf{P})$ defines only a proposal for a movement that can be confirmed by $RH(\mathbf{P}, \mathbf{P}^*)$, for this reason it is usually called the proposal or auxiliary probability density function. The Metropolis–Hastings sampling algorithm can be implemented by repeating the following steps:

1. Boot up the iterations counter of the chain $i = 0$ and assign an initial value $\mathbf{P}(0)$.
2. Generate a candidate value $\mathbf{P}^*$ of the distribution $q(\mathbf{P}^*|\mathbf{P})$.
3. Calculate the probability of acceptance of the candidate value $RH(\mathbf{P}, \mathbf{P}^*)$ by

$$RH(\mathbf{P},\mathbf{P}^*) = \min\left[1, \frac{p(\mathbf{P}^*|\mathbf{Y})\,q(\mathbf{P}^*|\mathbf{P})}{p(\mathbf{P}|\mathbf{Y})\,q(\mathbf{P}|\mathbf{P}^*)}\right] \tag{14.29}$$

4. Generate a random number u with uniform distribution, i.e., $u \sim U(0, 1)$,
5. If $u \le RH(\mathbf{P}, \mathbf{P}^*)$ then the new value is accepted and we let $\mathbf{P}(i+1) = \mathbf{P}^*$. Otherwise, the new value is rejected and we let$\mathbf{P}(i+1) = \mathbf{P}(i)$.
6. Increase the counter i to $i + 1$ and return to step 2.

In this way, a sequence is generated to represent the posterior distribution and inference on this distribution is obtained from inference on the samples $\{\mathbf{P}(1), \mathbf{P}(2), \ldots, \mathbf{P}(i), \mathbf{P}(i+1), \ldots\}$. We note that values of $\mathbf{P}(i)$ must be ignored until the chain has converged to equilibrium (the burn-in period). The success of the method depends on not so low acceptance rates and proposals that are easy to simulate. The method replaces a difficult to generate $p(\mathbf{P}|\mathbf{Y})$ by several generations of the proposal $q(\mathbf{P}^*|\mathbf{P})$. In this study we have chosen to adopt symmetrical chains, i.e., $q(\mathbf{P}^*|\mathbf{P}) = q(\mathbf{P}|\mathbf{P}^*)$ for all $(\mathbf{P}^*,\mathbf{P})$. In this case the ratio of Hastings, $RH(\mathbf{P}, \mathbf{P}^*)$ reduces to the ratio of the posterior densities calculated at the previous and proposed chain positions, and does not depend on $q(\mathbf{P}^*|\mathbf{P})$. For more details on theoretical aspects of the Metropolis–Hastings algorithm and MCMC methods, the reader should consult references [19,20].

Before addressing the estimation of the unknown parameters, the behavior of the determinant of the matrix $\mathbf{J}^T\mathbf{J}$ [29] needs to be analyzed in order to inspect the influence of the number of parameters to be estimated in the solution of the inverse problem. The sensitivity matrix $\mathbf{J}$ is defined as

$$\mathbf{J}(\mathbf{P}) = \left[\frac{\partial \mathbf{T}^T(\mathbf{P})}{\partial \mathbf{P}}\right]^T = \begin{bmatrix} \frac{\partial T_1}{\partial P_1} & \frac{\partial T_1}{\partial P_2} & \frac{\partial T_1}{\partial P_3} & \cdots & \frac{\partial T_1}{\partial P_{N_p}} \\ \frac{\partial T_2}{\partial P_1} & \frac{\partial T_2}{\partial P_2} & \frac{\partial T_2}{\partial P_3} & \cdots & \frac{\partial T_2}{\partial P_{N_P}} \\ \vdots & \vdots & \vdots & & \vdots \\ \frac{\partial TN_{meas}}{\partial P_1} & \frac{\partial TN_{meas}}{\partial P_2} & \frac{\partial TN_{meas}}{\partial P_3} & \cdots & \frac{\partial TN_{meas}}{\partial P_{N_p}} \end{bmatrix} \quad (14.30)$$

The sensitivity coefficients $J_{ij} = \partial T_i/\partial P_j$ give the sensitivity of T_i (solution of the direct problem) with respect to changes in the parameter P_j. A small value of the magnitude of J_{ij} indicates that large changes in P_j yield small changes in T_i. It can be easily noticed that the estimation of the parameter P_j is extremely difficult in such cases, because basically the same value for T_i would be obtained for a wide range of values of P_j. In fact, when the sensitivity coefficients are small, $|\mathbf{J}^T\mathbf{J}| \approx 0$ the inverse problem is said to be ill-conditioned. It can also be shown that $|\mathbf{J}^T\mathbf{J}|$ is null if any column of $\mathbf{J}$ can be expressed as a linear combination of other columns [29]. Therefore, it is desirable to have linearly-independent sensitivity coefficients J_{ij} with large magnitudes, so that the parameter estimation problem is not very sensitive to measurement errors, and accurate estimates of the parameters can be obtained. The comparison of the magnitude of the sensitivity coefficients, as well as the analysis of possible linear dependence, is more easily performed by using the reduced sensitivity coefficients instead of the original ones. The reduced sensitivity coefficients are obtained by multiplying the original sensitivity coefficients, J_{ij}, by the parameters they refer to. Therefore, they have units of the measured variables, which is used as a basis of comparison.

14.4 APPLICATIONS

14.4.1 Slip Flow Regime

The consideration of slip flow in the case of gas heat and fluid flow in microchannels, with its corresponding temperature jump condition for convective heat transfer, accounting for noncontinuum effects at the fluid-surface interactions, has been the subject of numerous investigations, and has been handled both analytically and numerically in previous works [26,27,30–34]. The usefulness of these solutions for the forced convection problem in microchannels, such as the accurate hybrid approach presented in Section 14.2, is highly dependent on the availability of reliable estimates for the momentum and thermal accommodation coefficients, as required by the slip and temperature jump boundary conditions. Fundamental experimental work on rarefied gas dynamics have offered measurements of the tangential momentum accommodation coefficient, requiring, for instance, high vacuum and molecular beams impinging on carefully prepared substrates, such as recently reviewed in [35], but a very few results are available for the actual conditions of the flow configuration within microchannels and their actual bounding walls [36,37]. The experiments indicate that the tangential momentum accommodation coefficient generally assumes values between 0.2 and 1.0, with the lower limit being associated with exceptionally smooth surfaces and the upper limit with very rough or highly oxidized surfaces [38]. Similar considerations are pertinent to the measurement of thermal accommodation coefficients [39], where an even more limited experimental database is available, and apparently no previous work seems to be available on the identification of this coefficient in actual heat and fluid flow conditions within specific pressure and temperature levels pertinent to microfluidics applications, and in addition for actual morphology and finishing of the microchannel walls.

Thus, the summarized and previously presented work in [23] takes advantage of the accuracy, robustness, and efficiency of the direct problem solution of Section 14.2, to tackle the associated inverse heat transfer problem, toward the simultaneous estimation of momentum and thermal accommodation coefficients in microchannel flows with velocity slip and temperature jump.

The example here considered corresponds to a parallel plates microchannel with the distance between the plates given by $2y_1$, average flow velocity u_{av}, thermal conductivity of the fluid k_f, thermal diffusivity of the fluid α_f, inlet flow temperature T_0, external environment temperature T_∞, and external heat transfer coefficient h_e. Conjugation effects are disregarded in this particular example and the Knudsen number, Kn, is within the slip flow regime bounds. The corresponding dimensionless groups are given by

$$Y = \frac{y}{y_1} \tag{14.31}$$

$$Z = \frac{\alpha_f z}{u_{av} y_1^2} \tag{14.32}$$

$$\theta(Y,Z) = \frac{T(y,z) - T_\infty}{T_0 - T_\infty} \tag{14.33}$$

$$W(Y) = \frac{u(y)}{u_{av}} \tag{14.34}$$

$$Bi = \frac{h_e y_1}{k_f} \tag{14.35}$$

$$Kn = \frac{\lambda}{2y_1} \tag{14.36}$$

$$Bi_{eff} = \frac{Bi}{1 + 2Kn\beta_t Bi} \tag{14.37}$$

and,

$$\beta_t = \frac{(2-\alpha_t)}{\alpha_t} \frac{2\gamma}{(\gamma+1)} \frac{1}{Pr} \tag{14.38}$$

is the wall temperature jump coefficient and α_t is the thermal accommodation coefficient, λ is the molecular mean free path, $\gamma = c_p/c_v$, while c_p is specific heat at constant pressure, c_v is specific heat at constant volume, and Pr is the Prandtl number. The dimensionless velocity profile is given as

$$W(Y) = \frac{6Kn\beta_v + 3(1-Y^2)/2}{1 + 6Kn\beta_v} \tag{14.39}$$

where

$$\beta_v = \frac{(2-\alpha_m)}{\alpha_m} \tag{14.40}$$

is the wall velocity slip coefficient and α_m is the tangential momentum accommodation coefficient. The ratio of the boundary conditions coefficients is also of interest and given as $\beta = \beta_t/\beta_v$.

The corresponding temperature jump is accounted for providing the temperature distribution at the wall:

$$\theta_w(Z) = \frac{1}{1+2KnBi\beta_t}\theta(1,Z) \tag{14.41}$$

The vector of parameters to be estimated in the slip flow example is then given by

$$\mathbf{P}^T = [\beta_v, \beta_t, Bi_{eff}] \tag{14.42}$$

14.4.2 Micromodel of Porous Media

Micromodels of porous media have been recently constructed and employed in the proposition and analysis of conceptual models for contaminated soils remediation and enhanced oil recovery from mature reservoirs [40–42]. Different microfabrication techniques have been utilized in constructing such pore scale 2D and 3D physical models of the porous medium, to be studied, also known as ROC (rock-on-a-chip) [43]. The aim is to obtain a fundamental and an integrated understanding of all the transport processes, to enable more efficient and cost-effective design of either remediation strategies for cleaning up a variety of recalcitrant chemicals in the subsurface or of the oil recovery technique from the petroleum reservoir. This is to be accomplished through a combination of theoretical pore-scale modeling of the processes involved, calibration and testing of the derived formulations using experimental data obtained with the physical micromodels, and upscaling of the pore-scale processes to the field scale [44]. The utilization of physical micromodels in the analysis of flow and transport in porous media then requires a careful characterization of the microfluidic chip, both in terms of the physical dimensional metrology and of the related hydraulic parameters in the macroscale modeling of the flow through this porous channel.

The physical dimensions and properties in the present example, previously presented in [24], are taken from an actually constructed micromodel [43,45], with regular porous structure for validation purposes, manufactured through micromilling on a PMMA substrate and sealed with a thermal pressing machine, which can be illustrated in Figure 14.1.

An infrared camera is then employed to capture the temperature distributions on the external surface of the microfluidic chip, while flowing hot water through the micromodel. Figure 14.2 illustrates the experimental apparatus and a typical infrared camera image from the external surface of the micromodel device.

Conjugated heat transfer is accounted for in terms of transversal heat conduction across the wall, and the direct problem solution provides the theoretical external surface temperatures, which are then employed in the inverse analysis through the MCMC approach, so as to identify the hydraulic and thermal parameters of the tested micromodel. The dimensionless quantities are given by

$$X = \frac{x}{H} \tag{14.43}$$

$$Y = \frac{y}{H} \tag{14.44}$$

14.5.2 Micromodel of Porous Media

The inverse problem analysis corresponding to the application in Section 14.4.2 was previously considered in Ref. [24], for hypothetical values of the parameters [S, Pe, Bi_{eff}] and under different experimental errors in the temperature-simulated measurements. As in the previous test case, more realistic conditions are here considered, for an actual experiment and simulation recently undertaken [43,45], involving a PMMA substrate and a porous channel microfabricated from a CT-scan image of an actual petroleum reservoir. The test case data obtained from the experiment with a hot water flow within the porous microchannel built on the PMMA substrate are summarized in Table 14.4.

Table 14.5 provides the exact values assumed for the three parameters, based on the conditions described in Table 14.4, corresponding to the water flow within the porous channel on the PMMA substrate. It also shows the bounding limits for the admissible interval for the parameters, as well as the corresponding initial guesses here adopted in the MCMC algorithm.

Based on the sensitivity analysis described in more details in [24], it has been observed that S has the smallest sensitivity coefficients of the three parameters, while Pe and Bi_{eff} are linearly dependent. Thus, some difficulty can be anticipated in the simultaneous estimation of all three parameters, unless some prior information can be provided. Therefore, since information can always be obtained for all three parameters, based on mass flow rate and pressure drop measurements and correlations for heat transfer coefficients, *a priori* in the form of a normal distribution with 30% standard deviation has been assumed for all three parameters, S, Pe, Bi_{eff}. Simulated wall temperature measurements were obtained with 50 terms in the eigenfunction expansions, with an experimental error of 2% of the maximum temperature value ($\sigma = 0.02$). Figure 14.6 illustrates the set of simulated experimental data (dot points) along the channel length and the exact temperature

TABLE 14.4
Input Data for the Porous Medium Micromodel Test Case

Parameter	Water-PMMA	Parameter	Water-PMMA
Pe	20	cp_s	1450 J/kg C
$M\ Da$	10^{-4}	k_f	0.598W/m C
Bi_{eff}	0.003	ρ_f	998.2 kg/m^3
e	2 mm	cp_f	4182 J/kg C
H	200 μm	ν_f	1.004×10^{-6}
L^*	1.5 cm	*Porosity*	30%
k_s	0.18 W/m C	N_{meas}	750
ρ_s	1190 kg/m^3		

TABLE 14.5
Exact, Minimum, and Maximum Bounds, and Initial Guesses for the Parameters in Porous Medium Micromodel Test Case

Parameter	Exact	Minimum	Maximum	Initial Guess
S	100	10	1000.0	160.0
Pe	20.0	2	200.0	101.0
Bi_{eff}	0.003	0.0003	0.03	0.01515

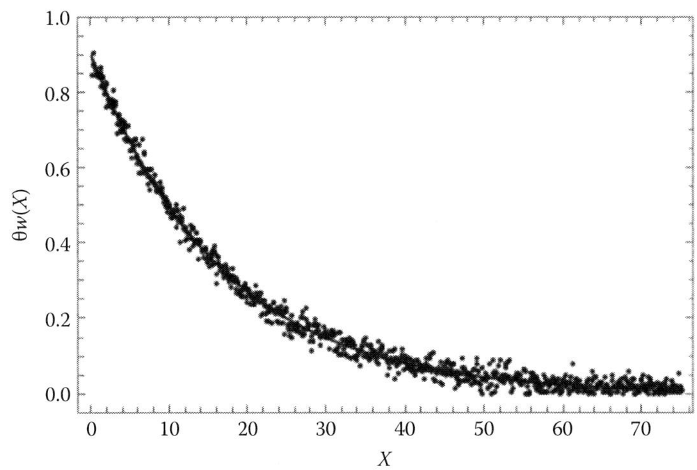

FIGURE 14.6 Simulated wall temperature measurements with 2% error.

curve (solid line), corresponding to the temperature distribution along the external face of the substrate wall.

The direct problem solution within the inversion algorithm is employed with just 10 terms, which is sufficient to achieve convergence in the temperature field for the range of dimensionless length employed.

Figure 14.7a through c illustrates the Markov chains evolution, for each of the three parameters to be identified, up to a total of 50,000 states in the MCMC algorithm. A burning up period of 6,000 states has been adopted. Clearly, the parameter S has the slowest converging chain, as anticipated by the sensitivity analysis presented in [24].

Table 14.6 provides the estimated results for the parameters, as well as the bounds for the 95% confidence interval for the estimations. It can be noticed that the estimates are satisfactory, in light of the realistic conditions imposed to the experimental error and to the priori's standard deviations. The wider confidence interval for the parameter S was somehow expected, due to the smaller sensitivity coefficients achieved for this parameter.

Finally, Figure 14.8 provides the velocity profiles reconstructed from the estimated values of the parameters (long dashed line), compared to the exact distributions (solid line). Also shown in each figure are the profiles obtained with the corresponding lower (dotted line) and upper (dashed-dotted line) bounds of the 95% confidence intervals for each of the parameters. One may observe the excellent agreement between the exact and estimated velocity profiles.

14.6 CONCLUSIONS

The direct-inverse problem analysis here presented involves the combined use of an accurate analytic-type direct problem solution, based on the Generalized Integral Transform Technique (GITT), with a robust Bayesian inference approach, based on the MCMC algorithm. A class of internal flow heat transfer problems in microchannels was here more closely analyzed, considering a gas slip flow model application and water flow in a micromodel of porous media. In each case, a set of three parameters was identified, considering simulated experimental measurements with added realistic noise, only for the external wall temperature longitudinal distribution, as obtained from a nonintrusive technique such as infrared camera thermography. The estimated parameters involve both fluid flow and heat transfer characteristics inherent to each specific application, such as momentum and thermal accommodation coefficients in the slip flow test case, and the product

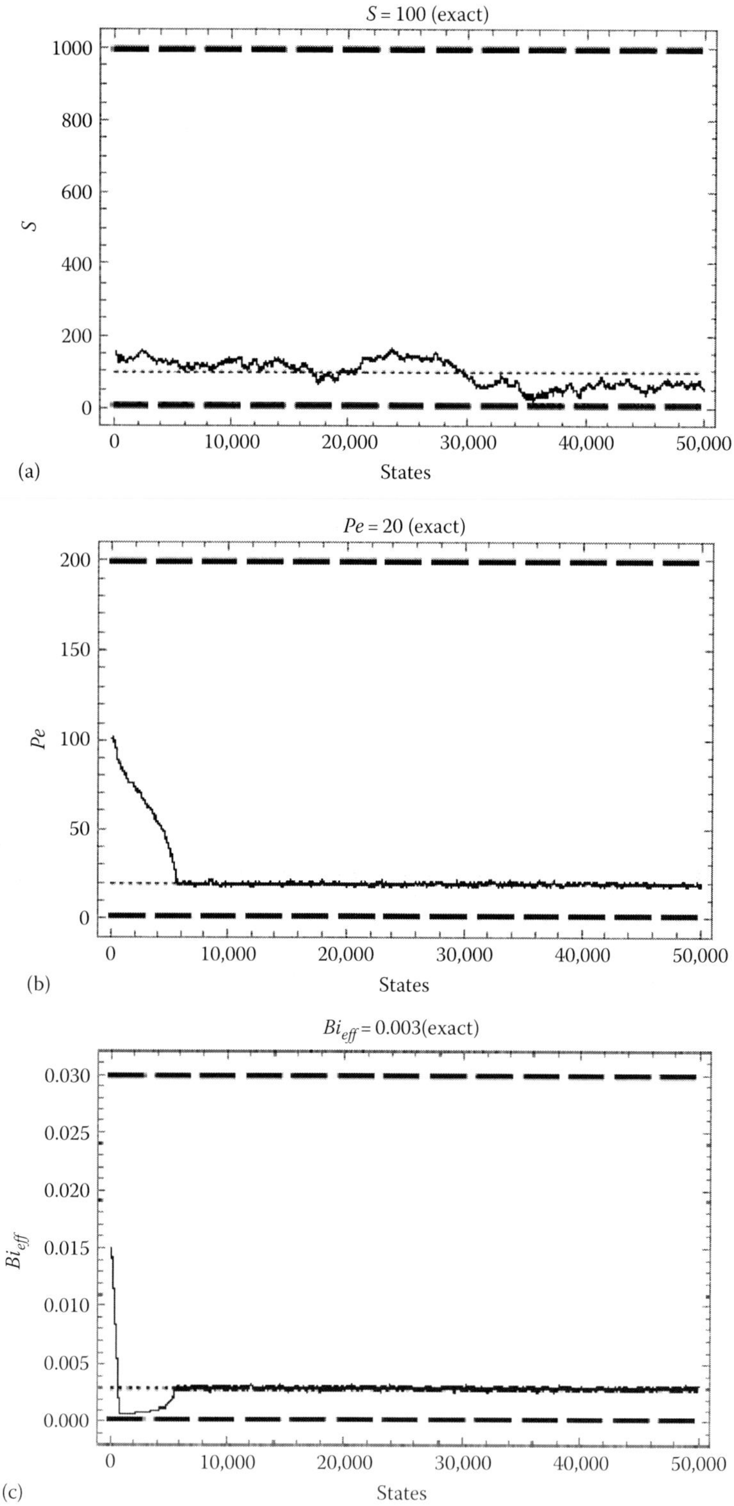

FIGURE 14.7 (a–c) Markov chains evolution for each of the three parameters in the porous medium micromodel test case [S, Pe, Bi_{eff}].

TABLE 14.6
Exact, Estimated, Minimum, and Maximum Bounds for Parameters in Porous Medium Micromodel Test Case

Parameter	Exact	Estimated	[Min, Max]
S	100	94.5363	[36.0403, 152.8300]
Pe	20.0	19.7876	[18.5264, 21.2139]
Bi_{eff}	0.003	0.003025	[0.002842, 0.003205]

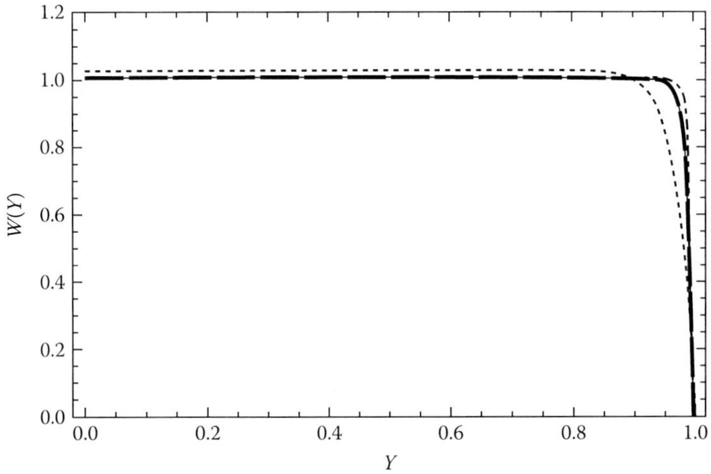

FIGURE 14.8 Reconstructed velocity profile with estimated parameters (long dashed line), with the lower (dotted line), upper (dashed-dotted line) bounds of confidence interval, and exact profile (solid line).

of effective viscosity ratio, Darcy number and Peclet number, for the porous micromodel test case. Despite the not so favorable sensitivity analysis, the three parameters were successfully estimated simultaneously in both test cases, for fairly realistic conditions, by employing prior information, when available, especially for those parameters with smaller sensitivity coefficients.

REFERENCES

1. G. Hetsroni, A. Mosyak, Z. Segal, G. Ziskind, A uniform temperature heat sink for cooling of electronic devices, *International Journal of Heat and Mass Transfer*, 45(2), 3275–3286 (2002).
2. V.A. Patil, V. Narayanan, Measurement of near-wall liquid temperatures in single phase flows through silicon microchannels, *Proceedings of the Third International Conference on Microchannels and Minichannels (ICMM 2005)*, Toronto, Ontario, Canada, 2005.
3. C. Pradère, M. Joanicot, J.-C. Batsale, J. Toutain, C. Gourdon, Processing of temperature field in chemical microreactors with infrared thermography, *QIRT Journal*, 3, 117–135 (2006).
4. J.M. Costa Junior, C.P. Naveira-Cotta, O. Fudym, Image processing of temperature fields from infrared thermography of micro-mixers with polymeric substrates, *Thermal Engineering (RETERM)*, 13(1), 71–79 (2014).
5. D.C. Knupp, C.P. Naveira-Cotta, R.M. Cotta, Theoretical-experimental analysis of conjugated heat transfer in nanocomposite heat spreaders with multiple micro-channels, *International Journal of Heat and Mass Transfer*, 74, 306–318 (2014).
6. D.C. Guerrieri, C.P. Naveira-Cotta, Experimental and theoretical analysis of a microchannel heat exchanger for high concentration photovoltaic cells, *International Symposium on Convective Heat and Mass Transfer*, CONV-14, Kusadasi, Turkey, June 2014.

7. C.P. Naveira-Cotta, H.R.B. Orlande, R.M. Cotta, Combining integral transforms and Bayesian inference in the simultaneous identification of variable thermal conductivity and thermal capacity in heterogeneous media, *ASME Journal of Heat Transfer*, 133(11), 1301–1311 (2011).
8. C.P. Naveira-Cotta, R.M. Cotta, H.R.B. Orlande, Inverse analysis with integral transformed temperature fields for identification of thermophysical properties functions in heterogeneous media, *International Journal of Heat and Mass Transfer*, 54(7–8), 1506–1519 (2011).
9. C.P. Naveira-Cotta, R.M. Cotta, H.R.B. Orlande, Estimation of space variable thermophysical properties, in: *Thermal Measurements and Inverse Techniques*, eds. H.R.B. Orlande, O. Fudym, D. Maillet, R.M. Cotta, CRC Press, Boca Raton, FL, Chapter 20, pp. 675–707, 2010.
10. D.C. Knupp, C.P. Naveira-Cotta, H.R.B. Orlande, R.M. Cotta, Experimental identification of thermophysical properties in heterogeneous materials with integral transformation of temperature measurements from infrared thermography, *Experimental Heat Transfer*, 26, 1–25 (2013).
11. D.C. Knupp, C.P. Naveira-Cotta, J.V.C. Ayres, R.M. Cotta, H.R.B. Orlande, Theoretical-experimental analysis of heat transfer in nonhomogeneous solids via improved lumped formulation, integral transforms and infrared thermography, *International Journal of Thermal Sciences*, 62, 71–84 (2012).
12. D.C. Knupp, C.P. Naveira-Cotta, J.V.C. Ayres, H.R.B. Orlande, R.M. Cotta, Space-variable thermophysical properties identification in nanocomposites via integral transforms, Bayesian inference and infrared thermography, *Inverse Problems in Science and Engineering*, 20(5), 609–637 (2012).
13. R.M. Cotta, Hybrid numerical-analytical approach to nonlinear diffusion problems, *Numerical Heat Transfer—Part B: Fundamentals*, 127, 217–226 (1990).
14. R.M. Cotta, *Integral Transforms in Computational Heat and Fluid Flow*, CRC Press, Boca Raton, FL, 1993.
15. R.M. Cotta, Benchmark results in computational heat and fluid flow: The integral transform method, *International Journal of Heat and Mass Transfer*, 37(Suppl. 1), 381–394 (1994).
16. R.M. Cotta, M.D. Mikhailov, *Heat Conduction: Lumped Analysis, Integral Transforms, Symbolic Computation*, Wiley-Interscience, Chichester, U.K., 1997.
17. R.M. Cotta, *The Integral Transform Method in Thermal and Fluids Sciences and Engineering*, Begell House, New York, 1998.
18. R.M. Cotta, M.D. Mikhailov, Hybrid methods and symbolic computations, in: *Handbook of Numerical Heat Transfer*, 2nd edn., eds. W.J. Minkowycz, E.M. Sparrow, J.Y. Murthy, John Wiley, New York, Chapter 16, 2006.
19. J. Kaipio, E. Somersalo, *Statistical and Computational Inverse Problems*, Springer-Verlag, New York, 2004.
20. D. Gamerman, H.F. Lopes, *Markov Chain Monte Carlo: Stochastic Simulation for Bayesian Inference*, 2nd edn., Chapman & Hall/CRC, Boca Raton, FL, 2006.
21. N. Zabaras, Inverse problems in heat transfer, in: *Handbook of Numerical Heat Transfer*, 2nd edn., eds. W.J. Minkowycz, E.M. Sparrow, J.Y. Murthy, Wiley, New York, pp. 525–557, 2006.
22. H.R.B. Orlande, M.J. Colaço, G.S. Dulikravich, Approximation of the likelihood function in the Bayesian technique for the solution of inverse problems, *Inverse Problems in Science and Engineering*, 16, 677–692 (2008).
23. C.P. Naveira-Cotta, H.R.B. Orlande, R.M. Cotta, Inverse analysis of forced convection in microchannels with slip flow via integral transforms and Bayesian inference, *International Journal of Thermal Sciences*, 49, 879–888 (2010).
24. C.P. Naveira-Cotta, Direct-inverse analysis of heat transfer n micro-models of porous media, *Proceedings of the 15th Int. Heat Transfer Conference*, IHTC, Kyoto, Japan, August 10–15, 2014.
25. C.P. Naveira-Cotta, R.M. Cotta, H.R.B. Orlande, O. Fudym, Eigenfunction expansions for transient diffusion in heterogeneous media, *International Journal of Heat and Mass Transfer*, 52, 5029–5039 (2009).
26. M.D. Mikhailov, R.M. Cotta, Mixed symbolic-numerical computation of convective heat transfer with slip flow in microchannels, *International Communications in Heat and Mass Transfer*, 32(3–4), 341–348 (2005).
27. R.M. Cotta, M.D. Mikhailov, S. Kakaç, Steady and periodic forced convection in microchannels, in: *Microscale Heat Transfer: Fundamentals and Applications*, eds. S. Kakaç et al., Springer, Dordrecht, the Netherlands, NATO Science Series II, pp. 193, 49–74, 2005.
28. S. Wolfram, *The Mathematica Book*, version 7.0, Cambridge-Wolfram Media, Champaign, IL, 2008.
29. M.N. Ozisik, H.R.B. Orlande, *Inverse Heat Transfer: Fundamentals and Applications*, Taylor & Francis, New York, 2000.
30. R.F. Barron, X. Wang, T.A. Ameel, R.O. Warrington, The Graetz problem extended to slip flow, *International Journal of Heat and Mass Transfer*, 40, 1817–1823 (1997).

31. F.E. Larrodé, C. Housiadas, Y. Drossinos, Slip flow heat transfer in circular tubes, *International Journal of Heat and Mass Transfer*, 43, 2669–2680 (2000).
32. G. Tunc, Y. Bayazitoglu, Heat transfer in microtubes with viscous dissipation, *International Journal of Heat and Mass Transfer*, 44, 2395–2403 (2001).
33. Y. Yener, S. Kakaç, M. Avelino, T. Okutucu, Single-phase forced convection in micro-channels—A state-of-the-art review, in: *Microscale Heat Transfer: Fundamentals and Applications*, eds. S. Kakaç, L.L. Vasiliev, Y. Bayazitoglu, Y. Yener, NATO ASI Series, Kluwer Academic Publishers, Dordrecht, the Netherlands, pp. 1–24, 2005.
34. B. Çetin, A. Gűvenç Yazicioḡlu, S. Kakaç, Slip-flow heat transfer in microtubes with axial conduction and viscous dissipation—An extended Graetz problem, *International Journal of Thermal Sciences*, 48, 1673–1678 (2009).
35. A. Agrawal, S.V. Prabhu, Survey on measurement of tangential momentum accommodation coefficient, *Journal of Vacuum Science & Technology A*, 26(4), 634–645 (2008).
36. E.B. Arkilic, K.S. Breuer, M.A. Schmidt, Mass flow and tangential momentum accommodation in silicon micromachined channels, *Journal of Fluid Mechanics*, 437, 29–43 (2001).
37. S.S. Hsieh, H.H. Tsai, C.Y. Lin, C.F. Huang, C.M. Chien, Gas flow in a long micro-channel, *International Journal of Heat and Mass Transfer*, 47, 3877–3887 (2004).
38. R.W. Barber, D.R. Emerson, Challenges in modeling gas-phase flow in microchannels: From slip to transition, *Heat Transfer Engineering*, 27(4), 3–12 (2006).
39. D.J. Rader, W.M. Trott, J.R. Torczynski, J.N. Castañeda, T.W. Grasser, *Measurements of Thermal Accommodation Coefficients*, Report SAND2005--6084, Sandia National Laboratories, Albuquerque, NM, 2005.
40. N.S. Kumar Gunda, B. Bera, N.K. Karadimitriou, S.K. Mitra, S.M. Hassanizadeh, Reservoir-on-a-Chip (ROC): A new paradigm in reservoir engineering, *Lab on a Chip*, 11, 3785–3792 (2011).
41. N.K. Karadimitriou, M. Musterd, P.J. Kleingeld, M.T. Kreutzer, S.M. Hassanizadeh, V. Joekar-Niasar, On the fabrication of PDMS micromodels by rapid prototyping, and their use in two-phase flow studies, *Water Resources Research*, 49, 2056–2067 (2013).
42. Q. Zhang, S.M. Hassanizadeh, A. Raoof, M.Th. van Genuchten, S.M. Roels, Modeling virus transport and remobilization during transient partially saturated flow, *Vadose Zone Journal*, 11(2), 2013.
43. L. Paré, C.P. Tostado, C.P. Naveira-Cotta, Characterization of micro-models of porous media fabricated using CO_2 laser machining, ENCIT 2014, *15th Brazilian Congress of Thermal Sciences and Engineering*, Belém, Brazil, November 10–13, 2014.
44. M.T. van Genuchten, C.P. Naveira-Cotta, T.H. Skaggs, A. Raoof, E.M. Pontedeiro, The use of numerical flow and transport models in environmental analyses, in: *Application of Soil Physics in Environmental Analyses*, eds. W.G. Teixeira et al., Springer International Publishing, Cham, Switzerland, Chapter 15, pp. 1–29, 2014.
45. G.S. Moura, C.P. Naveira-Cotta, Theoretical analysis of porous media micro-models, ENCIT 2014, *15th Brazilian Congress of Thermal Sciences and Engineering*, Belém, Brazil, November 10–13, 2014.
46. A. Nakayama, H. Koyama, F. Kuwahara, An analysis on forced convection in a channel filled with a Brinkman-Darcy porous medium: Exact and approximate solutions, *Warme und Stoffubertragung*, 23, 291–295 (1988).
47. D.A. Nield, A.V. Kuznetsov, M. Xiong, Thermally developing forced convection in a porous medium: Parallel-plate channel or circular tube with isothermal walls, *Journal of Porous Media*, 7(1), 19–27 (2004).

15 Application of the Lattice Boltzmann Method (LBM) to Thermal Microflows

A.A. Mohamad and Khaled M. Issa

CONTENTS

15.1 INTRODUCTION

The study of fluid flow and heat transfer at the microscale plays an important role in the development of micro-electromechanical systems (MEMS) [1,2] and their far reaching applications [3–6]. The challenge of reliable experimental measurements in microscale devices makes it difficult to discern how the various physical parameters affect the flow and thermal transport processes. Therefore, much insight can be gained from numerical simulations in the design and improvement of such devices. The reduction of channel dimensions in MEMS, however, can give rise to noncontinuum effects that cannot be appropriately modeled using the Navier–Stokes–Fourier (NSF) equations for fluid flow and heat transfer. Liquid flows in microchannels, for example, experience an increase in surface-to-volume ratio and interfacial phenomena becomes markedly important. In the presence of strong molecular interactions with the wall surface and an applied electric field, it can give rise to electrokinetic or electro-osmotic forces [7] sufficient to drive the flow. For microscale gas flows (referred to as *microflows* from this point onward), the mean free path (MFP) λ, which is the average distance a molecule travels between collisions, provides a good measure of the noncontinuum effects. In liquids, λ is close in size to the molecular diameter, whereas in gases λ can be comparable to the characteristic length of the flow geometry (H). Microflows, which are the focus of this chapter, are thus better categorized using the Knudsen number ($Kn = \lambda/H$). For low values ($Kn \leq 0.01$) the gas is considered to follow the continuum hypothesis and the NSF equations are applicable throughout the flow domain [2]. When $Kn \geq 0(10)$, the gas enters the free molecular regime where the NSF equations are inapplicable, requiring the use of one of the other appropriate techniques such as the molecular dynamics (MD) simulation. Between those two Kn limits, the flow can be further divided into two regimes: (1) slip flow ($0.01 < Kn < 0.1$) and (2) transition flow ($0.1 < Kn < 10$). In the slip flow regime, the NSF equations are still applicable in the bulk flow region with the use of appropriate velocity slip and temperature jump boundary conditions. For the transition flow regime, which lies between the slip and free molecular regimes, the kinetic boundary (or Knudsen layer) between

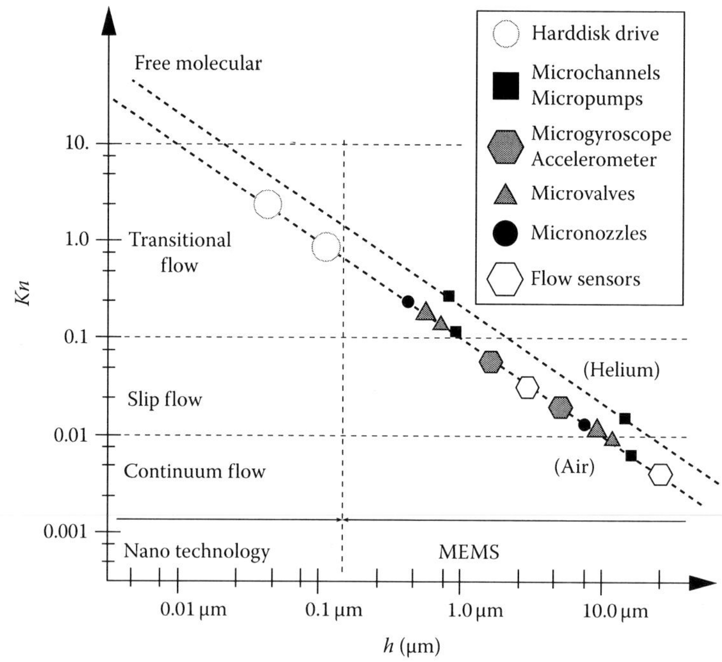

FIGURE 15.1 Flow regimes and *Kn* ranges for some typical MEMS applications. (From Karniadakis, G. et al., *Microflows and Nanoflows: Fundamentals and Simulation*, Springer, New York, 2005.)

the wall and bulk flow becomes more dominant. Flow parameters can undergo considerable variations within the Knudsen layer which must be accounted for in the numerical model to improve the accuracy of simulation results. Presence of the various flow regimes and their corresponding *Kn* in MEMS is shown in Figure 15.1.

Although particle methods such as MD [8] and discrete simulation Monte Carlo (DSMC) [9] are able to simulate fluid phenomena at the microscale (and/or nanoscale) and beyond the continuum hypothesis, they remain computationally intensive for practical applications. On the other hand, mesoscopic methods, such as the Lattice Boltzmann method (LBM), model the fluid as fictitious particles each consisting of a large number of molecules and evolving according to the particle density distribution function. In this chapter, we discuss the application of the LBM to thermal microflows based on the double-distribution approach. After introducing the thermal LBM (TLBM) in Section 15.2, we examine the boundary conditions used in the literature to describe the slip thermal flow regime in Section 15.3, followed by the transition thermal flow regime in Section 15.4. Finally, we summarize our findings and make recommendations in Section 15.5.

15.2 THERMAL LATTICE BOLTZMANN METHOD (TLBM)

The LBM evolved from Lattice Gas Automata (LGA) [10], in which a fluid is represented by fictitious particles hopping from one node to another on a lattice in discrete time and with discrete velocities. In LGA, lattice velocities connect nodes with their neighbors and determine how particles move on the lattice during a propagation collision-free step. Lattice nodes can be either empty or have a single particle residing on them. When more than one particle arrives at a node following the propagation (streaming) step, they collide based on predetermined collision rules that conserve mass, momentum, and energy. Some of the main disadvantages in LGA include the statistical noise inherent in the technique and lack of Galilean invariance, in addition to feasibility issues when it comes to extending the method to 3D space [11]. Thus, the initial intent behind

LBM was to overcome the statistical noise in LGA by replacing the Boolean particle number and discrete collisions with its average density distribution function and continuous collision operator, respectively [10]. The NSF equations can still be recovered from this system through a Chapman–Enskog multiscale analysis. It has also been shown [12,13] that the LBM can be derived from the continuous Boltzmann equation by special discretization of the latter in time and phase space.

One of the initial limitations of applying LBM to model fluid dynamics was the lack of an adequate thermal model. Initial attempts at a TLBM can be split into two approaches. The first is referred to as the multispeed approach [14–16] and utilizes a single density distribution function with additional lattice speeds and higher-order velocity terms in the equilibrium distribution function to recover the macroscopic temperature field. The multispeed approach to TLBM suffers from numerical instability and is limited in the range of temperature variations [17,18] that it can simulate. The second approach to TLBM is a passive-scalar one and relies on an additional distribution function to model the temperature evolution in the absence of viscous heat dissipation and/or pressure compression work [19,20]. This passive-scalar approach represents a TLBM with enhanced numerical stability compared to the multispeed counterpart. It was later improved [21] to incorporate viscous heat dissipation and pressure compression work effects. This was accomplished by applying the same concepts in the priori derivation of LBM from the Boltzmann equation [12] to an internal energy distribution function. This TLBM, also referred to as the double-distribution approach, has proven popular among researchers looking to extend LBM to thermal flows. The aim of this chapter is to outline this TLBM and the boundary conditions implemented in its extension to slip and transition flow regimes.

15.3 APPLICATION OF TLBM TO THERMAL MICROFLOWS IN THE CONTINUUM LIMIT ($Kn \leq 0.01$)

The double-distribution approach to TLBM was derived in [21] to produce the following two discrete lattice evolution equations in the absence of external forces:

$$\tilde{f}_i(\mathbf{x}+\mathbf{c}_i\Delta t, t+\Delta t) - \tilde{f}_i(\mathbf{x},t) = -\frac{\Delta t}{\tau_f + 0.5\Delta t}\left[\tilde{f}_i(\mathbf{x},t) - f_i^{eq}(\mathbf{x},t)\right] \tag{15.1}$$

$$\tilde{g}_i(\mathbf{x}+\mathbf{c}_i\Delta t, t+\Delta t) - \tilde{g}_i(\mathbf{x},t) = -\frac{\Delta t}{\tau_g + 0.5\Delta t}\left[\tilde{g}_i(\mathbf{x},t) - g_i^{eq}(\mathbf{x},t)\right] - \frac{\tau_g \Delta t}{\tau_g + 0.5\Delta t} f_i Z_i \tag{15. 2}$$

where

$$\tilde{f}_i = f_i + \frac{\Delta t}{2\tau_f}\left(f_i - f_i^{eq}\right) \tag{15.3}$$

$$\tilde{g}_i = g_i + \frac{\Delta t}{2\tau_g}\left(g_i - g_i^{eq}\right) + \frac{\Delta t}{2} f_i Z_i \tag{15.4}$$

Δt is the time step
τ_f and τ_g are the relaxation times for the density (f) and internal energy distribution functions (g), respectively
$\mathbf{c}_i$ is the lattice velocity along direction i and for the 2D D2Q9 lattice shown in Figure 15.2, the velocities are given by

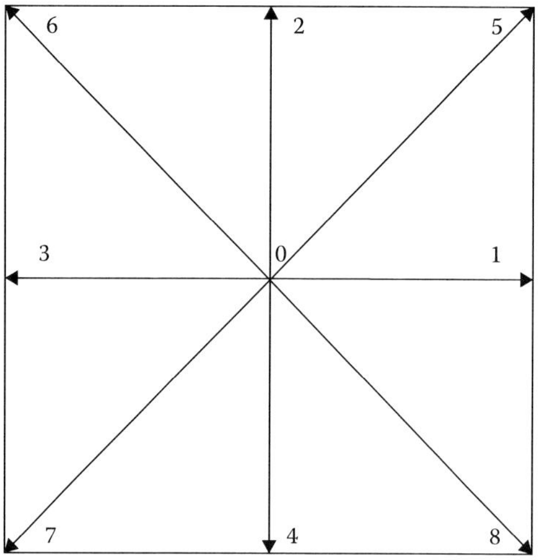

FIGURE 15.2 Two-dimensional D2Q9 lattice velocities.

$$\mathbf{c}_i = \begin{cases} (0,0) & i=0 \\ \left(\cos\left[(i-1)\frac{\pi}{2}\right], \sin\left[(i-1)\frac{\pi}{2}\right]\right)c & i=1-4 \\ \sqrt{2}\left(\cos\left[(i-5)\frac{\pi}{2}+\frac{\pi}{4}\right], \sin\left[(i-5)\frac{\pi}{2}+\frac{\pi}{4}\right]\right)c & i=5-8 \end{cases} \tag{15.5}$$

with $c=\sqrt{3RT}=\Delta x/\Delta t=1$. The equilibrium distribution functions for the density and internal energy are given by

$$f_i^{eq} = w_i\rho\left[1+\frac{3(\mathbf{c}_i\cdot\mathbf{u})}{c^2}+\frac{9(\mathbf{c}_i\cdot\mathbf{u})^2}{2c^4}-\frac{3(\mathbf{u}\cdot\mathbf{u})}{2c^2}\right] \tag{15.6}$$

$$g_i^{eq} = e f^{eq} \tag{15.7}$$

where
- **u** is the macroscopic velocity
- the internal energy $e = DRT/2$ (D,R, and T stand for dimension, the universal gas constant, and temperature, respectively)

The weighting factors are $w_0 = 4/9$, $w_{1-4} = 1/9$, and $w_{5-8} = 1/36$. The Z_i term in Equation 15.2 represents the viscous heating and compression work and is given by

$$Z_i = (\mathbf{c}_i-\mathbf{u})\cdot\left[\frac{\partial\mathbf{u}}{\partial t}+(\mathbf{c}_i\cdot\nabla)\mathbf{u}\right] \tag{15.8}$$

The relevant macroscopic variables and transport properties are calculated as follows: $\rho = \Sigma \tilde{f}_i$ (density), $\rho \mathbf{u} = \Sigma \mathbf{c}_i \tilde{f}$ (velocity), $\rho e = \Sigma \tilde{g}_i - 0.5\Delta t \Sigma f_i Z_i$ (internal energy per unit mass), $\upsilon = \tau_f c_s^2$ (kinematic viscosity, $c_s^2 = c^2/3$), and $D = 2\tau_g c_s^2$ (thermal diffusivity).

Unless mentioned otherwise, the analysis in the following text is based on the double-distribution TLBM 2D D2Q9 lattice. In terms of its application to macroscopic problems, some examples include modeling natural convection of a Boussinesq fluid in a square cavity [22], extending the method to circular boundaries to simulate natural convection in a concentric annulus [23], and in [24] it was used to study backward-facing step flows. In order to model thermal microflows in the slip and transition regimes, the boundary conditions in the TLBM need to be modified to account for noncontinuum effects. In the following sections, we describe such implementations.

An important step in the TLBM application to microflows is the definition of *Kn* in terms of the relaxation time.

15.4 TLBM IN THE SLIP FLOW REGIME (0.01 < *Kn* < 0.1)

When the gas MFP (λ) becomes comparable to the flow characteristic length (*H*), a slip flow regime begins to develop with a slip velocity and a temperature jump at the channel surface. These effects cannot be directly resolved by the NSF equations, which are still applicable for the bulk flow. Therefore, special boundary conditions are required to account for such rarefication effects on the thermal microflow. A review of microflow slip models can be found in [25]. One of the earliest attempts at modeling thermal microflows in the slip flow regime, using the double-distribution TLBM [21], utilized a diffuse scattering boundary condition (DSBC) [26–28] to capture the velocity slip and temperature jump at the wall [29,30]. The DSBC is based on the kinetic theory of gas-surface interactions and can be generally expressed as follows:

$$\left|(\mathbf{c}_i - \mathbf{u}_w)\cdot\mathbf{n}\right|\varphi_i = \sum_{(\mathbf{c}_{i'} - \mathbf{u}_w)\cdot\mathbf{n}<0} \left|(\mathbf{c}_{i'} - \mathbf{u}_w)\cdot\mathbf{n}\right|\Gamma_\varphi(\mathbf{c}_{i'} \to \mathbf{c}_i)\varphi_{i'} \tag{15.9}$$

where

$\varphi = f$ or $\varphi = g$

i and i' are the directions of the incident and reflected particles, respectively

$\mathbf{u}_w$ is the wall velocity

$\mathbf{n}$ is an inward pointing unit normal vector

The scattering kernel $\Gamma\varphi$ is given by

$$\Gamma_f(\mathbf{c}_{i'} \to \mathbf{c}_i) = \frac{A_N}{\rho_w}[(\mathbf{c}_i - \mathbf{u}_w)\cdot\mathbf{n}]f_i^{eq}\big|_{\mathbf{u}=\mathbf{u}_w} \tag{15.10}$$

$$\Gamma_g(\mathbf{c}_{i'} \to \mathbf{c}_i) = \frac{B_N}{\rho_w e}[(\mathbf{c}_i - \mathbf{u}_w)\cdot\mathbf{n}]g_i^{eq}\big|_{\mathbf{u}=\mathbf{u}_w, e=e_w} \tag{15.11}$$

with A_N and B_N acting as normalization coefficients that guarantee zero mass flow through the wall and a continuous energy flux across it, respectively. These coefficients can be expressed as follows:

$$A_N = \rho_w \frac{\sum_i \left|(\mathbf{c}_i - \mathbf{u}_w)\cdot\mathbf{n}\right| f_i}{\left|(\mathbf{c}_i - \mathbf{u}_w)\cdot\mathbf{n}\right| f_i^{eq}\big|_{\mathbf{u}=\mathbf{u}_w} \sum_i \left|(\mathbf{c}_{i'} - \mathbf{u}_w)\cdot\mathbf{n}\right| f_{i'}} \tag{15.12}$$

$$B_N = \rho_w e \frac{\sum_i \left|(\mathbf{c}_i - \mathbf{u}_w)\cdot\mathbf{n}\right| g_i}{\left|(\mathbf{c}_i - \mathbf{u}_w)\cdot\mathbf{n}\right| g_i^{eq}\Big|_{\mathbf{u}=\mathbf{u}_w, e=e_w} \sum_{i'} \left|(\mathbf{c}_{i'} - \mathbf{u}_w)\cdot\mathbf{n}\right| g_{i'}} \tag{15.13}$$

where ρ_w and e_w are the wall density and internal energy, respectively. In the DSBC, it is assumed that a particle loses its information about its state prior to colliding with the wall surface, and that the result of the collision is a reassignment of the particle state based on the Maxwellian distribution. The link between the DSBC and higher-order slip models [2] for velocity and temperature was also derived in [29,30] by applying Equation 15.10 to the boundary nodes using the distribution functions f (for the velocity slip) and g (for the temperature jump). To simplify the application of their model, the heat dissipation term (Z_i) was ignored. Conditions that justify this assumption for TLBM can be found in [31]. The model was first used to simulate a 2D thermal Couette flow. The TLBM results for the velocity slip and temperature jump were found to be in good agreement with Maxwell's theory [32], in addition to the results from MD and DSMC [33]. Although simulations were conducted for Kn = 0.01–0.5, there was an increased deviation from the reference data beyond the slip flow regime (Kn>0.1) as shown in Figure 15.3. The authors also simulated a developing

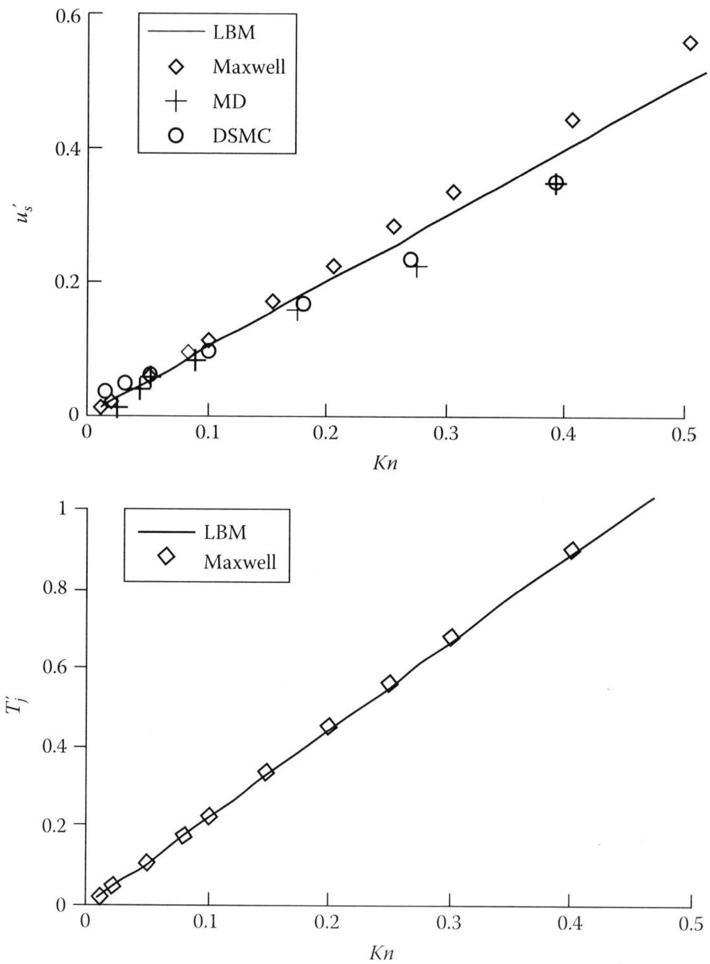

FIGURE 15.3 Dimensionless velocity slip (u_s') and temperature jump (T_j') for thermal micro-Couette flow. (From Niu, X.D. et al., *Comput. Fluids*, 36, 273, 2007.)

thermal Poiseuille flow in a microchannel within the slip flow regime. The results showed a drop in the Nusselt number ($Nu = 2H(\partial T/\partial y)_w/(T_w - T_o)$, where H is the channel height along the y-axis, and T_o is the bulk temperature) and friction coefficient ($C_f = \mu(\partial u/\partial y)_w/(0.5\rho u_o^2)$, where u_o is the bulk velocity with increased Kn, at a Reynold's number of Re = 0.01. The results compare well with numerical solutions of the compressible momentum and energy equations, using velocity slip and temperature jump boundary conditions [34]. The extension of the TLBM with DSBC to three dimensions was later demonstrated in the slip flow regime by the same authors [35].

In other early work on TLBM for microflows in the slip regime [36], the velocity slip and temperature jump were modeled using a Maxwell first-order slip boundary condition [1]. The viscous heating term Z_i from Equation 15.8 was retained in the double-distribution TLBM model [21]. For channel flow of height H along the y-axis, the corresponding velocity slip and temperature jump boundary conditions can be expressed as follows:

$$u_{y=0}^{s} = u_{y=0} - u_w = \sigma Kn\left(\frac{\partial u}{\partial y}\right)_{y=0} \tag{15.14}$$

$$u_{y=H}^{s} = u_w - u_{y=H} = \sigma Kn\left(\frac{\partial u}{\partial y}\right)_{y=H} \tag{15.15}$$

$$T_{y=0}^{j} = T_{y=0} - T_w = \alpha\left[\frac{2\gamma}{(\gamma+1)}\right]\left(\frac{Kn}{Pr}\right)\left(\frac{\partial T}{\partial y}\right)_{y=0} \tag{15.16}$$

$$T_{y=H}^{j} = T_w - T_{y=H} = \alpha\left[\frac{2\gamma}{(\gamma+1)}\right]\left(\frac{Kn}{Pr}\right)\left(\frac{\partial T}{\partial y}\right)_{y=H} \tag{15.17}$$

with superscripts s and j referring to "slip" and "jump," respectively, $\sigma = (2-\sigma_v)/\sigma_v$, where σ_v is the tangential momentum accommodation coefficient and represents the fraction of molecules that are reflected diffusively, and $\alpha = (2-\alpha_T)/\alpha_T$, with α_T as the thermal accommodation coefficient. Both coefficients are taken to be equal to 1.0, rendering the wall reflections as completely diffusive. Other parameters in the slip model include the specific heat ratio γ and the Prandtl number Pr.

To test the slip model, results from the TLBM were compared to analytical solutions of the NSF equations derived by applying the first-order slip model of Equations 15.14 through 15.17. This was conducted for thermal micro-Couette flows and good agreement was found for the velocity slip and temperature jumps at the wall boundary with Kn = 0.01–0.1, that is, in the slip flow regime. The accuracy of the first-order slip boundary condition implementation into TLBM was later demonstrated by applying it to pressure-driven thermal micro-Poiseuille flows [37] in two dimensions. As can be seen in Figure 15.4, results were once again validated with the Maxwell boundary condition solution of the NSF equations. A similar study was more recently reported in [38].

For the Maxwell slip model in Equations 15.14 through 15.17, the velocity slip and temperature jump are heavily influenced by the accommodation coefficients, Kn, and the corresponding velocity and temperature gradients at the wall. Hence, a main disadvantage in this slip model is that the choice of the accommodation coefficients is based on the notion of a diffusive reflection. Other inadequacies of the Maxwell slip model include the velocity slip becoming unbounded under certain conditions [39] in addition to deviations from benchmark results [40]. To improve on their velocity slip and temperature jump model, the work in [36,37] was later adopted to include a Langmuir slip model [41] based on adsorption theory [42]. In Langmuir's theory, the intermolecular forces

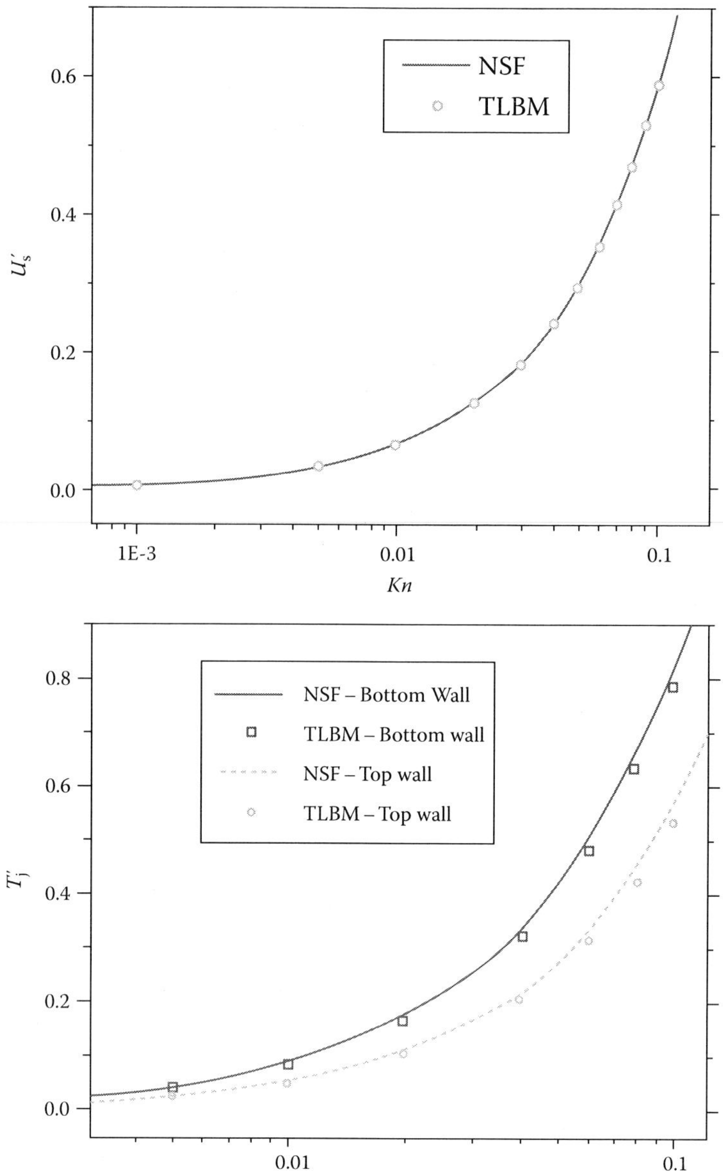

FIGURE 15.4 TLBM with first-order slip model boundary condition comparison with NSF solutions. Velocity slip (u_s') and temperature jump (T_j') for thermal micro-Poiseuille flow [37] with the top wall at a higher temperature.

between the gas and surface molecules are long-range attractive in nature and the gas molecules reside on the surface for a period of time before being reflected, that is, desorbed. The Langmuir slip model is given by [43]

$$u^s = (1 - \alpha_L)u_g + \alpha_L u_w \tag{15.18}$$

$$T^j = (1 - \alpha_L)T_g + \alpha_L T_w \tag{15.19}$$

where the subscript g denotes fluid values adjacent to the wall (or other reference value such as free-stream). Here the term α_L is the fraction of the surface covered by adsorbed molecules at thermal equilibrium and is computed by

$$\alpha_L = \frac{\beta p}{1+\beta p} \tag{15.20}$$

with p as the pressure, and [43]

$$\beta = \frac{1}{4\omega Kn} \tag{15.21}$$

with

$$\omega = \omega_o(v)\left(\frac{T_w}{T_r}\right)^{1+2/(v-1)} \exp\left(-\frac{D_e}{k_b T_w}\right) \tag{15.22}$$

and

$$\omega_o(v) = \frac{8\sqrt{2}}{5\pi} A'(v)\Gamma\left[4-\frac{2}{v-1}\right] \tag{15.23}$$

where

the subscript r denotes a reference value
D_e is heat of adsorption
k_b is the Boltzmann constant
v is the exponent of an inverse power-law molecular force
the corresponding values for ω_o and A' can be found tabulated in [43]

The variable ω plays a similar role to the accommodation coefficient from the Maxwell slip model, except its value is determined now on a more physical basis through the parameters mentioned earlier. From Equation 15.21, it can be seen that as $Kn \rightarrow 0$ one gets $\alpha_L \rightarrow 1$ in Equation 15.20, which results in macroscopic boundary conditions. Unlike the Maxwell model, the Langmuir velocity slip and temperature jump expressions from Equations 15.18 and 15.19 do not require the calculation of velocity and temperature gradients. The Langmuir slip model was validated by simulating thermal micro-Couette flows and was found to be in good agreement with the solutions from NSF with first-order slip boundary conditions imposed.

15.5 TLBM IN THE TRANSITION FLOW REGIME ($0.1 < Kn < 10$)

The application of the double-distribution TLBM in the transition flow regime was carried out in [44] using a Knudsen layer correction to the MFP (λ). The effects of the Knudsen layer, of thickness $\sim O(\lambda)$, become increasingly prominent in the transition flow regime, and the relatively simpler slip models discussed before are insufficient in capturing the complex physics. Compared to the flow in the bulk region, there is a reduction in the MFP within the Knudsen layer as a result of the molecular interactions between the rarefied gas and wall surface. In [44], the Knudsen layer correction was applied via an effective MFP (λ_e) given by

$$\lambda_e = \frac{\lambda}{1+\psi(x_n/\lambda)} \tag{15.24}$$

where

x_n is the normal distance to the wall
μ is the viscosity
p is pressure
R is the gas constant
MFP is given by $\lambda = \mu/p\sqrt{\pi RT/2}$

The correction factor function ψ is chosen to depend on the distance to the wall and to diminish beyond the Knudsen layer. Noting the lack of information on ψ, the study in [44] used a version that is applicable to Kramer's problem of sheared gas flow bounded by one surface [45]

$$\psi\left(\frac{x_n}{\lambda}\right) = \frac{7}{10}\exp\left(-\frac{x_n}{\lambda}\right) \tag{15.25}$$

The profile of ψ is shown in Figure 15.5 along with the resulting λ_e using $\lambda = 1.0$. The effective MFP can be seen to approximately recover λ around $x_n \approx 5\lambda$.

Additionally, the effects of temperature and density variations on the effective MFP (λ_e) were incorporated through λ. From kinetic theory, the MFP is related to viscosity by [46]

$$\mu = a\rho c\lambda \tag{15.26}$$

where

a is a constant
$c = \sqrt{8RT/\pi}$

If the viscosity (μ_{ref}) is known at a given reference temperature (T_{ref}), it can be approximated at a different temperature using

$$\mu = \mu_{\text{ref}}\left(\frac{T}{T_{\text{ref}}}\right)^w \tag{15.27}$$

where w depends on the type of molecular interactions and varies from 0.5 to 1.0 for hard-sphere and Maxwellian interactions, respectively [46]. Similarly, Equation 15.26 at a reference temperature

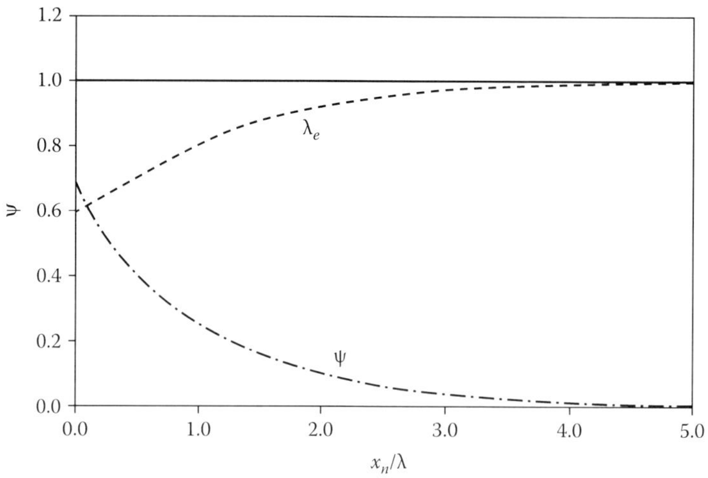

FIGURE 15.5 Correction function (ψ) and effective MFP from Equations 15.24 and 15.25 for $\lambda = 1.0$.

gives $\mu_{ref} = a\rho_{ref}c\lambda_{ref}$. Therefore, the combined effect of temperature and density variations on the MFP can be derived using Equation 15.28 as follows:

$$\frac{\mu}{\mu_{ref}} = \frac{a\rho\lambda\sqrt{8R/\pi}}{a\rho_{ref}\lambda_{ref}\sqrt{8R/\pi}}\left(\frac{T}{T_{ref}}\right)^{0.5} = \left(\frac{T}{T_{ref}}\right)^{w} \tag{15.28}$$

solving for λ/λ_{ref}

$$\frac{\lambda}{\lambda_{ref}} = \frac{\rho_{ref}}{\rho}\left(\frac{T}{T_{ref}}\right)^{w-0.5} \tag{15.29}$$

Expressing Equation 15.29 in terms of the effective MFP (λ_e) from Equation 15.24 gives the final expression:

$$\frac{\lambda_e}{\lambda_{ref}} = \frac{1}{1+\psi(x_n/\lambda)}\frac{\rho_{ref}}{\rho}\left(\frac{T}{T_{ref}}\right)^{w-0.5} \tag{15.30}$$

These effects were incorporated into the relaxation times by first expressing the Knudsen number as [47] follows:

$$Kn = \frac{\sqrt{8/3\pi}(\tau-0.5)}{N_H} \tag{15.31}$$

where N_H is the number of lattice sites spanning H. An "effective" Knudsen number can then be written as $Kn_e = \lambda_e/H$, which when substituted into Equation 15.31 results in $\tau = \sqrt{3\pi/8}N_H Kn_e + 0.5$. Using Equation 15.30, the density relaxation time can be expressed as follows:

$$\tau = \sqrt{\frac{3\pi}{8}}\frac{KnN_H}{1+\psi(x_n/\lambda)}\frac{\rho_{ref}}{\rho}\left(\frac{T}{T_{ref}}\right)^{w-0.5} \tag{15.32}$$

where now $Kn = \lambda_{ref}/H$. The thermal relaxation time becomes

$$\tau_g = \frac{(\tau-0.5)}{Pr} + 0.5 \tag{15.33}$$

This TLBM was tested using Maxwellian gas molecular interactions ($w = 1.0$) and DSBC [29] at the wall surface. The results in this case were limited to thermal transport in a gas confined between two parallel plates (Fourier flow) and were compared to DSMC values from a previous study [48]. As shown in Figure 15.6, the results demonstrated an improved agreement between TLBM and DSMC when the Knudsen layer correction was included for $Kn = 0.158$, 0.475, and 1.58. However, the deviation of the results from DSMC were found to increase with increasing Kn.

A similar approach applying TLBM with DSBC to the transition flow regime was also demonstrated in [49]. A D2Q13 lattice was used for which the density relaxation time can be expressed as [50]

$$\tau_f = \chi\sqrt{\frac{\pi}{4}}KnN_H + 0.5 \tag{15.34}$$

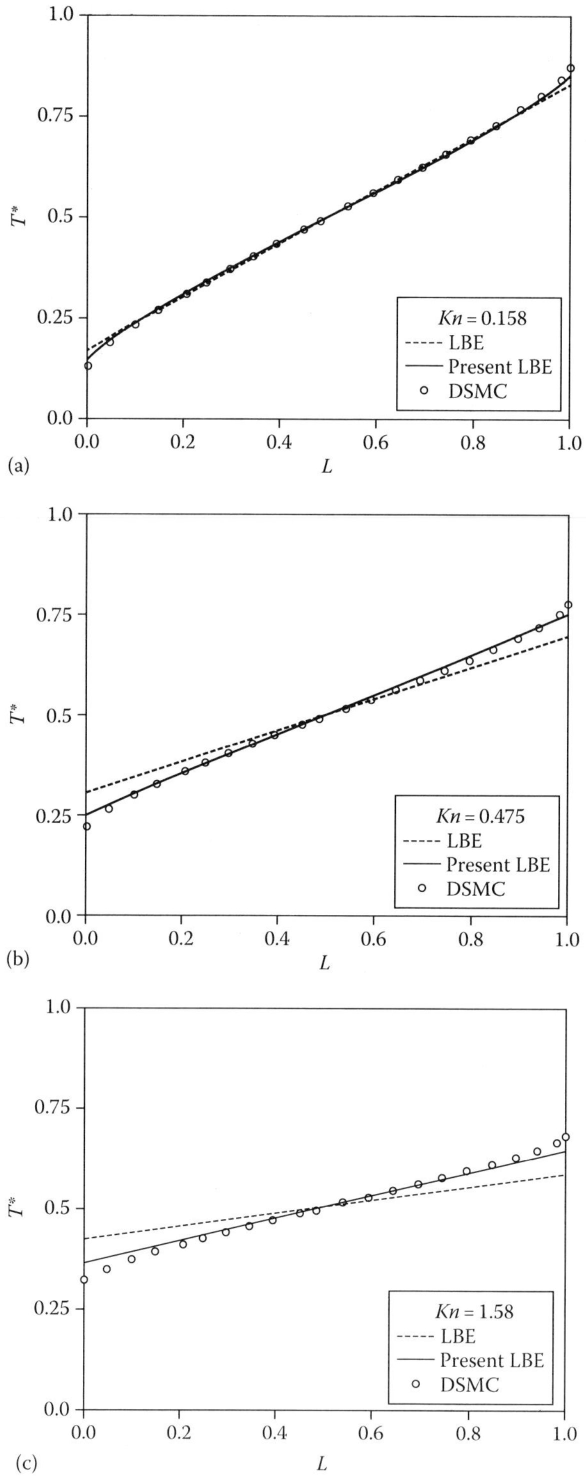

FIGURE 15.6 Nondimensional temperature jumps for rarefied gas heat conduction between two plates for (a) *Kn* = 0.158, (b) *Kn* = 0.475, and (c) *Kn* = 1.58. "Present LBE" refers to the TLBM with MFP corrections compared to "LBE" where no such corrections are applied. (From Zhang, Y.-H. et al., *Europhys. Lett.*, 77, 30003 p1–p5, 2007.)

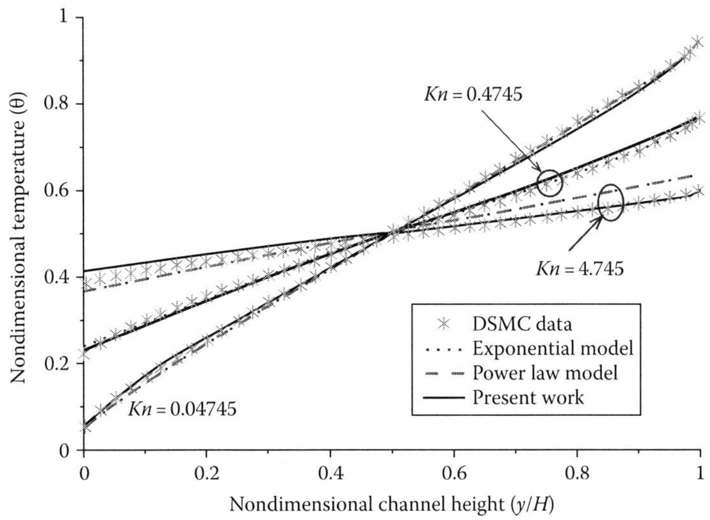

FIGURE 15.7 Temperature profile and jumps for rarefied gas Fourier flow between two plates with $\theta = (T - T_{cold})/(T - T_{hot})$. (From Lopez, P. and Bayazitoglu, Y., *Int. J. Heat Mass Transfer*, 65, 374, 2013.)

where $\chi = \lambda_e/\lambda$ is the normalized effective MFP. Functions for the amplitude and profile of χ were fitted using MD simulation data on rarefied gases [51]. Both Fourier and Couette flows were simulated. For the Fourier case, temperature profiles for $Kn = 0.04745$, 0.4745, and 4.745 were compared to DSMC data [48] and found to be in good quantitative agreement as given in Figure 15.7. For the isothermal Couette flow case studied, velocity profiles for Kn up to 11.28 were compared to linearized Boltzmann data [44] with good accuracy being reported well within the transition regime. For the thermal Couette flow however, the validation was qualitative with no reference data to compare to.

15.6 CONCLUSION

The accurate prediction of fluid flow and heat transfer phenomena in microflows is an important component in the development of microdevices. At the microscale, a gas can undergo different flow regimes with their corresponding unique characteristics as Kn increases beyond the continuum flow limit. The effects of gas rarefaction on the flow and thermal transport physics can ideally be captured using particle methods such as MD, where individual molecules are tracked. However, this approach is not computationally feasible for systems of relevant sizes. The TLBM on the other hand tracks the evolution of a large group of molecules at a time via the density distribution function and kinetic theory. While TLBM is capable of solving the NSF equations on a lattice, special boundary conditions are needed to incorporate the noncontinuum effects in the slip and transition flow regimes. For the former, both DSBC and Maxwell's first-order slip model have been demonstrated to produce fairly accurate results. The boundary conditions needed in the transition regime are relatively more complex in order to account for the effects on the fluid properties. This correction in the Knudsen layer, though, relies on data from more accurate simulations such as MD which are not necessarily available depending on the problem configuration. Still, this approach has shown good potential in modeling thermal microflows in the transition regime. The focus in most of the studies has been on simple channel flow geometries and more work is needed to extend TLBM for microflows in more complex flow paths, which is one of the main advantages in using the LBM. Furthermore, the use of a Knudsen layer correction function can be tricky depending on the available information and in that regard multiscale modeling can play a critical role, especially based on domain decomposition techniques.

REFERENCES

1. M. Gad-el-Hak, The fluid mechanics of microdevices—The Freeman scholar lecture, *Journal of Fluids Engineering*, 121, 5–33 (1999).
2. G. Karniadakis, A. Beskok, and N. Aluru, *Microflows and Nanoflows: Fundamentals and Simulation*, Springer, New York (2005).
3. J. de Jong, R.G.H. Lammertink, and M. Wessling, Membranes and microfluidics: A review, *Lab on a Chip*, 6, 1125–1139 (2006).
4. C. Habre, Microfluidics in commercial applications: An industry perspective, *Lab on a Chip*, 6, 1118–1121 (2006).
5. A.A. Rostami, A.S. Mujumdar, and N. Saniei, Flow and heat transfer for gas flowing in microchannels: A review, *Heat and Mass Transfer*, 38, 359–367 (2002).
6. P. Rosa, T.G. Karayiannis, and M.W. Collins, Single-phase heat transfer in microchannels: The importance of scaling effects, *Applied Thermal Engineering*, 29, 3447–3468 (2009).
7. S. Kandlikar, S. Garimella, D. Li, S. Colin, and M.R. King, *Heat Transfer and Fluid Flow in Minichannels and Microchannels*, Elsevier, Oxford, U.K. (2006).
8. D. Frenkel, and B. Smit, *Understanding Molecular Simulation*, Academic Press, San Diego, CA (2002).
9. G.A. Bird, *Molecular Gas Dynamics and the Direct Simulation of Gas Flows*, Oxford University Press, Oxford, U.K. (1994).
10. D.A. Wolf-Gladrow, *Lattice-Gas Cellular Automata and Lattice Boltzmann Models—An Introduction*, Springer-Verlag, Berlin, Heidelberg (2005).
11. S. Succi, *The Lattice Boltzmann Equation: For Fluid Dynamics and Beyond*, Oxford University Press, Oxford, U.K. (2001).
12. X. He, and L-.S. Lou, A priori derivation of the lattice Boltzmann equation, *Physical Review E*, 55, R6333–R6336 (1997).
13. T. Abe, Derivation of the lattice Boltzmann method by means of the discrete ordinate method for the Boltzmann equation, *Journal of Computational Physics*, 131, 241–246 (1997).
14. G. McNamara, and B. Alder, Analysis of the lattice Boltzmann treatment of hydrodynamics, *Physica A*, 194, 218–228 (1993).
15. F.J. Alexander, S. Chen, and J.D. Sterling, Lattice Boltzmann thermohydrodynamics, *Physical Review E*, 47, R2249–R2252 (1993).
16. Y. Chen, H. Ohashi, and M. Akiyama, Thermal lattice Bhatnagar-Gross-Krook model without nonlinear deviations in macrodynamic equations, *Physical Review E*, 50, 2776–2783 (1994).
17. P. Pavlo, G. Vahala, L. Vahala, and M. Soe, Linear-stability analysis of thermos-lattice Boltzmann models, *Journal of Computational Physics*, 139, 79–91 (1998).
18. G.R. McNamara, A.L. Garcia, and B.J. Alder, Stabilization of thermal lattice Boltzmann models, *Journal of Statistical Physics*, 81, 395–408 (1995).
19. A. Bartoloni, C. Battista, S. Cabasino et al., LBE simulations of Rayleigh-Benard convection on the APE100 parallel processor, *Journal of Modern Physics C*, 4, 993–1006 (1993).
20. X. Shan, Simulation of Rayleigh-Benard convection using a lattice Boltzmann method, *Physical Review E*, 55, 2780–2788 (1997).
21. X. He, S. Chen, and G.D. Doolen, A novel thermal model for the lattice Boltzmann method in incompressible limit, *Journal of Computational Physics*, 146, 282–300 (1998).
22. H.N. Dixit, and V. Babu, Simulation oh high Rayleigh number natural convection in a square cavity using the lattice Boltzmann method, *International Journal of Heat and Mass Transfer*, 49, 727–739 (2006).
23. H. Huang, T.S. Lee, and C. Shu, Thermal curved boundary treatment for the thermal lattice Boltzmann equation, *International Journal of Modern Physics C*, 17, 631–643 (2006).
24. C.-K. Chen, T.-S. Yen, and Y.-T. Yang, Lattice Boltzmann method simulation of backward-facing step flow with double plates aligned at angle to flow direction, *Journal of Heat Transfer*, 128, 1176–1184 (2006).
25. W.M-. Zhang, G. Meng, and X. Wei, A review on slip models for gas microflows, *Microfluidics and Nanofluidics*, 13, 845–882 (2012).
26. C. Cercignani, *Mathematical Methods in Kinetic Theory*, Springer Science+Business Media, New York (1969).
27. F. Sharipov and V. Seleznev, Data on rarefied gas flows, *Journal of Physical and Chemical Reference Data*, 27, 657–706 (1998).

28. K. Suga, Lattice Boltzmann methods for complex micro-flows: Applicability and limitations for practical applications, *Fluid Dynamics Research*, 45, 034501 1–31 (2013).
29. C. Shu, X.D. Niu, and Y.T. Chew, A lattice Boltzmann kinetic model for microflow and heat transfer, *Journal of Statistical Physics*, 121, 239–255 (2005).
30. X.D. Niu, C. Shu, and Y.T. Chew, A thermal lattice Boltzmann model with diffuse scattering boundary condition for micro thermal flows, *Computers & Fluids*, 36, 273–281 (2007).
31. Y. Shi, T.S. Zhao, and Z.L. Guo, Thermal lattice Bhatnagar-Gross-Krook model for flows with viscous heat dissipation in the incompressible limit, *Physical Review E*, 70, 066310 1–10 (2004).
32. D.R. Willis, Comparison of kinetic theory analyses of linearized Couette flow, *Physics of Fluids*, 5, 127–135 (1962).
33. D.L. Morris, L. Hannon, and A.L. Garcia, Slip length in a dilute gas, *Physical Review A*, 46, 5279–5281 (1992).
34. H.P. Kavehpour, M. Faghri, and Y. Asako, Effects of compressibility and rarefaction on gaseous flows in microchannels, *Numerical Heat Transfer Part A*, 32, 677–696 (1997).
35. Y.T. Chew, X.D. Niu, and C. Shu, Three-dimensional lattice Boltzmann BGK model and its application to flows with heat transfer in a rectangular microchannel, *International Journal for Numerical Methods in Fluids*, 50, 1321–1334 (2006).
36. Z.-W. Tian, C. Zou, Z.-H. Liu, Z.-L. Guo, H.-J. Liu, and C.-G. Zheng, Lattice Boltzmann method in simulation of thermal micro-flow with temperature jump, *International Journal of Modern Physics C*, 17, 603–614 (2006).
37. Z.-W. Tian, C. Zou, H.-J. Liu, Z.-L. Guo, Z.-H. Liu, and C.-G. Zheng, Lattice Boltzmann scheme for simulating thermal micro-flow, *Physica A*, 385, 59–68 (2007).
38. S. Gokaltun and G.S. Dulikravich, Lattice Boltzmann method for rarefied channel flows with heat transfer, *International Journal of Heat and Mass Transfer*, 78, 796–804 (2014).
39. H.-I. Choi, D.-H. Lee, and D. Lee, Complex microscale flow simulations using Langmuir slip condition, *Numerical Heat Transfer Part A: Applications*, 48, 407–425 (2005).
40. H.M. Kim, D. Kim, W.T. Kim, P.S. Chung, and M.S. Jhon, Langmuir slip model for air bearing simulation using the lattice Boltzmann method, *IEEE Transactions on Magnetics*, 43, 2244–2246 (2007).
41. S. Chen, and Z. Tian, Simulation of thermal micro-flow using lattice Boltzmann with Langmuir slip model, *International Journal of Heat and Fluid Flow*, 31, 227–235 (2010).
42. I. Langmuir, Surface chemistry, *Chemical Reviews*, 13, 147–191 (1933).
43. R.S. Myong, D.A. Lockerby, and J.M. Reese, The effect of gaseous slip on microscale heat transfer: An extended Graetz problem, *International Journal of Heat and Mass Transfer*, 49, 2502–2513 (2006).
44. Y.-H. Zhang, X.J. Gu, R.W. Barber, and D.R. Emerson, Modelling thermal flow in the transition regime using a lattice Boltzmann approach, *Europhysics Letters*, 77, 30003 p1–p5 (2007).
45. D.A. Lockerby, J.M. Reese, and M.A. Gallis, The usefulness of higher-order constitutive relations for describing the Knudsen layer, *Physics of Fluids*, 17, 100609 1–6 (2005).
46. S. Chapman and T.G. Cowling, *The Mathematical Theory of Non-Uniform Gases*, Cambridge University, Cambridge, U.K. (1971).
47. Y.H. Zhang, R.S. Qin, Y.H. Sun, R.W. Barber, and D.R. Emerson, Gas flow in microchannels—A lattice Boltzmann approach, *Journal of Statistical Physics*, 121, 257–267 (2005).
48. M.A. Gallis, D.J. Rader, and J.R. Torcynski, Calculations of the near-wall thermophoretic force in rarefied gas flow, *Physics of Fluids*, 14, 4290–4301 (2002).
49. P. Lopez and Y. Bayazitoglu, An extended thermal lattice Boltzmann model for transition flow, *International Journal of Heat and Mass Transfer*, 65, 374–380 (2013).
50. P. Lopez and Y. Bayazitoglu, A note on modeling of nano-scale thermal flow via the lattice Boltzmann method, *ASME 2012 International Mechanical Engineering Congress and Exposition*, Houston, TX, vol. 7, pp. 3007–3008 (2012).
51. N. Dongari, Y. Zhang, and J.M. Reese, Molecular free path distribution in rarefied gases, *Journal of Physics D: Applied Physics*, 44, 125502 1–6 (2011).

16 Heat and Fluid Flow of Gases in Porous Media with Micropores: Slip Flow Regime

Moghtada Mobedi, Murat Barisik, and A. Nakayama

CONTENTS

16.1 INTRODUCTION

Heat and fluid flow in porous media have found wide applications ranging from nature to industry. The flow of air into a human lung, flow of water in soil, fluid flow in underground geothermal fields, flow of crude oil in underground reservoirs, and even air flow between tree trunks in forests are some examples of transport in a porous environment. The flow of hot air for drying in fixed packed beds, the use of metal foams to enhance heat transfer, and fluid flow in membranes are examples of industrial applications. Due to these extensive uses, the number of studies that focuses on porous transport field has increased exponentially in recent years.

The mechanism of heat and fluid transport through the pores is very complex such that determining the velocity, pressure, and temperature fields is a challenging task. There are two main theoretical approaches to analyze heat and fluid flow in a porous structure known as pore scale method (PSM) and volume-averaged method (VAM). In PSM, the conservation of mass and momentum is solved for the fluid flowing through the voids of the porous medium. The energy equation can be solved for both solid and fluid flow in the pores to find temperature field both in solid and fluid regions. The PSM yields accurate distributions for velocity, pressure, and temperature. However, its practical application on porous media, which are generally heterogeneous with numerous pores and voids, is troublesome. This difficulty with PSM leads researchers to develop and employ VAM. VAM is developed based on volume average theory. The results of VAM may not provide an exact view for temperature, pressure, and velocity in the porous media since spaced-averaged values are used. However, the method is practical and the accuracy of results is generally in acceptable ranges. Porous media consists of solid and fluid phases considered as a continuum domain and the heat and fluid flow equations are established for this imaginary continuum domain. Establishment of VAM governing equations requires the definition of the volume average velocity, pressure, and temperatures as main dependent variables. Taking volume integral of the continuity, momentum and energy equations over a representative elementary volume (REV) of a porous structure results in additional unknown parameters such as permeability, Forchheimer coefficient, interfacial heat transfer coefficient, and thermal dispersion coefficients. These parameters are known as VAM transport parameters and the application of VAM requires the knowledge of values of VAM transport parameter. The values of permeability tensor depend on the geometrical parameters of the porous media. However, the values of Forchheimer, interfacial convective heat transfer coefficient, and thermal dispersion tensor depend on geometrical parameters, character of flow in the pores, etc. It should be mentioned that, recently, Nakayama et al. [1] showed that in addition to the aforementioned macroscopic transport properties, thermal tortuosity also plays an important role in determining an accurate temperature field in porous media. Consequently, a VAM transport parameter relating to thermal tortuosity should also be added to the aforementioned unknown transport parameters.

The two methods for determining VAM transport parameters are the experimental and theoretical methods. Determination of permeability and thermal dispersion by experimental methods were discussed in the review studies of Sharma and Siginer [2] and Ozgumus et al. [3] in detail. Pore scale simulation of heat and fluid flow for a REV of the porous media has become popular in recent years due to developments both in computer technology and computational methods. Multiple studies on pore scale simulation of heat and fluid flow in porous media can be found in the literature. For heat and fluid flow in isotropic periodic porous media, examples such as the studies of Nakayama et al. [4] on 2D porous media with square rods, Saito and de Lemos [5] on the determination of interfacial heat transfer coefficient for 2D porous media with square rods, and Nakayma and Kuwahara [6] on the modeling of 3D PSM for a regular and periodic porous media can be given. For nonisotropic but periodic porous media structure, studies were conducted by Ozgumus and Mobedi [7] for 2D porous media with rectangular rods and Nakayama et al. [8] for 2D porous media with different directional cell. For real heterogeneous porous media that can be represented in computer environment by using methods such as tomography or MRI, the studies of Peszynska and Trykozko [9], Akolkar and Petrasch [10], and Ucar et al. [11] should be noted.

The aforementioned studies were performed to determine VAM transport properties for porous media of pore sizes greater than micron scales in which gas flows at standard conditions permit the use of continuum assumptions and constitutive laws. However, there exist many porous media with micron-scale pores (such as microfiltration devices, packed beds with micron-scale particles, or micron-scale size fiber porous structure) and low-pressure environments (vacuum absorption applications) where nonequilibrium gas behavior develops known as the rarefaction. Most of the existing correlations suggested for the determination of VAM transport properties are developed without considering this rarefaction effect and may not be valid for the porous media having pores in micron scale and/or operated under low pressures. For such cases, the Knudsen number defined as the ratio of mean free path to the characteristic length is employed as a measure of the degree of rarefaction. For a certain level of rarefaction, Navier–Stokes equations are found to be valid with slip boundary conditions. Also known as the slip flow regime, Maxwell defined that velocity slip and temperature jump conditions can incorporate the rarefaction formed as a result of small scale and/or low pressure. New correlations or diagrams for the determination of VAM transport properties should be investigated or the suggested correlations in literature should be revised as a function of rarefaction.

Literature survey showed that the number of PSM studies examining the VAM transport parameters of porous media with microchannels is limited compared to studies performed for macro channels. Most of the previous studies are on the determination of permeability rather than thermal VAM transport properties, such as the study of Jeong [12] in which lattice Boltzmann method was used to determine the permeability for microporous structures. For thermal VAM transport properties, the study of Vu et al. [13] on forced convection of air through networks of square rods or cylinders can be given as an example. They found permeability values for 2D porous media with staggered squared rods. Furthermore, Kuwahara et al. [14] studied the heat and fluid flow for 2D porous media with square rods and microscale channels and determined the values of permeability and tortuosity. Xu et al. [15] studied the rarefied effects on internal heat transfer coefficients in microporous media for sintered bronze porous media with average particle diameters from 11 to 225 μm.

The aim of the present work is to perform a numerical study on 2D porous media consisting of square rods in micron size in order to determine VAM transport properties, including permeability and interfacial heat transfer coefficient in slip flow regime. Brief information on the VAM motion and heat transfer equations and the concept of heat and fluid flow in microscale channel with rarefaction effect in slipping region is provided. Then, the obtained results for permeability and interfacial heat transfer coefficient are presented for a 2D porous medium.

16.2 VAM MOTION AND HEAT TRANSFER EQUATIONS

A porous medium consists of solid particles and the voids between solid particles are shown in Figure 16.1. In order to obtain the pore scale temperature, velocity, and pressure fields, an REV is considered in the porous medium, then continuity, momentum and energy equations can be solved for the considered REV.

If a steady-state single phase and incompressible fluid flow exists in the voids between particles, the governing equations can be expressed as

$$
\begin{aligned}
&\vec{\nabla}\cdot\vec{V}=0\\
&\left(\vec{\nabla}.\vec{V}\right)\vec{V}=-\frac{1}{\rho}\vec{\nabla}p+\nu\vec{\nabla}^2\vec{V}\\
&(\rho c_p)_f(\vec{V}.\vec{\nabla})T_f=k_f\vec{\nabla}^2T_f\\
&\vec{\nabla}^2T_s=0
\end{aligned}
\tag{16.1}
$$

The solution of these equations with appropriate boundary conditions yields velocity, pressure, and the solid and fluid phase temperature distributions for the REV of porous media. For an isotropic porous media, the obtained microscopic results are valid for the entire porous media. For nonisotropic or heterogeneous porous media, different REVs at different locations should be considered.

For VAM, the porous medium consists of solid and voids in which the fluid flowing is replaced by an imaginary continuum medium. The derivation of VAM-governing equation requires the

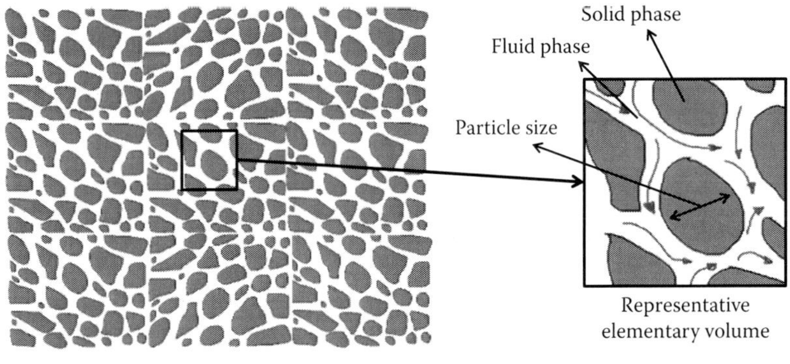

FIGURE 16.1 The illustration of a porous medium and considered representative elementary volume (REV).

definition of volume average. For an REV shown in Figure 16.1, two types of volume average as total volume average of quantity φ and the intrinsic volume average value of φ can be defined.

$$\langle \varphi \rangle = \frac{1}{V}\int_V \varphi \, dV$$
$$\langle \varphi \rangle^x = \frac{1}{V_x}\int_{V_x} \varphi \, dV \qquad (16.2)$$

where
- V is the total volume of REV
- V_x is the volume of considered phase in REV (s or f stands for solid or fluid phases, respectively).

The variable of φ can be velocity, pressure, or temperature in the present study. The following relation exists between the values of real and volume averaged of quantity of φ:

$$\langle \varphi \rangle = \varphi - \varphi' \qquad (16.3)$$

where φ′ refers to the deviation of volume-averaged value from the real value of φ for any location in REV. Taking volume integral from the continuity, momentum, and energy equations over the REV and neglecting thermal tortuosity effect in REV yield the macroscopic motion and energy equations:

$$\vec{\nabla}\cdot\langle \vec{V} \rangle = 0$$
$$\frac{1}{\varepsilon}\frac{\partial \langle \vec{V} \rangle}{\partial t} + \frac{1}{\varepsilon^2}\left(\langle \vec{V} \rangle \cdot \vec{\nabla}\right)\langle \vec{V} \rangle = -\frac{1}{\rho_f}\nabla \langle p \rangle^f + \frac{\nu}{\varepsilon}\nabla^2 \langle \vec{V} \rangle - \frac{\mu}{\rho_f K}\langle \vec{V} \rangle - \frac{C}{K^{1/2}}\left|\vec{V}\right|\vec{V}$$
$$(\rho c_p)_s (1-\varepsilon)\frac{\partial \langle T \rangle^s}{\partial t} = k_s (1-\varepsilon)\vec{\nabla}^2 \langle T \rangle^s + h_{sf}\frac{A_{sf}}{V}\left(\langle T \rangle^s - \langle T \rangle^f\right) \qquad (16.4)$$
$$(\rho c_p)_f \varepsilon \frac{\partial \langle T \rangle^f}{\partial t} + \rho_f c_{p_f}\langle \vec{V} \rangle \cdot \nabla \langle T \rangle^f = k_f \varepsilon \vec{\nabla}^2 \langle T \rangle^f + h_{sf}\frac{A_{sf}}{V}\left(\langle T \rangle^s - \langle T \rangle^f\right) + k_{dis}\vec{\nabla}^2 \langle T \rangle^f$$

where
- $\langle \vec{V} \rangle$ is the local average velocity
- $\langle p \rangle^f$ and $\langle T \rangle^f$ are the intrinsic average pressure and temperature of the fluid phase while $\langle T \rangle^s$ is the average temperature in the solid phase in REV

If these equations are compared with the pore-level governing equations, it can be seen that there are new terms in the VAM transport equation involving new parameters called as VAM transport parameters. The new unknown VAM transport parameters are K, h_{sf}, C, and k_{dis} named as permeability, interfacial heat transfer coefficient, Forchheimer coefficient, and thermal dispersion coefficient, respectively. As mentioned previously, these macroscopic transport parameters should be known in order to obtain volume-averaged temperature, velocity, and pressure fields.

16.3 HEAT AND FLUID FLOW IN MICROSCALE CHANNELS

The effect of pore size along with gas pressure is especially important since most of the porous media contains sub-micron-size pores in addition to existing low-pressure vacuum absorption applications. As the pore size becomes comparable to the mean free path of the gas molecules, nonequilibrium gas behavior occurs known as the rarefaction effects leading to the breakdown of the continuum assumptions and constitutive laws.

Nonequilibrium gas flows are classified by the Knudsen number ($Kn = \lambda/H$), which is the ratio of the local gas mean free path (λ) to the characteristic length scale (H), such as the channel height or pore size. Depending on Kn, transport is considered in the continuum ($Kn \leq 0.01$), slip ($0.01 \leq Kn \leq 0.1$), transition ($0.1 \leq Kn \leq 10$), and free-molecular ($Kn > 10$) flow regimes [16]. Within the context of continuum fluid dynamics, constitutive laws used in the definition of stress tensor and heat flux vector and the no-slip boundary condition break down with increased Knudsen number. For example in a 1 μm conduit, gas at standard pressure and temperature develops transport in the slip flow regime. For such cases, the continuum no-slip boundary condition becomes invalid due to gas rarefaction, but the Navier–Stokes equations can still be employed with an appropriate boundary conditions defined on the surface. The fluid particles adjacent to the boundary surface are not in thermodynamic equilibrium with the wall that there would be slip velocity and temperature jump at the channel wall (For a more detailed discussion on these, readers are referred to the textbook by Gad-el-Hak [17]). For the slip flow regime ($0.01 < Kn < 0.1$), slip velocity and temperature jump boundary conditions for a microtube can be defined as follows [18]:

$$u_n = -\frac{2-\sigma_m}{\sigma_m}\lambda\left.\frac{\partial u}{\partial n}\right|_{wall} \tag{16.5}$$

$$T_f - T_w = -\frac{2-\sigma_t}{\sigma_t}\frac{2\gamma}{\gamma+1}\frac{\lambda}{Pr}\left.\frac{\partial T_f}{\partial y}\right|_{wall} \tag{16.6}$$

where

σ_m is the tangential momentum accommodation coefficient
σ_t is the thermal accommodation coefficient
γ is the specific heat ratio

These slip flow models are successfully employed to consider the effect of rarefaction on microscale flow [19] while good agreements are obtained with experimental measurements [20].

16.4 THE CONSIDERED TWO-DIMENSIONAL POROUS MEDIA

The schematic of the considered porous medium is shown in Figure 16.2. The porous medium is an infinite media consisting of square rods in inline arrangement. Considering the periodicity of the porous structure, an REV with the dimensions $H \times H$ is employed to investigate the effect of Knudsen number on permeability and interfacial convective heat transfer coefficient. The dimension of the REV and the square blocks are constant for all studied cases. The height of REV (i.e., H) is 670 μm while the size of square block (i.e., D) is 335 μm. The porosity of porous media is unchanged and the constant is $\varepsilon = 0.75$.

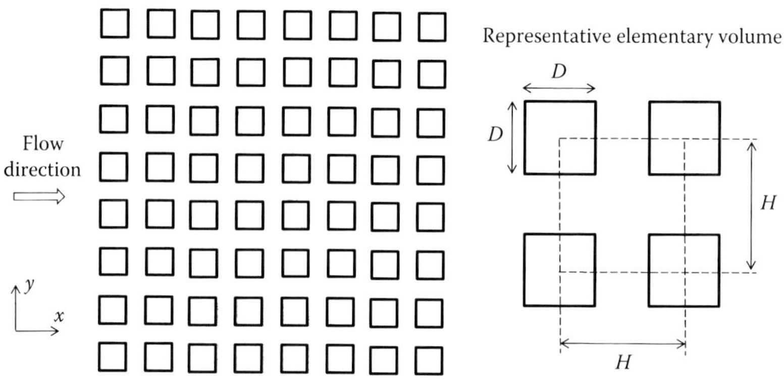

FIGURE 16.2 The studied 2D porous media.

TABLE 16.1
Cases Studied for the Determination of Permeability (T = 300 K)

Cases Studied	Pressure (Pa)	Density (kg/m³)	Mean Free Path (μm)	Knudsen Number
Case 1	101325	1.17659	6.70006E–08	0.0001
Case 2	10132.5	0.11766	6.70006E–07	0.001
Case 3	1013.25	0.011766	6.70006E–06	0.01
Case 4	506.625	0.005883	1.34001E–05	0.02
Case 5	202.65	0.002353	3.35003E–05	0.05
Case 6	101.325	0.001177	1.177000E–03	0.10

The fluid flowing through the porous media is air. For determination of permeability for the considered porous media, the dynamic viscosity of air is taken as 18.21×10^{-6} kg/ms. The air pressure is changed from 101325 to 101.325 Pa while its temperature is constant at 300 K. Based on the assumed temperature and pressure, the air density changes and different mean free path and consequently different Knudsen number are achieved. Table 16.1 shows the studied cases with corresponding density, mean free path, and Knudsen number. As shown in Table 16.1, in order to investigate permeability change with Knudsen number, six cases are studied. It should be mentioned that the permeability values are obtained in very low Re number as 0.001 to remove the inertia effect in the flow field.

In order to calculate the interfacial convective heat transfer coefficient, the solid particle is assumed at constant temperature of 310 K. A study is performed for two Reynolds numbers of 10 and 100 and six cases are considered for each Re number. The density of the air and mean free path are calculated according to the mean temperature which is the arithmetic average of periodic inlet and outlet temperatures and surface temperature.

16.5 GOVERNING EQUATIONS, BOUNDARY CONDITIONS, AND SOLUTION METHOD

The flow in the voids between the particles is assumed incompressible and steady. The fluid (i.e., air) is Newtonian fluid with constant thermophysical properties. The steady form of the continuity and momentum equations (Equation 16.1) is solved to determine the velocity and pressure fields for the fluid flow in the voids between the blocks. After obtaining the velocity field in the porous media, the energy equation for fluid phase is solved to obtain the temperature distributions in the REV. Considering REV in Figure 16.2, the boundary conditions for the microscopic equations are chosen as symmetry for the top and the bottom of the REV. The periodic velocity and temperature profiles are used for the inlet and outlet boundaries. The velocity and temperature gradients at the outlet boundary are assumed zero, hence no diffusion transport exists. The discretization interval is chosen as 0.02 mm which makes the number of grids as 500 × 500 for the entire domain. The Fluent commercial code based on the finite volume method is used to solve the governing equations. SIMPLE method is used for handling the pressure–velocity coupling. The power law scheme is employed for the discretization of the convection terms in the momentum and energy equations. The approximate errors are set to 10^{-9} for flow equations and 10^{-12} for temperature field.

16.5.1 Calculation of Permeability

As it is well known, permeability is a tensor quantity and based on Darcy's law for a 2D flow in Cartesian coordinate, and it can be defined as

$$\begin{pmatrix} \langle u \rangle \\ \langle v \rangle \end{pmatrix} = \frac{1}{\mu} \begin{pmatrix} K_{xx} & K_{yx} \\ K_{xy} & K_{yy} \end{pmatrix} \begin{pmatrix} \dfrac{\partial p}{\partial x} \\ \dfrac{\partial p}{\partial y} \end{pmatrix} \tag{16.7}$$

where K_{xy}, K_{yx}, K_{xy}, and K_{xy} are components of the permeability tensor. In the present problem, the permeability is calculated only for x direction since the structure of porous media is symmetrical not only in x and y directions but also with respect to xy and yx diagonals. The permeability is calculated based on velocity and pressure fields obtained from the continuity and momentum equations. Considering the REV shown in Figure 16.2, the following boundary conditions are used with σ_m=1 assumption:

$$\begin{aligned} &\text{On solid walls:} \quad u_n = -\lambda \left.\frac{\partial u}{\partial n}\right|_{wall} \; ; \quad u_t = 0 \\ &\text{For inlet boundary:} \quad u(0,y) = f(y),\ v(0,y) = 0 \\ &\text{For outlet boundaries:} \quad \frac{\partial u(H,y)}{\partial x} = \frac{\partial v(H,y)}{\partial x} = 0 \end{aligned} \tag{16.8}$$

where u_n and u_t are normal and tangential components of the velocity vector at the solid wall and they can be u or v components according to the position of wall. The mean free path is shown by λ in the earlier equation. The profile velocity of $f(y)$ is obtained by an iterative procedure. A uniform velocity distribution is assigned for the inlet velocity, and the outlet velocity is obtained by solving the continuity and momentum equations according to the aforementioned boundary conditions. Then, the obtained outlet velocity is used as inlet velocity profile and the same equations are solved again. The procedure continues until the inlet and outlet velocity profiles become identical (i.e., $f(y)$).

The Darcy velocity and pressure gradient in x direction for flow through the REV are calculated by the following relation:

$$\langle u \rangle = \frac{1}{H^2} \int_0^H \int_0^H u \, dx \, dy \tag{16.9}$$

$$\frac{d\langle p\rangle^f}{dx} = -\frac{1}{H(H-D)}\left[\int_{D/2}^{H-D/2} p\big|_{x=0}\,dy - \int_{D/2}^{H-D/2} p\big|_{x=H}\,dy\right] \tag{16.10}$$

Hence the permeability for x direction can be obtained from the following Darcy equation:

$$\langle u_f\rangle = -\frac{k_{xx}}{\mu}\frac{d\langle p\rangle^f}{dx} \tag{16.11}$$

Considering Figure 16.2, the permeability tensor for the present problem takes the following form:

$$\begin{pmatrix}\langle u_f\rangle \\ \langle v_f\rangle\end{pmatrix} = \frac{1}{\mu}\begin{pmatrix}K_{xx} & 0 \\ 0 & K_{yy}\end{pmatrix}\begin{pmatrix}\dfrac{\partial p}{\partial x} \\ \dfrac{\partial p}{\partial y}\end{pmatrix} \tag{16.12}$$

For the studied cases, the values of K_{xx} and K_{yy} are equal to each other due to symmetrical geometry of REV.

16.5.2 Calculation of Interfacial Heat Transfer Coefficient

For a thermal nonequilibrium condition, the heat transfer between the solid phase surface and the fluid flowing in the voids can be calculated by using the interfacial convective heat transfer concept. Mathematically, the convective heat transfer between the solid and fluid can be calculated by the following relation:

$$h_{sf}A_{ss}(\langle T_s\rangle - \langle T_f\rangle) = \frac{k_f}{V}\int_{A_{sf}} \vec{n}\cdot\vec{\nabla}T\,dA \tag{16.13}$$

where

h_{sf} is the interfacial convective heat transfer coefficient
A_{sf} is the solid–fluid interface area
A_{ss} is the specific solid–fluid interface area (i.e., $A_{ss} = A_{sf}/V$).

Correspondingly, the interfacial Nusselt number can be defined as follows:

$$\mathrm{Nu}_{sf} = \frac{h_{sf}H}{k_f} \tag{16.14}$$

where H represents the length and height of the REV. Although different characteristic lengths such as hydraulic diameter and the block height may be employed to define Nusselt number for the studied porous structure, the dimension of the REV is selected in this study. The same characteristic

length is also selected in many reported studies [4,5,7]. Considering the REV in Figure 16.2, the thermal boundary conditions for the PSM equations are as follows:
On the solid walls

$$T_f - T_w = -\kappa\lambda \left.\frac{\partial T_f}{\partial y}\right|_{wall}$$

On the top and bottom

$$\frac{\partial T_f}{\partial y} = 0$$

For inlet boundary

$$T(0, y) = g(y) \tag{16.15}$$

For outlet boundary

$$\frac{\partial T_f}{\partial x} = 0$$

where κ is a parameter that represents the degree of temperature jump, defined from the temperature jump boundary condition, $\kappa = ((2-\sigma_t)/\sigma_t)(2\gamma/(\gamma+1))(1/\text{Pr})$, and $\kappa = 0$ corresponds to no temperature jump at the wall, while $\kappa = 1.667$ is a typical value for air, which is the working fluid in many engineering applications and is taken so in this study. As seen in Equation 16.15, a temperature jump is assumed at the interface between the solid and fluid phases. The function of $g(y)$ is the temperature profile that provides thermal periodicity for the inlet and outlet boundaries of the REV. In order to determine the function of $g(y)$, it is assumed that a thermally fully developed convection heat transfer is valid which means that no change of the dimensionless temperature should be observed in sequential REVs in the flow direction. A uniform temperature profile which is different from the solid temperature is defined for the fluid inlet boundary and then the temperature field for the entire domain and consequently for the outlet boundary is obtained. The temperature at the inlet is determined from the dimensionless temperature profile at the outlet boundary. The iterative process continues until no change in the dimensionless temperature distribution at the inlet and outlet is observed. The dimensionless temperature equality for achieving thermally periodic flow is

$$\left.\frac{T - T_w}{T_b - T_w}\right|_{inlet} = \left.\frac{T - T_w}{T_b - T_w}\right|_{outlet} \tag{16.16}$$

where T_b is the bulk temperature and it can be obtained by the following relation for the considered REV:

$$T_b = \frac{\int_A uT dA}{\int_A dA} \tag{16.17}$$

Furthermore, following dimensionless temperature distribution is defined to provide temperature distribution of different REV in the same range:

$$\theta^* = \frac{T - T_{\min}}{T_{\max} - T_{\min}} \tag{16.18}$$

where T_{min} and T_{max} are the minimum and maximum temperatures in the considered temperature field.

16.6 RESULTS AND DISCUSSION

In this section, the results for permeability and interfacial heat transfer coefficient are presented separately.

16.6.1 Permeability

As discussed previously, a commercial code is used to solve the microscopic governing equations for the considered porous media. In order to define slipping boundary conditions for solid walls, a user defined function (UDF) is written and adapted to program. The validation of written UDF is checked with fully developed velocity profiles in a channel obtained analytically. Figure 16.3 shows the dimensionless velocity profile obtained by the employed code and the corresponding analytical results for a straight channel for three different Knudsen numbers. As shown, there is a good agreement between the analytical solution and the obtained computational results illustrating the correctness of the written UDF code.

Figure 16.4 shows the distribution of longitudinal dimensionless velocity component for two Knudsen numbers of 0.0001 and 0.1. The dimensionless velocity is defined as the ratio of velocity component to the mean inlet velocity value. A high-velocity gradient at the solid horizontal surface for Kn = 0.0001 can be observed. By increasing the Knudsen number from 0.0001 to 0.1, the gradient of velocity at the solid surface decreases considerably and this causes a lower pressure drop through the REV. Hence, it may be expected that the increase in Knudsen number increases the permeability of fluid through the porous medium. The values of permeability for the cases given in Table 16.1 are obtained and nondimensionalized with H^2. The results are presented in Figure 16.5.

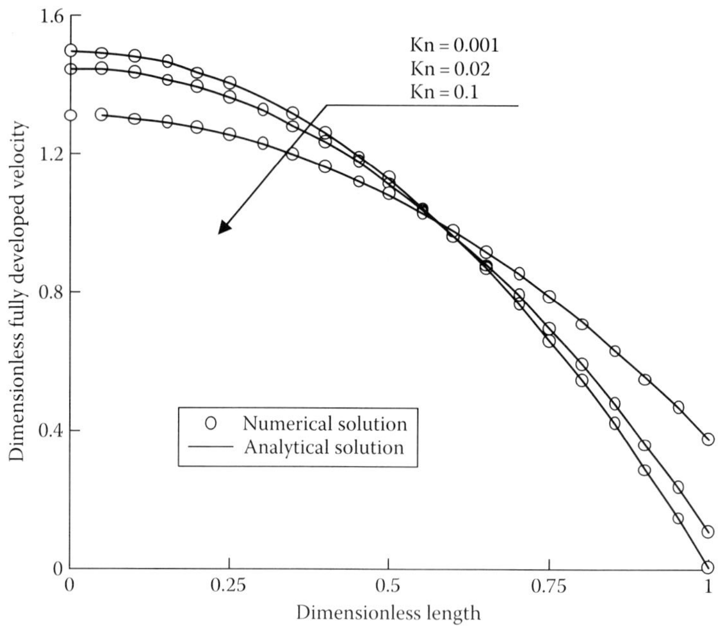

FIGURE 16.3 Comparison of computational results with the analytical solution.

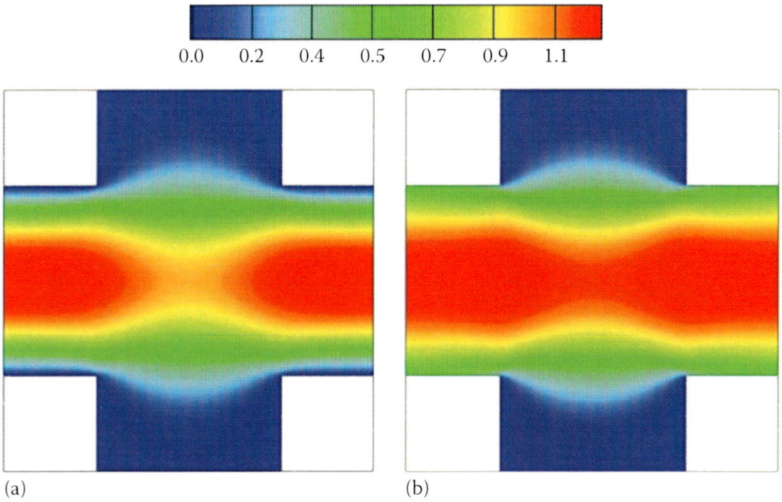

FIGURE 16.4 Distribution of dimensionless longitudinal velocity component for flow in REV: (a) Kn = 0.0001 and (b) Kn = 0.1.

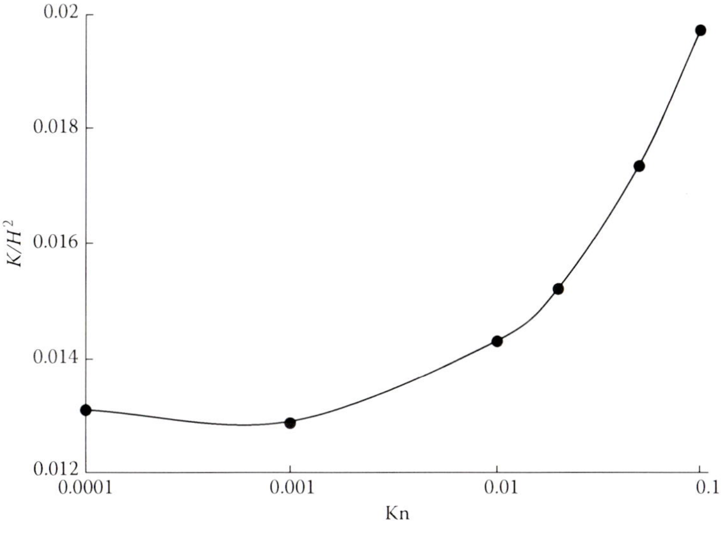

FIGURE 16.5 The change of dimensionless permeability with Kn number.

The permeability value increases considerably with Knudsen number indicating high permeability at lower Knudsen number.

Although many equations were suggested in the literature for the determination of permeability under rarefied effect, it seems that the relationship presented by Kuwahara and Nakayama [14] is the most suitable equation for the studied porous media in this work.

$$\frac{K}{K_{ns}} = 1 + C\,\mathrm{Kn} \tag{16.19}$$

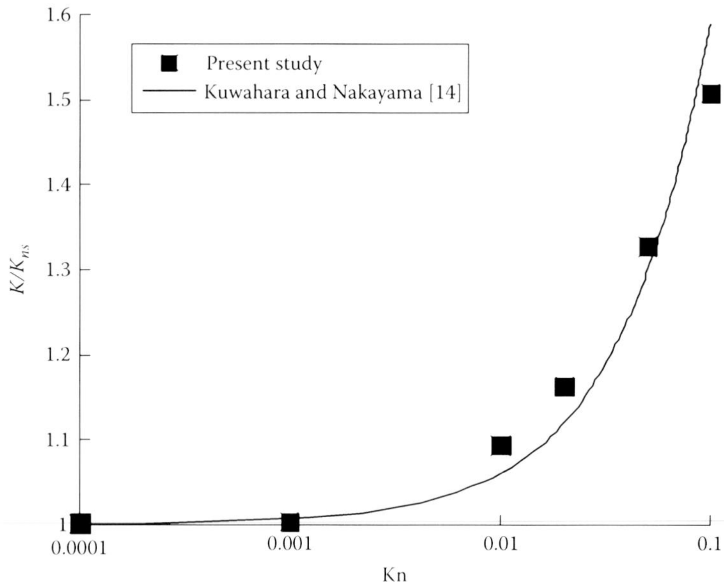

FIGURE 16.6 The comparison between the obtained numerical results with the correlation suggested by Kuwahara and Nakayama [14].

where K_{ns} is the permeability of the porous media for no-slip boundary condition. The coefficient of C is defined as

$$C = 9\frac{(1-\varepsilon)^{1/2}}{\varepsilon} \tag{16.20}$$

Figure 16.6 shows the comparison of the obtained results with correlation suggested by Kuwahara and Nakayama. As shown, good agreement between the correlation and the numerical results obtained in this study is observed.

16.6.2 Interfacial Convective Heat Transfer

Similar to the calculation of permeability, UDF codes are written to provide temperature jump for horizontal and vertical interfaces between solid and fluid boundaries. The written code and program are checked by comparing numerical results with Refs. [21,22] for a channel and it is shown in Figure 16.7. A good agreement between the computational results of present and reported studies can be observed.

Figure 16.8 shows the dimensionless temperature distribution for two porous media with two different Reynolds and Knudsen numbers. Figure 16.8a shows the temperature contours in porous media for Re = 10. It should be mentioned that the dimensionless temperature at the solid–fluid is 1 (i.e., $\theta^* = 1$). For Kn = 0.0001, the temperature of fluid at the region close to the solid surface is around 1 (i.e., $\theta^* = 1$) due to negligible temperature jump at the solid surface. By increasing Kn number from 0.0001 to 0.1, the effect of the slipping boundary condition can be observed clearly. The interface temperature for fluid region is considerably reduced and this causes fluid flowing in

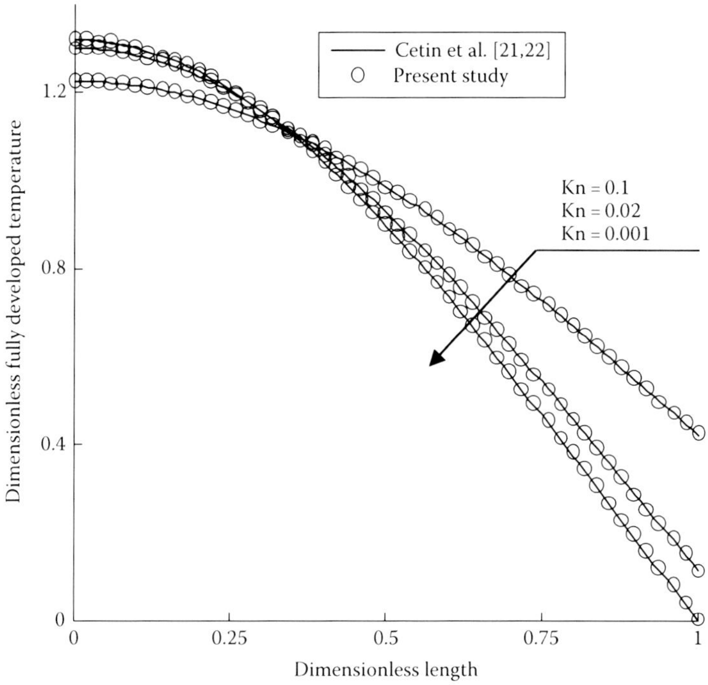

FIGURE 16.7 Comparison of obtained computational and reported temperature profiles for a straight channel.

the voids which cannot be heated. Figure 16.8b shows the same porous media; however, the Re number is increased to 100. The same behavior for temperature changes observed for Re = 10 can also be seen for fluid flowing in the voids with Re = 100. The Kn number has an important effect on the temperature field of fluid and it reduces the heat transfer rate in the voids. Hence, a reduction in interfacial heat transfer coefficient is expected.

The change in interfacial Nusselt number with Knudsen for two Reynolds number of Re = 10 and 100 is presented in Figure 16.9. As expected, the Nu number considerably decreases with Kn number for both Re = 10 and 100 flows. The interesting point of Figure 16.9 is that the difference between Nu number of flows with Re = 10 and 100 depends on the Kn number. By increasing Kn number, the difference between Nu numbers is reduced.

Our study shows that it is possible to express a relation for interfacial heat transfer coefficient based on no-slip Nusselt number. A relation for the determination of Nu_{sf} for porous media with $\varepsilon = 0.75$ is found as follows:

$$\frac{Nu_{sf}}{Nu_{ns,sf}} = 0.9915 - 11.757Kn + 57.83Kn^2 \tag{16.21}$$

where $Nu_{ns,sf}$ is the interfacial Nusselt number for porous media with no-slip boundary condition. Figure 16.10 shows the comparison of the suggested relation with the obtained numerical results.

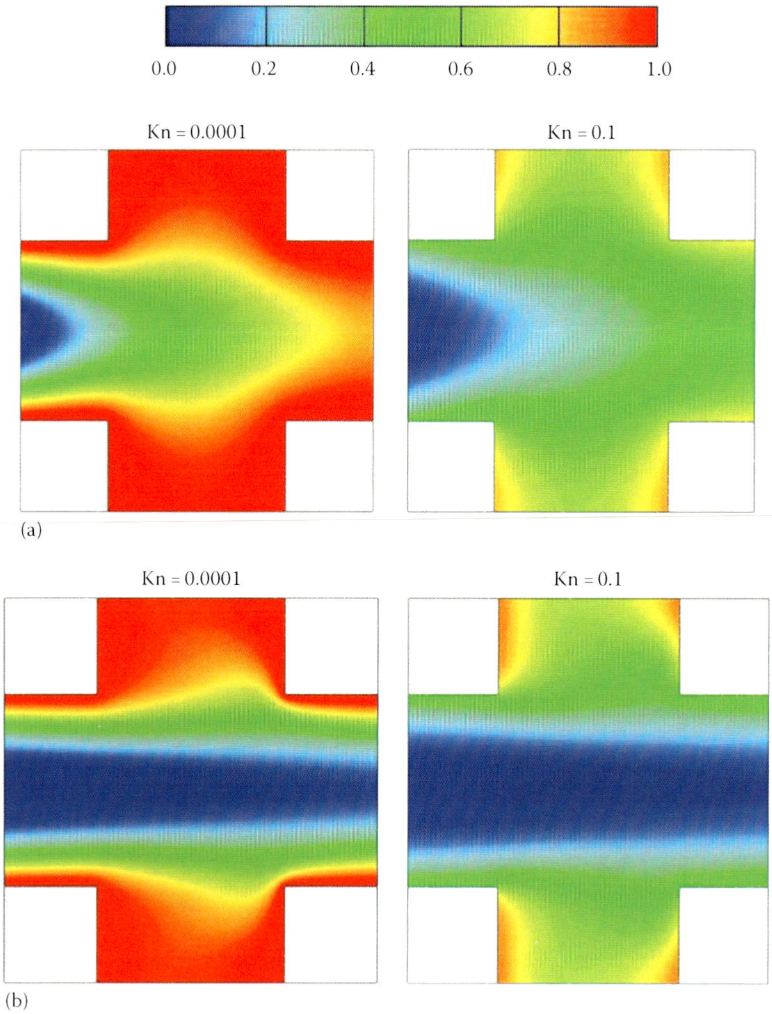

FIGURE 16.8 Dimensionless temperature distribution in the porous media: (a) Re = 10 and (b) Re = 100.

16.7 CONCLUSION

A numerical study on heat and fluid flow in a 2D porous media with microchannel is performed. The related PSM-governing equations and boundary conditions are solved and both permeability and interfacial heat transfer coefficient are calculated by using volume-averaged method. It is observed that for porous media with microchannel particularly at low gas pressure, the Kn number has a significant effect on both permeability and interfacial heat transfer coefficient. Hence, the traditional corrections reported in the literature for permeability and interfacial heat transfer coefficient cannot be used. Two-dimensional porous media studies showed that an increase in Kn number from 0.0001 to 0.1 increased the permeability by 51%. The reduction of interfacial heat transfer coefficient with Kn number depends on Re number. The interfacial heat transfer coefficients become closer to each other with an increase in the Kn number.

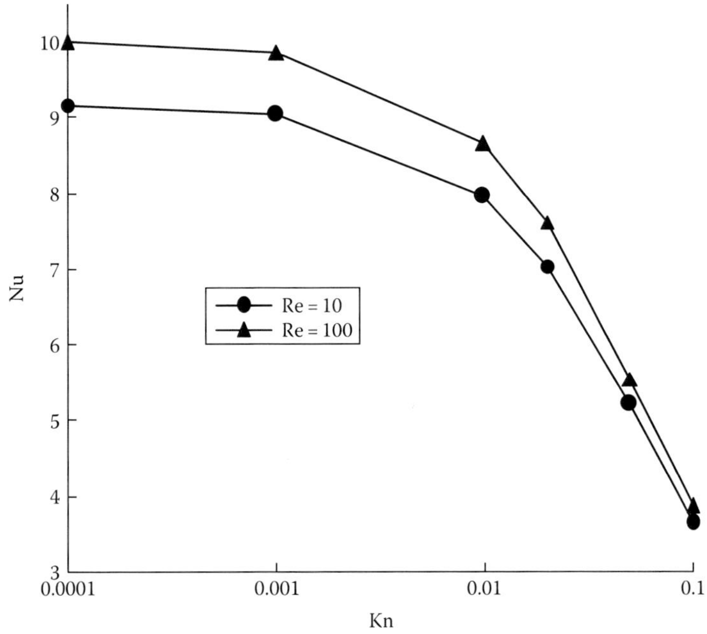

FIGURE 16.9 The change in Nu number with Kn for two Re numbers of 10 and 100.

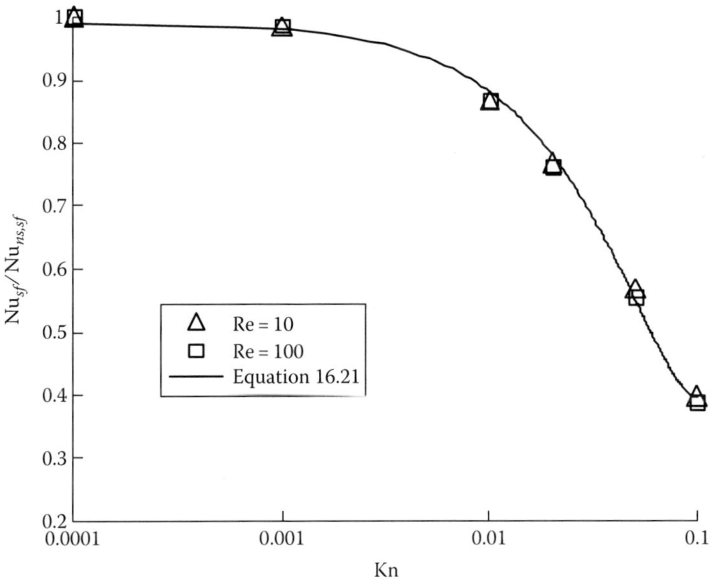

FIGURE 16.10 Comparison of the suggested correlation with obtained numerical results.

ACKNOWLEDGMENT

The authors want to thank M.Sc. student Safa Sabet for his great effort in obtaining computational results and it would have been difficult to finish this study on time without his help.

REFERENCES

1. C. Yang, A. Nakayama, A synthesis of tortuosity and dispersion in effective thermal conductivity of porous media, *Int. J. Heat Mass Transfer*, 53 (15), 3222–3230 (2010).
2. S. Sharma, D.A. Siginer, Permeability measurement methods in porous media of fiber reinforced composites, *Appl. Mech. Rev.*, 63 (2), 020802–020802-19 (2010).
3. T. Ozgumus, M. Mobedi, U. Ozkol, A. Nakayama, Thermal dispersion in porous media; a review on the experimental studies for packed beds, *Appl. Mech. Rev.*, 65 (3), 031001–031001-19 (2013).
4. F. Kuwahara, A. Nakayama, Numerical determination of thermal dispersion coefficients using a periodic porous structure, *J. Heat Transfer*, 121 (1), 160–163 (1999).
5. M.B. Saito, M.J.S. de Lemos, A correlation for interfacial heat transfer coefficient for turbulent flow over an array of square rods, *J. Heat Transfer*, 128, 444–452 (2006).
6. A. Nakayma, F. Kuwahara, Numerical modeling of convective heat transfer in porous media using microscopic structure, *Hand Book of Porous Media*, Edited by Vafai K., Marcel Dekker Inc., New York, pp. 441–487 (2000).
7. T. Ozgumus, M. Mobedi, Effect of pore to throat size ratio on interfacial heat transfer coefficient of porous media, *J. Heat Transfer*, 137 (1), 012602 (2014).
8. A. Nakayama, F. Kuwahara, T. Umemoto, T. Hayashi, Heat and fluid flow within an anisotropic porous medium, *J. Heat Transfer*, 124, 746–753 (2002).
9. M. Peszynska, A. Trykozko, Pore-to-core simulations of flow with large velocities using continuum models and imaging data, *Comput. Geosci.*, 17, 623–645, (2013).
10. A. Akolkar, J. Petrasch, Tomography-based characterization and optimization of fluid flow through porous media, *Transp. Porous Med.*, 95, 535–550 (2012).
11. E. Ucar, M. Mobedi, G. Altıntas, E. Glatt, Effect of voxel size in flow direction on permeability and Forchheimer coefficients determined by using micro-tomography images of a porous medium, *Progress in Computational Fluid Dynamics* (in press).
12. N. Jeoong, Advanced Study about the permeability for micro-porous structures using the Lattice Boltzmann method, *Transp. Porous Med.*, 83, 271–288 (2010).
13. T.L. Vu, G. Lauriat, O. Manca, Forced convection of air through networks of square rods or cylinders embedded in microchannels, *Microfluid Nanofluid*, 16, 287–304 (2014).
14. F. Kuwahara, A. Nakayama, Numerical simulation of rarefied gas flow through a porous media, *Trans. Jpn. Soc. Mech. Eng.*, 69, 189–195 (1999).
15. R.N. Xu, Y.L. Huang, P.X. Jiyang, B.X. Wang, Internal heat transfer coefficients in microporous media with rarefaction effects, *Sci. China, Technol. Sci.*, 55 (10), 2869–2876 (2012).
16. A. Beskok, G.E. Karniadakis, Simulation of heat and momentum transfer in complex micro-geometries. *J. Thermophys. Heat Transfer*, 8, 355–370 (1994).
17. M. Gad-el-Hak, *The MEMS Handbook*, CRC Press, New York (2001).
18. G.E. Karniadakis, A. Beskok, N. Aluru, *Micro Flows and Nano Flows: Fundamentals and Simulation*, Springer-Verlag, New York, 2005.
19. H.P. Kavehpour, M. Faghri, Y. Asako, Effects of compressibility and rarefaction on gaseous flows in microchannels. *Numer. Heat Transfer, Part A*, 32, 677–696 (1997).
20. E.B. Arkilic, K.S. Breuer, M.A. Schmidt, Gaseous flow in microchannels. Application of microfabrication to fluid mechanics. *ASME FED*, 197, 57–66 (1994).
21. B. Cetin, Analysis of single phase convective heat transfer in microtubes and microchannels, Master thesis, Middle East Technical University, 2005.
22. B. Cetin, H. Yuncu, S. Kakac, Gaseous flow in microchannels with viscous dissipation. *Int. J. Transp. Phenom.*, 8, 297–315 (2006).

17 Numerical Analysis of Subcooled Convective Boiling in Microchannels

Mourad Rebay, Ghazali Mebarki, Nadim El Wakil, and Anis Hamza

CONTENTS

17.1 INTRODUCTION

This chapter covers a numerical method for the resolution of the problem of subcooled convective boiling flows in microchannels heat sink. The focus is on the numerical procedure for tracking or capturing interface/surface shape. Fundamentals of the boiling phenomena and the thermophysical properties with dimensionless numbers usually used in the resolution of the two-phase flow problems are presented in Chapter 18. In particular, an analysis of the transition of phase change pattern between conventional macrochannels and mini/microchannels is given in that chapter. Since the focus of this chapter is the numerical modeling rather than the fundamentals of the boiling process, analysis and interpretation are limited to the concerned subject. Analysis of some experimental research works in the literature is also given in Chapter 18. For an in-depth understanding of fundamentals and analysis of convective boiling heat transfer in microscale, readers are referred to different excellent sources available in the literature. One can quote Refs. [1–13] among a long list of interesting studies.

Mini- and microchannel heat sinks have become known as one of the effective cooling techniques. The advantage of such devices is their ability to dissipate large heat flux in relatively small volumes, which makes them very efficient cooling systems. Mini- and microchannel heat sinks combine the attributes of very high surface-area-to-volume ratio, large convective heat transfer coefficient (even in single-phase flows), and small coolant fluid quantity. These attributes make mini/microchannels heat sinks very suitable for cooling devices with high dissipating energy like integrated circuits, microprocessors, and high-energy laser mirrors.

Boiling occurs when the temperature of the heater surface exceeds the saturation temperature. If the bulk temperature of the liquid is below saturation, the process is known as subcooled boiling. In heat sinks with phase change, a better understanding of two-phase flows occurring in phase change allows leading to an optimal design of compact heat exchangers and micro/minichannels heat sinks. Such design allows reducing manufacturing cost and energy consumption. Two-phase flows with phase change are known to be a very complex phenomenon. Indeed, in addition to classical effects such as buoyancy, viscosity, and pressure forces involved in single-phase flows, two-phase flows are also influenced by interfacial tension forces and the exchange of momentum between the liquid and the vapor phases. Analysis of physics and mechanism of nucleation, growth of bubbles, and convective heat transfer enhancement in micro/minichannels are presented in Refs. [1,2,10–13] among others.

Flow boiling in small hydraulic diameter channels is becoming increasingly important in many diverse applications. Flow boiling has been studied extensively for many decades. Correlations and experiments at the microscale have been concerned with the determination of the average heat transfer coefficient and average pressure drop over the length of the vaporization section. But, experiments have their own limitations, and because of the difficulty of having stable conditions in multiphase flows, the measurement inaccuracies can lead to a large error. Hence, there is a need to numerically model the two-phase flow through mini/microchannels, which would help to interpret better the available experimental data and also explain the underlying physics. A direct numerical simulation of the boiling phenomena is one of the promising approaches in order to clarify its heat transfer characteristics and discuss their mechanism. Therefore, many direct numerical simulation (DNS) procedures have been developed based on recent highly advanced computer technologies [14–17]. The tracking or capturing interface/surface shape such as the front tracking method, level set method, and volume of fluid treatments are briefly reviewed here.

Typically, there are several methods that are used for the simulation of boiling heat transfer. One can cite the following:

- *Marker and Cell method* [18,19]:
  The interface is marked by many massless particles that are convectively transported with the Eulerian velocity field and can be used to reconstruct the interface position on a fixed mesh. This method showed some difficulties in the determination of the precise position of the interface from the scattered particles after transporting the interface.
- *Front tracking method* [20,21]:
  This method is an extension of the marker and cell method and is thus based on massless particles that follow the motion of the interface driven by the Eulerian velocity fields. The interface is tracked by a Lagrangian frame, and the interface position can be tracked precisely, but it is necessary to introduce some artificial rules to treat the break-up interface and also coalescence of interfaces during computation. It has been shown that this method is very accurate in particular in terms of curvature calculation which is important for the simulation of very small bubbles.
- *Arbitrary Lagrangian–Eulerian method* [22]:
  This method is based on a dynamic mesh that follows the motion of the interface. Thus, the interface coincides with a boundary of the computational domain at all times. According to this feature, the center of the control volume may not coincide often with the center of mass, and eventually the introduction of large artificial viscosity will be inevitable. A remeshing procedure is often required to fit the boundary-fitted coordinate of each phase to the phase interface. This procedure is a time-consuming process for computation. Despite this disadvantage, this method presents an important advantage compared to the aforementioned methods; it treats the liquid–vapor interface as a boundary of the computational domain, which is a very important task when

the bubble–liquid interface is highly dependent on its curvature angle, as it is usually occurring in boiling in mini/microchannels.

- *Level-Set method* [23,24]:
  In this method, the distance of the numerical cell to the interface, which is convectively transported with the Eulerian velocity fields, is continuously tracked. The interface is represented by the zero-contour line of the level-set field. Analysis and various applications of this method have been given by Dhir and coworkers in Refs. [23,24].
- *Volume-of-Fluid (VOF) method* [25]:
  The VOF method is based on the use of information contained in the field of the volume fraction of one of the phases in a numerical cell and which is convectively transported with the Eulerian velocity fields. The VOF method allows the interface reconstruction between neighborhood cells and the precise calculation of interface curvature. The well-known problem of the VOF procedures is how to remove the "spurious velocity" caused by the surface tension calculation. Many studies have been conducted on this problem. Since most VOF procedures based on the continuum surface force [26] do not show high accuracy of the curvature calculation by using the VOF values in the neighboring cells, more accurate methods have been proposed to strongly suppress the spurious velocity such as the method by using the estimator function [27] and the second-order approximation of the local interface shape (parabolic reconstruction of surface tension) [28].

The computational fluid dynamics (CFD) is a powerful tool for fluid dynamics and thermal design in industrial applications, as well as in academic research activities. For that purpose, numerical simulations are widely used to investigate two-phase flows with vaporization in the commercialized CFD package, such as Fluent. Several methods have been proposed to model the two-phase flows. As it is one of the lowest computation time procedures, but precise interface tracking, VOF method has been found more suitable for simulating interface between two fluids or more. Thus, appropriate source terms need to be added in both momentum and energy equations, in order to take into account the heat absorption during vaporization or the heat release during condensation. VOF phase-change model has been used by Wu et al. [29] for phase change in a serpentine tube for refrigerant, and the numerical simulation procedure has been validated by an experimental study. More recently, Asghar et al. [30] have used a VOF method to study the two-phase flows with simultaneous vaporization and condensation phenomena in a thermosiphon tube. A good agreement with experimental results has been observed.

In this chapter, subcooled convective boiling problem has been numerically studied in a microchannel heat sink. A validation study has been carried out by comparison with the literature correlation and some typical results are displayed and discussed. A specific in-house program (UDF) using VOF method has been implemented in the commercialized CFD software (Fluent). We present here also an application of the developed method in convective boiling in microchannels heat sink used in the cooling of the light absorber in a Synchrotron beamline. This Euler–Euler multiphase approach, known as VOF method, is employed as most of bubble sizes are in the order of the channel dimensions (0.7 mm in width).

17.2 NUMERICAL SIMULATION OF SUBCOOLED CONVECTIVE BOILING

17.2.1 Volume of Fluid Modeling

The VOF formulation relies on the fact that two or more phases are not interpenetrating. The fields for all variables and properties are shared by the phases and represent volume-averaged values, as long as the volume fraction of each of the phases is known at each location. Thus, depending on the

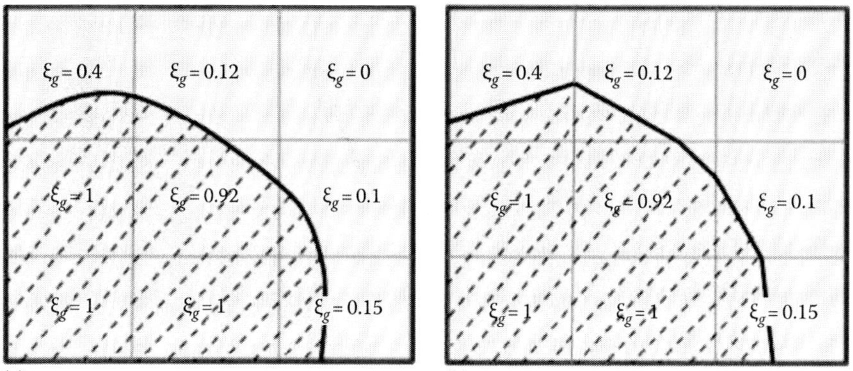

(a) (b)

FIGURE 17.1 Interface shape represented by the geometric reconstruction piecewise linear scheme. (a) Real interface and (b) PLIC reconstruction method.

volume fraction values, the variables and properties in any given cell are either purely representative of one of the two phases or a mixture of them. The vapor–liquid interface is tracked, thanks to the distribution of the vapor volume fraction ε_g in a computational cell. Vapor fraction varies from 0 to 1 ($\varepsilon_g = 0$ in the liquid phase and 1 in the vapor phase). Therefore, the vapor–liquid interface exists in the cell when ε_g has a value between 0 and 1. In the VOF model, a single set of transport equations is solved and the volume fraction of each fluid throughout the numerical cell is tracked. Appropriate jump boundary conditions are imposed at the liquid–vapor interface. The only drawback of VOF method is the so-called artificial (numerical) coalescence of gas bubbles which occurs when their mutual distances are less than the size of the computational cell. This fact also makes this approach memory consuming for the simulation of dispersed multiphase flows such as in large thermal systems [31]. When the cell is adjacent to the interface between two phases, the geometric reconstruction scheme, which is based on the piecewise linear interface calculation (PLIC) method, is applied to reconstruct the interface between fluids. This technique assumes that the interface between two fluids has a linear slope within each cell and uses this linear shape for the calculation of the fluid advection through the cell faces as represented in Figure 17.1.

The first step in this reconstruction schema is the positioning of the linear interface relative to the center of each partially filled cell, based on information concerning the vapor volume fraction and its derivatives in the cell. The second step is the calculation of the fluid advection amount through each interface using the computed linear interface representation and information about the normal and tangential velocity distributions at the interface. As a last step, the volume fraction in each cell is determined using the balance of fluid mass fluxes which were calculated previously [32].

Vaporization model: The vaporization, within the Fluent software, was modeled by an in-house program (UDF), for which source terms have been added in the continuity and energy equations. For that purpose, equations proposed by De Schepper et al. [33] were used to calculate the source terms (the mass transfer from liquid to vapor and inversely). Phase change was considered to occur at the saturation temperature T_{sat}. When the liquid phase temperature T_l exceeds this temperature, a mass with a mass flow rate S_M transfers from the liquid phase to the vapor phase and an amount of energy is then absorbed. In this case, two expressions describing the mass transfer in both phases (liquid and vapor) are required:
Liquid phase:

$$S_{M,l-g} = -t_{l-g}\varepsilon_l\rho_l \frac{T_l - T_{sat}}{T_{sat}} \tag{17.1}$$

Vapor phase:

$$S_{M,l-g} = t_{l-g}\varepsilon_l\rho_l \frac{T_l - T_{sat}}{T_{sat}} \quad (17.2)$$

where
- ε is the void fraction
- the subscripts l and g are referred to liquid and gas, respectively
- ρ is the density
- t will be defined later

The vaporization of liquid leads to a reduction in the liquid volume fraction and an increase in the vapor one. Local condensation of vapor that leaves the wall surface can also occur in the region in which $T_g < T_{sat}$. The mass source term is then calculated for each phase:

Liquid phase:

$$S_{M,g-l} = t_{g-l}\varepsilon_g\rho_g \frac{T_{sat} - T_g}{T_{sat}} \quad (17.3)$$

Vapor phase:

$$S_{M,g-l} = -t_{g-l}\varepsilon_g\rho_g \frac{T_{sat} - T_g}{T_{sat}} \quad (17.4)$$

The heat transfer corresponding to the vaporization or condensation process is determined by multiplying the calculated mass transfer with the latent heat of vaporization h_{lg}. The resulting source term S_E should be taken into account in the energy equation:

Vaporization:

$$S_E = -t_{l-g}\varepsilon_l\rho_l \frac{T_l - T_{sat}}{T_{sat}} h_{lg} \quad (17.5)$$

Condensation:

$$S_E = t_{g-l}\varepsilon_g\rho_g \frac{T_{sat} - T_g}{T_{sat}} h_{lg} \quad (17.6)$$

In the previous equations, t_{l-g} and t_{g-l} are empirical coefficients called mass-transfer intensity factors (mass transfer time relaxation parameters) with s^{-1} units. They determine the vaporization and condensation rates, respectively. Their values are chosen depending on the studied problems. In order to keep the interface temperature within $T_{sat} \pm 1$ K, they have been taken equal to 0.1 by different authors like Wu et al. [29], De Schepper et al. [32,33], Asghar et al. [30], and Sun et al. [34]. It has also been shown that excessively large values of t_{g-l} and t_{l-g} may cause divergence of the numerical procedure. On the other hand, it has been proven that too small values lead to a significant deviation between the interfacial and saturation temperatures.

17.2.2 Governing Equations

The mass, momentum, and energy conservation equations were applied for each volume of fluid. The equations are as follows:

Continuity:

$$\frac{\partial \rho}{\partial t}+\sum_{j=1}^{3}\frac{\partial(\rho \bar{u}_i)}{\partial x_i}=S_M \tag{17.7}$$

Momentum:

$$\frac{\partial(\rho \bar{u}_i)}{\partial t}+\sum_{j=1}^{3}\frac{\partial(\rho \bar{u}_i \bar{u}_j)}{\partial x_j}=-\frac{\partial p}{\partial x_i}+\sum_{j=1}^{3}\frac{\partial}{\partial x_j}\left[\mu\left(\frac{\partial \bar{u}_i}{\partial x_j}+\frac{\partial \bar{u}_j}{\partial x_i}-\frac{2}{3}\delta_{ij}\sum_{l=1}^{3}\frac{\partial \bar{u}_l}{\partial x_l}\right)\right]+\sum_{j=1}^{3}\frac{\partial}{\partial x_j}(-\rho\overline{u_i' u_j'})+F_V \tag{17.8}$$

Energy:

$$\frac{\partial(\rho E)}{\partial t}+\sum_{j=1}^{3}\frac{\partial(\rho E \bar{u}_j)}{\partial x_j}=\sum_{i=1}^{3}\sum_{j=1}^{3}\left(\frac{\partial}{\partial x_j}(\tau_{ij}-\rho\overline{u_i' u_j'})\bar{u}_i\right)-\sum_{j=1}^{3}\frac{\partial}{\partial x_j}q_j+S_E \tag{17.9}$$

Reynolds stress equation:

$$-\rho\overline{u_i' u_j'}=\mu_t\left(\frac{\partial \bar{u}_i}{\partial x_j}+\frac{\partial \bar{u}_j}{\partial x_i}\right)-\frac{2}{3}\left(\rho k+\mu_t\sum_{l=1}^{3}\frac{\partial \bar{u}_l}{\partial x_l}\right)\delta_{ij} \tag{17.10}$$

where
- δ_{ij} is the Kronecker function
- τ_{ij} is the strain tensor

Since the formation of bubbles, their growing up, agglomeration, and diffusion in the main flow involve different scales of motion through changes to the production term; the renormalization group (RNG) k-ε turbulence model is used for closing the equations. Indeed, RNG methods used to normalize the Navier–Stokes equations allow the account of the effects from smaller to larger scales of motion that contributes to the turbulent diffusion. The RNG approach is a mathematical technique that can be used to derive a turbulence model similar to the k-ε equation. With the turbulent viscosity being calculated in the same manner as with the standard k-ε model, the RNG k-ε turbulence model is governed by the following equations:

$$\frac{\partial(\rho k)}{\partial t}+\sum_{j=1}^{3}\frac{\partial(\rho k \bar{u}_j)}{\partial x_j}=\sum_{j=1}^{3}\frac{\partial}{\partial x_j}\left[\alpha_k \mu_{eff}\frac{\partial k}{\partial x_j}\right]+\bar{p}_k-\rho\varepsilon \tag{17.11}$$

$$\frac{\partial(\rho \varepsilon)}{\partial t}+\sum_{j=1}^{3}\frac{\partial(\rho \varepsilon \bar{u}_j)}{\partial x_j}=\sum_{j=1}^{3}\frac{\partial}{\partial x_j}\left[\alpha_\varepsilon \mu_{eff}\frac{\partial \varepsilon}{\partial x_j}\right]+C_{1\varepsilon}\frac{\varepsilon}{k}\bar{p}_k-C_{2\varepsilon}\rho\frac{\varepsilon^2}{k}-R_\varepsilon \tag{17.12}$$

with

$$\bar{p}_k = \sum_{i=1}^{3}\sum_{j=1}^{3}\left[\left(\mu_t\left(\frac{\partial \bar{u}_i}{\partial x_j}+\frac{\partial \bar{u}_j}{\partial x_i}-\delta_{ij}\sum_{l=1}^{3}\frac{2}{3}\frac{\partial \bar{u}_l}{\partial x_l}\right)-\frac{2}{3}\rho k\delta_{ij}\right)\frac{\partial \bar{u}_j}{\partial x_i}\right] \quad (17.13)$$

and

$$\mu_t = \rho C_\mu \frac{k^2}{\varepsilon} \quad (17.14)$$

$$R_\varepsilon = \frac{C_\mu \rho \eta^3 (1-\eta/\eta_0)}{1+\beta\eta^3}\frac{\varepsilon^2}{k} \quad (17.15)$$

The constant values are explicitly derived in the RNG k-ε turbulence model and they are given by

$$C_{1\varepsilon}=1.42,\ C_{2\varepsilon}=1.68,\ C_\mu=0.0845,\ \text{and}\ \alpha_k=\alpha_\varepsilon\approx 1.393,\ \eta_0=4.38,\ \text{and}\ \beta=0.012$$

In the earlier equations, μ is the dynamic viscosity and k is the thermal conductivity of the fluid mixture. They depend on the volume fractions of each phase (liquid or gas) and are calculated by using the following equations:

$$\rho = \varepsilon_l \rho_l + \varepsilon_g \rho_g \quad (17.16)$$

$$\mu = \varepsilon_l \mu_l + \varepsilon_g \mu_g \quad (17.17)$$

$$k = \varepsilon_l k_l + \varepsilon_g k_g \quad (17.18)$$

The energy E is considered as a weighted mass-averaged variable. It is expressed as

$$E = \frac{\varepsilon_l \rho_l E_l + \varepsilon_g \rho_g E_g}{\varepsilon_l \rho_l + \varepsilon_g \rho_g} \quad (17.19)$$

Therefore, for liquid and gaseous phases, its expression is, respectively,

$$E_l = C_{v,l}(T - T_{shared})\ \text{and}\ E_g = C_{v,g}(T - T_{shared}) \quad (17.20)$$

where
- C_v is the heat capacity under constant volume
- T_{shared} is the shared temperature at the interface between the two phases

The tracking of the interface between the phases has been accomplished by solving the continuity equation for the volume fraction of the two phases, using the following equations:

$$\frac{\partial \varepsilon_l}{\partial t} + \sum_{j=1}^{3}\frac{\partial(\varepsilon_l \bar{u}_j)}{\partial x_j} = \frac{S_{M,l-g}}{\rho_l} \quad (17.21)$$

$$\frac{\partial \varepsilon_g}{\partial t} + \sum_{j=1}^{3}\frac{\partial(\varepsilon_g \bar{u}_j)}{\partial x_j} = \frac{S_{M,g-l}}{\rho_g} \quad (17.22)$$

17.2.3 Surface Tension

A formulation of the continuum surface force model developed by Brackbill et al. [26] is used in the computation of surface tension. In this model, surface tension is considered as a continuous function taking into account 3D effects across an interface rather than as a boundary value condition on it. Thus, the surface tension σ is a surface force that is converted into a volume force F_V acting as a source term in the momentum equation (Equation 17.8) and it is written as follows:

$$F_V = \sigma \frac{\varepsilon_l \rho_l K_g \nabla\varepsilon_g + \varepsilon_g \rho_g K_l \nabla\varepsilon_l}{\frac{1}{2}(\rho_l + \rho_g)} \tag{17.23}$$

where K_l and K_g are the interface curvatures of liquid and vapor, respectively, which have the following expressions:

$$K_l = \frac{\Delta\varepsilon_l}{\left|\nabla\varepsilon_l\right|}, \quad K_g = \frac{\Delta\varepsilon_g}{\left|\nabla\varepsilon_g\right|} \tag{17.24}$$

The surface tension of water, in contact with its vapor, has been introduced using the commonly used equation given by Linhard and Linhard [35]:

$$\sigma = 235.8\left(1 - \frac{T_{sat}}{647.096}\right)^{1.256}\left[1 - 0.625\left(1 - \frac{T_{sat}}{647.096}\right)\right] \tag{17.25}$$

17.3 VALIDATION OF THE NUMERICAL SIMULATION IN CONVECTIVE BOILING IN CYLINDRICAL TUBE

In order to validate the developed numerical simulation procedure, both vapor fraction and heat transfer coefficient are computed. The results are compared with Kandlikar correlation given in Ref. [36] in the case of two-phase flow with vaporization in a vertical cylindrical tube. The tube material is copper. It is 500 mm in length, 16 mm in diameter, and 1 mm in thickness. The tube is subjected to a uniform heat flux density (φ = 100 kW/m^2). The flow is from downward to upward with a Reynolds number Re = 480 of the full liquid (water) entering in the tube. Concerning the heat transfer coefficient correlation, the presence of both nucleate and convective boiling terms has been considered in the correlation introduced by Kandlikar and Balasubramanian [37] as follows:
For $\mathrm{Re}_{LO} > 100$

$$h_{TP} = \max(h_{TP,NBD}, h_{TP,CBD}) \tag{17.26}$$

with

$$h_{TP,NBD} = 0.6683\,Co^{-0.2}(1-x)^{0.8}h_{LO} + 1058\,Bo^{0.7}(1-x)^{0.8}F_{Fl}\,h_{LO} \tag{17.27}$$

as the two-phase flow nucleate boiling heat transfer coefficient, and

$$h_{TP,CBD} = 1.136Co^{-0.9}(1-x)^{0.8}h_{LO} + 667.2Bo^{0.7}(1-x)^{0.8}F_{Fl}h_{LO} \tag{17.28}$$

represents the two-phase convective boiling heat transfer coefficient.
The coefficient

$$Co = \left[(1-x)/x\right]^{0.8}\left(\rho_g/\rho_l\right)^{0.5} \tag{17.29}$$

is the convection dimensionless number, and

$$Bo = \varphi/\dot{m}h_{lg} \tag{17.30}$$

is the boiling dimensionless number; where

- x is the mass quality
- φ is the applied heat flux density
- $\dot{m}$ is the mass flow rate.
- F_{Fl} is a fluid surface parameter and is tabulated in Ref. [4] for different fluid-surface combinations.

For single phase, the heat transfer coefficient h_{LO} is given by Gnielisnki correlation [38] in the turbulent flow. For laminar flow, Kandlikar and Steinke [36] suggested:

$$h_{LO} = \frac{(\text{Re}_{LO} - 1000)\,\text{Pr}_l(f/2)(k_l/D)}{1 + 12.7(\text{Pr}_l^{2/3} - 1)(f/2)^{0.5}} \tag{17.31}$$

For $3000 \le \text{Re}_{LO} \le 10^4$, and

$$h_{LO} = \frac{\text{Nu}_{LO}k_l}{D} \tag{17.32}$$

For $100 \le \text{Re}_{LO} \le 1600$.

In the transition region (Reynolds numbers from 1600 to 3000), a linear interpolation is suggested for h_{LO}.

In our numerical simulation procedure, the heat transfer rate has been calculated as follows:

$$h = \frac{\varphi}{T_w - T_b} \tag{17.33}$$

The fluid bulk temperature T_b is defined as

$$T_b = \frac{1}{A_c u_m} \int_{A_c} uT\,dA \tag{17.34}$$

where

- u_m is the fluid mixture mean velocity
- u and T are the axial velocity and local fluid temperature, respectively
- T_w is the wall temperature
- A_c is the cross-section of the tube

The tube has been divided into several sections, and the average values of φ, T_w, T_b, and h have been computed for each section.

Void fractions have been calculated using our numerical procedure and compared to those obtained by Chen [39] and Hamersma and Hart [40] correlations, which can be used for any flow regime. These correlations are as follows:

Chen correlation:

$$\varepsilon = \left(1 + 0.18\left(\frac{1-x}{x}\right)^{0.6}\left(\frac{\rho_g}{\rho_l}\right)^{0.33}\left(\frac{\mu_l}{\mu_g}\right)^{0.07}\right)^{-1} \tag{17.35}$$

Hamersma and Hart correlation:

$$\varepsilon = \left(1 + 0.26\left(\frac{1-x}{x}\right)^{0.67}\left(\frac{\rho_g}{\rho_l}\right)^{0.33}\right)^{-1} \tag{17.36}$$

Figures 17.2 and 17.3 depict the vapor volume fraction in the tube at the beginning of the boiling process at times $t = 1.4$ s and $t = 1.7$ s, respectively. The zooms on the entrance and exit regions in the tube show that, as expected, bubble formation takes place at first on the inner surface of the heated wall. In the entrance, the individual bubbles with small diameters are transported from the wall by the liquid flow. Some bubbles condense as soon as they leave the wall, because they are surrounded by liquid at a temperature lower than the temperature of saturation. While moving downstream in the tube, the fluid temperature increases, and as a consequence, smaller bubbles grow, but not enough to be agglomerated. The bubbles volume enlarges and some bubbles coalesce. In a steady-state regime, as the tube diameter is much higher than the bubble dimensions, the classical flow pattern occurs from the inlet to the exit section of the tube. As shown in Figure 17.4, in a steady-state regime, near the inlet where the liquid subcooling is too high to permit bubble nucleation, the flow

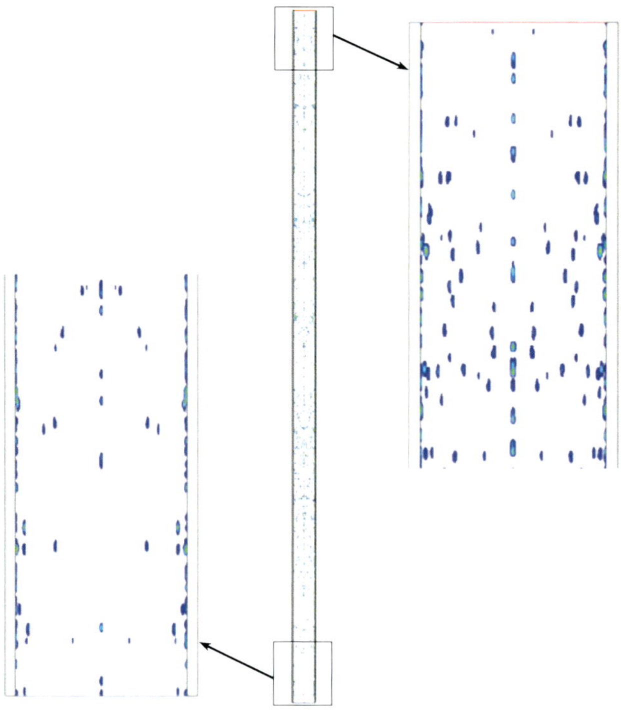

FIGURE 17.2 Contours of vapor volume fraction at $t = 1.4$ s.

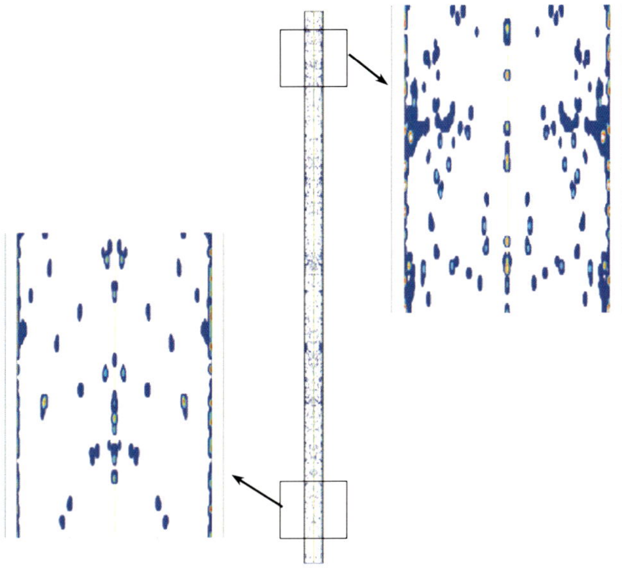

FIGURE 17.3 Contours of vapor volume fraction at $t = 1.7$ s.

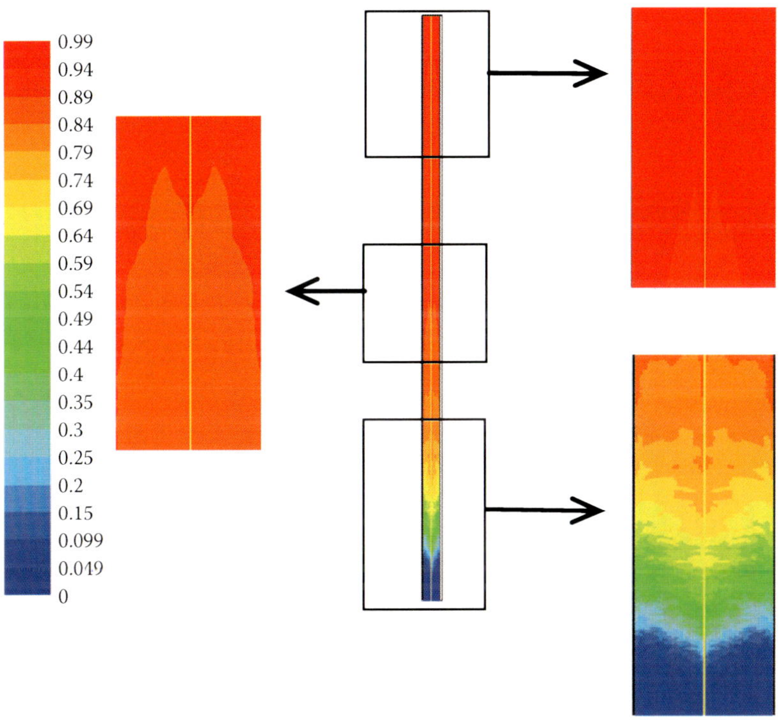

FIGURE 17.4 Contours of vapor volume fraction (Re = 480, $\varphi = 100$ kW/m^2).

regime is single-phase liquid (vapor volume fraction is zero). Following the initiation of boiling, the flow becomes bubbly with individual bubbles dispersed in the liquid. This region is followed by slug flow region where the bubbles grow enough to be agglomerated. Continuous growth of the bubbles makes the flow regime predominantly annular. Mist flow is therefore attained. The vapor volume fraction reaches saturation at 100% in the dry-mist flow at the exit of the tube.

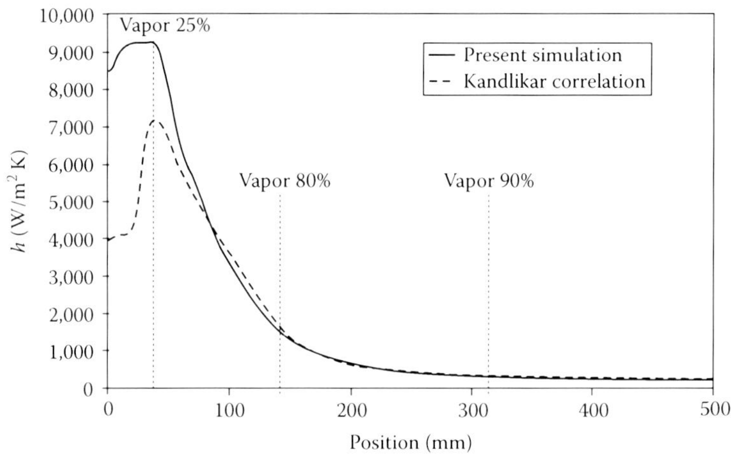

FIGURE 17.5 Comparison of the computed heat transfer coefficient with Kandlikar correlation (Re = 480, φ = 100 kW/m^2).

In Figure 17.5, the heat transfer coefficient obtained using the present numerical simulation is shown as a function of the position. The heat transfer coefficient obtained by the Kandlikar correlation (Equations 17.26 through 17.32) is also shown in that figure. It should be noted that in Equations 17.27 and 17.28, the introduced values of x have been computed using our numerical procedure. It is obvious that a very fair agreement has been reached between the two approaches especially for higher positions along the tube, for which the flow regime is well established and disturbances due to the entrance effects disappear. The difference in the entrance region is probably due to the fact that the liquid flow is considered not fully developed in the present simulation, which it is not the case in the Kandlikar correlation.

Void fraction evolution along the channel length is compared to the results of Chen [39] (correlation 1) and Hamersma and Hart [40] (correlation 2) in Figure 17.6. As for the heat transfer coefficient, the values of x obtained by our numerical procedure have been used for the computations of ε using Equations 17.35 and 17.36. It is observed from the figure that the numerical simulation has the same general tendency as Chen and Hamaresma correlations. Relative difference is lower than 10% on the exit section.

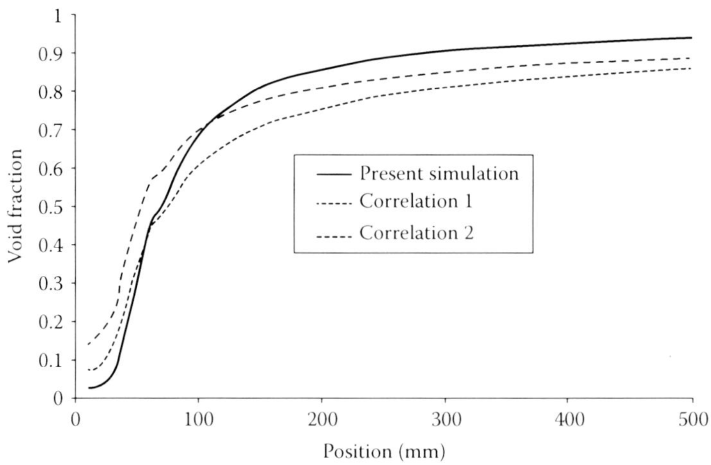

FIGURE 17.6 Comparison of the numerical void fraction with Chen (correlation 1) and Hamersma and Hart (correlation 2).

17.4 APPLICATION TO THE COOLING OF SYNCHROTRON LIGHT ABSORBER

17.4.1 Synchrotron Light Source Description

Numerical modeling developed here has been applied for the design of a microchannel heat sink for the cooling of the light absorber used in the synchrotron light source (SOLEIL) situated in Paris region in France [41]. Figures 17.7 and 17.8 are the schematic illustrations of a synchrotron light source. SOLEIL's nominal energy is 2.75 GeV. The light spins in a storage ring 354 m in circumference. Synchrotron radiation is a light emitted by electrons of very high energy when their trajectory is curved with a magnetic field. This is done tangentially to the trajectory. Figure 17.7 illustrates the functioning principle of the synchrotron source: a beam of electrons, emitted by an electron gun, is firstly accelerated in a 16 m linear accelerator, the LINAC. The electrons reach the speed of light at an initial energy level of 100 MeV.

- After this initial acceleration, the electron beam is directed toward a second accelerator, a circular one called the Booster, which brings the energy level up to the operating value (i.e., 2.75 GeV in Synchrotron SOLEIL).
- At this energy level, the electrons are injected into the 354 m circumference storage ring

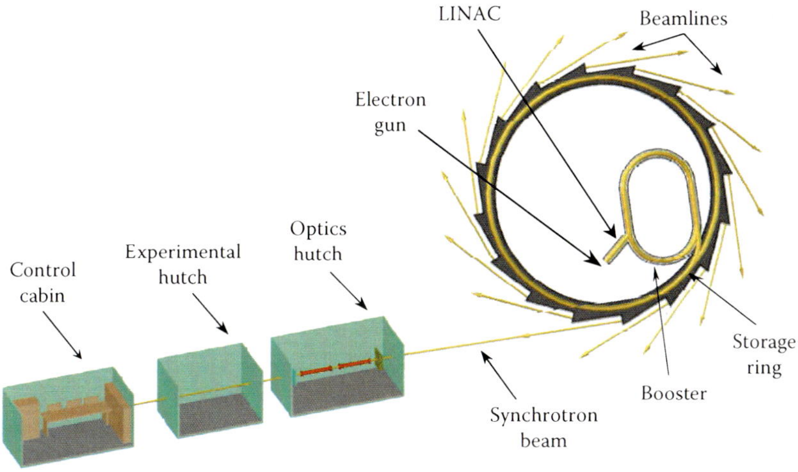

FIGURE 17.7 General description of the synchrotron light source.

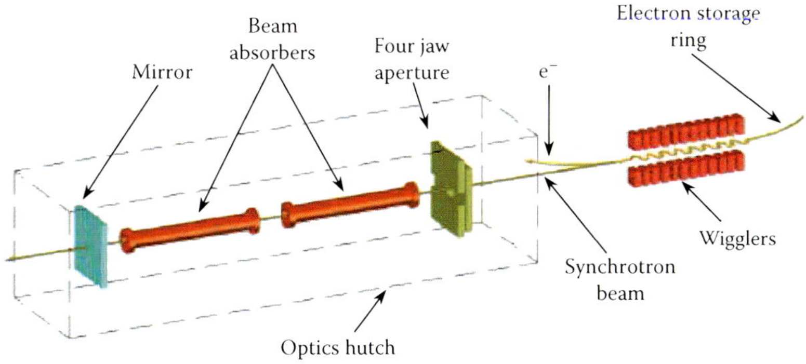

FIGURE 17.8 Generation of the synchrotron radiation.

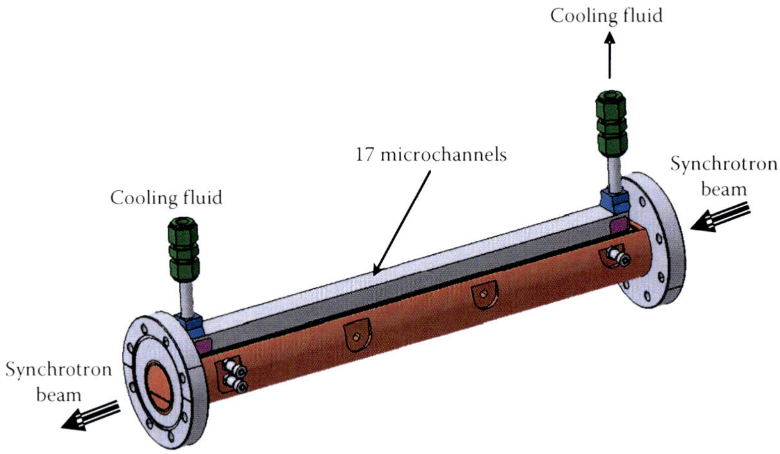

FIGURE 17.9 Schema of the light absorber equipped with 17 minichannels heat sink.

- In the storage ring, magnetic devices (curving magnets dipoles) and wigglers (a succession of alternating magnets) control the trajectory. The electrons lose energy in the form of light (the synchrotron radiation).
- The produced synchrotron radiation is directed toward the experimental stations called beamlines.
- Each beamline, with a specific wavelength radiation varying from infrared to x-ray radiations, allows the material characterization of samples in different fields of material sciences (e.g., biology, medical, chemistry, and earth sciences)

Implementation of the light absorbers on the beamline is illustrated in Figure 17.8. Light absorber, schematized in Figure 17.9, can absorb (totally or partially) the synchrotron light. Indeed, the absorber can be monitored for partially absorbing the light and thus it confines the beamline. It can also totally absorb the beamline and plays the role of a front end of the beam when it is necessary to stop it for preparing the sample or replacing any component of the beam line. Throughout its path, the beam passes through numerous components that are used to confine, define, monitor, intercept or absorb, transmit, reflect, or monochromatize the beam. The overall functions of the beamline components are personnel protection and safety, equipment protection, beam diagnostics, and beam characteristics modification (by selecting a spatial or spectral part of the beam).

17.4.2 Light Absorber Cooler

The performed prototype of the heat sink microchannels studied here is intended to cool the light absorber mounted on the beamline PSICHE in Synchrotron-SOLEIL source. Efficient cooling of the absorber allows improving the quality of the light beam and ensures equipment protection. The heat flux density chosen for the numerical simulation (700 kW/m^2) is expected to be reached in the future development of PSICHE beamline. It is then important to design the microchannels heat sink for the effective and efficient cooling of the absorber without the risk of deformation or melting the beam conditioners. In fact, thermal stress (high temperature or high nonuniform temperature) can induce mechanical deformation of the beam conditioners (e.g., mirrors, absorbers, and gear boxes), which results in a dramatic reduction of the quality of the output beam from the beamline. One of the most promising solutions is the cooling liquid flows inside the mini- and microchannels. The interest of this work lies in taking into account the conjugate heat transfer in minichannels etched in 3D complex geometry [42].

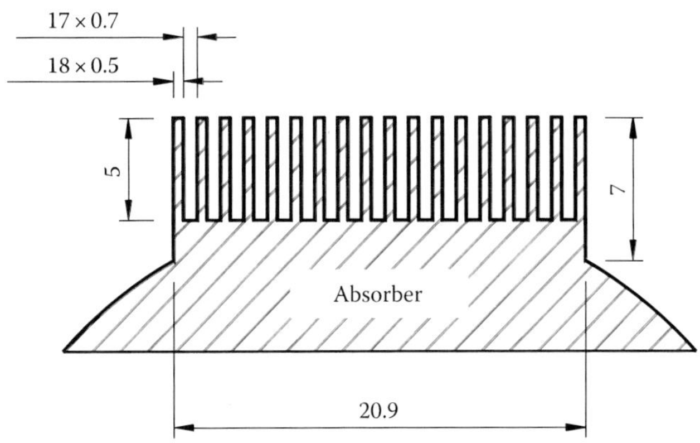

FIGURE 17.10 Transversal cross section of the minichannels.

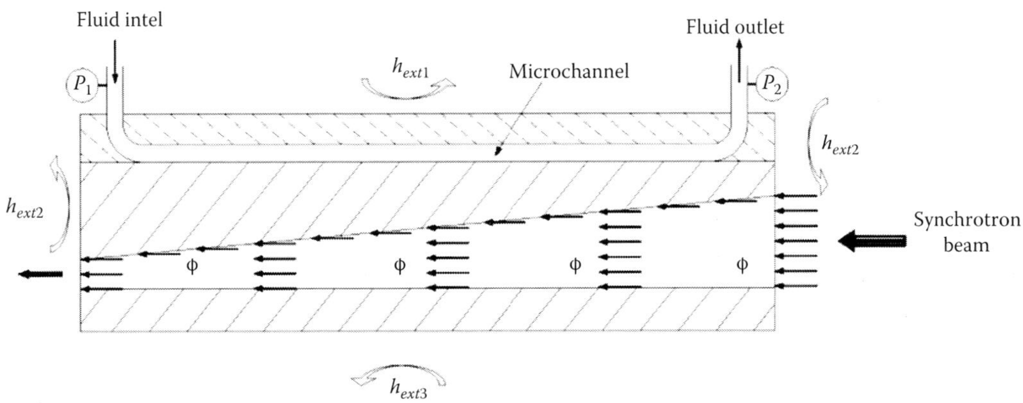

FIGURE 17.11 Schema of the problem.

The light absorber studied here was made of copper and has a cylindrical shape as shown in Figure 17.9. Seventeen channels of rectangular cross-section of 700 μm wide, 5 mm height, and 420 mm length were engraved on the upper surface of the absorber; they constitute the microchannels heat sink (Figure 17.10). The hydraulic diameter D_h = 1.228 mm. The average roughness of the channels ε = 6 μm; the working fluid is deionized water which prevents the fluid-wall electrostatic interaction. All the walls except the heating surface are exposed to external natural convection-radiation with the surrounding air and surfaces. The corresponding heat transfer coefficient are h_{ext1} on the upper surface of the absorber, h_{ext2} on their vertical right and left sides, and h_{ext3} on the lower surface of the absorber (Figure 17.11). The total inlet mass flow rate varied from 0.27 to 2.95 L/min, which corresponds to an inlet Reynolds number, based on the hydraulic diameter of the fluid cross-section, ranging from 500 to 6000. The outlet pressure was fixed at 3 bars since this is the network pressure in Synchrotron-SOLEIL facility. As it can be seen in Figure 17.11, the heat load from the synchrotron beamline is distributed on the inclined surface in the inner passage of the absorber. This is for increasing the solid surface subjected to the heat flux which results in a decreasing of the heat flux density.

17.4.3 Single-Phase Flow Cooling

Experimental measurements of the pressure drop P_1–P_2 versus water flow rate were performed on the microchannels heat sink in single-phase flows of deionized water. The simulation results have been compared to the experimental measurements and to the Kandlikar correlation [4,36] in Figure 17.12. The Kandlikar correlation gives the total pressure drop across the inlet and outlet

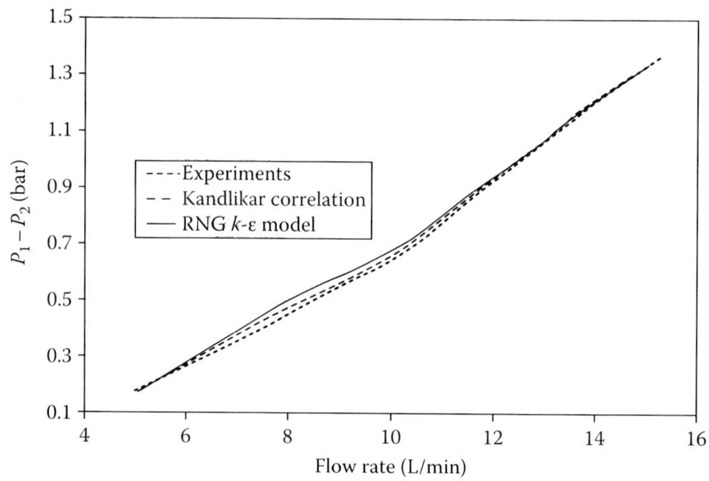

FIGURE 17.12 Total pressure drop in the heat sink.

collectors taking into account the losses at the entrance, the bends, the exit as well as the frictional losses. This correlation can be written as the sum of different terms in the following form:

$$\Delta P = \frac{\rho u_m^2}{2}\left[2K_{90}\left(\frac{A_c}{A_p}\right)^2 + \left(K_c + K_e\right) + \frac{4 f_{app} L}{D_h}\right] \quad (17.37)$$

where

- ρ is the fluid density
- u_m is the fluid mean velocity at the channel inlet
- L is the channel length
- D_h is the hydraulic diameter
- A_c and A_p are the total channel area and the total plenum cross-section area, respectively
- K_{90} is the pressure drop coefficient at the 90° bends which is recommended by Phillips [43] to be approximately 1.2
- K_c and K_e represent the contraction and expansion pressure drop coefficients due to the sectional cross changes and can be obtained from the diagrams derived from Phillips [43] and Kays and London [44]
- f_{app} includes the combined effects of frictional losses and the additional losses in the developing flow region

For turbulent regime, Phillips [43] proposed for f_{app} a correlation that covers both the developing and fully developed flow regions as follows:

$$f_{app} = A\mathrm{Re}_{eq}^{B} \quad (17.38)$$

$$A = 0.0929 + \frac{1.01612}{x/D_h} \quad \text{and} \quad B = -0.2680 + \frac{0.3293}{x/D_h} \quad (17.39)$$

where

- x is the length from the entrance
- Re_{eq} is the laminar-equivalent Reynolds number for rectangular channel geometries, given by

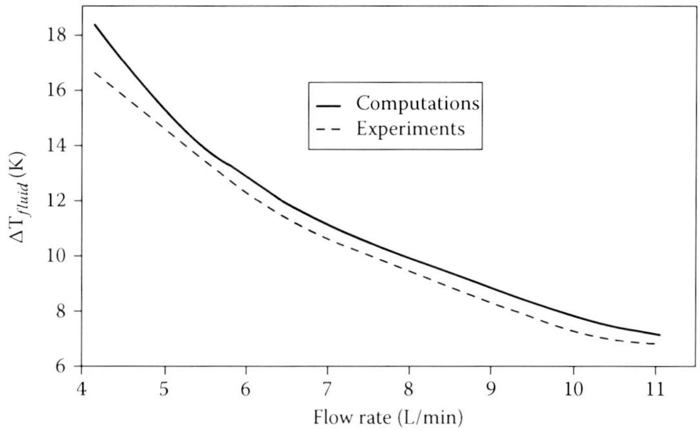

FIGURE 17.13 Comparisons of measured and calculated temperatures of the cooled fluid.

$$Re_{eq} = \frac{\rho u_m D_{le}}{\mu} \tag{17.40}$$

$$D_{le} = \left[\frac{2}{3} + \frac{11}{24}\alpha_c\left(2 - \alpha_c\right)\right] D_h \tag{17.41}$$

α_c is the channel aspect ratio defined as $\alpha_c = a/b$ for a rectangular channel with short side "*a*" and long side "*b*."

As it can be noted from Figure 17.12, the CFD simulation results are in good agreement with those obtained by both experiment and Kandlikar correlation.

Figure 17.13 shows the evolution of the bulk fluid temperature variation from the entry to the exit of the heat sink for different flow rates. The simulations results are compared to the measured temperatures of the cooling fluid. For different flow rates, the measured temperature variations are about 0.5°C lower than the simulation results. This is partially due to the measurement uncertainties (estimated to 0.2°C) and to the heat losses from the connecting conduits of the network water distribution to the heat sink, which is not taken into account in the computations.

17.4.4 Two-Phase Flow Cooling

17.4.4.1 Flow Pattern

Some typical results of the subcooled flow boiling in rectangular microchannels heat sink are given in the following text for the Reynolds number Re = 685 and Re = 1056 of the inlet liquid flow and for a heat flux density ϕ = 700 kW/m^2.

Bubble formation, coalescence, formation of slugs or plugs, and local dry-out conditions are all important in understanding the heat transfer phenomena, especially in microchannel flow. As it has been noted by different researchers [1,4], distinction should be made between parallel conventional channels and multiple microchannels with common inlet and outlet plena, manifolds, or headers. The flow misdistribution, mixing default, and oscillations can lead to instability in the boiling flow. Moreover, the evaporation of the liquid phase affects the flow in two ways. Firstly, it alters the pressure drop characteristics by introducing an acceleration pressure drop component that can be quite large at higher heat fluxes. Secondly, when the channel dimension is quite small, the effects of surface tension forces become more important in defining the two-phase structure. The confinement resulting from the small size of the channel can affect the bubble dynamics.

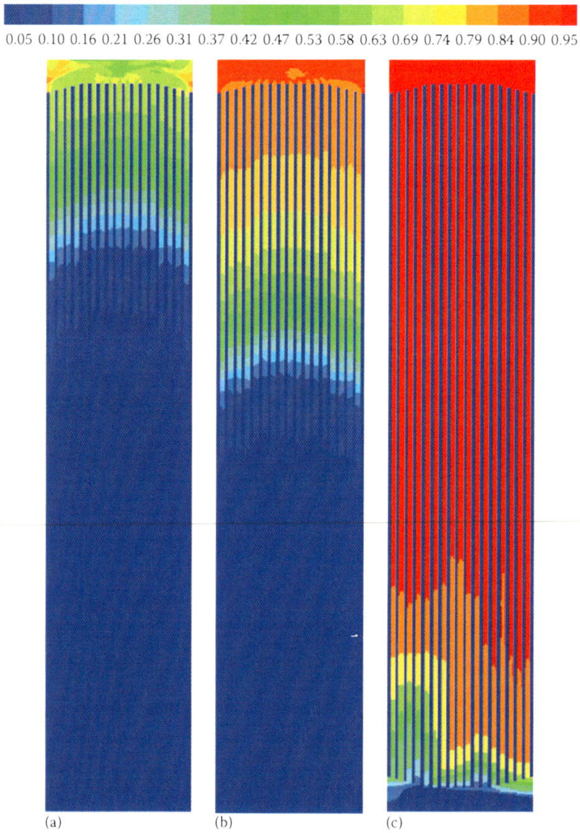

FIGURE 17.14 Vapor fraction on the mid-height of the channels at different time Re = 1056: (a) t = 25 s, (b) t = 45 s, and (c) t = 155 s.

It has been shown in the literature that there are only three major flow regimes for flow boiling in microchannels: isolated bubble flow, confined bubble flow (similar to plug flow in conventional channel), and annular slug flow. Confined flow is obtained when the bubbles span the entire width of the microchannel. It can lead to the collapse and dispersal of some part of the liquid slugs that separate the neighboring bubbles. Such behavior generates bubble growth in nucleate boiling and intermittent dry-out flow with pressure fluctuations and instabilities in microscale flow boiling. These patterns have been observed in the present numerical results in both steady-state and transient-state regimes [45,46]. Readers can refer Chapter 18 for a detailed description of the physics of boiling in microchannels.

Figure 17.14 shows, for Re = 1056, the vapor fraction evolution along the heat sink at the channel's mid-height (y = 2.5 mm) and at different time. Blue-color regions correspond to the microchannels' walls or to 0% vapor fraction volume and the red regions correspond to 100% vapor fraction. In the simulation conditions, it is obvious that all the liquid at the inlet becomes a vapor at the outlet of the heat sink after 155 s (Figure 17.14c). Also conduction in the absorber is with great importance since the heat transfer is conjugated in the boiling flow in microchannels and in the absorber.

17.4.4.2 Heat Transfer Coefficient

Variation of heat transfer coefficient along the central microchannel for both single-phase flow and two-phase flow with vaporization is illustrated in Figure 17.15. From the entrance to the location x = 120 mm, the flow is single-phase liquid, and the heat transfer coefficient matched well with the conventional correlation in single-phase flow with uniform heat flux. Beyond this location, the

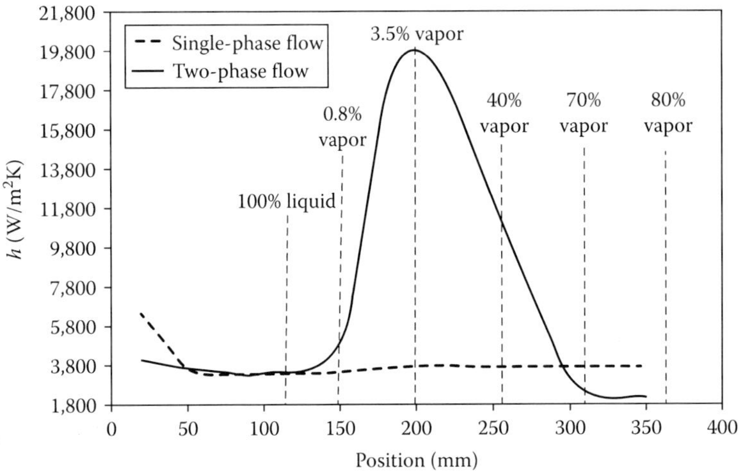

FIGURE 17.15 Comparison between two-phase flow and single-phase flow cooling.

vapor bubbles are starting to develop and the flow is becoming two phases. The two-phase flow heat transfer coefficient increases significantly because of the bubble nucleation and the bubbly regime. Indeed, a maximum is reached at $x = 200$ mm, then it decreases to become lower than the value corresponding to single-phase liquid flow at the heat sink exit (vapor region). The same evolution of the heat transfer coefficient has been noted by Sobierska et al. [47].

REFERENCES

1. J.R. Thome, Boiling in micro-channels: A review of experiment and theory, *International Journal of Heat and Fluid Flow*, 25, 128–139 (2004).
2. L.P. Yarin, A. Mosyak, and G. Hetsroni, *Fluid Flow, Heat Transfer and Boiling in Micro-Channels*, Springer, New York (2008).
3. S.M. Ghiaasiaan, *Two-Phase Flow, Boiling and Condensation in Conventional and Miniature Systems*, Cambridge University Press, Cambridge, U.K. (2008).
4. S.G. Kandlikar, S. Garimella, D. Li, S. Colin, and M.R. King, *Heat Transfer and Fluid Flow in Minichannels and Microchannels*, Elsevier, Oxford, UK (2006).
5. K.A. Triplett, S.M. Ghiaasiaan, S.I. Abdel-Khalik, and D.L Sadowski, Gas-liquid two-phase flow in micro-channels Part 1: Two-phase flow patterns, *International Journal of Multiphase Flow*, 25, 377–394 (1999).
6. S. Kandlikar, An extension of the flow boiling correlation to transition, laminar, and deep laminar flows in mini-channels and micro-channels, *Heat Transfer Engineering*, 25(3), 86–93 (2004).
7. Y. Chen, C. Zhang, M. Shi, and J. Wu, Three-dimensional numerical simulation of heat and fluid flow in noncircular microchannel heat sinks, *International Communications in Heat and Mass Transfer*, 36, 917–920 (2009).
8. W. Li and Z. Wu, A general criterion for evaporative heat transfer in micro/mini-channels, *International Journal of Heat and Mass Transfer*, 53, 1967–1976 (2010).
9. Z. Yang, X.F. Peng, and P. Ye, Numerical and experimental investigation of two phase flow during boiling in a coiled tube, *International Journal of Heat and Mass Transfer*, 51, 1003–1016 (2008).
10. A. Bontemps, B. Agostini, and N. Caney, Flow boiling in mini-channels, in *Micro-Scale Heat Transfer: Fundamentals and Applications*, pp. 217–230, S. Kakaç, L.L. Vasiliev, Y. Bayazitoğlu, and Y. Yener (eds), Springer, Dordrecht, the Netherlands (2005); G. Kumaraguruparan, T. Sornakumar, Development and testing of aluminum micro channel heat sink, *Journal of Thermal Science*, 19(3), 245–252 (2010).
11. M. Asadi, G. Xie, and B. Sunden, A review of heat transfer and pressure drop characteristics of single and two-phase micro-channels, *International Journal of Heat and Mass Transfer*, 79, 34–53 (2014).
12. Y. Wang and Z.-G. Wang, An overview of liquid-vapor phase-change, flow and heat transfer in mini- and micro-channels, *International Journal of Thermal Sciences*, 86, 214–247 (2014).

13. S.M. Kim and I. Mudawar, Universal approach to predicting saturated flow boiling heat transfer in mini/micro-channels—Part II. Two-phase heat transfer coefficient, *International Journal of Heat and Mass Transfer* 64, 1239–1256 (2013).
14. A. Esmaeeli and G. Tryggvason, Computations of film boiling. Part I: Numerical method, *International Journal of Heat and Mass Transfer*, 47, 5451 (2004).
15. T. Kunugi, Brief review of latest direct numerical simulation on pool and film boiling, *Nuclear Engineering and Technology*, 44, 852–858 (2012).
16. T. Fuchs, J. Kern, and P. Stephan, A transient nucleate boiling model including microscale effects and wall heat transfer, *Journal of Heat Transfer*, 128, 1257 (2006).
17. T. Harirchian and S.V. Garimella, Flow regime-base modelling of heat transfer and pressure drop in micro-channel flow boiling, *International Journal of Heat and Mass Transfer*, 55, 1946–1960 (2012).
18. R. Zhuan and W. Wang, Flow pattern of boiling in micro-channel by numerical simulation, *International Journal of Heat and Mass Transfer*, 55, 1741 (2012).
19. S. Madhaven and R. Mesler, A study of vapour bubble growth on surfaces, *Heat Transfer*, 5, B2.6 (1970).
20. G. Tryggvason, B. Bunner, A. Esmaeeli, D. Juric, N. Al-Rawahi, W. Tauber, J. Han, S. Nas, and Y.-J. Jan, A front-tracking method for the computations of multiphase flow, *Journal of Computational Physics*, 169, 708 (2001)
21. S.O. Unverdi and G. Tryggvason, A front-tracking method for viscous, incompressible, multi-fluid flows, *Journal of Computational Physics*, 100, 25 (1992).
22. C.W. Hirt, A.A. Amsden, and J.L. Cook, An arbitrary Lagrangian-Eulerian computing method for all speeds, *Journal of Computational Physics*, 14, 227 (1974).
23. G. Son and V.K. Dhir, A level set method for analysis of film boiling on an immersed solid surface, *Numerical Heat Transfer Part B*, 52, 153 (2007).
24. J. Wu and V.K. Dhir, Numerical simulations of the dynamics and heat transfer associated with a single bubble in subcooled pool boiling, *Journal of Heat Transfer*, 132, 11501 (2010).
25. C.W. Hirt and D.B. Nichols, Volume of fluid (VOF) method for the dynamics of free boundaries, *Journal of Computational Physics*, 39, 201 (1981).
26. J.U.D. Brackbill, D.B. Kothe, and C. Zemach, A continuum method for modeling surface tension, *Journal of Computational Physics*, 100, 335–354 (1992).
27. M. Meier, G. Yadigaroglu, and B.L. Smith, A novel technique for including surface tension in PLIC-VOF methods, *European Journal of Mechanics—B/Fluids*, 21, 61 (2002).
28. Y. Renardy and M. Renardy, Prost: A parabolic reconstruction of surface tension for the volume-of-fluid method, *Journal of Computational Physics*, 183, 400 (2002).
29. H.L. Wu, X.F. Peng, P. Yea, and Y.E. Gong, Simulation of refrigerant flow boiling in serpentine tubes, *International Journal of Heat and Mass Transfer*, 50, 1186–1195 (2007).
30. A. Asghar, M. Rahimi, and A. Alsairafi, CFD modelling of flow and heat transfer in a thermosyphon, *International Communications in Heat and Mass Transfer*, 37, 312–318 (2010).
31. V.S. Annaland, M., N.G. Deen, and J.A.M. Kuipers, Numerical simulation of gas bubbles behaviour using a three-dimensional volume of fluid method, *Chemical Engineering Science*, 60 (11), 2999–3011 (2005).
32. S.C.K. De Schepper, G.J. Heynderick, and G.B. Marin, CFD modelling of all gas–liquid and vapor–liquid flow regimes predicted by the Baker Chart, *Chemical Engineering Journal*, 138(3), 349–357 (2008).
33. S.C.K. De Schepper, G.J. Heynderick, and G.B. Marin, Modelling the vaporization of a hydrocarbon feedstock in the convection section of a steam cracker, *Computers & Chemical Engineering*, 33, 122–132 (2009).
34. D.-L. Sun, J.-L. Xu, and L. Wang, Development of a vapor–liquid phase change model for volume of fluid method in fluent, *International Communications in Heat and Mass Transfer*, 39, 1101–1106 (2012).
35. J.H. Lienhard IV and J.H. Lienhard V., *A Heat Transfer Textbook*, Phlogiston Press, Cambridge, MA (2003).
36. S.G. Kandlikar and M.E. Steinke, Predicting heat transfer during flow boiling in minichannels and microchannels, *ASHRAE Transactions*, 109(1), 1–9 (2003).
37. S.G. Kandlikar and P. Balasubramanian, An extension of the flow boiling correlation to transition laminar and deep laminar flows in minichannels and microchannels, *Heat Transfer Engineering*, 25(3), 86–93 (2004).
38. V. Gnielinski, New equations for heat and mass transfer in turbulent pipe and channel flow, *International Chemical Engineering*, 16(2), 359–368 (1976).

39. J. Chen, A further examination of void-fraction in annular two-phase flow, *International Journal of Heat Mass Transfer*, 29, 1760–1763 (1986).
40. P.J. Hamersma and J. Hart, A pressure drop correlation for gas/liquid pipe flow with a small liquid holdup, *Chemical Engineering Science*, 42, 1187–1196 (1987).
41. Synchrotron SOLEIL website, http://www.synchrotron-soleil.fr.
42. A. Hamza, *Amélioration du transfert de chaleur à l'interface solide-fluide dans les échangeurs de SOLEIL*, PhD thesis, University of Reims Champagne-Ardenne, Reims, France, 2013.
43. R.J. Phillips, *Microchannel heat sinks Advances in Thermal Modeling of Electronic Components and Systems*, Hemisphere Publishing Corporation, New York (1990).
44. W.M. Kays and A.L. London, *Compact Heat Exchangers*, McGraw-Hill, New York (1984).
45. G. Mebarki, A. Hamza, J.P. Daguerre, J. Dasilva Castro, and M. Rebay, Etude d'un dissipateur thermique à mini-canaux destiné au conditionnement du faisceau sur le synchrotron, In *Proceedings of the 20th French Congress of Mechanics* (*Congrès Français de Méc.*), Besançon, France, 2011, pp. 18–24.
46. G. Mebarki, M. Rebay, N. El Wakil, A. Hamza, and S. Rahal, Numerical simulation of vaporization in mini channels using multiphase VOF model, In *Proceedings of the International Symposium on Convective Heat and Mass Transfer*, Kusadasi-Turkey, June 8–13, 2014.
47. E. Sobierska, R. Kulenovic, and R. Mertz, Heat transfer mechanism and flow pattern during flow boiling of water in a vertical narrow channel—Experimental results, *International Journal of Thermal Sciences*, 46, 1172–1181 (2007).

18 Two-Phase Flow in Microchannels

Nadia Caney, Hasna Louahlia-Gualous, and André Bontemps

CONTENTS

18.1 INTRODUCTION

Small channels are being increasingly considered for usage involving phase change processes in existing or completely new applications. Information on boiling and condensation in mini- or microchannels is needed to improve heat transfer performances for ultra-high heat flux equipment and miniaturized systems. Compared to single-phase processes, phase change enables exchanging more energy by unit volume with a relatively lower jump of temperature at the wall.

Two-phase flow heat transfer in microchannels is an emerging field and the treatment given in this chapter must be considered as an introduction to future developments. Heat transfer being strongly related to flow patterns, knowledge of occurrence of different flow regimes as a function of fluid properties, channel size, vapor quality, and experimental conditions is necessary for understanding and modeling. In the first part of this chapter, we define the main physical properties useful to describe the phenomena, together with dimensionless numbers of interest which allow us to evaluate prominent driving forces. In particular, the criteria defining transition between conventional macrochannels and mini/microchannels are given. In these two types of channels, flow regimes are described while noting their differences. Specific attention is paid to surface properties as wettability plays an important role.

Boiling phenomena are treated in considering heterogeneous boiling. It is the rapid vaporization of liquid associated with bubble formation on a solid surface. It may be analyzed into two categories, namely nucleate boiling and flow boiling, the latter being also referred to as convective boiling. Heterogeneous boiling is significantly affected by interactions at the solid–liquid interface whose two main properties are surface wettability and topography.

The usual two-phase flow regimes encountered are reviewed and it is shown that those occurring in conventional channels are not realistic. The main flow boiling heat transfer results and heat transfer mechanisms are given and discussed. Current theoretical models are presented. Semiempirical correlations are presented to predict heat transfer coefficients. An analysis of pressure drop is presented by considering both the core friction and the acceleration components.

For condensation inside microchannels, flow regimes identified for adiabatic and condensate flows are shown to be different from those observed in conventional channels. Condensate flow patterns are analyzed for single and parallel multimicrochannel. Several transition criteria are defined for various fluids, operating conditions, and channel sizes. In particular, the cyclic and developed condensate flows in microchannels are described. Our attention is focused on the investigation of condensation heat transfer and hydrodynamic characteristics of slug flow and annular bubbly flow because they are the basic two-phase flow patterns in microchannels. Within flow regimes, models have been proposed for heat transfer and pressure drop. Condensation in noncircular channels has been studied and modeled, highlighting the prominent role of surface tension, microchannel shapes, and aspect ratio. This role is also observed in studies on channels with hydrophilic or hydrophobic surfaces. Existing correlations for condensation heat transfer in macroscale channels have been evaluated for use in the microscale channels.

18.2 FUNDAMENTAL DEFINITIONS

Before describing the processes of boiling and condensation, it is necessary to first introduce some important parameters that characterize two-phase flows.

18.2.1 Geometrical and Physical Parameters

18.2.1.1 Hydraulic Diameter

This parameter is defined from the ratio of the cross-sectional area of the fluid A_f to the wetted perimeter P_l, by

$$D_h = \frac{4A_f}{P_l} \tag{18.1}$$

18.2.1.2 Vapor Quality

Through the channel, the total mass flow rate $\dot{m}$ is equal to the sum of the mass flow rates of vapor $\dot{m}_v$ and liquid $\dot{m}_l$:

$$\dot{m} = \dot{m}_v + \dot{m}_l \tag{18.2}$$

The vapor quality x is defined as the ratio of the vapor mass flow rate to the total mass flow rate:

$$x = \frac{\dot{m}_v}{\dot{m}} \tag{18.3}$$

Due to the difficulty in determining this quantity, equilibrium (or thermodynamic) quality is often used by assuming that both phases are flowing under saturation conditions:

$$x = x_e = \frac{H - H_l}{H_{lv}} \tag{18.4}$$

where
- H is the total specific enthalpy of the liquid–vapor mixture
- H_l is the specific enthalpy of the liquid
- H_{lv} is the latent heat of vaporization

18.2.1.3 Mass Velocity

The total mass flux or mass velocity G is the total mass flow rate divided by the total channel cross-sectional area:

$$G = \frac{\dot{m}}{A_f} \tag{18.5}$$

18.2.1.4 Superficial Velocities

The superficial velocities of vapor and liquid phases, j_v and j_l, are defined, respectively, by

$$j_v = \frac{Gx}{\rho_v} \tag{18.6}$$

$$j_l = \frac{G(1-x)}{\rho_l} \tag{18.7}$$

nanoscale transition. The threshold to confined bubble flow [5] should be the most convenient criterion, since the confinement number Co is used as a criterion to differentiate between macro- and microscale flow boiling. This number can be defined as the ratio between the capillary length (or Laplace constant) l_c and the hydraulic diameter:

$$l_c = \left(\frac{\sigma}{g(\rho_l - \rho_v)} \right)^{1/2} \tag{18.19}$$

and

$$Co = \frac{l_c}{D_h} = \left(\frac{\sigma}{g(\rho_l - \rho_v) D_h^2} \right)^{1/2} \tag{18.20}$$

Beyond a critical value of the confinement number, heat transfer and flow characteristics will be significantly different from those observed in macroscale. Cornwell and Kew [5] suggested the criterion of $Co \geq 0.50$, but the criterion suggested by Ullmann and Brauner [6] is $Co \geq 0.79$.

However, Cheng and Wu [7] have proposed the Bond number as a sufficient classification criterion:

- Microchannel for $Bo \leq 0.5$ where surface tension is dominant compared to gravity.
- Minichannel for $0.5 < Bo \leq 3$ where surface tension and gravity have an equivalent impact on the flow.
- Macrochannel for $Bo > 3$ where gravity is the dominant force.

The most recent evaluations are those of Li and Wu [8] and Harirchian and Garimella [9] who link the transition both to confinement, inertial and friction forces and to flow regime. The former have defined a convective confinement number (we will call it Cov) by

$$Cov = Bo^{0.5} Re_l = \frac{Re_l}{Co} \tag{18.21}$$

where

$$Re_l = \frac{G(1-x)D_h}{\mu} \tag{18.22}$$

They propose the following simple correlation: when $Cov \leq 200$, the channel can be considered as a microchannel.

The latter have included the flow regimes (see Section 18.2.3) in a study on parallel microchannels. They have introduced a similar convective confinement number in using the following Re number:

$$Re = \frac{GD}{\mu} \tag{18.23}$$

where

$$D = \sqrt{A_f} \tag{18.24}$$

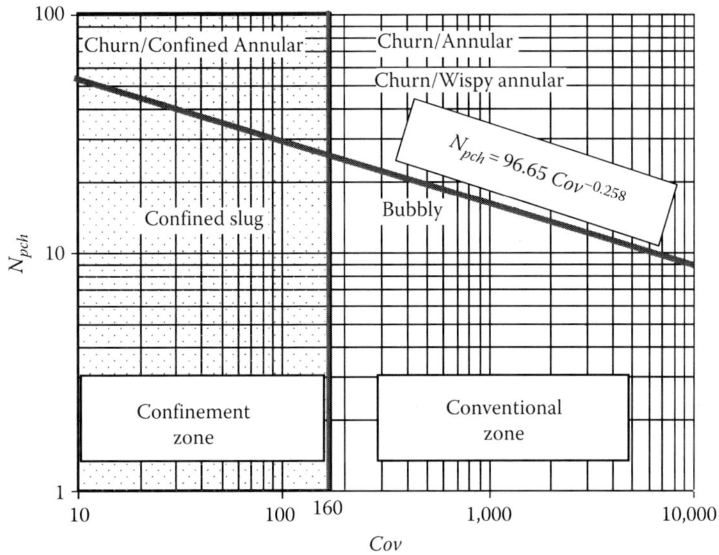

FIGURE 18.1 Confinement and flow regimes zones. (Adapted from Harirchian, T. and Garimella, S.V., *Int. J. Heat Mass Transfer*, 55, 1946, 2012.)

They propose that $Cov = 160$ represents the transition to confined flow. Another transition criterion is proposed by defining a specific phase change number N_{pch} to determine the flow regimes. This number is a modified boiling number:

$$N_{pch} = Bl \frac{L_H}{D_H} \frac{\rho_l - \rho_v}{\rho_v} \tag{18.25}$$

where
L_H is the axial heated length
D_H the heated diameter

The results are summarized in Figure 18.1.

In spite of the attempts to identify the macro- to microtransition threshold summarized earlier, a clarified universal criterion is still not well established due to lack of reliable experimental data in various microchannel geometries. It is interesting to note the comparison made by Charnay et al. [10] between different selection criteria as a function of the saturation temperature of R-245fa. This highlights discrepancies between the different estimations.

Further experimental investigations on flow boiling as well as on condensation heat transfer in some abnormal channel geometries are essential for building a more comprehensive database and deducing a more global macro- to microtransition criterion.

18.2.3 Flow Regimes

18.2.3.1 Conventional Channels

At very low quality, bubbly flows are often observed. Due to buoyancy, bubbles flow mainly in the upper portion of the channel. Increasing the quality, coalescence of small bubbles occurs, producing larger plug-type bubbles. This is referred to as the plug flow regime (cf. Figure 18.2).

At low flow rates and somewhat higher qualities, stratified flow may be observed when liquid and vapor are separated. In the regime of the stratified flow, if the flow rate is high enough to make

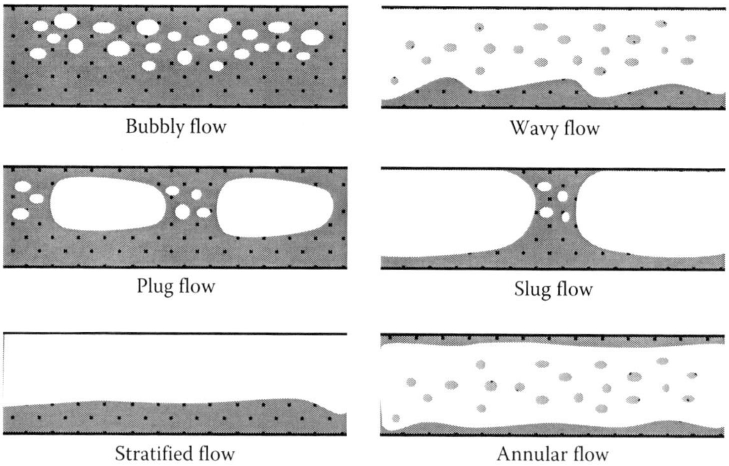

FIGURE 18.2 Schematic representation of regimes of flow boiling in horizontal channels.

the liquid–vapor interface unstable, the interface becomes wavy. This type of flow is categorized as wavy flow, which has strong vapor shear on the interface ejecting liquid droplets in the vapor core flow. At higher liquid flow rates, the amplitude of the waves may grow, forming large slug-type bubbles. This is referred to as the slug flow.

At high velocity of vapor and moderate velocity of liquid, annular flow may be observed. This regime is characterized by a thin liquid film on the top of the channel and a thicker liquid film at the bottom.

18.2.3.2 Microchannels

In microchannels, capillary effects become important, reducing stratification of the liquid and hence, the channel orientation is mostly of secondary importance. In general, two-phase flows in microchannels can have four principal flow patterns. The schematic view of these patterns is presented in Figure 18.3.

In bubbly flow, discrete bubbles move inside the liquid phase. Generally, these bubbles are smaller in size than the hydrodynamic diameter of the channel. This flow pattern covers a very small range of vapor quality. In slug flow, the bubbles have approximately the same diameter as the channel. The bubble nose is hemispherical and the vapor phase is separated from the wall by a thin liquid film. These bubbles are called *elongated bubbles* since they become greater in length due to the confinement effect. The liquid phase is contained mostly in the liquid slugs, which separate successive vapor bubbles.

By increasing the vapor quality, semiannular flow may be observed. This flow pattern refers to as a continuous vapor phase at the channel core together with some churning liquid zones at the

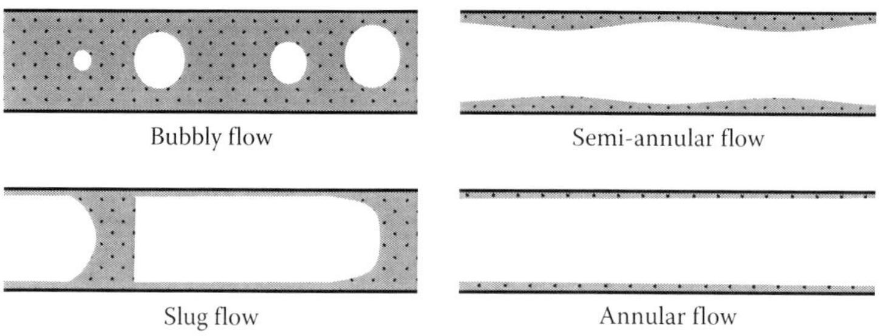

FIGURE 18.3 Schematic representation of regimes of flow boiling in horizontal microchannels.

TABLE 18.1
Comparison of Physical Properties of Air–Water and R134a at Saturation

Air/Water at 20°C and 1 Bar			R134a at Saturation, 20°C and 5.7 Bars		
ρ_l/ρ_v	μ_l (Pa s)	σ_l (N/m)	ρ_l/ρ_v	μ_l (Pa s)	σ_l (N/m)
846	10^{-3}	0.072	44.2	21.1×10^{-5}	0.0088

wall. Indeed, the churning liquid zones are created by deformation and coalescence of the previous elongated bubbles. When the churning liquid zones disappear, the flow regime is categorized as annular flow.

Most of experiments of the literature are concerned with adiabatic two-phase flow of air–water at atmospheric pressure. However, properties of a refrigerant (e.g., R134a at 20°C and 5.7 bars) in boiling conditions are quite different from those of an adiabatic mixture (cf. Table 18.1).

The study of Revellin and Thome [11] in horizontal microtubes of 0.5 and 0.8 mm with R134a and R245fa concludes that

- Classical transitions between flow regimes in macrochannels are not valid for microchannels.
- Maps established with adiabatic conditions (air–water) are not valid in convective boiling conditions.

To illustrate these conclusions, we present a figure (Figure 18.4) in which three flow pattern maps are superimposed:

- The map of Mandhane et al. [12] in adiabatic conditions deduced from the results of air–water mixtures flowing in conventional channels (tubes with diameters from 12.7 to 165.1 mm). The type of flow pattern is in boxed text preceded by *M* and the transition boundaries are in full lines.

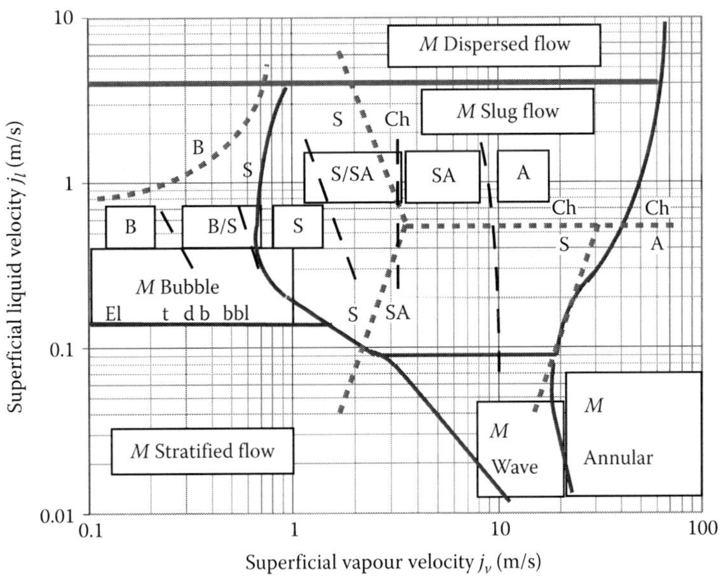

FIGURE 18.4 Comparison of three flow pattern maps (see text): Air–water from Mandhane et al. [12] with flow pattern boundaries in solid lines; air–water from Triplett et al. [13] with boundaries in dashed lines (short dashes), and boiling R-245fa from Charnay et al. [10] with boundaries in dashed lines (long dashes).

- The map of Triplett et al. [13] in adiabatic conditions deduced from the results of air–water mixtures flowing in microchannels (tube of 1.1 mm diameter). The type of flow patterns is indicated on both sides of a transition boundary (B for bubbly, S for slug, Ch for churn, and A for annular). The transitions are in dashed lines (short dashes).
- The map of Charnay et al. [10] in diabatic conditions deduced from results for R-245fa during convective boiling in a 3 mm inner diameter tube. The flow patterns are indicated with their boxed initial letters and the transitions are in dashed lines (long dashes).

It is clearly seen that it is difficult to extrapolate the flow regime with a map not corresponding to geometrical and thermophysical conditions.

18.2.4 Wetting Phenomenon

In heterogeneous boiling, the type of contact between the liquid, vapor, and the heated surface can strongly impact the heat and mass transfers. At small scale, this contact is mainly dominated by surface tension forces and is referred to as wetting phenomena. Understanding of wetting laws is crucial in interpreting some boiling physics, such as bubble growth in nucleate boiling and intermittent dry-out in microscale flow boiling.

If a liquid drop is small enough to neglect the flattening action of gravity, its sticking on a solid surface is referred to as surface wettability. An important parameter used to characterize the surface wettability is the contact angle θ between the liquid and the solid surface at the vapor/liquid/solid line (triple line). The solid surface is wetted if $\theta < 90°$ (cf. Figure 18.5a) and unwetted if $\theta \geq 90°$ (cf. Figure 18.5b). In the case of water, wetted surface is called "hydrophilic" and unwetted surface is called "hydrophobic."

18.2.4.1 Young's Relation

The value of the contact angle was first discussed by Young [14]. At the base of the liquid droplet, the three phases: solid, liquid, and vapor coexist in the so-called triple contact line (TCL). Each interface draws the TCL so as to minimize the corresponding surface area. Thus, balancing the surface tensions on the direction of potential motion (i.e., the horizontal) enables the determining of the contact angle as

$$\cos\theta = \frac{\sigma_{sv} - \sigma_{sl}}{\sigma_{lv}} \tag{18.26}$$

where σ_{sv}, σ_{sl}, and σ_{lv} are the surface tensions (i.e., energy per unit surface) of the interface of solid–vapor, solid–liquid, and liquid–vapor, respectively.

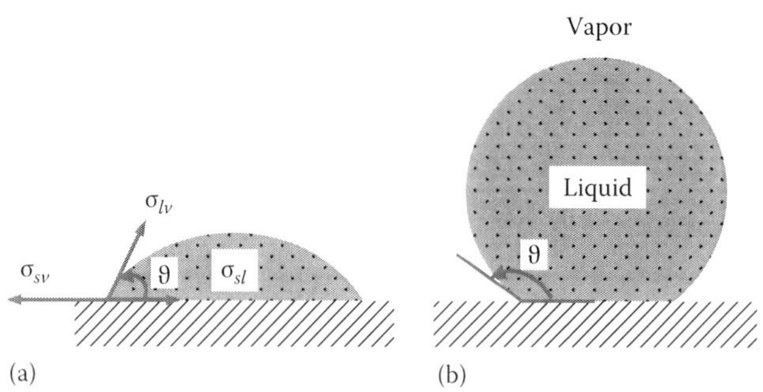

FIGURE 18.5 Liquid droplets on (a) wetted surface and (b) unwetted surface.

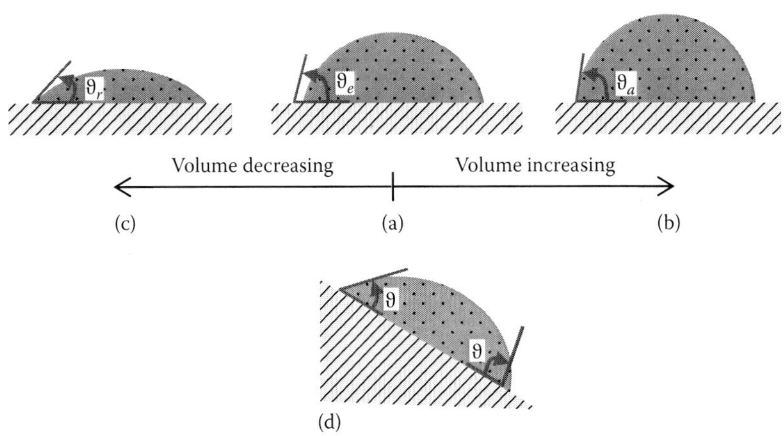

FIGURE 18.6 Contact angle hysteresis: (a) static contact angle θ_e, (b) advancing contact angle θ_a, (c) receding contact angle θ_r, and (d) contact angles of droplet on inclined surface.

18.2.4.2 Contact Angle Hysteresis

Solid surface is generally imperfect due to heterogeneity of surface roughness and chemistry. It can also be polluted by deposition of contaminants floating in air. These defects allow the TCL to pin on the surface, generating multiple values for the observed contact angle. This is the so-called contact angle hysteresis (cf. Figure 18.6), which can be quantified by measurement of the contact angle when varying the droplet volume.

First, a droplet is deposited on the solid surface. The measured contact angle is equal to the static contact angle θ_e (cf. Figure 18.6a). Afterwards, the droplet is slowly fed by using a syringe. If the contact angle is unique, the diameter of the TCL will gradually increase. However, the TCL frequently remains pinned, and the angle is seen to increase. If the droplet volume stops increasing, the contact angle remains stuck at its last value, which is one of the many possible observable angles. There is a critical value of the droplet volume, beyond which the TCL suddenly widens and the contact angle reaches the largest value, called the "advancing contact angle" θ_a (cf. Figure 18.6b). Conversely, decreasing the droplet volume enables the observing of the smallest contact angle, below which the TCL reduces; this angle is called the "receding contact angle" θ_r (cf. Figure 18.6c). The contact angle hysteresis can also be observed on an inclined solid surface (cf. Figure 18.6d). The angle is larger at the front than at the rear of the droplet.

18.2.4.3 Adhesion Energy

If the contact angle was unique, due to gravity a droplet would necessarily run down on an inclined surface. However, it is still stuck on the surface up to a threshold of the inclined angle. Indeed, the contact angle hysteresis generates a capillary force that opposes the gravity. This force reaches its maximum value at the front and rear angles of θ_a and θ_r, respectively. To characterize this adhesion of a liquid droplet on a solid surface, Dupré [15] defined the solid–liquid adhesion energy W_{sl}^a as

$$W_{sl}^a = \sigma_{sv} + \sigma_{lv} - \sigma_{sl} \tag{18.27}$$

Combining Equations 18.26 and 18.27, we obtain the well-known Young–Dupré equation as

$$W_{sl}^a = \sigma_{lv}(1 + \cos\theta) \tag{18.28}$$

18.2.5 Theories of Solid Surface Tension: Zisman Theory and Owens–Wendt Theory

As presented in the previous section, the contact angle is a result of minimization of the system energy and can be determined by a balance of the surface tensions. Determining the surface tension of liquid–vapor interfaces, in short called liquid surface tension, is simple and this can be achieved by a variety of techniques. However, that of solid–vapor interface, in short called solid surface tension, is more complicated since it cannot be directly measured. In general, the values of the solid surface tension are calculated from a set of contact angles, developed by bringing various liquids in contact with the solid surface, but there is no universal agreement in the literature about its definition. The following paragraphs present two of the most well-known theories.

18.2.5.1 Zisman Theory

The theory of Zisman [16] gives probably the most widely used definition of solid surface tension. Zisman defines the solid surface tension as the highest liquid–vapor surface tension that a liquid should have to completely wet the solid surface, with a contact angle of 0°. This comes from the fact that generally, the contact angle decreases if the liquid surface tension decreases.

In surface science, the solid–liquid surface interactions are categorized into polar and dispersive types. The polar interactions account for dipole–dipole, dipole-induced dipole, hydrogen bonding, and other site-specific interactions. The dispersive interactions account for Van der Waals and other nonsite-specific interactions. A surface is called polar if it is capable of polar-type interactions with liquids. In general, the Zisman theory works best for nonpolar surfaces, for example, poly(ethylene) and poly(propylene), but it becomes inadequate for polar surfaces, for example, glasses, ceramics, and metals. This is because the Zisman theory is a one-parameter model since it only considers the overall values of the surface tensions.

18.2.5.2 Owens–Wendt Theory

Owens and Wendt [17] developed a theory that accounts for polar-type interactions between solid surfaces and liquids. The Owens–Wendt theory is typically the most applicable to surfaces that have low surface charge and are moderately polar in nature. This theory has limitation since it leaves the choice of liquids used up to the experimenter.

Wetting of a liquid on a solid surface is impacted not only by the surface chemistry but also by the surface geometry. Two distinct models, developed independently by Wenzel [18] and Cassie and Baxter [19] (named as the Cassie model), are commonly used to explain the effect of surface roughness on the apparent contact angle of liquid drops.

18.2.5.3 Wenzel Model

Wenzel [18] was possibly the first person to analyze the effect of surface roughness on the static contact angle. He observed that surface roughness caused a wetted surface to behave as if it were more wetted and an unwetted surface to behave as if it were more unwetted. Hence, he suggested modifying Young's equation by multiplying the numerator of the right side by a roughness factor, r, which is ratio of the actual area to the apparent area of the contact surface:

$$\cos\theta = r\cos\theta_{smooth} \tag{18.29}$$

where θ_{smooth} is the contact angle measured on a smooth surface of the same material, given by Young's equation as $\cos\theta_{smooth} = (\sigma_{sv} - \sigma_{sl})/\sigma_{lv}$.

18.2.5.4 Cassie Model

The Cassie model [19], on the other hand, postulates that the hydrophobic nature of a rough surface is caused by microscopic pockets of air remaining trapped below the liquid droplet, leading to a composite interface. If Φ_s is the fraction of the solid in contact with the liquid, the Cassie equation yields

$$\cos\theta = -1 + \Phi_s(1 + \cos\theta_{smooth}) \quad (18.30)$$

In contrast to the Wenzel relation, the Cassie model enables the possibility of $\theta > 90°$, even with $\theta_{smooth} < 90°$. The transition from the Wenzel to the Cassie state was observed [20,21] at a critical contact angle θ_c, which is obtained by equating Equations 18.29 and 18.30:

$$\cos\theta_c = \frac{\Phi_s - 1}{r - \Phi_s} \quad (18.31)$$

The value of the critical contact angle is necessarily greater than 90° as $r > 1 > \Phi_s$. Thus, it may be anticipated that for $\theta < 90°$, a surface cannot exist in the Cassie state or that the creation of highly nonwetting surfaces ($\theta \gg 90°$) requires $\theta > \theta_c > 90°$.

18.3 BOILING

18.3.1 Industrial Applications

With the increasing demand of cooling ultra-high heat flux equipment and miniaturization, several studies have been carried out to improve heat transfer performances by flow boiling. Industrial applications with phase change typically use water-based fluids or refrigerants, depending on considerations of the amount of heat to be transferred, operating conditions, and interactions of the fluid with components. Nowadays, the need has arisen for very high heat flux cooling of computer chips. Heat flux dissipation rates have risen from about 30 W/cm^2 just a few years ago to 100 W/cm^2 today. Other applications of boiling and two-phase flow in small channels are related to automotive and domestic air conditioning systems.

18.3.2 Heat Transfer

It has been shown through a number of experiments that boiling is controlled by two additive components: nucleate boiling and convective boiling. Nucleate boiling is due to nucleating bubbles and their subsequent growth and removal from the heated surface. Convective boiling is due to heated fluid moving from the heated surface to the flow core. These two mechanisms cannot be separated with any precision since they are closely interconnected. Some experimental studies show a significant dependence on heat flux of the heat transfer coefficient and a minor dependence on mass flow rate and vapor quality [22–26]. These studies illustrate a predominance of nucleate boiling on heat transfer. Other studies [27–28] observe classical behavior of flow boiling: nucleate boiling occurrence and heat flux dependence for low vapor qualities and convective boiling and mass flow rate dependence for higher vapor qualities leading to dry-out. Some studies observe a simple dependence on mass flow rate and vapor qualities [29–31].

18.3.2.1 Confinement Influence

Different studies of the literature give conclusion about confinement effect on flow boiling. Tran et al. study [32] revealed that geometry has no influence on heat transfer. However, in this study, the test section was a tube of more than 2 mm hydraulic diameter with a confinement number of 0.32. According to Kew and Cornwell criterion [5], the channel is not confined. The study of Harirchian and Garimella [33] shows the existence of a critical section value (0.089 mm^2) from which heat transfer is intensified. The study of Yen et al. [34] shows that in nucleate boiling regime, heat transfer is better for a squared surface than for a rectangular one as the corners promote nucleation. Recently, the results of Soupremanien et al. [35] show that the augmentation of the channel aspect ratio favors the onset of boiling for a lower heat flux.

18.3.2.2 Intermittent Dry-Out

In microchannels flow boiling, the local heat transfer coefficient increases with heat flux up to a critical value of the vapor quality beyond which its deterioration occurs. Deterioration of the heat transfer coefficient is believed to be caused by the intermittent dry-out, which refers to an unstable breakdown of the liquid film in contact with the wall. It seems that intermittent dry-out is rather due to mechanical instabilities than thermal fluctuations. High shear force can make the liquid–vapor interface wavy and cause intermittent dry-out if the wave amplitude becomes greater than the liquid film thickness.

This phenomenon has been observed in microchannels for small values of vapor quality [24] and is a consequence of the channel confinement; the reduction of hydraulic diameter augments heat transfer coefficient as the liquid film is thinner.

Another phenomenon experimentally identified in flow boiling is the oscillatory nature of the flow. Some intermittent local dry-out can occur in confined spaces. This occurrence certainly influences the evolution of the heat transfer coefficient at high vapor quality. Experimental investigations of two-phase flow instabilities in narrow channels showed vapor slug formation blocking the two-phase flow and pushing it back to the inlet. Periodic slug flow with quick dry-out and rewetting was observed too. This phenomenon occurs faster and lingers longer when the heat flux increases. However, the exact influence on the heat transfer coefficient has not been quantified yet. Consequently, there is no clear indication that boiling phenomena in small diameter channels are either dominated by nucleate boiling or convective boiling or any new mechanism. Yet, there is strong evidence that these mechanisms are not interconnected as in conventional tubes.

18.3.2.3 Models for Heat Transfer

A universal method for predicting saturated boiling heat transfer must be able to tackle both nucleate boiling–dominated and convective boiling–dominated regimes up to the location of incipient dry-out. Various empirical correlations have been proposed for flow boiling in microchannels. Those most often cited in the literature are presented.

Kandlikar [36] developed a general correlation adapted to minichannels which gives very good results for low qualities but fails to take dry-out into account, as noted by the author himself. Tran et al. [37], Aritomi et al. [38], and Kew and Cornwell [39] established correlations of heat transfer coefficient for various refrigerants. All these authors noted that the local heat transfer coefficient was only dependent on the heat flux. Accordingly, they concluded that the governing mechanism was nucleate boiling and no dependence on quality was considered. Huo et al. [40] studied boiling of refrigerant R134a in minichannels and highlighted the prevalence of nucleate boiling and the occurrence of dry-out at low vapor quality. This result is confirmed by Bontemps et al. [41] who show that the effect of confinement on dry-out is found to be decreased in the critical quality. Some experimental studies in similar geometries show a dependence of the heat transfer coefficient on vapor quality. The Oh et al. [42] and Kandlikar [36] correlations illustrate a clear evolution of the heat transfer coefficient with vapor quality. These works indicate that nucleate boiling may not be the only mechanism governing boiling in minichannels and that new mechanism may occur too.

In an effort to develop a generalized correlation for saturated flow boiling, Gungor and Winterton [43] and Liu and Winterton [44] proposed correlations based on fairly large databases for macrochannels, and Bertsch et al. [45] and Li and Wu [46] for mini/microchannels.

Recently, a universal method for predicting the two-phase heat transfer coefficient for saturated flow boiling in mini/microchannels was developed by Kim and Mudawar [47]. A rather complete but not conclusive review has been made by Wang and Wang [48].

Thome [49] proposed a semiempirical three-zone model, which predicts the trends observed in minichannels. In this model, the dominant boiling mechanism is the evaporation of the liquid film lying between confined bubbles and the wall. The calculation implies the optimization of minimum and maximum liquid layer thickness and bubble generation frequency parameters. The three-zone model has been successfully compared to different data of the literature. However, it significantly underpredicts data at very low vapor qualities where the flow regime is bubbly flow and where heat transfer is controlled by nucleate boiling mechanism.

18.3.2.4 Heat Transfer Coefficient

Many empirical correlations have been proposed to determine heat transfer coefficients. Recently, Li and Wu [46] have analyzed experimental results in single as well as multiple mini/microchannels. They obtained a correlation using the boiling number, Bond number, and Reynolds number. The correlation is based on 3744 data points, covering a wide range of working fluids, operating conditions, and different microchannel dimensions. The correlation can be used for 0.19 mm < D_h < 2.01 mm and is written as

$$h = 334Bl^{0.3}\left(Bo\,Re_l^{0.36}\right)^{0.4}\frac{\lambda_l}{D_h} \tag{18.32}$$

18.3.3 Pressure Drop

Two-phase pressure drop is generally expressed as the sum of frictional, gravitational, and acceleration (or deceleration) components:

$$\left(\frac{dP}{dz}\right)_{TP} = \left(\frac{dP}{dz}\right)_{frictional} + \left(\frac{dP}{dz}\right)_{acceleration} + \left(\frac{dP}{dz}\right)_{gravitational} \tag{18.33}$$

The acceleration component is positive for boiling flows because of flow acceleration along the stream-wise direction, negative for condensing flows because of flow deceleration, and negligible for adiabatic flows. Unlike macrochannel flows, where gravitational effects can be significant, especially for low mass velocities, these effects are far less significant for mini/microchannel flows, which are dominated by surface tension forces and large shear stresses.

The two main approaches to predict frictional pressure drop in microchannels are the homogeneous and the separated flow models. The first one assumes that the two-phase fluid behaves as a single-phase fluid. It is not depending on the size of the channel. The second one considers that the phases are separated and interact through their interface. It depends strongly on the flow regime.

In the homogeneous model, the three terms of the total pressure gradient can be expressed as

$$\left(\frac{dP}{dz}\right)_{frictional} = \frac{2f_{TP}G^2}{\rho_{TP}D_h} \tag{18.34}$$

$$\left(\frac{dP}{dz}\right)_{acceleration} = G^2\frac{d}{dz}\left(\frac{1}{\rho_{TP}}\right) \tag{18.35}$$

$$\left(\frac{dP}{dz}\right)_{gravitational} = \rho_{TP}g\sin\theta \tag{18.36}$$

where

f_{TP} is a two-phase friction factor

G and ρ_{TP} are the mass velocity and an averaged two-phase density, respectively

θ is the angle between the channel and a horizontal plane

The two-phase friction factor f_{TP} can be written as

$$f_{TP} = \frac{16}{Re_{TP}} \quad Re_{TP} \leq 2300$$
$$f_{TP} = 0.079 Re_{TP}^{-0.25} \quad Re_{TP} > 2300 \tag{18.37}$$

with

$$Re_{TP} = \frac{GD_h}{\mu_{TP}} \tag{18.38}$$

and where μ_{TP} is an averaged two-phase viscosity.
In this model, two questions arise:

- What is the appropriate homogeneous density?
- What is the appropriate homogeneous viscosity?

The homogeneous density can be found from the phase densities by

$$\rho_{TP} = \left(\frac{x}{\rho_v} + \frac{1-x}{\rho_l}\right)^{-1} \tag{18.39}$$

For the two-phase viscosity, numerous models have been introduced. The readers can refer to the publication of Asadi et al. [50] for a general review of two-phase viscosity models.

It is often found that the frictional term is preponderant. Recent studies summarize the best design method to calculate this term and it is often concluded that the homogeneous model method can be fairly reliable and that separated flow methods perform poorly. As a consequence, no recommendation can be established for microchannels and the authors invite the readers to refer to Chapter 20 of the engineering data book [49] and to a recent review of predictive methods for microchannels [50].

18.3.4 Boiling on Enhanced Surfaces

18.3.4.1 Boiling on Conventional Enhanced Surfaces

This section provides a general overview of boiling on enhanced surfaces. In general, enhanced surfaces cover coated, rough, and extended surfaces. The coated surfaces are manufactured by depositing various materials on the working surfaces. Typical deposit materials are, for example, Teflon and metal powders. A fine-scale porous coating may lead to enhancement of nucleate boiling. Surface roughness is usually formed by machining or restructuring the surface. Rough surfaces contain artificial nucleation sites, which provide higher performance than a plain surface. Extended surfaces are the most widely employed ones in heat exchangers. They provide area increase and thereby improve the heat transfer efficiency. Table 18.2 presents a summary of techniques of surface treatments that have been used to develop enhanced surfaces.

In summary, many advances have been made to create enhanced surfaces that have better performance of boiling heat transfer than natural surfaces. The two basic techniques to fabricate enhanced surfaces are coating and structuring. The boiling process from enhanced surfaces is especially related to the geometry and the spacing of the nucleation sites. Today, researches on this subject focus on creating enhanced surfaces at lower scales (micro- and nanoscale).

TABLE 18.2
Techniques to Fabricate Enhanced Surfaces

Technique	Description
Abrasive treatment	Abrasively roughening the surface
Chemical gravure	Exposing the surface to an etching bath
Electroplating	Using electrical current to reduce the cations of a desired material from a solution and coat the surface with a thin layer of the material
Open grooves	Forming parallel grooves by sharp pointed scriber
Painting	Mixing the particles with a paint and applying on the surface
Sintering	High-temperature method to deposit a layer of metal particles or metal fiber on the surface
Spraying	Directing the molten metal on to the surface using a spray gun
Three-dimensional cavities	Pressing cylindrical or conical cavities into the surface

18.3.4.2 Boiling on Surfaces Structured at Micro- and Nanoscale

Recent interest of boiling with micro- and nanostructured surfaces has mainly derived from research on nanofluid boiling. Nanofluids are fluids containing suspended nanoparticles, which account for particles having a size smaller than 100 nm. Compared to the milli- and microfluids, nanofluids are more stable since they undergo little erosion and gravitational deposition over long time periods. Boiling of nanofluids has become an attractive subject over the past decade. A number of researchers observed deposition of nanoparticles on the heated surface. This intensified the surface roughness and wettability. The critical heat flux was also found to be significantly improved. However, controversial results were obtained for the heat transfer coefficient. Most studies reported no change of heat transfer and, even worse, the presence of heat transfer deterioration. A few others reported a heat transfer enhancement.

The enhancement of CHF was also noticed and was related to the enhancement of surface wettability and liquid spreadability. The effect of microstructure on CHF enhancement was not explained. The greatest CHF enhancement was observed with the surface containing both the micro and nanostructures. Each surface also showed some enhancement in boiling heat transfer.

18.3.4.3 Effects of Surface Wettability on Boiling Heat Transfer

Some experimental data have recently emerged questioning the role of surface wettability on nucleate boiling mechanism. Different types of coating to modify surface wettability have already been investigated and have exhibited interesting performances.

Coatings are provided at micrometric or nanometric scales either by particle depositions [51,52] or by an artificial creation of cavities to generate controlled nucleation sites [53]. Recent studies show that the influential parameters for enhancing heat transfer during boiling are bubbles size, emission frequency, and nucleation sites density. Surface wettability is also involved by connecting the contribution of those different parameters. For example, hydrophilic surfaces make bubbles detachment easier. The bubble is created in an active cavity. When liquid moves toward that cavity, the gravitational force stretches the bubble vertically. Consequently, the latter phenomenon is intensified for hydrophilic surfaces. However, for such surfaces, bubble diameter departure increases, due to the augmentation of bubble growth time. On the contrary, hydrophobic surfaces promote vapor confinement which easily initiates bubble nucleation; bubble growth time is then smaller (compared to that of the hydrophilic case) but the detachment stage is reduced leading to bubbles coalescence and rapid occurrence of critical heat flux.

A few studies have tried to develop experiments combining complementary properties of hydrophilic and hydrophobic surfaces. Jo et al. [54] have investigated both types of experiments: a hydrophilic surface including hydrophobic sites and the inverse pattern. They have observed a better enhancement of heat transfer and a more efficient heat removal for the hydrophilic surface including hydrophobic sites. The authors conclude that thermal performances of the surface are dependent on the size of hydrophobic sites and on the distance between each spot. Recently, Bertossi et al. [55] have reported the demonstration on thermal reversible switching behavior between hydrophilicity and hydrophobicity. Physicochemical techniques were used to obtain a coating switching near the saturation temperature of water at atmospheric pressure. It was found that the coated surfaces perform well and exhibit better thermal performances than the uncoated ones (average boiling enhancement about 20%).

One promising technique of enhanced surfaces is carried out by using carbon nanotubes. Carbon nanotubes are very thin cylindrical tubes whose diameter ranges from 1 to 100 nm and the length of which can reach several micrometers. This type of structures possesses some very specific interesting physical properties such as a very high mechanical resistance or very high electrical and thermal conductivities. For several years, thermal studies concerning pool boiling have been carried out on heated surfaces coated with nanotubes carpets [56–58]. Most of authors explain performance enhancement by capillary phenomena and creation of numerous nucleation sites; thanks to the wetted carbon nanotubes gathering.

As a conclusion, an ideal boiling surface has complex requirements on its wettability; it requires hydrophobicity to promote nucleation and hydrophilicity to promote bubbles detachment. Such a surface which juxtaposes regions of high and low wettability or which can temporally mix these properties offers an elegant solution to this dilemma. However, all the experimental results of the literature are valuable for nucleate boiling field and not proven in convective conditions.

18.4 CONDENSATION

18.4.1 Industrial Applications

Condensation in microchannels has received considerable attention because it offers appreciable advantages such as high heat transfer coefficient, high compactness, and low working fluid charge. Due to the ability to improve component and system energy efficiency, condensation in microchannels is considered as the appropriate option for reducing greenhouse gas emissions. Condensation in microchannels has been used in wide applications such as compact heat exchangers for automotive and refrigeration applications, electronic cooling, microfuel cells, microreactors, etc.

In previous works, numerous experimental studies on condensation flow patterns in microchannels show new two-phase flow mechanisms in submillimeter scale channels [7,59–63]. Therefore, condensation heat transfer in microchannels could not be predicted by using the known correlations proposed for large conventional channels because of the difference in surface tension magnitude, interfacial shear stress, and gravitational forces.

18.4.2 Condensation Flow Regimes

18.4.2.1 Condensation Flow Regimes in Conventional Channels

Condensation in macrochannels has been widely studied and is well understood compared to condensation in microchannels which requires more attention. Several classifications for transition from microscale to macroscale condensation heat transfer have been proposed taking into account working fluid temperature, pressure, and physical properties. For condensation flow, surface tension plays a predominant effect on two-phase flow structure, more specifically, in the microchannel corners. It could be used as an index to design the microchannel condensation heat transfer. Many criteria have been proposed (Section 18.2.2) and it can be concluded that condensation heat transfer in microchannels cannot be predicted using the known correlations proposed for larger channels. The reason is that new two-phase flow mechanisms in submillimeter scale channels are not observed in macrochannels.

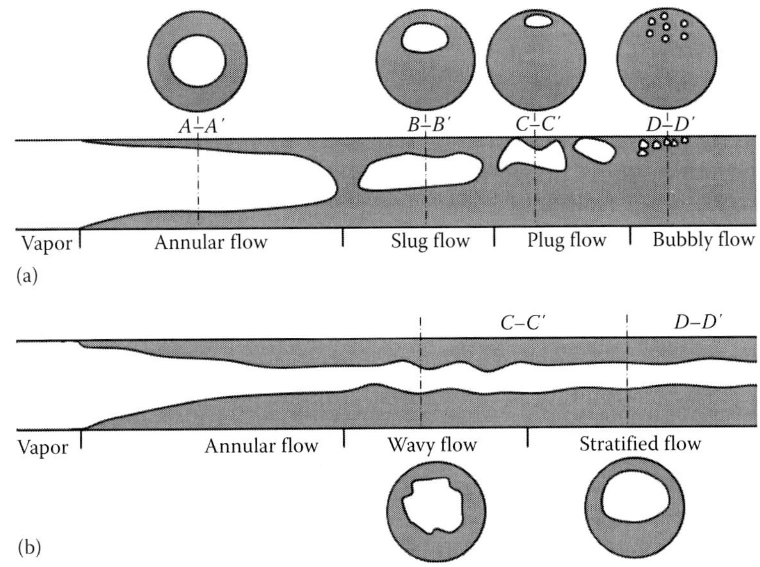

FIGURE 18.7 Condensation in macrochannels: flow regimes. (a) Annular/slug/plug/bubbly flow, (b) annular/wavy/stratified flow.

Different flow regimes are identified for condensation in macrochannels: annular flow, wavy flow, slug flow, plug flow, and bubbly flow (Figure 18.7a). At the channel inlet, vapor flows at saturated pressure and temperature. It starts to change its phase under channel cooling effect. Condensation occurs and a microliquid film is formed on the heat exchange surface. Film thickness increases along the vapor flow, forming the annular flow at the channel upstream. During the condensation process, vapor velocity decreases while the liquid velocity increases along the channel length. Condensate film flows under effect of shear stress at the liquid–vapor interface, liquid gravity, and wall shear stress that tends to retain condensate film at the surface. For annular flow, interfacial shear stress is the most dominant force on the liquid. At the end of the annular flow, accumulation of the condensate liquid occupies the major part of the channel cross-section. Transition from annular to slug/plug flows occurs where the interfacial shear stress has no significant effect on the two-phase flow compared to the gravity. For plug flow, bubbles sizes are lower than for slug flow. Under buoyancy forces, bubbles flow toward the channel top side where vapor condensation takes place. Along the channel axial direction, bubble sizes decrease because of the continuous condensation process. Bubbles are transported by liquid flow and moved to the channel exit. During their movement, condensation of vapor occurs at liquid/vapor interface of each bubble and their sizes are reduced continually to form discrete bubbly flow near the channel exit after which total condensation is obtained where the flow becomes in a liquid single phase.

At low mass vapor velocity, three condensate flow regimes are identified in a horizontal macrochannel: annular, wavy, and stratified flow (Figure 18.7b). Stratification phenomena are more important in the case when interfacial shear stress becomes very low. In this case, condensate film accumulates in the bottom side of the macrochannel and a thin liquid film remains on its upper side.

18.4.2.2 Condensation Flow Regimes in a Single Microchannel

Structure of the two-phase flow in a microchannel is mainly influenced by the size and the shape of the channel. Little information is available in the open literature about hydrodynamic and thermal characterization of condensation flow patterns inside a single minichannel or microchannel. Two-phase flow regimes in the channels with hydraulic diameter lower than 1 mm have been studied by several authors mainly for adiabatic air–water flows [64–66]. All these studies concern flow pattern maps with transition lines in terms of gas superficial velocity versus liquid superficial velocity. Compared to these studies, fewer investigations have been conducted on condensation heat transfer and flow regimes in a single microchannel. Compared to condensation in a macrochannel, stratification effect becomes insignificant in the microchannel because surface tension effect increasingly counteracts gravitational effect, thereby extending the size of the annular flow pattern.

18.4.2.2.1 Condensation in Circular Microchannels

Condensation flow structure in a circular microchannel has not been widely studied. Various flow patterns were identified for steam condensation inside a single glass circular microtube where the mass flow rate is controlled at the channel inlet. The condensate flow structure depends mainly on time and location in the microtube. The basic condensate flow regimes identified in a single circular microchannel constitute a succession of annular/oscillatory/wavy/bubbly and slug flows as shown in Figure 18.8. Figure 18.9 shows wavy annular flow with bubbles injection in a single microtube where the flow direction is from the right to the left and when only vapor is entering the

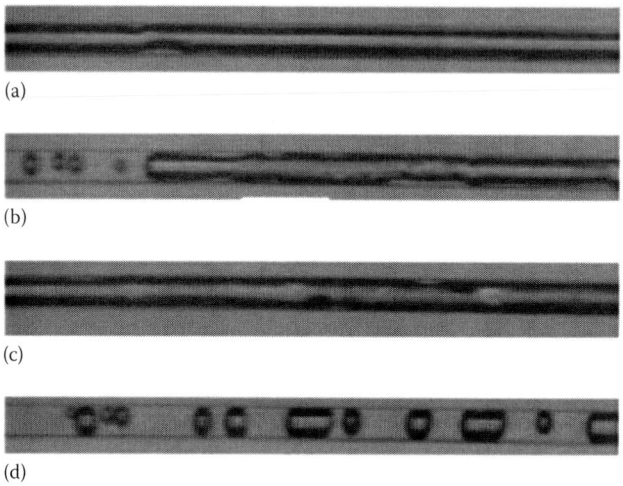

FIGURE 18.8 Flow regimes: (a) annular, (b) wavy/bubbly, (c) oscillatory, and (d) slug flow.

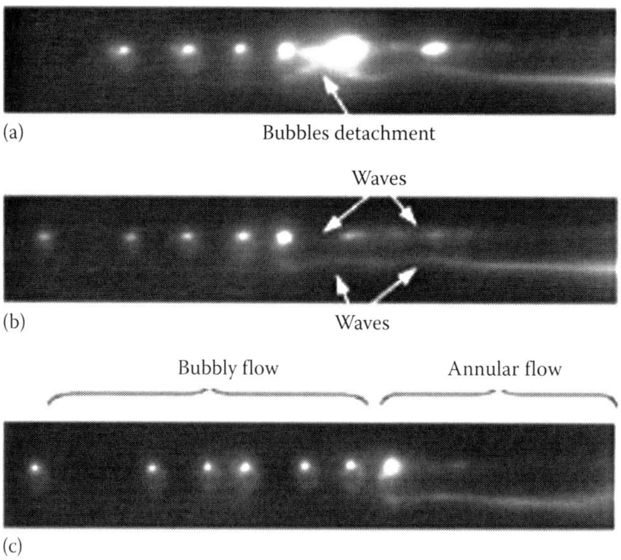

FIGURE 18.9 Annular wavy/bubbly flow. (a) $t = 0$ ms, (b) $t = 16$ ms, (c) $t = 96$ ms.

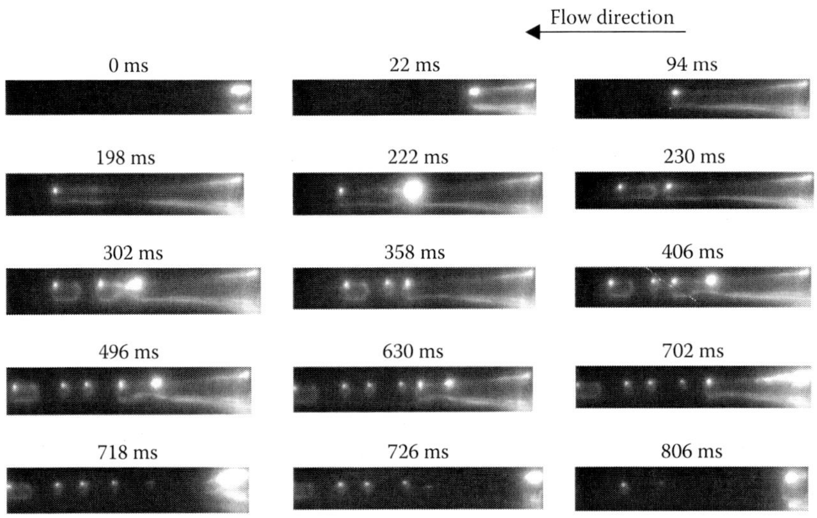

FIGURE 18.10 The complete sequence of the smooth annular and slug bubbly flow.

microtube. Vapor is condensed along the cooling surface forming a thin liquid film on the surface. Under Kelvin–Helmholtz instabilities, waves appear on the film surface resulting from interfacial shear between liquid and vapor phases moving at different velocities. Generally, waves grow with time due to liquid accumulation and reach a limit amplitude equivalent to the microtube radius, for which waves around the vapor flow perimeter collapse causing strangulation of vapor flow and bubble formation (Figure 18.9a). After detachment of each bubble, annular flow size is reduced and it moves back from the detachment zone to the microtube entrance during a short time period. Under effect of the incoming vapor and diminution of the friction force by condensation process, liquid film velocity decreases and vapor core diameter is reduced inducing transition from annular to the wavy flow (Figure 18.9b). After this step, bubbly flow occurs under successive strangulations of vapor core due to wave growth. As shown by Zhang et al. [67], vapor velocity varies along the front vapor interface and achieves a maximum value at the channel centerline. This explains why the breakup point of the vapor core occurs at the center of the microchannel.

Figure 18.10 shows a sketch of flow patterns inside a single microtube where the upstream flow is annular and the downstream flow is a bubbly/slug flow. The condensate flow is periodic during time. The first images show that annular flow length increases during time because of incoming vapor. The first bubble (called Taylor's bubble) having an elongated form is injected. After this step, different spherical bubbles are injected resulting from the condensation process and instabilities in the microtube. Bubbles move to the microtube exit near the channel top because of the buoyancy forces. When Taylor's bubble reaches microtube exit, it causes an increase of the pressure and a flow reversal (image at $t = 718$ ms). A hemispherical meniscus is formed (image at $t = 726$ ms) blocking the microtube inlet and increasing the vapor flow pressure allowing evacuation of bubbles to the microtube exit (image at $t = 806$ ms).

18.4.2.2.2 Condensation in Noncircular Microchannel

One of the important mechanisms enhancing condensation heat transfer in a noncircular microchannel is the condensate flow distribution in the microchannel cross-section under the surface tension effect. The major part of the condensate film is generally retained in the microchannel corners and a very thin film is found in all the microchannel sides between the corners.

Generally, there are two types of condensation flow patterns in noncircular microchannel: developed flows that are stable during time and periodic flows. As identified by numerous authors, the periodic condensate flow in a microchannel is a succession of different flow regimes on the entire

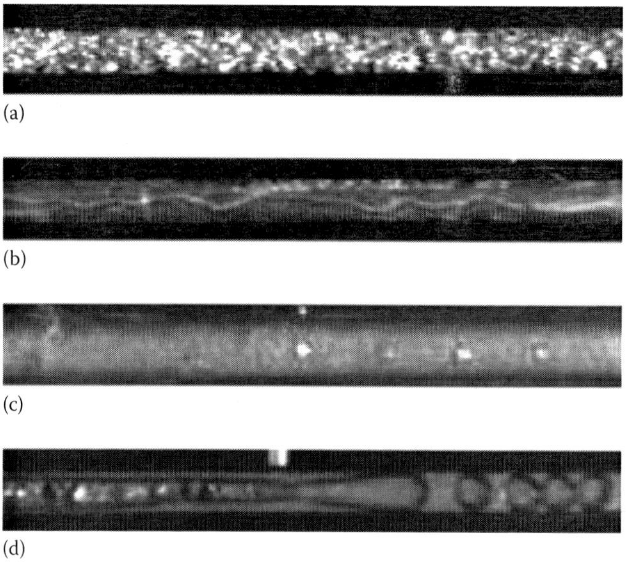

FIGURE 18.11 Flow regimes: (a) mist flow, (b) churn flow, (c) annular flow, and (d) annular/bubbly flow.

length or only in the microchannel half length. These flows are mist flow, churn flow, annular flow, liquid slug/bubbly flow, liquid ring flow, and annular/bubbly flow as shown by Figure 18.11. Mist flow is observed only in microchannel upstream and the churn flow is formed at the microchannel downstream (Figure 18.11a and b). The mist flow is a mixed two-phase flow formed by steam and microdroplets. It is obtained for turbulent superheated steam flows. Creation of microdroplets in mist flow is mainly due to surface tension that is predominant in noncircular microchannels. Churn flow is characterized by an oscillatory displacement of vapor flow in the microchannel. Continuous incoming vapor at the microchannel entrance increases the interfacial shear until it exceeds capillary force. Therefore, transition from mist flow to churn and annular flow occurs. For annular flow (Figure 18.11c), vapor has enough energy to move up the condensate film. When the surface tension and viscous force between vapor and liquid phases are sufficient to overcome the inertial force, interfacial instability is established. Therefore, wave formation leads to bubbles emission with sizes depending on cooling heat flux, flow velocity, and inlet pressure (Figure 18.11d). This makes transition from annular flow to slug flow which takes place generally in the entire microchannel. The incoming vapor at the channel inlet leads to the large bubble sizes and small liquid slugs. Liquid slugs appear when the front of the vapor plug breaks apart and forms the elongated bubbles. After this step, mist flow reappears near the channel inlet because the incoming vapor leads to the confined bubbles near the exit which merge and cause apparition of mist flow. Mist flow moves along the microchannel in order to occupy the entire channel length. After this, the cycle of condensate flow is repeated (Figure 18.12). As discussed by Triplett et al. [13], the transition from elongated bubbly flow to mist flow occurs at high velocity when bubbles become unstable near their trailing end which leads to their disappearance.

There are two types of flows which are considered as the base condensate flows in the noncircular microchannel: the upstream mist flow followed by a set of confined bubbles (Figure 18.13a) and the slug flow (Figure 18.13b). A microliquid film formed on the surface, whose thickness increases as we move away from the entrance due to friction forces at the liquid–steam interface and the wall that holds the liquid and amortizes the flow. These forces lead to a difference in speed with the liquid–steam interface, thus promoting the formation of interfacial waves. The wave amplitude enclosing the liquid increases over time, leading to a bottleneck (injection zone) of this flow and a bubble ejection. The latter moves toward the channel exit while condensing. Slug flow consists

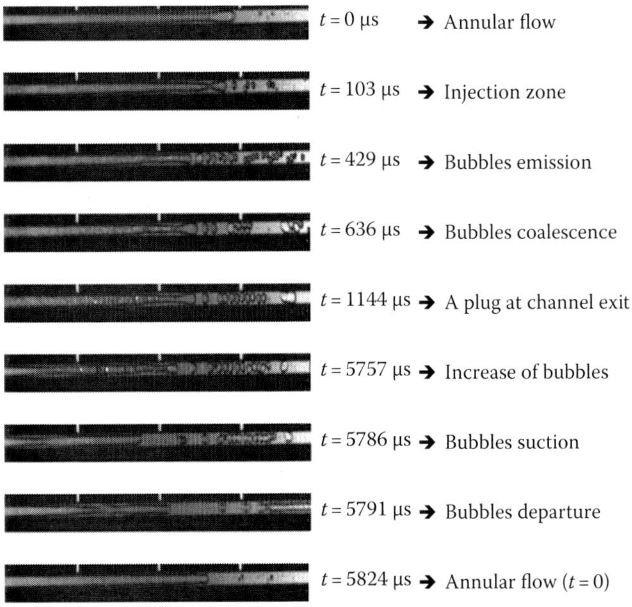

FIGURE 18.12 Sketch of flow patterns.

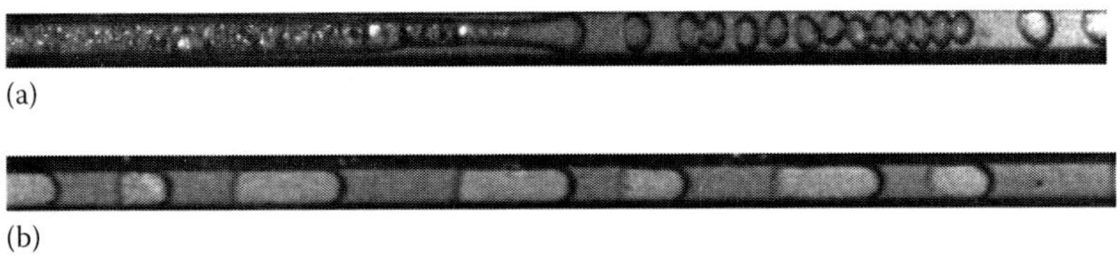

FIGURE 18.13 The base flow regimes in noncircular microchannels. (a) Mist/bubbly flow, (b) slug flow.

of an elongated bubble train with liquid slug separating one bubble from the other as shown in Figure 18.13b. Bubbles are separated from one another by liquid slugs resulting from condensation of vapor inside the microchannel. Elongated bubbles and liquid slugs having different sizes travel the microchannel from the entrance to the exit. Bubble size may be reduced and liquid slug size may be increased because of the condensation process on the bubble surfaces. This affects bubble velocity and leads to the coalescence phenomenon between the neighboring bubbles.

18.4.2.3 Condensation Flow Regimes in Hydrophobic Surface of Microchannels

Wettability is the ability of a liquid to maintain contact with a surface (Figure 18.14a). It is characterized by a low contact angle that is considered as a potential parameter affecting condensation

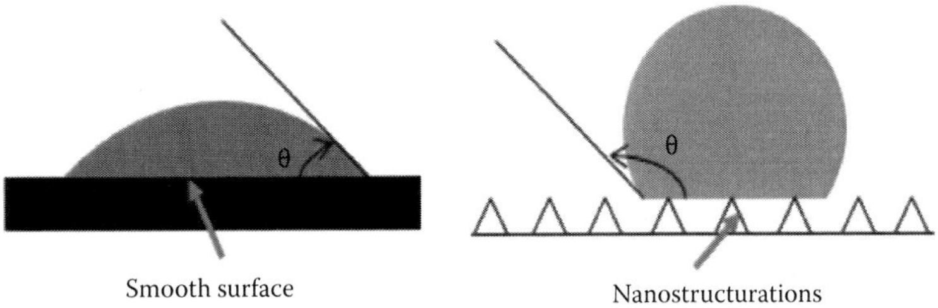

FIGURE 18.14 Hydrophobicity: (a) hydrophilic surface and (b) hydrophobic surface.

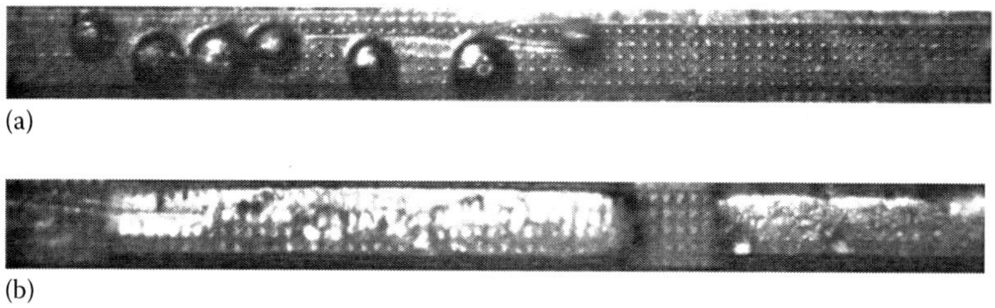

FIGURE 18.15 Condensate flow regimes in a single nanostructured microchannel. (a) Rotational bubbles, (b) elongated bubbles.

heat transfer efficiency. Conversely, it is extremely difficult to wet a hydrophobic surface whose contact angle is high (Section 18.2.4). Hydrophobicity can be used as a mean to enhance condensation heat transfer. In previous works, it was shown that the condensate length increases when the contact angle decreases [68] and the average condensation heat transfer is highly enhanced by increasing the contact angle [69]. Hydrophobicity of a heat exchange surface could be modified by material deposition or nanostructuring microfabrication process (Figure 18.14b) as discussed by numerous authors [70]. It is also recommended as an innovative method to remove passively the liquid condensate from the surface.

Generally, dropwise condensation only occurs on a nonwetting surface with less ability to retain liquid. Heat transfer coefficient for dropwise condensation is higher than for film condensation because liquid film thermal resistance reduces heat transfer at the surface. Hydrophobic surface has an advantage to remove condensate droplets from the surface in the absence of gravity. This contributes to reduce thermal resistance of droplets and accelerates vapor condensation process. Droplets velocity is related to the surface hydrophobicity degree and is controlled by hydrophobicity gradient on the surface. Figure 18.15 shows flow regimes observed for steam condensation inside a microchannel with hydrophobic surface realized by nanostructurations. For a microchannel with hydrophobic surface, bubbles move away from the microchannel with high-speed velocity under shear stress and the effect of the surface structuration that reduces the wettability effect. The rotational movement of the bubbles under surface hydrophobic effect is highlighted in Figure 18.15a. It contributes in speeding up bubble movement and coalescence, thereby forming elongated bubbles (Figure 18.15b). Condensate flow is more accelerated using nanostructured surface compared to smooth surface.

18.4.3 Condensation Heat Transfer: Models and Correlations

For flow with phase change, local and average heat transfer coefficients are related to condensate flow regimes. So far, there is no information available in the literature about the relationship between distribution of the condensate flow structure in microchannel and heat transfer coefficient resulting from this distribution. The problem is very complex depending on numerous physical parameters such as fluid, velocity, diameter, shape of the section, and channel orientation.

Local condensation heat transfer is mainly controlled by liquid film thickness distribution on the heat exchange surface where heat flux is transferred to the coolant. Condensate film is very thin and measurement of temperature gradient on the normal surface is complex. Several correlations and numerical models have been proposed in previous works for condensation heat transfer [71–73] in order to predict local and average heat transfer coefficients. However, each model was limited to a validity domain such as the scale of hydraulic diameter, pure vapor, pressure, velocity, etc. Few investigations about condensation heat transfer inside a microchannel have been found in the literature. It has been indicated by several authors that the capillary forces affected the flow pattern and the proposed correlations for condensation heat transfer in macrochannel could not be available to estimate condensation heat transfer in microchannels.

18.4.3.1 Modeling of Condensation in a Circular Microchannel

Several studies were conducted for condensation using the numerical procedure. Zhang et al. [74] used the volume of fluid (VOF) method to study condensation in horizontal miniature tube with diameter ranging from 1 to 3 mm. Begg et al. [75] proposed an analytical model for annular film condensation in a miniature tube of diameters 1.6, 2.3, and 3.4 mm. They neglected the vapor temperature gradient, the inertia terms in the condensate film, and the gravitational body force.

Analysis of annular condensation flow inside a vertical microtube could be applied to the condensation inside horizontal tube when the gravitational force was negligible compared to the frictional force at the liquid–vapor interface. Generally, physical models comprise a miniature tube with a diameter D cooled with a uniform wall heat flux (Figure 18.16). Saturated pure or mixture vapor flow enters the microtube with a uniform velocity, mass fraction, and temperature. The condensate film flows along the axial direction under the effects of pressure, surface tension, gravity, and shear stress forces. Considering a smooth liquid film and 2D flow, condensation heat and mass transfer could be treated numerically by solution of the coupled equations.

During condensation inside a single microtube, velocities, temperature, pressure, film thickness, and species mass fraction are changed continuously in the liquid and vapor flows. For condensation inside a single microtube, governing equations for both the liquid ($k = l$) and vapor ($k = v$) phases are used in cylindrical coordinates.

Mass conservation equation:

$$\frac{\partial U_k}{\partial z} + \frac{1}{r}\frac{\partial}{\partial r}(rV_k) = 0 \tag{18.40}$$

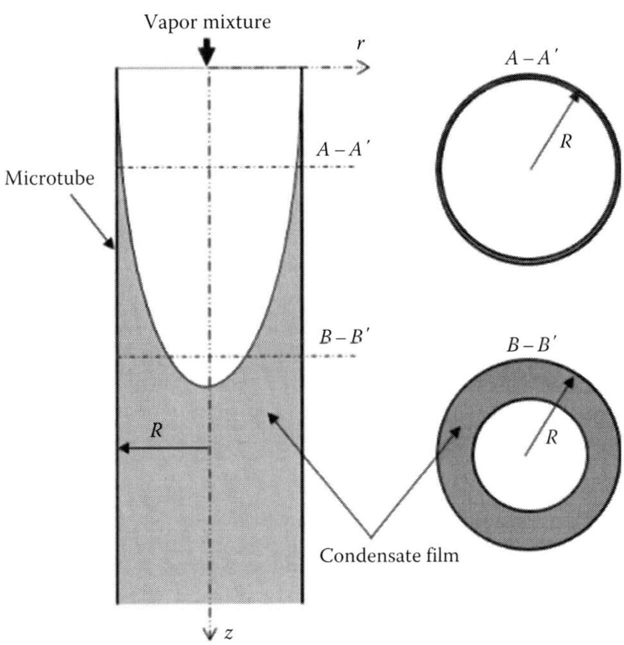

FIGURE 18.16 Physical model for condensation in a microtube.

Momentum conservation equation:

$$U_k\frac{\partial U_k}{\partial z}+V_k\frac{\partial U_k}{\partial r}=-\frac{1}{\rho_k}\frac{dP_k}{dz}+g+\frac{1}{\rho_k}\frac{1}{r}\frac{\partial}{\partial r}\left(\mu_k r\frac{\partial U_k}{\partial r}\right) \tag{18.41}$$

Energy conservation equation:

$$\rho_k c_{p,k}\left(U_k\frac{\partial T_k}{\partial z}+V_k\frac{\partial T_k}{\partial r}\right)=\frac{1}{r}\frac{\partial}{\partial r}\left(\lambda_k r\frac{\partial T_k}{\partial r}\right)+\rho_k D_k(c_{p,1}-c_{p,2})\frac{\partial T_k}{\partial r}\frac{\partial C_{1,k}}{\partial r} \tag{18.42}$$

Mass species conservation equations:

$$U_k\frac{\partial C_{1,k}}{\partial z}+V_k\frac{\partial C_{1,k}}{\partial r}=\frac{1}{r}\frac{\partial}{\partial r}\left(D_k r\frac{\partial C_{1,k}}{\partial r}\right) \tag{18.43}$$

The mixture velocities V_k, U_k, and the temperature T_k are computed from Equations 18.40 through 18.42, respectively. The mass fraction of the more volatile component in the mixture ($C_{1,k}$) is determined from Equation 18.43.

Closure of the physical model is obtained through a heat balance in the liquid film (thickness δ) as

$$-\lambda_l R\left.\frac{\partial T_l}{\partial r}\right|_{wall}=\lambda_v(R-\delta)\left.\frac{\partial T_v}{\partial r}\right|_{R-\delta}-\frac{d}{dz}\int_{R-\delta}^{R}\rho_l U_l(H_{lv}+c_{p,l}(T_{int}-T_l))r\,dr \tag{18.44}$$

The mass balance for each z location is given by

$$\dot{m}_{in}(z=0)=\dot{m}_v(z)+\dot{m}_l(z) \tag{18.45}$$

where

$\dot{m}_v(z)$ is the vapor mass flux

$\dot{m}_l(z)$ is the condensate mass flux at each location z

Equations 18.40 through 18.43 are subjected to the classical boundary conditions at the wall and tube axis where the flow symmetry is assumed. At the liquid–vapor interface, continuity of the longitudinal velocity and temperature must be adopted. The shear stress at the liquid–vapor interface is determined by taking into account the surface tension gradient related to the interfacial temperature gradient along the microtube:

$$\mu_l\left.\frac{\partial U_l}{\partial r}\right|_{int}=\mu_v\left.\frac{\partial U_v}{\partial r}\right|_{int}+\frac{d\sigma}{dT}\frac{dT_{int}}{dz} \tag{18.46}$$

As defined by the known Young–Laplace equation, the wetting characteristics and the surface tension result in a pressure difference between the liquid and vapor flows, which can be shown as

$$P_{v,int}-P_{l,int}=\sigma\left\{\frac{d^2\delta}{dz^2}\left[1+\left(\frac{d\delta}{dz}\right)^2\right]^{-3/2}+\frac{1}{R-\delta}\cos\left(a\tan\frac{d\delta}{dz}\right)\right\}-P_d \tag{18.47}$$

Numerous solution procedures have been proposed in the previous literature. The numerical model has advantages in estimating locally the main physical parameters characterizing condensation heat transfer such as film thickness distribution, profiles of temperature, velocity and mass fraction of species, heat transfer coefficient, and pressure drop.

18.4.3.2 Modeling of Condensation in a Noncircular Microchannel

Condensation heat transfer in a noncircular microchannel is a fundamental thermal process used in many industrial applications. A physical model of annular film condensation inside a noncircular microchannel was investigated by numerous authors [76–78].

As shown in Figure 18.17, vapor flow enters the microchannel with a fixed mass flux, pressure, and temperature. It condenses inside the microchannel having a wall temperature lower than vapor saturation temperature. The heat transfer from the cooling water to the condensate flow is assumed to be one dimensional. The film thickness on the heat exchange surface varies with vapor quality depending on the microchannel axial position. Hemispherical meniscus of condensate film end is one of the unknown parameters in the physical model.

During condensation inside a noncircular microchannel, governing equations for both liquid and vapor phases separately are used in Cartesian coordinates. Continuity, momentum, and energy equations are considered in the liquid and vapor phases including the Laplace–Young equation to define condensate film meniscus curvature:

$$\frac{dP_l}{dz} = \frac{dP_v}{dz} + \frac{\sigma}{R^2}\frac{dR}{dz} \tag{18.48}$$

$$\frac{dP_v}{dz} = 2\frac{\rho_v U_v^2}{A_v R}\frac{A_l dR}{dz} - 2U_v\rho_v\frac{dU_v}{dz} - 2\rho_v\frac{U_v^2}{A_v}\frac{dA_v}{dz} - \frac{f_{vL}\,\rho_v\,U_v^2\mathcal{P}_{int}}{2\,A_v} \tag{18.49}$$

$$\frac{dU_l(z)}{dz} = \frac{q\mathcal{P}}{A_l\rho_l\,H_{lv}} - 2\frac{U_l}{R}\frac{dR}{dz} \tag{18.50}$$

$$\frac{dU_v}{dz} = -\frac{q\mathcal{P}}{\rho_v A_v\,H_{lv}} - \frac{2\,A_l\dot{m}_v}{\rho_v A_v^2 R}\frac{dR}{dz} \tag{18.51}$$

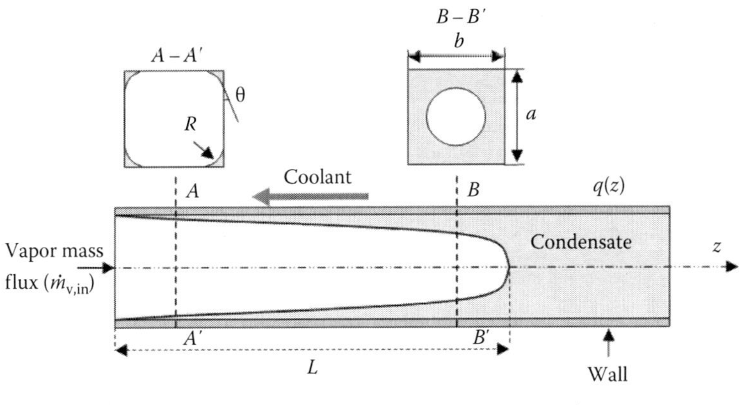

FIGURE 18.17 Geometrical configuration of the annular condensate flow.

$$\frac{dR}{dz} = \frac{0.5\, f_{int}\, \rho_v\, \mathcal{P}_{int} U_v^2 - 0.5\, f_l\, \rho_l \mathcal{P}\, U_l^2 + 0.5\, f_{int}\, \rho_v\, \mathcal{P}_{int}(A_l/A_v)U_v^2 - (2q\,\mathcal{P}A_l/H_{lv})[(U_v/A_v) + (U_l/A_l)]}{\sigma(A_l/R^2) + 2(\rho_v\, U_v^2 A_l^2/A_v R) - (2\rho_l A_l/R)U_l^2} \quad (18.52)$$

where

R is the curvature radius
U_l and U_v are the liquid and vapor velocities
P_l and P_v are the liquid and vapor pressure
$\dot{m}_v$ is the vapor mass flow rate
A_v and A_l are the vapor and liquid cross section
$\mathcal{P}_{int}$ is the liquid–vapor perimeter
$\mathcal{P}$ is the microtube perimeter
f_{int} is the liquid–vapor interface friction factor
q is the wall heat flux

18.4.3.3 Correlations of Condensation Heat Transfer in Microchannel

Very few correlations giving the Nusselt number for condensation inside microchannels have been published until now. Some experimental studies on the measurement of condensation heat transfer coefficients have been carried out in previous works. Cavallini et al. [79] studied condensation in multiport minichannels (D = 1.4 mm) in directly measuring wall temperature by thermocouples. They reported that the Wilson plot technique usually used in heat transfer measurements gives high uncertainty. Kim et al. [80] have proposed a new technique to measure condensation heat transfer of R134a inside a smooth single microtube (D = 691 μm). They found that the measured local heat transfer coefficients are close to the values predicted by the correlation of Shah [81] for the lower mass fluxes. Wang et al. [82] studied condensation of R134a in a port rectangular extrusion with hydraulic diameter of 1.46 mm and showed that their data were poorly predicted by the conventional large tube correlations and the best predictions were obtained using the correlation of Akers and Rosson [83]. Numerous studies [84] were conducted on condensation heat transfer and pressure drop in multichannels because it is easy to make measurements with more accuracy. However, more attention has been focused on the development of new correlations for predicting condensation heat transfer in microchannels. Evaluation of the usual correlations of conventional macrochannel in the case of microchannel was conducted by El Mghari et al. [69]. Average condensation heat transfer obtained from his numerical model was compared to those predicted by seven correlations proposed by Dobson et al. [85], Wang et al. [82], Koyama et al. [86], Shah [81], Akers and Rosson [83], and Wang and Rose [87]. Among these predictive correlations, those for condensation in microchannels were proposed by Wang et al. [82] and Koyama et al. [86]. All these correlations are listed in Table 18.3. Figure 18.18 shows comparison of steam condensation heat transfer in a square microchannel with hydraulic diameter of 110 μm. It is shown that the best predictions are obtained by the correlations of Dobson et al. [85] and Koyama et al. [86] for which the mean relative deviation is about 7%. It can be seen that the correlation of Traviss and Rohsenow [88] overpredicts highly the results of Reference 69 for low film Reynolds number, whereas the Akers's correlation [83] strongly underpredicts them for the whole range of the tested Reynolds numbers. Additionally, the correlation of Shah [81] proposed for condensation in channels with large hydraulic diameter from 7 to 40 mm and that of Wang et al. [82] given for condensation heat transfer inside a horizontal small rectangular channel with a hydraulic diameter of 1.46 mm are reported. Average Nusselt numbers obtained by these both correlations give the same trend and reasonable results compared to those predicted numerically. A summary of the maximum relative deviations is given in Figure 18.19.

Shah [84] proposed a new correlation to predict condensation heat transfer coefficient in micro- and minichannels. This correlation has been validated with data for eight fluids, including

TABLE 18.3
Correlations for Condensation Flow Heat Transfer

Authors	Correlations	
Dobson et al. [85]	$Nu = 0.023Re_l^{0.8}Pr^{0.3}\left(\frac{2.61}{X_{tt}^{0.805}}\right)$	(18.53)
	$Pr = \frac{\mu_L C_p}{\lambda_l},\ X_{tt} = \left(\frac{1-x}{x}\right)^{0.9}\left(\frac{\mu_l}{\mu_v}\right)^{0.1}\left(\frac{\rho_v}{\rho_l}\right)^{0.5}$	(18.54)
Wang et al. [82]	$Nu = 0.0274Pr\ Re_l^{0.6792}x^{0.2208}\left(\frac{1.376+8X_{tt}^{1.655}}{X_{tt}^2}\right)^{0.5}$	(18.55)
Traviss and Rohsenow [88]	$Nu = \left(\frac{1}{X_{tt}}+\frac{2.89}{X_{tt}^{0.478}}\right)\left(\frac{0.15Pr\ Re_l^{0.9}}{F}\right)$	(18.56)
	$F = \begin{cases} 0.707Pr\ Re_l^{0.5} & \text{for } Re_l < 50 \\ 5Pr + 5\ln(1+5Pr) + \ln\left(0.09636Re_l^{0.812} - 1\right) & \text{for } Re > 50 \\ 5Pr + 5\ln(1+5Pr) + 2.5\ln\left(0.00313Re_l^{0.812}\right) & \text{for } Re_l > 5 \end{cases}$	
Koyama et al. [86]	$Nu = 0.0152\left(1+0.6Pr_l^{0.8}\right)\left(\frac{\phi_v}{X_{tt}}\right)Re_l^{0.77}$	(18.57)
	$\phi_v^2 = 1 + 21(1-e^{-0.319D})X_{tt} + X_{tt}^2$	(18.58)
Shah [81]	$Nu = 0.023Re_l^{0.8}\ Pr^{0.33}\left[1+3.8\left(\frac{x}{1-x}\right)^{0.76}\left(\frac{P}{Pcr}\right)^{-0.38}\right]$	(18.59)
Akers et al. [83]	$Nu = CRe_{TP}^n(Pr)^{1/4}$	(18.60)
	$Re_{TP} = G\left[\left(1-x\right)+x\left(\frac{\rho_L}{\rho_v}\right)^{0.5}\right]\frac{D_h}{\mu_l}$	(18.61)
	$C = \begin{cases} 0.0256,\ n = 0.8 & \text{for } Re_{TP} > 50{,}000 \\ 5.03,\ n = \frac{1}{3} & \text{for } Re_{TP} < 50{,}000 \end{cases}$	

tube diameters from 0.49 to 5.3 mm, flow rates from 50 to 1400 kg/ms, and Bond numbers from 0.4 to 72. It predicts 500 data points from 15 studies on condensation in small diameter channels with a mean deviation of 15.9% including single circular and rectangular channels as well as multiport channels.

Shah has proposed correlations for condensation heat transfer considering two regimes: turbulent and mixed regimes using the dimensionless vapor velocity J_v as an index:

$$J_v = xG(\rho_v(\rho_l - \rho_v)gD)^{-0.5} \tag{18.62}$$

For turbulent regime:

$$J_v \geq 0.98(Z + 0.263)^{-0.62} \tag{18.63}$$

$$h_{TP} = h_{turb} = h_{lS}\left(1 + \frac{3.8}{Z^{0.95}}\right)\left(\frac{\mu_l}{\mu_v}\right)^{(0.0058+0.557P_r)} \tag{18.64}$$

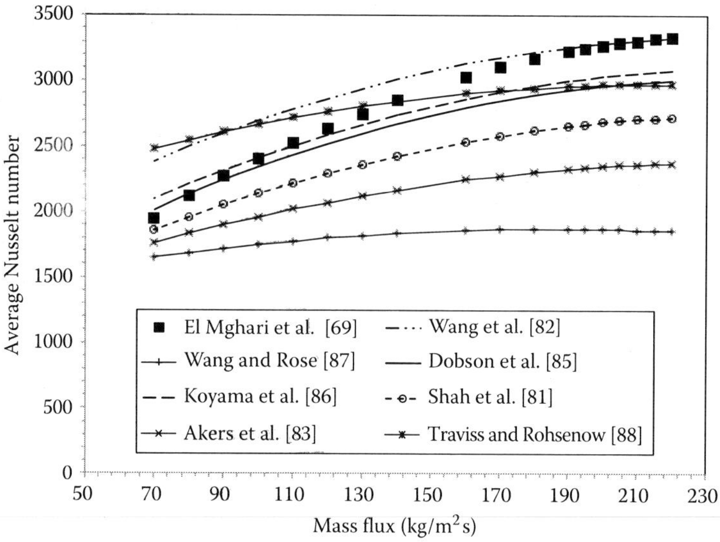

FIGURE 18.18 Comparison of the numerical results [69] with predictions of the available literature correlations.

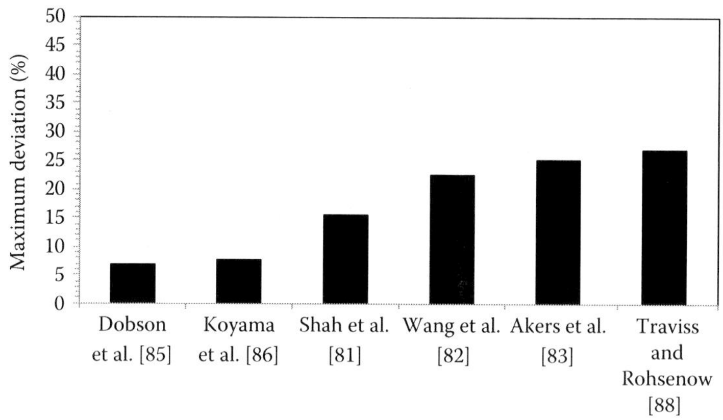

FIGURE 18.19 Comparison of various correlations for condensation heat transfer.

where

$$Z = \left(\frac{1-x}{x}\right)^{0.8} Pr_l^{0.4} \tag{18.65}$$

Heat transfer coefficient for single liquid phase is estimated using the following correlation (Dittus Boelter):

$$h_{lS} = 0.023 Re_{lS}^{0.8} Pr_l^{0.4} \frac{\lambda_l}{D_h} \tag{18.66}$$

Condensation heat transfer coefficient is defined using a modified correlation of Nusselt for laminar film condensation:

$$h_{TP} = h_{turb} + h_{Nu} \tag{18.67}$$

where

$$h_{Nu} = 1.32 Re_{lS}^{-1/3} \left(\frac{\rho_l(\rho_l - \rho_v) g \lambda_l^3}{\mu_l} \right)^{1/3} \quad (18.68)$$

REFERENCES

1. J.R. Thome, Boiling in micro-channels: A review of experiment and theory, *Int. J. Heat Fluid Flow* 25, 128–139 (2004).
2. S.S. Mehendale, A.M. Jacobi, and R.K. Shah, Fluid flow and heat transfer at micro- and meso-scales with application to heat exchanger design, *Appl. Mech. Rev.* 53, 175–193 (2000).
3. S.G. Kandlikar and W.J. Grande, Evolution of micro-channel flow passages: Thermohydraulic performance and fabrication technology, *Heat Transfer Eng.* 24, 3–17 (2003).
4. L. Lin and A.P. Pisano, Bubble forming on a micro line heater, *Micromech. Sens. Actuators Syst., ASME DSC*, Atlanta, GA, 32, 147–164 (1991).
5. K. Cornwell and P.A. Kew, Boiling in small parallel channels, in *Energy Efficiency in Process Technology*, P.A. Pilavachi (ed.), pp. 624–638, Springer, the Netherlands (1993).
6. A. Ullmann and N. Brauner, The prediction of flow pattern maps in mini-channels, *Multiphase Sci. Technol.* 19, 49–73 (2007).
7. P. Cheng and H.Y. Wu, Mesoscale and micro-scale phase-change heat transfer, *Adv. Heat Transfer* 39, 469–573 (2006).
8. W. Li and Z. Wu, A general criterion for evaporative heat transfer in micro/mini-channels, *Int. J. Heat Mass Transfer* 53, 1967–1976 (2010).
9. T. Harirchian and S.V. Garimella, Flow regime-base modelling of heat transfer and pressure drop in micro-channel flow boiling, *Int. J. Heat Mass Transfer* 55, 1946–1960 (2012).
10. R. Charnay, J. Bonjour, and R. Revellin, Experimental investigation of R-245fa flow boiling in minichannels at high saturation temperature: Flow patterns and flow pattern maps, *Int. J. Heat Fluid Flow* 46, 1–16 (2014).
11. R. Revellin and J.R. Thome, Experimental investigation of R-134a and R-245fa two-phase flow in micro-channel for different flow conditions, *Int. J. Heat Mass Transfer* 28, 63–71 (2007).
12. J.M. Mandhane, G.A. Gregory, and K. Aziz, A flow pattern map for gas–liquid flow in horizontal pipes, *Int. J. Multiphase Flow* 1, 537–553 (1974).
13. K.A. Triplett, S.M. Ghiaasiaan, S.I. Abdel-Khalik, and D.L Sadowski, Gas–liquid two-phase flow in micro-channels Part 1: Two-phase flow patterns, *Int. J. Multiphase Flow* 25, 377–394 (1999).
14. T. Young, An essay on the cohesion of fluids, *Phil. Trans. R. Soc. Lond.* 95, 65–87 (1805).
15. A. Dupré, Théorie mécanique de la Chaleur, p. 369, Paris, France: Gauthier-Villars (1869).
16. W.A. Zisman, Relation of the equilibrium contact angle to liquid and solid constitution, *Adv. Chem. Ser.* 43, 1–51 (1964).
17. D.K. Owens and R.C. Wendt, Estimation of the surface free energy of polymers, *J. Appl. Polym. Sci.* 13, 1741–1747 (1969).
18. R.N. Wenzel, Resistance of solid surfaces to wetting by water, *Ind. Eng. Chem.* 28, 7426–7431(1936).
19. A.B.D. Cassie and S. Baxter, Wettability of porous surfaces, *Trans. Faraday Soc.* 40, 546–551 (1944).
20. B. He, N.A. Patankar, and J. Lee, Multiple equilibrium droplet shapes and design criterion for rough hydrophobic surfaces, *Langmuir* 19, 4999–5003 (2003).
21. A. Lafuma and D. Quéré, Superhydrophobic states, *Nat. Mater.* 2, 457–460 (2003).
22. G.M. Lazarek and S.H. Black, Evaporative heat transfer, pressure drop and critical heat flux in a small vertical tube with R113, *Int. J. Heat Mass Transfer* 25(7), 945–960 (1982).
23. M.W. Wambsganss, D.M. France, J.A. Jendrzejczyk, and T.N. Tran, Boiling heat transfer in a horizontal small-diameter tube, *J. Heat Transfer* 115, 963–972 (1993).
24. B. Agostini, A. Bontemps, Vertical flow boiling of refrigerant 134a in small channels, *Int. J. Heat Fluid Flow* 26(2), 296–306 (2005).
25. A. Cavallini, S. Bortolin, D. Del Col, M. Matckovic, and L. Rossetto, Flow boiling inside a single circular mini-channel: Measurement of local heat transfer coefficient, in *Proceedings of ASME-JSME Thermal Engineering Summer Heat Transfer Conference*, Vancouver, British Columbia, Canada (2007).

26. D. Shiferaw, X. Huo, T.G. Karayiannis, and D.B.R. Kenning, Examination of heat transfer correlations and a model for flow boiling of R134a in small diameter tubes, *Int. J. Heat Mass Transfer* 50(25–26), 5177–5193 (2007).
27. D.S. Wen, Y. Yan, and D.B.R. Kenning, Saturated flow boiling of water in a narrow channel: Time-averaged heat transfer coefficients and correlations, *Appl. Therm. Eng.* 24, 1207–1223 (2004).
28. K.I. Choi, A.S. Pamitran, and J.T. Oh, Two-phase flow heat transfer of CO_2 vaporization in smooth horizontal mini-channels, *Int. J. Refrig.* 30, 767–777 (2007).
29. H. Oh, M. Katsuta, and K. Shibata, Heat transfer characteristics of R134a in a capillary tube heat exchanger, in *Proceedings of 11th International Heat Transfer Conference*, Kyongju, Korea, Vol. 6, pp. 473–478 (1998).
30. L. Qi, P. Zhang, Z. Wang, and L.X. Xu, Flow boiling of liquid nitrogen in micro-tubes: Part II—Heat transfer characteristics and critical heat flux, *Int. J. Heat Mass Transfer* 50(25–26), 5017–5030 (2007); *Int. J. Heat Mass Transfer* 52(3–4), 1080–1080 (2009).
31. F. Madrid, N. Caney, and P. Marty, Study of a vertical boiling flow in rectangular mini-channels, *Heat Transfer Eng.* 28(8), 1–9 (2007).
32. T.N. Tran, M.W. Wambsganss, and D.M. France, Small circular- and rectangular-channel boiling with two refrigerants, *Int. J. Multiphase Flow* 22, 485–498 (1996).
33. T. Harirchian and S.V. Garimella, Effects of channel dimension, heat flux, and mass flux on flow boiling regimes in micro-channels, *Int. J. Multiphase Flow* 35(4), 349–362 (2009).
34. T.H. Yen, N. Kasahi, and Y. Suzuki, Forced convective boiling heat transfer in microtubes at low mass and heat fluxes, *Int. J. Multiphase Flow* 29, 1771–1792 (2003).
35. U. Soupremanien, S. Le Person, M. Favre-Marinet, and Y. Bultel, Influence of the aspect ratio on boiling flows in rectangular mini-channels, *Exp. Therm. Fluid Sci.* 35, 797–809 (2011).
36. S. Kandlikar, An extension of the flow boiling correlation to transition, laminar, and deep laminar flows in mini-channels and micro-channels, *Heat Transfer Eng.* 25(3), 86–93 (2004).
37. T. Tran, M. Wambsganss, D. France, and J. Jendrzejczyk, Boiling heat transfer in a small horizontal rectangular channel, Heat Transfer Atlanta, *AIChE*, 89, 253–261 (1993).
38. M. Aritomi, T. Miyata, M. Horiguchi, and S. Sudi, Thermohydraulics of boiling two-phase flow in high conversion light water reactors (thermohydraulics at low velocities), *Int. J. Multiphase Flow* 19(1), 51–63 (1993).
39. P.A. Kew and K. Cornwell, Correlations for the prediction of boiling heat transfer in small diameter channels, *Appl. Therm. Eng.* 17(8–10), 705–715 (1997).
40. X. Huo, L. Chen, Y. Tian, and T. Karayiannis, Flow boiling and flow regimes in small diameter tubes, *Appl. Therm. Eng.* 24, 1225–1239 (2004).
41. A. Bontemps, B. Agostini, and N. Caney, Flow boiling in mini-channels, in *Micro-Scale Heat Transfer: Fundamentals and Applications*, S. Kakaç, L.L. Vasiliev, Y. Bayazitoğlu, and Y. Yener (eds.), pp. 217–230, Dordrecht, the Netherlands: Springer (2005).
42. H. Oh, M. Katsuta, and K. Shibata, Heat transfer characteristics of R134a in a capillary tube heat exchanger, in *Proceedings of 11th IHTC*, Vol. 6, pp. 131–136 (1998).
43. K.E. Gungor and R.H.S. Winterton, A general correlation for flow boiling in tubes and annuli, *Int. J. Heat Mass Transfer* 29, 351–358 (1986).
44. Z. Liu and R.H.S. Winterton, A general correlation for saturated and subcooled flow boiling in tubes and annuli, based on a nucleate pool boiling equation, *Int. J. Heat Mass Transfer* 34, 2759–2766 (1991).
45. S.S. Bertsch, E.A. Groll, and S.V. Garimella, A composite heat transfer correlation for saturated flow boiling in small channels, *Int. J. Heat Mass Transfer* 52, 2110–2118 (2009).
46. W. Li and Z. Wu, A general correlation for evaporative heat transfer in micro/mini-channels, *Int. J. Heat Mass Transfer* 53, 1778–1787 (2010).
47. S.M. Kim and I. Mudawar, Universal approach to predicting saturated flow boiling heat transfer in mini/micro-channels—Part II. Two-phase heat transfer coefficient, *Int. J. Heat Mass Transfer* 64, 1239–1256 (2013).
48. Y. Wang and Zh.-G. Wang, An overview of liquid–vapor phase-change, flow and heat transfer in mini- and micro-channels, *Int. J. Therm. Sci.* 86, 214–247 (2014).
49. J.R. Thome, *Wolverine Engineering Data Books III*, Wolverine Tube Inc. (2004).
50. M. Asadi, G. Xie, and B. Sunden, A review of heat transfer and pressure drop characteristics of single and two-phase micro-channels, *Int. J. Heat Mass Transfer* 79, 34–53 (2014).
51. H.T. Phan, N. Caney, P. Marty, S. Colasson, J. Gavillet, Surface wettability control by nanocoating: The effects on pool boiling heat transfer and nucleation mechanism, *Int. J. Heat Mass Transfer* 52, 5459–5471 (2009).

52. T. Hendricks, S. Krishnan, C. Choi, C. Chang, and B. Paul, Enhancement of pool boiling heat transfer using nanostructured surfaces on aluminium and copper, *Int. J. Heat Mass Transfer* 53, 3357–3365 (2010).
53. Y. Nam, E. Aktinol, V.K. Dhir, and Y.S. Ju, Single bubble dynamics on a superhydrophilic surface with artificial nucleation sites, *Int. J. Heat Mass Transfer* 54, 1572–1577 (2011).
54. H. Jo, H.S. Ahn, S. Kang, and M.H. Kim, A study of nucleate boiling heat transfer on hydrophilic, hydrophobic and heterogeneous wetting surfaces, *Int. J. Heat Mass Transfer* 54, 5643–5652 (2011).
55. R. Bertossi, N. Caney, J.A. Gruss, and O. Poncelet, Pool boiling enhancement using switchable polymers coating, *Appl. Therm. Eng.* 77, 121–126 (2015).
56. Z. Yao, Y.W. Lu, and S.G. Kandlikar, Effects of nanowires height on pool boiling performance of water on silicon chips, *Int. J. Therm. Sci.* 50, 2084–2090 (2011).
57. S. Ujereh, T. Fisher, and I. Mudawar, Effects of carbon nanotube arrays on nucleate pool boiling, *Int. J. Heat Mass Transfer* 50, 4023–4038 (2007).
58. R. Bertossi, N. Caney, J.A. Gruss, J. Dijon, A. Fournier, and P. Marty, Influence of carbon nanotubes on deionized water pool boiling performances, *Exp. Therm. Fluid Sci.* 61, 187–193 (2015).
59. A. Agarwal, T.M. Bandhauer, and S. Garimella, Measurement and modelling heat transfer in noncircular micro-channel, *Int. J. Refrig.* 33, 1160–1179 (2010).
60. A. Odaymet, H. Louahlia-Gualous, and M. De Labachelerie, Local heat transfer and flow patterns during condensation in a single silicon micro-channel, *Nano-scale Micro-scale Thermophys. Eng.* 16, 220–241(2012).
61. H. Louahlia-Gualous and B. Mecheri, Unsteady steam condensation flow pattern inside a miniature tube, *Appl. Therm. Eng.* 27, 1225–1235 (2007).
62. A. Odaymet and H. Louahlia-Gualous, Experimental study of slug flow for condensation in a single square micro-channel, *Exp. Therm. Fluid Sci.* 38, 1–13 (2012).
63. Q. Su, G.X. Yu, H.Sh. Wang, and J.W. Rose, Micro-channel condensation: Correlations and theory, *Int. J. Refrig.* 32, 1149–1152 (2009).
64. P.M.Y. Chung and M. Kawaji, The effect of channel diameter on adiabatic two-phase flow characteristics in micro-channels, *Int. J. Multiphase Flow* 30, 735–761 (2004).
65. A. Kawahara, P.M.Y. Chung, and M. Kawaji, Investigation of flow pattern, void fraction and pressure drop in a micro-channel. *Int. J. Multiphase Flow* 28, 1411–1435 (2002).
66. Z. Feng and A. Serizawa, Visualization of two-phase flow patterns in an ultra-small tube, in *Proceedings of the 18th Multiphase Flow Symposium of Japan*, Suita, Osaka, Japan, pp. 33–36 (1999).
67. W. Zhang, J.L. Xu, J.R. Thome, Periodic bubble emission and appearance of an ordered bubble sequence (train) during condensation in a single micro-channel. *Int. J. Heat Mass Transfer* 51, 3420–3433 (2008).
68. J.W. Rose, Surface tension effects and enhancement of condensation heat transfer, *Chem. Eng. Res. Des.* 82, 419–429 (2004).
69. H. El Mghari, M. Asbik, H. Louahlia-Gualous, and I. Voicu, Condensation heat transfer enhancement in a horizontal non-circular micro-channel, *Appl. Therm. Eng.* 64, 358–370 (2014).
70. B.S. Sikarwar, S. Khandekar, and K. Muralidhar, Effect of surface hydrophobicity on heat transfer during dropwise condensation: A numerical study, in *ECI Eighth International Conference on Boiling and Condensation Heat Transfer*, June 3–7, Lausanne, Switzerland (2012).
71. S. Nebuloni and J.R. Thome, Numerical modeling of laminar annular film condensation for different channel shapes, *Int. J. Heat Mass Transfer* 53, 2615–2627 (2010).
72. H. Louahlia-Gualous and M. Asbik, Numerical modeling of annular film condensation inside a miniature tube, *Numer. Heat Transfer A: Appl.* 52 (3) 251–273 (2007).
73. M. Miscevic, P. Lavieille, and B. Piaud, Numerical study of convective flow with condensation of a pure fluid in capillary regime, *Int. J. Heat Mass Transfer* 52, 5130–5140 (2009).
74. Y. Zhang, A. Faghri, and M.B. Shafii, Capillary blocking in forced convection condensation in horizontal miniature channels, *J. Heat Transfer Trans. ASME* 123, 501–511 (2001).
75. E. Begg, D. Khrustalev, and A. Faghri, Complete condensation of forced convection two-phase flow in a miniature tube, *J. Heat Transfer Trans. ASME* 121, 904–915 (1999).
76. Y. Chen, J. Wu, M. Shi, and G. Peterson, Numerical simulation for steady annular condensation flow in triangular micro-channels. *Int. Commun. Heat Mass Transfer* 35, 805–809 (2008).
77. Y. Chen, X. Li, J. Wu, and M. Shi, One dimensional numerical simulation for steady annular condensation flow in rectangular micro-channels, *Heat Mass Transfer* 46, 75–82 (2009).
78. T.S. Zhao and Q. Liao, Theoretical analysis of film condensation heat transfer inside vertical mini-triangular channels, *Int. J. Heat Mass Transfer* 45, 2829–2842 (2002).

79. A. Cavallini, G. Censi, D. Del Col, L. Doretti, G.A. Longo, and L. Rosseto, Experimental investigation on condensation heat transfer and pressure drop of new HFC refrigerants (R134a, R125, R32, R410a, R236ea) in a horizontal smooth tube. *Int. J. Heat Mass Transfer* 24, 73–87 (2001).
80. M.H. Kim, J.S. Shin, C. Huh, T.J. Kim, and K.W. Seo, A study of condensation heat transfer in a single mini-tube and a review of Korean micro and mini channel studies, in *First International Conference on Microchannels and Minichannels*, April 24–25, Rochester, NY (2003).
81. M.M. Shah, A general correlation for heat transfer during film condensation inside pipes. *Int. J. Heat Mass Transfer* 22, 547–556 (1979).
82. W.W.W. Wang, T.D. Radcliff, and R.N. Christensen, A condensation heat transfer correlation for millimeter-scale tubing with flow regime transition. *Exp. Thermal Fluid Sci.* 26, 473–485 (2002).
83. W.W. Akers and H.F. Rosson, Condensation inside a horizontal tube, *Chem. Eng. Progr. Symp. Ser.* 56(30), 145–149 (1959).
84. M.M. Shah, Heat transfer during condensation inside small channels: Applicability of general correlation for macro-channels, in *Proceedings of the 14th International Heat Transfer Conference*, IHTC 14, Washington, DC (2014).
85. M.K. Dobson, J.C. Chato, D.K. Hinde, and S.P. Wang, Experimental evaluation of internal condensation of refrigerants R-12 and R-134a, in *Proceedings of the ASHRAE Winter Meeting*, New Orleans, LA, pp. 744–754 (1994).
86. S. Koyama, K. Kuwahara, and K. Nakashita, Condensation of refrigerant in a multi-port channel, *First International Conference on Micro-channels and Mini-channels*, Rochester, NY, pp. 193–205 (2003).
87. H.S. Wang and J.W. Rose, Theory of heat transfer during condensation in micro-channels, *Int. J. Heat Mass Transfer* 54, 2525–2534 (2011).
88. D.P. Traviss and W.M. Rohsenow, Flow regimes in horizontal two-phase flow with condensation, *ASHRAE Trans.* 79 (2) 31–39 (1973).

Index

D

E

F

U

V